TRANSITS OF VENUS: NEW VIEWS OF THE SOLAR SYSTEM AND GALAXY

IAU COLLOQUIUM No. 196

COVER ILLUSTRATION:

The painting by Dupain (1847–1933) that decorates the ceiling of the octagonal west rotunda of the Perrault Building of the Paris Observatory showing an allegory of the passage of Venus over the sun. See S. Débarbat, page 47. Copyright Paris Observatory.

INTERNATIONAL ASTRONOMICAL UNION
UNION ASTRONOMIQUE INTERNATIONALE

TRANSITS OF VENUS: NEW VIEWS OF THE SOLAR SYSTEM AND GALAXY

PROCEEDINGS OF THE 196th COLLOQUIUM OF THE
INTERNATIONAL ASTRONOMICAL UNION
HELD IN PRESTON, LANCASHIRE,
UNITED KINGDOM
7 – 11 JUNE 2004

Edited by

D. W. KURTZ
Centre for Astrophysics, University of Central Lancashire, Preston, UK

CAMBRIDGE UNIVERSITY PRESS
The Edinburgh Building, Cambridge CB2 2RU, UK
40 West 20th Street, New York, NY 10011 4211, USA
477 Williamstown Road , Port Melbourne, VIC 3207, Australia
Ruiz de Alarcón 13, 28014 Madrid, Spain
Dock House, The Waterfront, Cape Town 8001, South Africa

First published 2005

Printed in the United Kingdom at the University Press, Cambridge

Typeset in System $\text{\LaTeX}\,2_\varepsilon$

A catalogue record for this book is available from the British Library

Library of Congress Cataloguing in Publication data

ISBN 0 521 84907 1 hardback
ISSN 1743-9213

Table of Contents

Part 1. TRANSITS OF VENUS: HISTORY, RESULTS AND LEGACY

Chairs: Steve Dick & Mary Brück

Part 4. THE JEREMIAH HORROCKS MEMORIAL PUBLIC LECTURE

Chair: Gordon Bromage

Part 5. NEW VIEWS OF THE GALAXY: PARALLAXES, DISTANCES AND IMPLICATIONS FOR ASTROPHYSICS

Chair: Naoteru Gouda

Part 6. NEW VIEWS OF THE GALAXY: FUTURE SPACE AND GROUND-BASED PROGRAMMES

Chair: Brian Warner

Part 7. MEETING SUMMARY

Part 8. TRANSIT OF VENUS: 8 JUNE 2004 – pictures from the transit and banquet

Preface

On 24 November 1639 (Julian Calendar) in the tiny Lancashire village of Much Hoole, Jeremiah Horrocks made the first observations of a Transit of Venus. He was one of the first Englishmen to appreciate the astronomical revolution going on in Europe following the works of Tycho, Galileo and Kepler. Horrocks, who died at the age of 22, can be considered to be the father of British astrophysics for the remarkable depth of his accomplishments. His legacy reverberates today.

Following the first relatively precise determination of the AU from the opposition of Mars in 1672 by Richer and Cassini, the great scientifically competitive expeditions to observe the transits of Venus in 1761 and 1769 were the first example of modern "big science"; those expeditions have given us some of the most colourful stories in all astronomy. With the length of the astronomical unit known, and with the discovery of stellar parallax in the 1830s, our view of the universe was fundamentally changed. It is fair to say that modern astrophysics blossomed from these determinations.

I had been captivated by the stories of the transit of Venus Expeditions of the 18th century from my graduate student days in the 1970s. So when I found myself a professor at the University of Central Lancashire (UCLan) in Preston at the beginning of 2001, I asked Gordon Bromage, Head of the Centre for Astrophysics (CfA), if he had any plans for the then-upcoming 8 June 2004 Transit. He did. With Much Hoole only 12 km from Preston, he had been thinking about this transit for many years.

He and I therefore began to plan and IAU C196 took shape. We wanted a meeting with history of astronomy mixed with fundamental astronomy and solar system astronomy. We received broad support for the meeting from two Divisions and many Commissions of the IAU, allowing us to build a strong, diverse SOC. With their advice, IAU C196 grew to be the eclectic meeting that you will find in these proceedings. For example, you will find here the amazing and surprising history of the English North-country astronomers of the 17th century; the definition of the AU at a precision of 1.4 m (not easy!); the explanation for the infamous black drop effect; a possible Mayan observation of a transit of Venus in the 13th century; a thorough discussion of the vexed question of leap seconds and time scales; history, distances, parallaxes, the solar system at exquisite precision and future space missions that will revolutionise fundamental astronomy yet again in the next decade.

We recorded the discussion of C196, thanks to the technical control of Hannah Worters and a team of UCLan graduate students. Nuala Jones then made a first transcription of the discussion and I went through it all again in detail. You will find, therefore, a real record of the discussion as it happened at the meeting. Of course, I cleaned up the inevitable ums and ahs, stops and starts, that occur in speech, but I stayed true to the style of the speaker. The speakers themselves were then given the chance to make corrections of fact to their discussion, but not to change the style of the actual discussion. You will thus find that the real flavour of the meeting comes through.

Gordon and I, and our team in the CfA at UCLan, wanted C196 to be a special event, as well as a scientific meeting. This was everyone's first transit of Venus; of course, and all would want to see it. On the day of the transit our programme began with an 05:15 BST (UT +1) departure from the University residences for the 20-minute trip to the UCLan's Alston Observatory where Arthur Missira with help from CfA staff and students had set up an array of telescopes with solar filters, with projection and (I have to admit that we did worry that it might be necessary) with real-time data projection of the transit from a guaranteed sunny site.

First contact was at 06:19:46 BST with the sun low in the east and cloud in the sky. We missed first contact. The clouds then broke about 1 minute later and the rest of the transit, including second contact, was easily observed by all. With a 6-hour transit in progress, people began to settle down to individual activities after the initial excitement, and we walked to Alston Hall, next to the Observatory, for a full breakfast. Pictures of the transit viewing and Alston Hall are in Part 8, and elsewhere in these proceedings.

With everyone very happy with the clear weather and the views of the transit, we then made the 30-minute drive to Much Hoole, where we toured Carr House and visited the room from which it is widely believed that Jeremiah Horrocks made his seminal observations in 1639. The BBC was there, too, broadcasting the transit live and interviewing Allan Chapman, one of the keynote speakers for C196. More telescopes had been set up in the grounds of Carr House by Robert Walsh and Stew Eyres of the CfA for continued viewing of the transit.

Following our tour of Carr House, the meeting participants then walked the kilometre to Much Hoole Church, where Horrocks had "higher duties" on the day of the transit in 1639. There Barbara Hassall and CfA graduate students had set up yet more telescopes and the public had turned out in huge numbers to experience the event and look through the telescopes.

Admittedly, we did have some problems with clouds on-and-off as third and fourth contact arrived, but we were able to see much of the end of the transit, and everyone was relaxed and happy by then. With fourth contact at 12:23:28 BST and more than 6 hours of transit viewing, we were well-rewarded for all the effort put in over the year of planning and organising of C196.

In addition to the professional astronomy, Gordon and I and our CfA team saw this transit as a marvellous opportunity for public outreach, as did all observatories in our hemisphere where it was observable. Robert Walsh organised a 6-month-long series of "Jeremiah Horrocks Memorial Lectures" that culminated in Michael Perryman's "Our Galaxy in three dimensions", given to both the meeting participants and the general public the evening after the transit. I had seen Michael give this presentation as an Invited Discourse at the Manchester IAU GA in 2000, and I knew we had to have it for C196; it makes one feel like a child again, seeing astronomy through new eyes. Michael was very willing, but warned me that I would need some specialised projection equipment that would cost a lot in both money and time. I blithely said "no problem, I'll do it", and the lecture was set. (But now I know what he meant!) Our 470-seat lecture hall was sold out for this event. You will find the lively discussion following the talk after Michael's paper; speakers who are "anonymous" were from the public.

Robert Walsh and Stew Eyres also included schools in our transit activities. Tarleton High School of Lancashire was with us at Carr House. They, Thomas Hardye School of Science in Dorchester and William Howard High School in Carlisle were given telescopes by us to observe the transit and calculate the AU. We also sent telescopes to South Africa through the SALT (Southern African Large Telescope) collateral benefits programme, run by Clifford Nxomani at the South African Astronomical Observatory. The CfA at UCLan is the leader of a UK consortium who are partners in SALT, and we took full advantage of this relationship to extend the schools' observations of the transits to the southern hemisphere. The Royal Society of New Zealand held a transit of Venus competition for high school students. The winners came on an expedition to the UK, and a group from Pakuranga College, led by Marilyn Head, also joined us for the meeting and transit viewing.

For our public outreach activities for the transit it is a pleasure to acknowledge support from ESO's VT2004 programme, from UCLan, from ESA and, of course, from the IAU.

With an early morning start on 8 June 2004 for participants and 6 hours of activities that morning, we then kept the afternoon of that day free for some rest, then held the conference banquet at the historic Hoghton Tower in the rolling Lancashire Hills just outside Preston. This fortified Manor House has been in the de Hoghton family since it was built in the 16th century; the land has been in their hands since it was given to them by William the Conqueror in 1066. It is widely believed that Shakespeare worked at the Tower in 1580-81, and the story is told that King James I was served at the high table in the Banqueting Hall a loin of beef he so liked that he knighted it on the spot, "Sir Loin". With the clear skies of the morning, the banquet was a real astronomical celebration.

During the planning for the meeting Mary Brück said to me, "You *are* going to arrange a tour of Stonyhurst Observatory for us, aren't you?" So I said "Of course we are!", and then I got busy to make it happen. Stonyhurst is a famous Catholic School many centuries old; Arthur Conan Doyle and Gerard Manly Hopkins are amongst its many illustrious Old Boys. And so is Chris Corbally of the Vatican Observatory. Chris kindly put me in touch with the current director of the Stonyhurst Observatory, Fintan O'Reilly, who also teaches at the school. Fintan gave us a tour of the observatory and some of the school gardens with great enthusiasm. Stonyhurst Observatory was built originally in 1838 and has had some distinguished directors: e.g. Father Cortie who led the British Solar Eclipse Expedition to the Tonga Islands in 1911, and Father Stephen Perry, FRS, who died in 1889 while making solar observations in the West Indies.

Stonyhurst College and Hurst Green village have another claim to fame: J.R.R. Tolkien's sons taught at the College and he spent a good deal of time there. It is locally believed that much of *The Lord of the Rings* was imagined during his stays at Stonyhurst. Names in the village support this, from Shire Lane to his favourite watering hole, The Shireburn Arms. It was at that pub that the participants had dinner following the tour of the Observatory.

The meeting was a great success with a delightful mixture of history, fundamental and solar system astronomy, with the observation of the transit of Venus, with outstanding public outreach, and with a unique social-astronomical programme to complement the science.

It is my pleasure to thank many people for the success of this meeting: I thank the owners of Carr House who so generously allowed all of us to visit their beautiful home (1611) during the transit and climb the winding stairs to the first floor room where Horrocks is thought to have made his 1639 observation – all while juggling the BBC's live broadcast with all of their equipment, cameras, microphones, wires and support crew. Arthur Missira of the CfA's Alston Observatory went far beyond the call of duty in setting up the Observatory for the transit viewing, with help from many of the staff and graduate students of the CfA. Rob Walsh and Stew Eyres worked day and night to set up of the telescopes at Carr House and get telescopes prepared and shipped to the high schools in the UK and South Africa. Hannah Worters and Nuala Jones were indispensable in making and transcribing the recording of the discussion at the meeting. The pictures in this volume were taken by Arthur Missira, Janet Strom and myself. Financial support was provided by the IAU, the University of Central Lancashire and ESA.

Finally, I would like to give special thanks to the IAU for supporting this meeting – with its unusual, successful mix of History and Science – as an IAU Colloquium.

D.W. Kurtz, co-chair SOC
Preston, Lancashire, UK

THE ORGANIZING COMMITTEES

Scientific

Don Kurtz (Co-chair, UK)
Gordon Bromage (Co-chair, UK)
Nicole Capitaine (France)
Dale Cruikshank (USA)
Suzanne Débarbat (France)
Steven Dick (USA)
Mike Feast (South Africa)

Julieta Fierro (Mexico)
Naoteru Gouda (Japan)
Anne Lemaitre (Belgium)
Mikhail Marov (Russia)
Wayne Orchiston (Australia)
Jay Pasachoff (USA)
Luisa Pigatto (Italy)

Local

Gordon Bromage (Chair)
Barbara Hassall
Steven Chapman
Peter Hingley
Nuala Jones

Don Kurtz
Paul Marston
Robert Walsh
Emma Woodward

Acknowledgements

This Colloquium was proposed by IAU Commissions 8 (Astrometry) and 16 (Physical Study of Planets and Satellites). The coordinating division was Division I (Fundamental Astronomy) and the supporting division was Division III (Planetary Systems Sciences). Supporting Commissions were: 4 (Ephemeredes), 7 (Celestial Mechanics and Dynamical Astronomy), 8 (Astrometry), 16 (Physical Study of Planets and Satellites), 19 (Rotation of the Earth), 33 (Structure and Dynamics of the Galactic System) and 41 (History of Astronomy).

Financial support was provided by the University of Central Lancashire, the IAU and ESA.

CONFERENCE PHOTOGRAPH

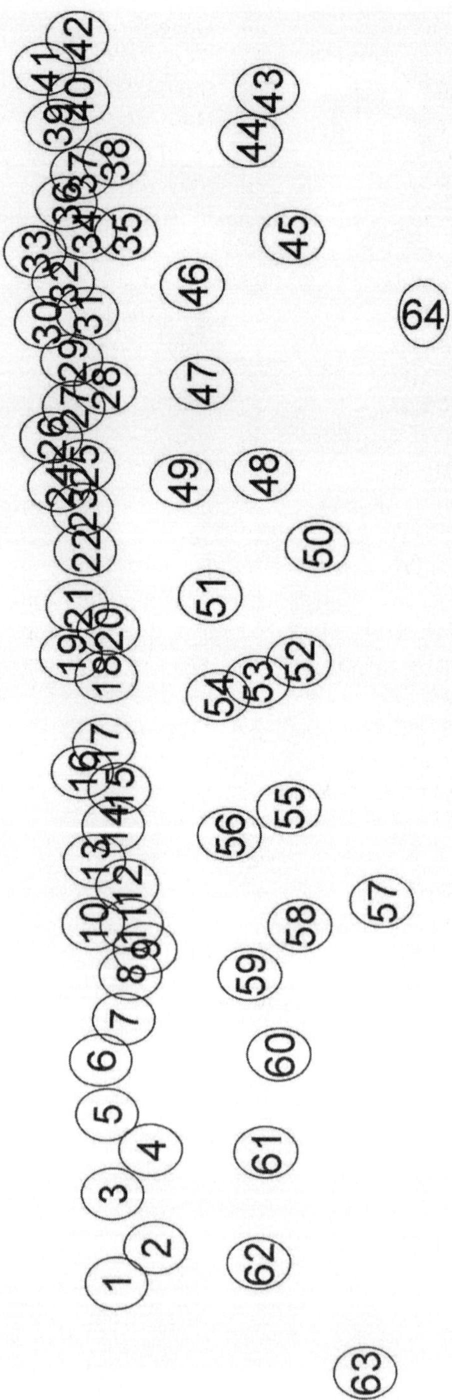

1	Brian Warner	33	Dave Monet
2	Brenda Shepherd	34	Alice Monet
3	Malcolm McVicar	35	Sylvia Monet
4	Margery Miller	36	Dennis McCarthy
5	Giovanni Gronchi	37	Diane McCarthy
6	Taihei Yano	38	Margaret Morris
7	Jon Posonby	39	Walter Brisken
8	Christine Allen	40	David Clarke
9	Anne Lemaitre	41	John Butler
10		42	Don Kurtz
11	Janet Strom	43	Allan Chapman
12	Jeannie Allen	44	
13	Richard Strom	45	Andreas Papageourgiou
14	Nicole Capitaine	46	Yukiyasu Kobayashi
15	Elena Pitjeva	47	Suzanne Debarbat
16		48	Kleomenis Tsiganis
17	Mikhail Marov	49	Jesús de Alba Martinez
18	Sandrine D'Hoedt	50	Barbara Bromage
19		51	Coryn Bailer-Jones
20	Mary Brück	52	Wayne Orchiston
21	Bernard de Saedeleer	53	Michael Perryman
22	Hideyuki Kobayashi	54	Gert Zech
23	Ed Budding	55	Steve Dick
24	Fritz Benedict	56	David Sellars
25	Yoshiyuki Yamada	57	Gordon Bromage
26	Myles Standish	58	Anne Samson
27	Floor van Leeuwen	59	Benoit Noyelles
28	Naoteru Gouda	60	Steven Chapman
29	Nick Kollerstrom	61	Jacqueline Mitton
30	Ricky Smart	62	Simon Mitton
31	Takuji Tsujimoto	63	Clive Davenhall
32	Seiji Ueda	64	Jaymie Matthews

Participants

Christine **Allen**, Instituto de Astronomia, UNAM, Mexico	chris@astroscu.unam.mx
Coryn **Bailer-Jones**, Max-Planck-Institut für Astronomie, Germany	calj@mpia.de
Steve **Bell**, Rutherford Appleton Laboratory, UK	sab@nao.rl.ac.uk
Fritz **Benedict**, McDonald Observatory, USA	fritz@astro.as.utexas.edu
David **Berry**, University of Central Lancashire, UK	dsb@ast.man.ac.uk
Martin **Birch**, University of Central Lancashire, UK	mjbirch@uclan.ac.uk
Kate **Bird**, University of Central Lancashire, UK	kate.bird@ghg.org.uk
Walter **Brisken**, National Radio Astronomy Observatory, USA	wbrisken@nrao.edu
Barbara **Bromage**, University of Central Lancashire, UK	bjibromage@uclan.ac.uk
Gordon **Bromage**, University of Central Lancashire, UK	gebromage@uclan.ac.uk
Mary **Brück**, University of Edinburgh, UK	mt.bruck@virgin.net
Ed **Budding**, COMU, Turkey and Carter Obs., New Zealand	ebudding@comu.edu.tr
Ian **Butchart**, University of Central Lancashire, UK	ibutchart@uclan.ac.uk
John **Butler**, Armagh Observatory, UK	cjb@arm.ac.uk
Nicole **Capitaine**, Observatoire de Paris, France	nicole.capitaine@obspm.fr
Allan **Chapman**, Modern History Faculty, Oxford University, UK	
Steven **Chapman**, University of Central Lancashire, UK	sachapman@uclan.ac.uk
David **Clarke**, Glasgow University, UK	d.clarke@astro.gla.ac.uk
Phil **Cox**, University of Central Lancashire, UK	pccox@uclan.ac.uk
Mark **Cropper**, MSSL/UCL, UK	msc@mssl.ucl.ac.uk
Clive **Davenhall**, University of Edinburgh, UK	acd@roe.ac.uk
Durruty Jesús **de Alba Martinez**, University of Guadalajara, Mexico	dalba@astro.iam.udg.mx
Suzanne **Débarbat**, Observatoire de Paris, France	suzanne.debarbat@obspm.fr
Ronaldo **de Freitas Mourão**, Museu de Astronomia do Rio de Janeiro	mourao@ronaldomourao.com
Bernard **De Saedeleer**, University of Namur (FUNDP), Belgium	Bernard.Desaedeleer@fundp.ac.be
Sandrine **D'Hoedt**, University of Namur (FUNDP), Belgium	sandrine.dhoedt@fundp.ac.be
Steven **Dick**, NASA HQ, USA	steven.j.dick@nasa.gov
Vladimir **Elkin**, University of Central Lancashire, UK	vge@sa1.star.uclan.ac.uk
Clive **Elphick**, Much Hoole, UK	clive.elphick@uuplc.co.uk
Stewart **Eyres**, University of Central Lancashire, UK	spseyres@uclan.ac.uk
Naoteru **Gouda**, National Astronomical Obs. of Japan, Japan	naoteru.gouda@nao.ac.jp
Giovanni **Gronchi**, University of Pisa, Italy	gronchi@dm.unipi.it
Barbara **Hassall**, University of Central Lancashire, UK	bjmhassall@uclan.ac.uk
Marilyn **Head**, Royal Society of New Zealand, New Zealand	debbie.woodhall@rsnz.org
Stephen **Higgins**, University of Central Lancashire, UK	swhiggins@uclan.ac.uk
Philip **Judge**, National Centre for Atmospheric Research, USA	judge@ucar.edu
Hideyuki **Kobayashi**, National Astronomical Obs. of Japan, Japan	hkobaya@hotaka.mtk.nao.ac.jp
Yukiyasu **Kobayashi**, National Astronomical Obs. of Japan, Japan	yuki@merope.mtk.nao.ac.jp
Nicholas **Kollerstrom**, University College London, UK	nk@astro3.demon.co.uk
Don **Kurtz**, University of Central Lancashire, UK	dwkurtz@uclan.ac.uk
Anne **Lemaitre**, University of Namur, Belgium	anne.lemaitre@fundp.ac.be
Mikhail **Marov**, M.V. Keldysh Institute, RAS, Russia	marov@spp.keldysh.ru
Michael **Marsh**, University of Central Lancashire, UK	msmarsh@uclan.ac.uk
Paul **Marston**, University of Central Lancashire, UK	pmarston@uclan.ac.uk
Jaymie **Matthews**, University of British Columbia, Canada	matthews@astro.ubc.ca
Dennis **McCarthy**, U.S. Naval Observatory, USA	mccarthy.dennis@usno.navy.mil
Jim **Message**, University of Liverpool, UK	sx20@liverpool.ac.uk
Arthur **Missira**, University of Central Lancashire, UK	amissira@uclan.ac.uk
Jacqueline **Mitton**, Royal Astronomical Society, UK	jmitton@dial.pipex.com
Simon **Mitton**, St Edmund's College, University of Cambridge, UK	smitton@cambridge.org
Dave **Monet**, US Naval Observatory, USA	dgm@nofs.navy.mil
Margaret **Morris**, University of Glasgow, UK	mmorris671@aol.com
Jane **Noglik**, University of Central Lancashire, UK	jbnoglik@uclan.ac.uk
Mark **Northeast**, University of Central Lancashire, UK	MSNortheast@uclan.ac.uk
Benoit **Noyelles**, IMCCE, France	Benoit.Noyelles@imcce.fr
Wayne **Orchiston**, Anglo-Australian Observatory, Australia	wo@aaoepp.aao.gov.au
Andreas **Papageourgiou**, University of Central Lancashire, UK	alp@sa1.star.uclan.ac.uk
Michael **Perryman**, European Space Agency, Netherlands	mperryma@rssd.esa.int
Luisa **Pigatto**, University of Padova, Italy	pigatto@pd.astro.it
Elena **Pitjeva**, Institute of Applied Astronomy RAS, Russia	evp@quasar.ipa.nw.ru
John **Ponsonby**, ex-Jodrell Bank, UK	johnpon@supanet.com
Dimitri **Pourbaix**, Universite Libre de Bruxelles, Belgium	pourbaix@astro.ulb.ac.be
Jon **Riley**, University of Central Lancashire, UK	jdriley@uclan.ac.uk
Anne **Sansom**, University of Central Lancashire, UK	AESansom@uclan.ac.uk
David **Sellers**, UK	David.Sellers@ntlworld.com
Ricky **Smart**, Torino Observatory, Italy	smart@to.astro.it
Ignas **Snellen**, Institute for Astronomy, Edinburgh, UK	ignas@roe.ac.uk
Myles **Standish**, JPL, USA	ems@smyles.jpl.nasa.gov
Richard **Strom**, ASTRON & University of Amsterdam, Netherlands	strom@astron.nl
John **Southworth**, Keele University, UK	jkt@astro.keele.ac.uk
Kleomenis **Tsiganis**, CNRS/Observatoire de Nice, France	tsiganis@obs-nice.fr
Takuji **Tsujimoto**, National Astronomical Obs. of Japan, Japan	taku.tsujimoto@nao.ac.jp
Seiji **Ueda**, Graduate University of Advanced Studies, Japan	seiji.ueda@nao.ac.jp
Stephane **Valk**, University of Namur (FUNDP), Belgium	stephanevalk@fundp.ac.be
Robert **van Gent**, Institute for History & Foundations of Mathematics, Netherlands	r.h.vangent@astro.uu.nl
Floor **van Leeuwen**, Institute of Astronomy, Cambridge, UK	fvl@ast.cam.ac.uk
Robert **Walsh**, University of Central Lancashire, UK	rwwalsh@uclan.ac.uk
John **Walton**, University of Central Lancashire, UK	jkwalton@uclan.ac.uk
Brian **Warner**, University of Cape Town, South Africa	warner@physci.uct.ac.za
Hannah **Worters**, University of Central Lancashire, UK	hlworters@uclan.ac.uk
Yoshiyuki **Yamada**, Kyoto University, Japan	yamada@amesh.org
Taihei **Yano**, National Astronomical Obs. of Japan, Japan	yano.t@nao.ac.jp
Gert **Zech**, Astronomisches Rechen-Institut, Germany	zech@ari.uni-heidelberg.de

Part 1

TRANSITS OF VENUS: HISTORY, RESULTS AND LEGACY

John Butler and Allan Chapman view the start of the transit of Venus,
8 June 2004

Transits of Venus: New Views of the Solar System and Galaxy
Proceedings IAU Colloquium No. 196, 2004
D.W. Kurtz, ed.

© 2004 International Astronomical Union
doi:10.1017/S1743921305001225

Jeremiah Horrocks, William Crabtree, and the Lancashire observations of the transit of Venus of 1639

Allan Chapman

Modern History Faculty, Wadham College, Oxford University, Oxford OX13BD, UK

Abstract. When Jeremiah Horrocks correctly predicted, and with his friend William Crabtree observed, the Venus transit of 24 November 1639, these two men became more than the first astronomers in history to witness a rare celestial phenomenon. For Horrocks's and Crabtree's achievement constituted in may ways the first major astronomical discovery to be made in Renaissance England. It is also clear from their writings, moreover, that the two men, working in the isolation of rural Lancashire and well away from London or the universities, were fully conversant with contemporary discoveries made in continental Europe by Tycho Brahe, Galileo, Kepler, Gassendi, and others. In many ways, therefore, their work begs more questions than can easily be answered, such as why the rural North-West produced not only Horrocks and Crabtree, but other contemporary astronomers such as the Lancastrians Charles Towneley, Jeremy Shakerley, and their Yorkshire friend William Gascoigne. Yet in addition to whatever regional circumstances might have been present, and how easy it might have been for an educated rural Lancastrian to be fully informed about what astronomers in Paris, Prague, or Florence were doing, what cannot be denied is the outstanding originality of their wider achievement. For Jeremiah Horrocks in particular was a physical scientist of genius. His correct determination of the elliptical shape of the lunar orbit by 1638 when he was about 20 and his wider work on planetary dynamics place him amongst the most creative researchers of the seventeenth century. Central to Horrocks's and Crabtree's achievement was Crabtree's realisation by 1636 that contemporary published astronomical tables were unreliable, and that if one wanted to do serious work in understanding the heavens, then one had to observe and measure them for oneself, and learn to draw original conclusions.

Standing in the early dawn light, at the University of Central Lancashire's Alston Observatory at Longridge, Preston, as the Sun rose on 8 June 2004, was a moving experience. For as it broke through the low cloud on that hot and sultry summer morning, we all saw the Sun as no human being for the last 120 years had seen it: with a distinct semi-circular bite taken out of its lower left limb (or upper right, as seen through a telescope). Venus, indeed, had begun to transit at 6.24 a.m. B.S.T., but first contact could not be observed, and not until 'Cythera', as the ancients called her, was already indenting the solar limb about 6.30, could we see with our own eyes that the 2004 transit of Venus was actually happening. Second contact, at which point Venus makes her first appearance as a distinct disk on the solar surface, was witnessed at 6.44 a.m., and after that we set about the leisurely process of watching the planet making its stately way across the Sun. Then after a hearty breakfast at the Hall, the IAU party was taken by bus to the Lancashire village of Much Hoole, 8 miles south-west of Preston, to observe Venus as she gently slid off the face of the Sun around lunch time.

Of course, all the astronomers at Alston Hall had state-of-the-art equipment. High-resolution catadioptric and reflecting telescopes, mylar and other filters, binoculars, and the great Wilfred Hall equatorial refractor, the 15-inch object-glass of which could produce a projected image of Venus that was as big as a tennis ball, were all freely available on site, thanks to the University. Very importantly, we were not only able to predict the elements of the transit with great accuracy, but also to know in advance how big the Venusian image would appear on the solar disk, and how fast it would move. For these last crucial nuggets of information we really had two men to thank. They were Jeremiah Horrocks and William Crabtree, who between them had successfully computed and observed the first recorded transit of Venus on 24 November 1639 (4 December New Style), and had told us what to expect. What brought the IAU to Preston and the North-West to observe and celebrate the 2004 transit was the fact that Horrocks and Crabtree were local lads, and subsequent British scientists came to recognise them as the founding fathers of research astronomy in these islands.

And just as astronomers had gathered in Alston Hall and Much Hoole, so some 30 members of the Salford Astronomical Society had brought telescopes to various parts of the city of Salford, largely under the organisation of Ken Irvine, Director of the Salford Observatory, where they both observed the transit scientifically and demonstrated it to the public. For it had been in Salford that Crabtree had observed the 1639 transit.

By the time of the 1639 Venus transit, what we might think of as 'modern' astronomy was already advancing rapidly. But within the science several major debates were taking place, not only about the structure of the heavens themselves, but also about mankind's capacity to unravel that structure. The classical philosophy of the Greeks, for instance, had been concerned with defining the nature of the absolute and eternal truths, such as those of geometry, and had largely worked on the assumption that the way to get at truth itself was through disciplined intellectual deduction, based upon precise axioms. Sense knowledge was mistrusted, largely because the senses could, on a certain level, be so capricious. For simple human perceptions could vary across a very wide spectrum, giving rise to endless opinion and debate about what exactly was seen by a given individual, whereas the intellectual truths of deductive geometry and mathematics seemed to be beyond dispute. Indeed, this perceptual problem lay at the heart of the scientific revolution, for much of the new data that were pouring out not only of astronomy, but also of the sciences of geography and anatomy, were sensory and not deductive in character. For who could have deduced the existence of the American continent from first principles laid down by Strabo, Ptolemy, and the other classical geographers; and how could William Harvey in 1628 have discovered the circulation of the blood in living creatures from a careful reading of Hippocrates and Galen? In fact, all of these discoveries were the products of new physical, exploratory, and experimental researches that led human experience into hitherto uncharted territories of understanding. For it seemed that if one could discipline the human senses by means of the right procedures and techniques – such as the use of specially designed instruments which enhanced human perception, and a system of public peer review whereby results could be verified across a community of observers – then one really could discover those new continents of knowledge of which the visionary philosopher Sir Francis Bacon had written.

Where the new scientific approach of that period which we call the European Renaissance differed from that of the classical and medieval worlds lay not just in its recognition of the true value of sense-knowledge, but in the way in which it developed that knowledge within a method that embodied an investigative or forensic attitude to nature. For as Bacon had recognised, and begun to explore in *The Advancement of Learning* (1605) and *Novum Organon* ('The New Method', 1620), the experimental method of science was capable of opening up and researching new fields of natural knowledge that had been closed to traditional understanding. For whereas classical and medieval science had been what might be called curatorial in its approach, passing on a given body of facts and interpretations, such as the geometry of Euclid and the deductive physics of Aristotle, the experimental method aimed to break into pastures new and go beyond the systems of the ancient philosophers. And in many ways, this is why the great geographical discoveries of Columbus and Magellan had exerted such a profoundly imaginative effect on Renaissance scientific thinking: because these continental discoveries were not only sensory in character, but also subject to independent verification by other navigators. It had not been for nothing that the allegorical title page of Bacon's *Novum Organon* had depicted a modern three-masted ship sailing out through those very Pillars of Hercules which the ancients believed marked the westernmost boundaries of human knowledge, to make new discoveries in that great ocean of wonders that lay beyond.

And in the same way that one could think of an explorer's ship as a scientific instrument that facilitated the discovery of natural truths which could never have been revealed without it, so this new forensic, experimental approach to nature quickly gave rise to new instruments that took us to and opened up hitherto unknown realms of nature. Those unprecedentedly accurate angle-measuring instruments devised by Tycho Brahe between 1572 and 1598, for instance, were contrived with the expectation of providing firm physical proof, either for or against the geocentric cosmology of Copernicus; and though failing in this original intention because the required accuracies were beyond the manufacturing capacities of the age, they nonetheless generated a vast new harvest of astronomical facts which were to provide the foundation for the researches of Kepler, Gassendi, Horrocks, Crabtree, and Gascoigne.

But it was the telescope, first used for scientific purposes by Galileo in 1609, which became arguably the most momentous scientific instrument of the Renaissance. For the telescope fundamentally changed mankind's understanding of the universe, showing the Moon to be a world possessing distinct topographical features, the planets to be rotating spheres and not mere points of light, and the stars to be countless in number and receding ever dimmer into a seeming infinity. By the mid-1630s, when Horrocks, Crabtree, and Gascoigne were beginning to make their own highly original contributions, continental European astronomy was firmly established as the fastest advancing and the most intellectually provocative of all the sciences. What seemed to give astronomy such intellectual authority was its susceptibility to instrumental quantification. For whilst medicine and the life sciences, in spite of major discoveries in anatomy, were still incapable of curing most illnesses, and chemists, because of their fundamentally incorrect understanding of the nature of matter, were unable to explain chemical change in a coherent fashion, astronomy seemed to go from strength to strength. Much of that development hinged upon the fact that, unlike medicine and chemistry, astronomy was not a comparative or qualitative science, but a quantitative, exact one. For the new Tychonic measuring instruments and post-Galilean telescopes made it possible to measure angles to an increasing level of precision, and thereby establish physical standards based upon the division of the 360° circle, to make astronomy a truly predictive science. Conversely, medicine, botany, chemistry, mineralogy, and meteorology still lacked any distinct criteria which rendered them amenable to precise quantification, so that they were left thrashing around in a myriad of discrete observed empirical facts that lacked any kind of coherent interpretative or predictive structure.

It would be wrong, however, to see astronomy's progress between 1570 and 1635 as simple, uncomplicated, and based upon self-evident hard facts. There were still big issues, both intellectual and technical, that awaited resolution, some of which are still with us today. One might say that three especial problems were felt to be in need of solution by the astronomers of the early seventeenth century. The first of these was to establish the truth or falsehood of the Copernican heliocentric theory. The second was to ascertain the size, or perhaps the infinity, of the universe. And the third was to understand the nature of the force which made the planets move through space with such unchanging regularity. For once Tycho Brahe's new star and cometary observations between 1572 and 1588 had seriously undermined the credibility of the ancient crystalline spheres, the question remained as to how planets and comets moved through a seeming void with no apparent means of support or impulse. Did they move under the action of some form of magnetism or unspecified occult force, or were their periods and velocities somehow governed by their individual ponderousness and distances from each other?

While none of these questions could be adequately answered even by 1650, it is important to bear in mind that they assumed a vital importance in the creative scientific imaginations of Horrocks, Crabtree, and Gascoigne; as their surviving writings so clearly testify, these three North-country astronomers wrestled with them mightily in that creative dynamic developing between practical observation and theory out of which all original science springs. Without any doubt, they were the first Englishmen to work consistently with true originality towards this end.

Why was work of this originality brought to life in a thinly-populated strip of northern England lying between Liverpool to the west, Leeds to the east, Preston in the north and Salford in the south? And how, one might well ask, did Jeremiah Horrocks, working perhaps as a teacher and Bible Clerk in Toxteth, Liverpool, and Much Hoole; William Crabtree, a Salford cloth dealer; and William Gascoigne, the son of a minor county gentleman of Middleton, Leeds, obtain the resources to actively advance the great intellectual movement going on in continental Europe? For in Europe, the great new discoveries in astronomy and the other sciences came out of rich cities, royal courts, universities or ecclesiastical chapters – of great centres of patronage in Florence, Paris, Copenhagen, and elsewhere. Why creative English astronomy took root in relatively remote and rural Lancashire and west Yorkshire is hard to tell, though it did initiate what would, in many ways, be something of an enduring tradition in British science: for more than a couple of centuries to come, and well into the nineteenth century, it was indeed the inspired, informed 'Grand Amateurs', often of modest yet independent circumstances, who were to form the intellectual cutting edge of British science, and who addressed themselves to the most profound questions of the age.

While Horrocks, Crabtree, and Gascoigne may have lived in a geographically remote part of Britain, as far as metropolitan or direct continental influences were concerned, one must not forget that each of these men had received a sound education: two of them had attended university, and all had enough spare cash and access to those lines of communication necessary to obtain copies of the works of Tycho, Kepler, Galileo, Descartes, Gassendi, Lansberg, and several other contemporary scientific researchers. For the names and ideas of these and other men are regularly mentioned and evaluated in the surviving correspondence of the North-country astronomers.

As Peter Aughton has recently shown, from surviving wills of the Horrocks and Aspinwall families, and from Oxford and Cambridge University registers, several of Jeremiah Horrocks's relatives had preceded him to university. For while the Horrockses and Aspinwalls (on his mother's side) were successful watch-makers and yeomen farmers active in the Toxteth area, they clearly valued education, making sure that clever sons went at least to grammar school, and often to university. While I have not been able to trace William Gascoigne's name in any registers, he nonetheless claimed in one of his surviving letters to have been at Oxford. William Crabtree made no university claims, it is true, but he is said to have attended the 'Manchester School', which was probably a grammar school attached to the Collegiate Church (which became Manchester Cathedral in 1847). This means, therefore, that all three men would have been Latin literate – the key to the world of learning in the early seventeenth century – and Chetham's College

Library (in the School founded by Humfrey Chetham in 1653, and still standing to the immediate north of Manchester Cathedral) later acquired one of William Gascoigne's notebooks, in which he had made several juvenile attempts to correctly spell his name in Latin! Without the Latin tongue, indeed, the works of Tycho, Kepler, Gassendi and others would have been closed to them.

Through his work in the textile trade, William Crabtree is also likely to have had connections in London, and the Crabtree and Horrocks correspondence contains references to Samuel Foster, Professor of Geometry at Gresham College, London. Samuel Foster, indeed, was one of the men whom Crabtree, no doubt at Horrocks's request, appears to have alerted to the impending transit of Venus in 1639, though Foster does not seem to have made a successful observation.

Yet while the scientific brilliance and originality of Horrocks and his friends in the north is beyond doubt, it would be entirely wrong to think of them as the only English people interested in modern astronomy in that age. Indeed, Thomas Harriot, the mathematical friend of Sir Walter Raleigh and the 'Wizard' Earl of Northumberland, was a convinced Copernican, as had been Thomas Digges a generation before in the 1570s. Harriot, moreover, was already using a newly-imported 'Dutch spyglass' to observe sunspots through the thick winter mists of the Thames valley as early as 8 December 1610, which was several months ahead of Galileo's 'official' discovery of the spots in 1611; though neither Harriot nor his Welsh friend Sir William Lower ever published their results, so that their original work only saw the wider light of day after 1833 when some of Harriot's surviving manuscripts were printed in Stephen P. Rigaud, 'Account of Harriot's Astronomical Papers', and followed by more detailed researches by twentieth-century Harriot scholars. And in addition, there were Samuel Foster's predecessors at Gresham College, London, which after its foundation in 1597 had produced a succession of illustrious Professors of Astronomy and Geometry: men such as Henry Briggs, Henry Gellibrand, and Edmund Gunter. In 1619, moreover, Sir Henry Savile endowed Oxford University with two scientific chairs, once again in Astronomy and Geometry; and if Gascoigne had in fact studied at Oxford, as he later claimed, it is likely that he could have attended the lectures of John Bainbridge, who was Savilian Professor of Astronomy throughout the 1620s and 1630s, and was probably the first man to publicly lecture on Galileo and Kepler in Oxford. Likewise, Gascoigne may have heard the lectures of Henry Briggs, who left Gresham College to become Oxford's first Savilian Professor of Geometry up to his death in 1631. Both Bainbridge and Briggs were convinced Copernicans.

How, therefore, can one claim such outstanding originality for Horrocks, Crabtree, and Gascoigne, if England already enjoyed not only this richness in mathematicians and astronomers, but also a flourishing London-based trade in mathematical instrument-making, as men like Humfrey Cole and Elias Allen manufactured large numbers of accurate quadrants, dials, and precision computational scales?

I would argue that what made the North-country astronomers so important was their attitude to research and the seeking out and testing of new astronomical knowledge, for the London and Oxford professors, in so many ways, had different areas of interest from the northern group. Indeed, their own especial interest in experimental research tended more to the study of geomagnetism than to planetary astronomy; while another long-term interest of the London and Oxford professors lay in the improvement of computational techniques – such as the development of logarithms, which had been pioneered by Henry Briggs – for the wider mathematical community. Yet when it came to astronomy itself, they still seemed happy to work within the wider ground rules laid down by Tycho Brahe, Kepler, and others, and to use printed volumes of planetary tables (often deriving from Tycho's original observations) as a way of calculating their celestial triangles.

None of these London or Oxford astronomers, therefore, seem to have been active, night-by-night practical observers of the heavens – and here lies the crunch. For if one simply uses a cross-staff, quadrant, or similar angle-measuring instrument to check occasional planetary angles against the standard tables, then one is more likely to be able to explain away errors on scale mis-graduation and similar grounds than if one is experiencing the mounting nightly frustration of finding such a proliferation of errors as to make one declare the tables false or useless, as William Crabtree and Jeremiah Horrocks declared the tables of Philip Lansberg to be by 1636!

Another important factor in assessing an astronomer's originality was how he used his telescope. Did he see the telescope as a way of merely confirming Galileo's discoveries, or did he want to devise new usages in order to extract a bit more research mileage out of the simple optical systems of the 1630s? For it is clear that, by the 1630s, telescope technology had hit an *impasse*, as object glasses and magnifications had not improved much since about 1615, and would not do so significantly till the new optical innovations of Giuseppi Campani and Eustachio Divini and the Huygens brothers in the 1650s. So considering the very limited capacity of contemporary telescopes, how might a skilled observer use his instrument to give him that extra mileage? By using it, perhaps, to watch the dark edge of the Moon snuff out the individual stars of the Pleiades and thence draw the conclusion that the stars were but points of light and not disks – as Horrocks and Crabtree did on the night of 19 March 1637 [*Op. Post.* 151]; by contriving to turn it into an angle-measuring micrometer (as Gascoigne did, as will be shown presently); or else by using it to observe the transit of Venus!

For these were some of the areas in which Horrocks and his friends displayed such astonishing scientific creativity in the late 1630s. To them, indeed, astronomy had ceased to be the essentially passive science of error correction, and became in their hands a more aggressive, forensic intellectual discipline that aimed to go beyond what the past had laid down, and work by a self-conscious process of inquisition. While the great figures of modern astronomy, such as Galileo, Kepler, and Gassendi, had discovered stunning new facts, and while Gassendi's published observations show that he was constantly putting Tycho to the test, it was in many respects the North-country astronomers who came to think of astronomy from an 'experimental' perspective. One wonders, indeed, whether these men were familiar, if only by repute, with the above-mentioned writings of Sir Francis Bacon, whose emphasis upon taking nothing in nature on trust, and on using 'experiment' as a tool whereby one might inquire forever deeper into the structures of the natural world, seemed so clearly reflected in what was going on in Preston, Salford, and Leeds.

There is nothing to suggest that, up to about 1640, Horrocks and his circle were using instruments that were innately superior to those being used in London, Oxford, or in the great cities of continental Europe. Even the best telescopes at this period rarely had effective object glasses of more than 1.5 inches in diameter, and a clear magnification of 40 or 50 times would have been doing very well for that time. And while one might have made guesses at very small angles, such as those displayed between the stars of the Pleiades, by comparing such angles to the known 30′ of the lunar diameter, it was not possible to use the telescope as a direct measuring, as opposed to a viewing, instrument. In fact, the only way in which one could make measurements through a telescope before 1640 was by using the telescope to project an image of the Sun, as a preliminary to ascertaining the relative positions of objects upon its surface, as Galileo, Father Christopher Scheiner, and others did with sunspots, Gassendi and Quietanus did with the Mercury transit of 1631, and Horrocks and Crabtree did with the Venus transit in 1639.

To measure larger angles around the sky, one used one or more instruments of a type that had been familiar to European astronomers for centuries. These could include the 90° graduated quadrant, for reading vertical angles. Or if one wanted to measure the horizontal (or 'Right Ascension') angles between, let us say, the Moon and a bright star, then one used a version of the 'Astronomical Radius' or 'Baculus'. Astronomical Radii were 'T'-shaped configurations, often made in wood, where the short end of the 'T' was graduated with a tangent scale against which the observer could measure his angles when the Radius was held up to his eye, in the manner of a crossbow. By the 1630s, indeed, European astronomical literature was rich in the details of how one might design and make a 'Radius' and a quadrant, and there is no reason to believe that the probably home-made angle-measuring instruments of Horrocks and Crabtree were any better than those described by their older Parisian contemporary, Pierre Gassendi, with whose published works they were clearly familiar. I would suggest that what made the northern astronomers so significant is the manner in which they used them.

Having become disillusioned with the reliability of contemporary astronomical tables by 1636, William Crabtree and then Jeremiah Horrocks took it upon themselves to measure celestial angles on a nightly basis after the manner of Tycho Brahe. The primary objects of concern were the Moon, Sun, and planets as they moved around the ecliptic against the background of the Zodiac stars, and their movements with relation to each other. As a consequence, Jeremiah Horrocks in particular began to draw some remarkable conclusions from the emerging data. For one thing, over a given period of time Saturn seemed to be slowing down while Jupiter was speeding up, in a way which was quite inexplicable in terms of existing planetary theory, but seemed to suggest that the planets exerted some kind of attraction upon each other. But more immediately significant was the behaviour or the Moon.

The dynamics of the lunar orbit around the Earth, and its relation to the Sun, had long been acknowledged as one of the most baffling in astronomy. That the Moon did not rotate around the Earth in a simple circular orbit, but in an eccentric circle, had been known for centuries, and aspects of 'transit' or eclipse phenomena had long been the indicator. Why, for instance, were some total eclipses of the Sun – when the Moon transits the solar disk – annular, while others were not? Could it be that the Earth-Moon distance varied in accordance with a complex cycle? And why did eclipses of the Moon, when the Earth, Moon, and Sun were on a straight axial line, come in such complex varieties? For instance, the terrestrial shadow could be bronze, black, or blood-red in colour, and cover the whole or only part of the lunar disk. Of course, Hipparchus back in the second century BC had developed his famous 'Diagram' of the shadows projected within the Earth, Moon, and Sun system, and had even used them to compute a remarkably accurate proportionate distance of the Moon from the Earth in terms of Earth radii; yet the dynamical astronomy of the Moon was still replete with unanswered questions.

By 1638 Jeremiah Horrocks, who had a detailed familiarity with the celestial mechanics of the 'Hipparchian Diagram', had come to an analytical understanding of the Earth-Moon system that was substantially in advance of anything that had been achieved by any preceding or contemporary astronomer. It was based, moreover, on his own and Crabtree's original observations, and went on to beg wider questions in planetary dynamics, such as the nature of the force that moved astronomical bodies around their centres of rotation. For by 1638, Horrocks had come to realise that the Moon moved around the Earth not in an eccentric circle – where the lunar orbit itself was a perfect circle, but where the Earth did not occupy the geometrical centre of that circle – but in an ellipse. In this respect, he became the first astronomer to go beyond Kepler's work on the elliptical orbit of Mars around the Sun, to demonstrate that another astronomical body in the solar system also moved in an elliptical orbit, with one of the foci of the lunar orbit within the Earth. Once such a realisation had been made, and the eccentric circle model abandoned, several long-standing anomalies about the lunar orbit became amenable to explanation.

Yet Horrocks had not merely realised that the lunar orbit around the Earth fitted neatly with a Keplerian ellipse, but he went on to draw an even more profound dynamical conclusion: namely, that the long axis of this ellipse – or apside line – itself possessed an independent rotation in space over a period of time. In short, the whole Earth-Moon system was rotating around a point in space, while the dynamics of the system seemed to relate the whole to the constantly changing position of the Sun. Horrocks's own interest in the tides further related to this problem, as variations in the Earth's tidal patterns seemed to be governed by close and further approaches by the Moon, and also by the Sun. Indeed, the complex dynamical relationship between the Earth, Moon, and Sun, and the moving elliptical planes operating within this system, which Jeremiah Horrocks had been the first astronomer to describe coherently, would later become immortalised as Sir Isaac Newton's 'three bodies problem', and would play a major role in the development of the theory of Universal Gravitation. And Newton would acknowledge his fellow-Cantabrigian in *Principia* (1687).

Horrocks's realisation of the ellipticity of the lunar orbit, and of the apsidal precession of that orbit, was very much the product of meticulous observation and independent analysis. By 1637 or so, one can see Horrocks looking for further cross-checks whereby to test his overall ideas, and this becomes particularly manifest in his concern with seasonal and systematic variations in the diameter of the Sun and Moon. Did the Moon's (and Sun's) diameter over a given period vary in accordance with predictions derived from elliptical orbital criteria? Of course, it had been known from classical times that the lunar and solar diameters were around $30'$, or $0°5$. What Horrocks needed, however, were very slight yet crucial variations around this figure to an accuracy of a few arc seconds.

There were two ways of attempting to obtain these values. The first, which had even been known to medieval astronomers, was to make a pinhole projection of the Sun into a darkened room. By measuring the exact projection distance between the pinhole and the solar image falling on a screen, along with the precise diameter of that solar image, one could obtain the angular diameter as a tangent function. Indeed, as early as 1636, Horrocks and Crabtree were monitoring changes in the solar diameter by means of this projection technique, and finding that on 12 December 1636 [*Op. Post.* 140], for example, with a pinhole aperture of 5 units, at a projection distance of 4100 units, the solar diameter was $42\frac{1}{2}$ units, or $31'\,26''$. (Seventeenth-century astronomers generally worked in geometrically-related proportions rather than specific units of measure, but if Horrocks's 5-unit projection aperture was around one-twenty-fourth of an inch across – '*quarum uncia nostra Anglica habet 24*' – then the projection distance would have been about 14 ft 3 inches, and the solar diameter about $1\frac{3}{4}$ inches.) A brilliant full Moon diameter could also be measured by the same technique. Horrocks was fully familiar with the projection method, though its main defect lay in the inevitable slight fuzziness which a pinhole image produced, leading to an inevitable margin of error when trying to work to a critical level of accuracy. In the Latin translation of Horrocks's letter to Crabtree which John Wallis produced in 1673, this pinhole technique was referred to as the 'Foramen' or aperture method.

The second method was really a refinement of the Astronomical Radius mentioned above. Horrocks referred to it as his instrument '*filis ferries*', or with two iron needles or fine pointers, on 4 January 1637 [*Op. Post.* 255]. A pair of metal pointers, needles, or stretched wires (? *filum*) would be mounted vertically and parallel to each other in the end of a long rod, with only a small part of an inch between them. Horrocks would view the Sun or Moon between these needles, and then gradually move them up or down the rod on a slider or transom until he reached a point where the angle subtended between the needles as viewed from the end of the rod was exactly the same as the angular diameter of the body under observation, in so far as it now filled the opening between the needles precisely. By knowing in advance the exact distance between the needles, in relationship to the feet, inches, and fractions or other units of proportion down the rod at which he had needed to place his eye to make the body fill the space, he could compute the precise angular diameter of the Sun or Moon. On 9 December 1636 [*Op. Post.* 252], he used such a pair of '*stylis ferries*', mounted on his 11-foot-long Astronomical Radius, to obtain a value of between 34′ and 35′ for the solar diameter. Then, a month later, on 4 January 1637 [*Op. Post.* 255], he used the same 11-foot Radius and '*filis ferries*' configuration to measure the angle subtended between Jupiter and Venus.

Even so, these were still naked-eye methods, for in 1637 neither Horrocks nor any other astronomer in Europe had yet devised a reliable way of measuring angular distances through the telescope. That fundamental invention was to be made in the late 1630s when William Gascoigne happened to make a lucky discovery. For Gascoigne, who was a keen inventor, had noticed that a spider had spun a thread through a Keplerian optical arrangement with which he was working. As luck would have it, the spider's web fell exactly at the combined focal points of the objective and eye glasses, so that when looking through the optical arrangement, Gascoigne saw the web bright and sharp within the field of view. Gascoigne quickly grasped the optical physics involved and started to use these reticules as marker points in the field of what was probably the first telescopic sextant. This sextant was five feet in radius, and modelled on the sextant of Tycho Brahe (though Tycho's sextant was only a naked-eye instrument), and was capable of measuring the distance between astronomical bodies – such as the angle between the Moon and a fixed star – to an unprecedented degree of accuracy.

Then, in addition to the telescopic sextant which reputedly measured down to $0°001$ against an engraved 60° scale, Gascoigne had the brilliant idea of putting two vertical marker points into the same telescopic field. Each marker point could be controlled by a fine pitched screw so that he could adjust them, be they marker points, two stretched threads, needles, or knife-edges, until they precisely enclosed the Sun, Moon, planetary bodies, or cluster of stars. Once they were enclosed, Gascoigne could use the known pitches of the screws to measure the size of the prime-focus image falling within the telescope to a very tiny fraction of an inch, and, knowing the focal length of the lens producing that image, calculate the angular size of the object to a hitherto unattainable degree of accuracy.

Indeed, Gascoigne's telescopic sights and micrometer were the ideal instruments for testing Horrocks's theories about elliptical orbits, and for establishing the exact angular diameter of the Moon at key stages in its orbit as the apsidal line turned around the Earth with relation to the Sun. The telescopic sight could monitor the Moon's nightly motion against the fixed stars to a level of accuracy impossible for any previous astronomer – and thereby make it feasible to establish precise variations of the Moon's velocity in different parts of its orbit – while the micrometer enabled the observer to correlate regular changes in the Moon's observed diameter to specific parts of its orbit. In short, Gascoigne's telescopic sights and micrometer became key instruments in establishing the elliptical, Keplerian orbit of the Moon, first realised by Horrocks from less accurate observations made about three years previously. The whole sequence of events, moreover, exemplifies that perceived relationship between theory, invention, observation, and refinement of data, which so characterised the working methods of Jeremiah Horrocks and his friends.

As soon as Gascoigne's inventions had been seen by Crabtree, probably on a visit to Yorkshire in the late summer or early autumn of 1640, their significance was immediately recognised. Writing to Gascoigne from Salford on 30 October 1640, Crabtree asked how he might obtain such instruments for himself, and, no doubt, for Horrocks: 'Could I purchase it with Travel, or procure it with Gold, I would not be without a Telescope for observing small Angles in the Heavens [the micrometer]; nor want the Use of your other Device of a Glass in a Cane [tube] upon the moveable Ruler of your Sextant [telescopic sight].' Observation of the changing lunar diameter with relation to orbital position, moreover, was clearly high on both men's intellectual agenda, for as Crabtree lamented in the same letter, 'I lost the little Paper, wherein I noted the Moon's Diameter, which we observed when I was with you: I pray you send it to me' [*Phil. Trans.* 1717, p. 607]. What is more, by the time of his writing this letter to Gascoigne, Crabtree was also making arrangements for a visit to Wigan, Lancashire, 'where much Brass is cast', to see if he could 'procure such an one cast', meaning, one presumes, a frame or small precision components for a sextant similar to Gascoigne's which Crabtree himself could complete and to which he could add telescopic sights.

Sadly, it is unlikely that Jeremiah Horrocks had time to use Gascoigne's telescopic sights and micrometer, if duplicates of Gascoigne's originals were only becoming available in Lancashire in the late autumn or winter of 1640, for Horrocks died suddenly on 3 January 1641. Horrocks's sudden death, moreover, was rendered all the more poignant by the excited expectation which he himself entertained of Gascoigne's instruments being used to independently substantiate his own discoveries, especially those of the lunar orbit; for as Crabtree wrote to Gascoigne on 28 December 1640, 'My Friend Mr Horrox professeth, that little Touch which I gave him, hath ravished his mind quite from itself, and left him in an Exstasie between Admiration and Amazement. I beseech you, Sir, slack not your Intentions for the Perfection of your begun Wonders.' [*Phil. Trans.* 1717, p. 608].

It is clear from surviving correspondence, however, that Crabtree and Gascoigne did try to use the new instruments to provide observational substantiation for Horrocks's work, for in a letter to Crabtree, dated Christmas Eve 1641, Gascoigne declares 'Mr Horrox his Theory of the Moon I shall be shortly furnished to try' [*Phil. Trans.* 1717, p. 605], while William Derham, who edited these letters for the *Philosophical Transactions* of the Royal Society in 1717, adds [p. 609] 'Then ... follows an Account of the Agreement of Mr Horroxs Theory of the Moon with Mr Gascoigne's Observations.' Maddeningly from our point of view, however, Derham fails to print these crucial sections of the correspondence: a correspondence which, while surviving in at least ten detailed letters in 1717, has subsequently vanished, so that all we have are the selections which Derham edited for the Royal Society.

Yet other fragments of what was clearly a very substantial correspondence between Crabtree and Gascoigne had entered the possession of the Revd John Flamsteed, for not only did Flamsteed see Horrocks, Crabtree, and Gascoigne as the founding fathers of British research astronomy, and the intellectual heirs of Galileo and Kepler, but he began his massive three folio volume *Historia Coelestis Britannica* (1725) by printing five pages of their surviving letters and observations, made between 1638 and 1643. Of the highest significance in our understanding of the three North-country astronomers, moreover, is a set of nearly 80 observations of the lunar diameter, made by Gascoigne with his micrometer, between January 1641 and December 1642, and printed by Flamsteed in his *Historia.* What is immediately clear from these micrometer observations is that they record orbital variations in the lunar diameter that have a precision which transcends all the previous results obtained by pinhole or iron stylus methods. For the gradual, often nightly, changes are expressed in arc *seconds*, and display clear flow patterns of diameter change over the course of individual lunations.

As a way of trying to make an objective assessment of the quality of these lunar diameter observations, I was delighted when some years ago astronomers at the Royal Greenwich Observatory (RGO), Herstmonceaux – an organisation subsequently destroyed by a short-sighted government – agreed to run Gascoigne's historical observations through their computer. When they had back-calculated the position of the Moon to those dates and times upon which Gascoigne made his observations, I was astonished at how good these early observations were. In addition, the former RGO was kind enough to run a set of Gascoigne's planetary diameter measurements through their computer as well, though as one might expect, Gascoigne's micrometric observations of the diameters of Jupiter, Mars, and Venus displayed wider margins of error, as these objects are so small when observed from the Earth, especially in the low-powered and aberrated telescopes of the early seventeenth century. Yet they were in a different league of accuracy when compared with previous planetary diameter measures made by European astronomers using needles and holes in conjunction with the naked eye. For example, Gascoigne's measurement of the apparent angular diameter of Jupiter for the night of 25 August 1640 was 51″, whereas the correct value calculated by the RGO for that date was 41″ [Chapman 1990, p. 43].

If results of this quality were being obtained by the infant micrometer of 1640, one can fully appreciate Horrocks's, Crabtree's, and Gascoigne's exhilaration at the prospect of providing physical substantiation for Horrocks's Lunar Theory. And while poor Horrocks died just as Gascoigne's micrometric observations were getting into their regular stride, at least Crabtree and Gascoigne – who both died in the summer of 1644 – lived long enough both to see Horrocks vindicated, and to know that they had pushed astronomy beyond what Kepler had achieved, as well as having gained insights into nature disclosed to no other men.

To understand Horrocks, Crabtree, and the transit of Venus of 1639, and to place their observation of that event into its proper historical and technical context, I believe that it has been necessary to take the somewhat circuitous route that we have followed so far. For Horrocks and Crabtree were not simply lucky amateurs who saw something that all the European professionals had missed, but were men working within a very clear research agenda of their own, who were acutely aware of the strengths and weaknesses of contemporary European astronomy, and who had developed their own very ingenious working methods.

Of course, the transit of Venus was very much of a type in many respects with the researches outlined above, in so far as both its observation and interpretation were rooted in the study of planetary dynamics, the passing of bodies in front of bodies, and the measurement of precise angles. It was also part and parcel of Horrocks's and Crabtree's wider agenda of showing the tabular calculation school of astronomy to be wrong, for while the *Prutenic Tables* of Erasmus Rheinhold and the *Tables* of Lansberg could perhaps have predicted a transit, those predictions would not have fallen on the correct day, while the tables of Largomontanus put the Venus conjunction position too far south for the planet to have passed across the Sun in any case.

It is likely that this confusion found amongst the tables, combined with Horrocks's own established experience as a planetary position observer, led him to the conclusion by 26 October 1639 [*Op. Post.*, p. 331] – on which date he wrote to Crabtree to alert him – that Venus would pass across the Sun's disk on 24 November. Horrocks seems, in addition to Crabtree, to have asked his brother Jonas, then living in Toxteth, to keep watch as well, which leaves us to conclude that Jeremiah was not the only astronomical Horrocks and possessor of a telescope, though Jonas Horrocks saw nothing. Crabtree also appears to have passed on notification of the event to Samuel Foster, Gresham Professor of Astronomy in London; though as in the case of Jonas, Horrocks subsequently cites no actual observation by Foster. Yet one man who would have been ideally suited to make the transit observation, and to whom no reference is made either in surviving correspondence or in Horrocks's subsequent treatise *Venus in sole visa*, is William Gascoigne. One is therefore left to assume that in October 1639 Gascoigne had not yet made the acquaintance of Crabtree, in spite of the fact that detailed astronomical observations by Gascoigne survive from 10 December 1638 (published by Flamsteed, *H.C.B.*, p. 1), suggesting that Gascoigne was already a well-equipped and experienced astronomer by that date. William Derham, however, in the Crabtree to Gascoigne letter dated 7 August 1640 which he published in *Philosophical Transactions* in 1711, was of the opinion that this letter was the first communication which passed between the two men, though the superscription 'To his Loving Friend Mr William Gascoigne, at his Father's House in or near Leeds in Yorkshire' and references to prior astronomical communications suggest that the two men had already come to know each other quite well by August 1640. Indeed, it is possible that they had known each other since the winter of 1639-40, for Flamsteed mentions a letter '*Ex alia Gascoignii ad Crabtraeum*' dated 22 March 1640 [*H.C.B.*, p. 3]; but as it is cited in a context of observations made in October and December 1640, one wonders whether this was a typographical error for 22 March 1640/41, falling as this date did immediately prior to the then legal New Year of Lady Day, 25 March. (On another occasion Flamsteed cites a Crabtree letter describing a lunar eclipse dated 18 March 1640/41: *H.C.B.*, p. 1.) On the other hand, Crabtree's description of himself as '*de facie ignotus*' ['though my face is unknown to you'] would imply that the two men had not yet met. But either way, it would be interesting to know who introduced Crabtree to Gascoigne.

One likely agent of introduction between the two men were the brothers Charles and Christopher Towneley, of Towneley Hall, Burnley. For the Towneleys, like Gascoigne, were intellectually-inclined Roman Catholic gentry living some 20 miles apart across the Lancashire and Yorkshire Pennines. What is more, William Crabtree was clearly on good terms with 'Mr Townley' already by this time, mentioning him twice in the August 1640 letter, and hoping in the near future to bring him, along with 'my Friend and second Self Mr Jeremiah Horrox, being near Preston . . . to see Yorkshire, and you'. Firm friendships, therefore, already seem to have been in place by the summer of 1640.

The long letter which Crabtree sent to Gascoigne on 7 August 1640 – it took up 10 pages when printed in the *Philosophical Transactions* in 1711 – was concerned with solar observations, and Crabtree's agreement with Galileo that sunspots were actually on the solar surface, as opposed to being little planetoids in orbit around the Sun, as Gassendi was arguing. Yet in addition to sunspots, Crabtree's letter to Gascoigne discusses two other topics, both on the same theme of seeing dark objects in silhouette upon the solar disk, and of trying to make measurements of them. One of these was the details of a recent solar eclipse, where Crabtree promised to send not only his own observations, made at Broughton, Salford, along with those of Jeremiah Horrocks made at Much Hoole 'between Liverpool and Preston', but also 'Mr Foster's at London' [*Phil. Trans.* 1711, p. 288]; Mr Foster seems to have been the Gresham College Professor.

The other solar-related topic in the 7 August 1640 letter is the 24 November 1639 transit of Venus. From Crabtree's wording, it is clear that Gascoigne had not seen the transit for himself – which is confirmed by the absence of Gascoigne's name in Horrocks's subsequent *Venus in sole visa* – although a prior discussion had clearly taken place between Crabtree and Gascoigne about the angular diameter of the transiting Venus on the Sun's disk, for it is mentioned in the 7 August letter. Crabtree describes his own partial clouding-out in Salford, though he did have time for at least one good observation. His friend Horrocks at Much Hoole, however, had been more fortunate, having seen Venus 'clearly from the time of its coming onto the Sun till the Sun's setting'. However, both Crabtree's and Horrocks's observations had been in close agreement as far as the main features of the transit were concerned, such as the small angular size of Venus when seen on the Sun's disk, and the Venusian orbital details extracted from the observation.

Central to Crabtree's and Horrocks's concern with the transit of Venus, however, had been Kepler's prediction of impending Venus and Mercury transits across the Sun's disk in 1631, in *Admonitio ad Astronomos* (1630). No one had been able to observe the Venus transit of 6 December 1631, because it occurred after sunset in most European locations. But as we say above, the Mercury transit, on 7 November 1631, had been successfully observed by Pierre Gassendi in Paris, who had drawn some significant facts from the event. The most important of these was the surprisingly small angular size of Mercury as it passed across the Sun: a mere $20''$, indeed, as against the several arc minutes traditionally ascribed to it by astronomers looking at Mercury as it shone in the evening or morning sky. Secondly, by being able to see Mercury on the disk of the Sun, one could fix the planet's position in the sky to a much higher level of accuracy than before, for the daily position of the Sun in the ecliptic was known more precisely than that of any other 'planetary' or wandering body. And from Mercury's position at a particular time on a given day with relation to the Sun's centre, it would be possible to apply precise corrections to the theoretical knowledge of Mercury's orbit. The technique which Gassendi had employed to observe the 1631 Mercury transit was that of telescopic projection on to a white screen – basically the same method as that used by Galileo in 1612, and which had been pretty well perfected by Christopher Scheiner for making a daily sunspot record and described and illustrated by him in *Rosa Ursina* (1626-1630).

By the late 1630s, Horrocks and Crabtree were familiar both with Kepler's Mercury and Venus transit predictions for 1631, and with Gassendi's observation. It is also likely that it was Gassendi's surprising discovery of Mercury's smallness when seen in transit which aroused Horrocks's suspicion that Venus might also subtend a much smaller angular diameter than the several arc minutes traditionally ascribed to it, and made him especially keen to observe the event when he realised that a transit would occur on 24 November 1639. And as we saw above, Horrocks came to the conclusion that a transit would occur on that day from disagreements about the exact day and time of Venus's inferior conjunction as displayed in the major astronomical tables, for it is at inferior conjunction, of course, that Venus passes between the Earth and the Sun. Kepler had been of the opinion that after the 1631 transit the Earth, Venus, and solar straight-line inferior conjunction would not re-align for another 130 years. But from his awareness of the errors of the tables, his own observations of the motions of Venus, and by that peculiar genius which made him one of the most brilliant planetary dynamicists who has ever lived, Jeremiah Horrocks came to grasp by mid-October 1639 that Kepler had been mistaken, and that Venus would once again cross the Sun on 24 November.

At the time of the transit, Horrocks was living in the Lancashire village of Much Hoole, some 8 miles south-west of Preston. Legend has it that he was curate at the parish church of St. Michael, at which the Revd Robert Fogg was Rector, but there are no contemporary references to his clerical status, and even Wallis, Flamsteed, and Derham, all of whom were Anglican clergymen, and who were to be such active publishers of his scientific importance over the next eighty years, never speak of Horrocks as an ordained colleague.

Most important of all in this respect, Jeremiah Horrocks, who seems to have been no more than 21 or 22 years old at most in 1639, was simply too young to have been ordained deacon, for the minimum age for an Anglican deacon, both in 1639 and today, is 23, and for a priest, 24. It is more likely that Horrocks was working as a schoolmaster or a private tutor in Much Hoole – perhaps in the employ of the Stones family of Carr House, who were not only local landowners but also merchants trading with London and Amsterdam – and assisting in church as a Bible Clerk on Sundays. For in this capacity, where he would have read the Psalms, Collects, and Old Testament Lessons, he would have gained invaluable experience for a future career as a clergyman. And every evidence is that Horrocks, who was a deeply devout Cambridge graduate, would have presented himself for ordination when he reached the right age.

Though Horrocks's calculations led him to believe that Venus would cross the Sun on Sunday 24 November 1639, he decided, just to be on the safe side, to keep watch on Saturday 23rd. After all, the standard tables did put inferior conjunction on the Saturday. And to make the observation, he used a technique similar to that used by Gassendi in 1631. A small refracting telescope was used to project an image of the Sun on to a six-inch-diameter circle drawn on to a sheet of white paper. The circle had been carefully divided into 360°, so that it would be possible to establish the coordinates of the transit with great accuracy [Horrox, *Venus*, p. 121]. Although Horrocks does not say so, one presumes that the telescope and projection screen must have been connected to each other by a wooden bar or shaft, to maintain a line of collimation.

The day of the transit was overcast, and while Horrocks mentions the 'higher' duties which occupied his time for parts of the day, and gave rise to the story that he was the parish curate, he nonetheless enjoyed sufficient snatches of leisure to keep an eye open for any breaks in the clouds. But not until 3.15 in the afternoon, and with less than 40 minutes to sunset, did Horrocks see Venus just entered upon the solar disk (ingress had probably occurred at 2.49 p.m.), at 62° 30′ from the top limb, seen as a telescopic reversal. He then managed to secure two more sightings, at 3.35 and 3.45 p.m., before he lost the Sun on the horizon. William Crabtree in Salford seems to have enjoyed the same short late-afternoon break in the clouds, later telling Gascoigne that 'The Clouds depriv'd me of part of the Observation', though his and Horrocks's 'Observations agreed, both in the Time and Diameter [of Venus], most precisely' [*Phil. Trans.* 1711, p. 288].

The central concern to both Horrocks and Crabtree was the apparent angular diameter of Venus. No doubt on the basis of Gassendi's Mercury diameter measurement in 1631, Horrocks confidently informed Crabtree on 26 October 1639 [*Op. Post.*, p. 331] that he expected the Venusian diameter to be about 1′. On the day of the transit, Horrocks found that the planet actually subtended 1′ 12″, or 1′ 16″ at most (after applying corrections), and not the 7′ supposedly attributed to it by Kepler or the 11′ of Lansberg [p. 331], while Crabtree independently measured Venus's diameter to 1′ 3″ [Horrox, *Venus*, pp. 131, 187]. Both figures were obtained by each man marking the size of Venus on his projection screen, and calculating it as a fraction of the known apparent solar diameter for the day. What is more, Crabtree and Horrocks independently found the planet Venus to be perfectly round and jet black in colour: clearly a totally opaque body, and not self-luminescent as certain ancient philosophers had believed the planets to be.

Though William Crabtree does not seem to have obtained more than a single observation of Venus in transit because of clouds, Jeremiah Horrocks obtained three. From these three Venus positions, he was able to calculate the planet's velocity, making it possible to work out when the transit would have started and ended. It was also possible, from the angle at which it passed across the Sun, to calculate the node (the point at which the orbit of Venus intersects the ecliptic) and other characteristics of the orbit to a new level of accuracy, and thereby substantially improve our understanding of the theory of the planet's motion.

The 20″ apparent diameter of Mercury observed by Gassendi and the 1′ 16″ corrected diameter of the transiting Venus led Horrocks to derive a new value for the horizontal solar parallax of 14″, which was very much smaller than the values ascribed to it by classical and modern astronomers. Yet impressive as Horrocks's figure was, one must not forget that he came to it via a line of celestial geometrical reasoning which we now know to have been false. For influenced as he was by Kepler's harmonic and geometrical reasoning, Horrocks had concluded that, when viewed from the Sun, both Mercury and Venus would subtend angular diameters of 28″ [Horrox, *Venus*, p. 208]. Knowing that the Earth was around 8000 miles in diameter (and presumably larger than the inner planets), he thought that it too would subtend 28″ when viewed from the Sun. Believing as he did that the planets (with the exception of Mars, which presented such a tiny visible expanse when viewed through the telescope) became physically larger and moved more slowly in orbits of increasing radii as their distance from the Sun increased, one comes to understand why Jeremiah Horrocks was so concerned with obtaining accurate measurements for the angular diameters of the planets: if Mercury, Venus, and Earth subtended 28″ when seen from the Sun, did Mars, Jupiter, and Saturn do likewise? Knowing the distance of the Sun was a key component in this exercise, as it would enable the astronomer to supply some physical dimensions to the proportions predicated by Keplerian theory. Saturn, for instance, thought to be right at the edge of the solar system in the seventeenth century, seemed to be especially large when viewed with simple telescopes at that time, for the ring system was often mistaken for planetary bulk. And as the line connecting the centre of the Sun to the centre of a planet bisects this 28″ diameter, thereby creating a pair of right-angled triangles across the Earth's equatorial diameter, Horrocks considered that the solar parallax must be 14″.

Erroneous in its fundamental cosmological assumptions as Horrocks's value for the solar parallax was, its very smallness with regard to all previous parallax values had the immediate effect of making the solar system seem incredibly large, with an Astronomical Unit of 15 000 Earth radii, or 59 600 000 miles. With Tycho Brahe's solar parallax value, however, the Sun was only about 4 600 000 miles away, while Kepler's 59″ parallax would still have put it at only around 10 000 000 miles. In this respect, therefore, Jeremiah Horrocks is not without significance as a cosmologist, and as an early proponent of cosmological vastness based not on philosophical speculation so much as upon (albeit misinterpreted) telescopic observations and measurements.

It must be fully understood, however, that there was no way in which Horrocks could have measured, or even thought of measuring, the solar parallax directly from the 24 November 1639 transit, as eighteenth- and nineteenth-century astronomers attempted to do from later transits. For the measurements taken in 1761, 1769, 1874, and 1882 came from a variety of observing stations scattered around the globe, as astronomers laid out great base lines across the Earth's surface. For to measure the solar parallax from trigonometrical observations of Venus on the Sun's disk, one needs at least two widely-spaced observing stations.

Looking back in hindsight, one can see how the 1639 transit formed a natural apotheosis to the short and staggeringly brilliant astronomical career of Jeremiah Horrocks. While he did follow-up work on the post-transit Venus positions, as the planet moved to eastern elongations as a morning star, and was, according to Crabtree (7 August) clearly fascinated by the news of Gascoigne's instruments, this would turn out to be his last great piece of research. Mercifully, he had sufficient time to write up his and Crabtree's findings into *Venus in sole visa*, and then he died suddenly on 3 January 1641. We do not know what he died of, at the age of 22 or 23, and sadly, our only surviving record of his death comes from the manuscript obituary penned by Crabtree himself:

> 'Mr Jeremiah Horrox's Letters to me in the years 1638, 1639, 1640, up to the day of his death very suddenly on the morning of the 3rd January [1641]; the day before he had arranged to come to me. Thus God puts an end to all worldly affairs. I have lost alas my most dear Horrox. Hinc illae lachrimae [thence those tears]. Irreparable loss!' [*Op. Post.*, p. 338; English, Whatton, Memoir, p. 58]

And if Horrocks was planning a journey from Toxteth (having returned to the family home from Much Hoole sometime in the summer or autumn of 1640), then one might assume that he was not suffering from a long-term disease. Perhaps he died in an accident, or from a sudden virulent winter infection. We simply do not know.

Judging from the letters printed by Flamsteed, it is clear that Crabtree and Gascoigne continued to correspond until 1642. But this correspondence seems to have ended with the onset of the Civil Wars in August 1642, for Gascoigne received a commission as Providore for Yorkshire in the army of King Charles I. There is no record of Crabtree having served in any army, though as the Salford and Manchester areas were supportive of Parliament, the two astronomical friends could have found themselves on different sides in the great conflict. Yet Crabtree, Gascoigne, and their mutual friend Charles Towneley senior all died within a month of each other: Towneley and Gascoigne falling in the Battle of Marston Moor, fought on 2 July 1644, and Crabtree making his will on 19 July, and being buried within the precincts of the Manchester Collegiate Church on 1 August 1644, close to where he had received his education. We have no information as to the cause of Crabtree's death.

After the deaths of Horrocks, Crabtree, and Gascoigne, various people tried to collect together and preserve their manuscripts, though it was the Pendle Forest, Lancashire, astronomer Jeremy Shakerley who first acknowledged Horrocks's achievement in print, in three books, which he published between 1649 and 1653. Then Christopher Towneley, Charles's younger brother, who survived the battle of Marston Moor and lived on until 1674, certainly obtained a body of the manuscripts, along with the 'carkasse' of Gascoigne's sextant, and these were studied by the young John Flamsteed after 1672. Flamsteed, deeply Protestant Anglican clergyman that he was, became a friend and long-standing correspondent of the Roman Catholic Richard Towneley, Charles's son and heir to the Towneley estates in Lancashire, while Richard Towneley himself was an active scientist, leaving a large collection of books, manuscripts, and scientific instruments in his will in 1706. Other bodies of manuscripts found their way to London, where Sir Jonas Moore (who as a young man in Lancashire had been part of the Towneley circle, and as a knighted Royalist in the 1660s was an early Fellow of the Royal Society and the patron of John Flamsteed) assisted in bringing them to the attention of Oxbridge and Metropolitan scientists. The early Royal Society resolved to see these national treasures published, and though an unspecified number were destroyed in the Great Fire of 1666 – in the hands of Nathaniel Brooks the printer – sufficient remained for John Wallis to edit into the *Opera Posthuma* of Jeremiah Horrox (*sic*) in 1673.

But it was Horrocks's account of the 1639 Venus transit which not only immortalised him and his friend Crabtree, but was also the only complete major work that he left at his death in January 1641. This manuscript was certainly copied by others over the next few years: by John Worthington of Manchester, perhaps, Horrocks's Emmanuel College student contemporary, who could well have introduced Horrocks to Crabtree in 1635 or 1636, when Jeremiah was staying over in Manchester on the last leg of his journey back to Liverpool from Cambridge, and who was clearly interested in his deceased friend's achievements. Worthington rose up to be a Head of House and a dignitary of Cambridge University in later life, and tried to find a publisher for '*Venus in Sole Visa*'. Samuel Hartlib, the expatriate German scholar living in England, is said to have obtained two manuscript copies of '*Venus*', one of which was studied by a member of the Mercator dynasty of Flemish cartographers and scientists. Then in April and May 1661, as the original Fellowship of the Royal Society was forming, the great Dutch astronomer Christiaan Huygens, who was elected a Fellow of the Society in 1663, was in London. Huygens was certainly interested in Horrocks, and also acquired a copy of the '*Venus*' manuscript from Sir Robert Moray, who in turn had got it from Sir Paul Neile. Via Huygens, the manuscript, or yet another copy of it, was passed on to his Polish friend Johannes Hevelius in Dantzig, and when Hevelius published his own account of the Mercury transit of 1661, he issued it in conjunction with a printing of Horrocks's Venus transit manuscript. Thus, in 1662, Horrocks's *Venus in Sole Visa* at last saw the light of day in an elegantly-printed volume under the imprimatur of one of Europe's most illustrious astronomers. Published in this way, it is hardly surprising that the work of Horrocks and Crabtree came to be blazoned across learned Europe. In the wake of this prestigious publication, the Royal Society's issuing of *Opera Posthuma* (in which Horrocks's and Crabtree's letters were translated into Latin for a clearly targeted international readership), and the publishing of fragments of the Crabtree-Gascoigne correspondence by Derham and then Flamsteed in the early eighteenth century, it is clear that the work of the North-country astronomers had entered into the full consciousness of the European scientific movement. Newton sang the praises of Horrocks in *Principia* (1687), while they were accorded recognition on a variety of levels: as original proponents of practical observation, as celestial mechanists of genius, and as inventors of crucial new scientific instruments, such as the telescopic sight and the micrometer.

By the nineteenth century, Horrocks and Crabtree in particular came to be seen as iconic figures by British astronomers. To Sir John Herschel, for instance, Horrocks was 'the pride and boast of British astronomy'. He was commemorated – at long last – in Much Hoole parish church, and in Westminster Abbey in 1874, while artists such as Eyre-Crowe, Ford Maddox Brown, and W. R. Lavender immortalised Horrocks and Crabtree at those short, critical moments when Venus briefly revealed herself in transit, in magnificent oil paintings. It is true that here Horrocks and Crabtree, of whose authentic appearances we know nothing whatsoever, were draped in the garb of the Victorian romantic imagination: both men are thin and emaciated, and Crabtree, who was 29 years of age in 1639, looks like a manic 70-year-old with an incongruously young wife and small children, while Horrocks has become a gaunt puritan! Their early deaths also excited the romantic imagination of Victorian scientific hagiographers: was not Gascoigne a dashing Cavalier, tragically meeting his death in battle at the age of 32, and could not Horrocks easily be seen as an intense genius, nervous and deeply devout; a starving curate, probably consumptive, whose frail hold on life finally snapped in his early twenties? It was not for nothing that in 1959 W. F. Bushell styled him 'The Keats of English Astronomy'. Yet in reality we know nothing about these men's appearance, physical build, health, or – with the exception of Gascoigne – cause of death. One can only assume that they enjoyed a considerable physical robustness, considering the amount of stunning achievement they packed into such short lives, at the same time as earning their living.

Yet historical *reality* never ceases to be full of surprises. Ten or twelve years ago, no one could have predicted that major Horrocks documents would emerge out of their hiding-place of centuries, and present themselves in the saleroom. Yet this is precisely what happened in the spring of 1995, when Mark Westwood, a scholar and academic book dealer of Hay-on-Wye and himself an Emmanuel College graduate, was commissioned by an undisclosed vendor to sell what was probably Jeremiah Horrocks's own handwritten Latin manuscript of '*Venus in Sole Visa*'. Both Mr Westwood and I agreed that the handwriting was very similar to that of Horrocks himself, while the leather boards into which a subsequent owner had bound it also contained, probably in a scribal hand, a fair copy of '*Venus*', along with the manuscript of Horrocks's '*Praeludium Astronomicum*'. Mercifully, when these manuscripts came up for auction later in 1995, it was possible for Cambridge University Library to purchase them, so that they will be curated in perpetuity by the academic institution in which Horrocks first acquired his extra-curricular fascination with modern astronomy.

Then equally remarkably, in November 2004, Horrocks's manuscript '*Observationes Astronomiae . . .*' for the years 1635 to 1637 suddenly came up for sale at Sotheby's as part of a larger collection of historical astronomical documents. This valuable manuscript, written over 36 leaves and almost certainly in Horrocks's own hand, shows that he was observing as early as June 1635, which may have been around the time that he met Crabtree. It also contains his original observations of Jupiter and Saturn for 1636, which were to be so significant in his emerging ideas of planetary attraction, as well as other important investigations. And as all of this work was being done at the Horrocks family home at Toxteth, Liverpool, between Jeremiah's return from Cambridge and his departure for Much Hoole in 1639, it represents the earliest major scientific research to have been undertaken in and adjusted to the geographical position of Liverpool, and therefore has enormous cultural significance for the North-West of England.

Unlike the '*Venus in Sole Visa*' manuscript, I was unable to examine or even see this '*Observationes*' manuscript, and can only speak of it from the details contained in the published Sotheby's catalogue, and from what my friend Alan Bowden of National Museums on Merseyside, Liverpool, could tell me from his own brief inspection. For sadly, the saleroom bidding for the manuscript escalated beyond the sums which had been pledged for Merseyside Museums, and it was knocked down to another purchaser. Attempts are currently in hand, January 2005, to prevent the manuscript from leaving Great Britain if any request for an export licence is made, or at the very least, of having its contents computer-scanned and made available to the international scholarly community.

Yet as we have seen above, Jeremiah Horrocks had no geometrical method, beyond his mistaken idea that the planets subtended 28″ diameters as viewed from the Sun, by which one could measure the solar parallax from the three brief views of Venus in transit that he obtained in November 1639, so Horrocks's and Crabtree's observations played no enduring part in establishing that vital astronomical quantity. Indeed, another 38 years would elapse before an astronomer realised how the Sun's parallax might truly be measured from an inner planet transit. And that realisation was to be made by the young Edmond Halley when, on 7 November 1677, he observed a transit of Mercury from the island of St Helena in the south Atlantic. It occurred to Halley, a geometer of genius, who curiously enough was in 1677 about the same age that Horrocks had been in 1639, that if astronomers in Europe were observing Mercury's passage across the Sun at the same time as he was in St Helena, then they would see it on a slightly different part of the Sun's surface. London or Paris and St Helena are around 56° apart on the Earth's surface, and after establishing exact trigonometrical base lines between observing stations (using Jupiter's satellites to determine longitudes and accurate telescopic quadrants to fix latitudes), it should be possible to use a transit of Mercury to obtain the solar parallax by triangulations and timings. This would be accomplished by the astronomer using a good telescope to observe the ingress and egress of the transiting planet, while timing the duration of passage with an accurate regulator clock. For observers stationed at two different latitudes on the Earth would indeed see Mercury enter and leave the Sun from slightly different parts of the limb, and describe lines of slightly different lengths upon it. If these differences of both position and time could be measured with sufficient accuracy, Halley argued, then the distance of Mercury, and by extrapolation of the Sun, could be extracted from them.

But Mercury turned out to be too small and fast-moving for practical measurement by this method, which led Halley to realise that Venus transits, with a larger planetary body and longer transit time, would be much more suitable. The only problem was that no individual around in 1677 was likely to witness the next Venus transit in 1761. Yet this did not prevent Halley from paying considerable attention to the celestial mechanics of a forthcoming Venus transit which he knew that he would never see, and over the next few decades and into the early eighteenth century, he published several papers which developed his method of observation. As 1761 drew closer other astronomers, learned societies, and governments began to make plans for 1761 and 1769. Joseph-Nicholas Delisle in France refined the data, published a *Mappemonde* of where in the world the whole or parts of the 1761 Venus transit would be seen, and improved Halley's originally-suggested observing technique. Delisle also drew attention to the fact that in 1753 Mercury would transit the Sun's disk, and that this event could act as something of a training ground for the Venus transits of 1761 and 1769. The eighteenth-century transits, moreover, assumed a new intellectual importance in the wake of Newton's Laws of Motion, for if the Astronomical Unit could at last be established with accuracy, then not only could the proportions of the solar system be ascribed definitive physical values, but gravitation theory could be further perfected.

Other scholars and historians of astronomy, however, will go on in the present volume to discuss the problems, adventures, and achievements of the observers of the 1761, 1769, 1874, and 1882 transits. It is hard to imagine that anyone, no matter where they were in the world, could have observed the Venus transit of 8 June 2004 without a sense of wonder and of great occasion. And those who had gathered at Broughton Spout, Salford, within a stone's throw from where William Crabtree had lived, or those, including the IAU Colloquium 196 Conference Party, who, after witnessing the start of the transit at Alston Hall, then assembled in the gardens of Carr House and St Michael's churchyard, Much Hoole, where Horrocks had lived in 1639, to watch its latter stages, could not have failed to appreciate the historical significance of time, place, and event. For it was from Salford and Much Hoole that, 364 years before, a profoundly far-reaching observation had been made by two English amateurs. It would subsequently blazon their names before the *savants* of Europe, and inaugurate for English research astronomy an enduring reputation for originality and excellence.

Acknowledgements

I wish to thank Dr Robert van Gent for his information, and for directing me to sources regarding the passage of Horrocks's manuscript of '*Venus in Sole Visa*' from London to Hevelius, via Christiaan Huygens in Holland. I also wish to thank Carl Barry and Lilian Fletcher of Salford for access to their genealogical researches into the Crabtree family, and also Kenneth Irvine, Director of the Salford Astronomical Society Observatory. Among them, these three have produced credible evidence to suggest that a house still standing in Broughton Spout, Salford, is likely to have been the one where Crabtree lived and from which he observed the 1639 transit.

References

Abbreviations

>Op. Post. – see Horrocks, *Opera Posthuma*
>
>H.C.B. – see Flamsteed, *Historia Coelestis Britannica*
>
>Phil. Trans. – see *Philosophical Transactions*

Primary sources: letters and publications

Jeremiah Horrox [Horrocks], *Jeremiae Horrocci Liverpoliensis Angli ex Palinatu Lancastriae Opera Posthuma* (London, 1673), for the letters of Jeremiah Horrocks and William Crabtree.

J. Horrocks, *Venus in Sole Pariter Visa Anno 1639*, published in Latin by Johannes Hevelius, as an addition to his *Mercurius in Sole Visa Gedani* (Dantzig, 1662).

J. Horrocks, *Venus in Transit Across the Sun*, English translation of *Venus in Sole Visa*, by Arundel Blount Whatton, and appended to 'A Memoir of his [Horrocks's] Life and Labours' (London, 1859).

J. Horrocks: known manuscripts:

'*Venus in Sole Visa*'; '*Venus in Sole Visa*'; '*Jeremiae Horroxj Praeludium second Astronomicum Liber primus De motu Solis*': three original manuscripts (two of 'Venus') bound into the same set of leather boards, and acquired at auction by Cambridge University Library, 1995.

'*Venus in Sole Visa*', acquired by the Royal Observatory, Greenwich, Library, 1838. RGO 76/1.

'*Venus in Sole Visa*', also in the Royal Observatory collection, RGO 68/C.

'Philosophical Exercises', RGO 68/B.

'*Explicatio brevis et perspicua Diagrammatis Hpparchi*', RGO 68/D.

Other MSS in RGO 68/E.

'Astronomical observations and notes for the years 1635-1638'. Autograph manuscript in eighteenth-century vellum-backed paper boards. Sold in auction at Sotheby's, London, 4 November 2004, to an unspecified vendor. Nothing currently known beyond details in Sotheby's sale catalogue.

John Flamsteed, *Historia Coelestis Britannica*, 3 vols. (London, 1725). Vol. I, pp. 1-5, for Crabtree-Gascoigne correspondence and observations, 1638-1643.
J. Flamsteed, *The 'Preface' to John Flamsteed, 'Historia Coelestis Britannica', or British Catalogue of the Heavens, 1725*, edited and introduced by A. Chapman, based on a translation by Alison Dione Johnson (National Maritime Museum Monograph No. 52, 1982).

William Crabtree, Letter to William Gascoigne on sunspots and solar observations, 7 August 1640, ed. William Derham, *Philosophical Transactions* 27 (1711), 279-90.

William Crabtree and William Gascoigne, selections from correspondence 1640-1, ed. William Derham, *Philosophical Transactions* 30 (1717), 604-10.

William Gascoigne to William Oughtred, undated letter but probably written in February 1641: Stephen J. Rigaud, *Correspondence of Scientific Men of the Seventeenth Century*, 1 (Oxford, 1841), 35-9. Also Gascoigne to Oughtred, 2 December 1640, Rigaud, *Correspondence* 1, 33-4. [Rigaud also reproduces letters by Flamsteed, John Collins, and others about Horrocks, Crabtree, and Gascoigne from the 1660s and 1670s.]

Edward Shereburne, *The Sphere of Marcus Manilius* (1675): see supplement 'Catalogue of famous astronomers' for early biographies of Horrocks, Crabtree & co.

J. Crossley, 'The Diary and Correspondence of Dr John Worthington', 3 vols. (Chetham's Society Publications, 13, 36, 114 (Manchester, 1847-1886).

Jeremy Shakerley Anatomy of "*Urania Practica*" (London, 1649).

Jeremy Shakerley *Synopsis Compendiana* (London, 1651).

Jeremy Shakerley *Tabulae Britannicae* (London, 1653).

Richard Towneley, 'A description of an instrument for dividing a foot into many thousand parts', *Philosophical Transactions*, 1667, 541.

Secondary sources

Peter Aughton, *The Transit of Venus. The Brief, Brilliant Life of Jeremiah Horrocks, Father of British Astronomy* (Weidenfeld and Nicolson, Windrus Press, London, 2004).

E. Axon, 'The Broughton Estate plans', *Transactions of the Lancashire and Cheshire Antiquarian Society*, 1905. Estate plan of 1637 drawn by William Crabtree.

John E. Bailey, 'The writings of Jeremiah Horrox and William Crabtree ... reprinted with additions', in the *Palatine Notebook*, Dec. 1882 and Jan. 1883 (1883).

Francis Baily, *An Account of the Revd. John Flamsteed* (1835).

Carl Barry and Lilian Fletcher, 'William Crabtree of Broughton. A hunt for his domicile' (Salford, 2004). A computer text of genealogical and topographical researches. Copy generously presented to A. Chapman, December 2004. Copy also on deposit in the Central Library, Peel Park, Salford.

W. F. Bushell, 'Jeremiah Horrocks, the Keats of English Astronomy', *Mathematical Gazette* 43, 343 (February 1959) 1–16; reprinted in the North West Astronomy Series by the Liverpool Astronomical Society, 1992

A. Chapman, 'Jeremiah Horrocks, the Transit of Venus, and the "New Astronomy" in early seventeenth-century England', *Quarterly Journal of the Royal Astronomical Society* 31 (1990), 333-57. Reprinted in Chapman, *Astronomical Instruments and their Users: Tycho Brahe to William Lassell* (Variorum, Ashgate Collected Studies, Aldershot, 1996).

A. Chapman, *Three North-Country Astronomers* (Manchester, 1982).

A. Chapman, *William Crabtree 1610-1644. Manchester's First Mathematician* (Manchester Statistical Society, 1995).

A. Chapman, 'The Astronomical Work of Thomas Harriot', *Quarterly Journal of the Royal Astronomical Society* 36 (1995), 97-107.

A. Chapman, 'Reconstructing the angle-measuring instruments of Pierre Gassendi', in *Learning, Language, and Invention: Essays presented to Francis Maddison*, edd. W. D. Hackman and A. J. Turner (Variorum, Aldershot, 1994). Reprinted in Chapman, *Astronomical Instruments and their Users*.

A. Chapman, *Dividing the Circle. The Development of Critical Angular Measurement in Astronomy, 1500-1850* (Praxis-Wiley, Chichester, 1990, 1995).

Chetham Society, *Miscellanies*, vol. 63, for Crabtree family.

Betty M. Davis, 'The astronomical work of Jeremiah Horrox', University of London M.Sc. thesis (1967).

Sidney B. Gaythorpe, 'A Galilean telescope made about 1640 by William Gascoigne ...', *Journal of the British Astronomical Association*, June 1929, 238-41.

S. B. Gaythorpe, 'Horrocks' observations of the Transit of Venus, 1639', *Journal of the British Astronomical Association*, Dec. 1936, 60-8; July 1954, 309-15.

S. B Gaythorpe, 'Jeremiah Horrocks: Date of Birth, Parentage, and Family Associations', *Transactions of the Historical Society of Lancashire and Cheshire* 106 (1954), 23-33.

S. B. Gaythorpe, 'Horrocks' Observations and Contemporary Ephemerides', *Journal of the British Astronomical Association* 47 (1937), 156-7.

David W. Hughes, 'Horrocks's bogus law', *Astronomy and Geophysics* vol. 46, Feb. 2005, 114-16.

David W. Hughes, 'Why did other European astronomers not see the Transit of Venus', these proceedings.

John Roche, 'The Radius Astronomicus in England', *Annals of Science* 8 (1977), 102-12.

John W. Shirley (ed.), *Thomas Harriot, Renaissance Scientist* (O.U.P., 1974).

F. J. M. Stratton, 'Jeremiah Horrox and the Transit of Venus', *Occasional Notes of the Royal Astronomical Society* 7 (1939).

G. H. Turnbull, 'Samuel Hartlib's Influence on the Early Royal Society', *Notes and Records of the Royal Society* 10 (1953), 101-30.

Charles Webster, 'Richard Towneley (1629-1707). The Towneley group and seventeenth-century science', *Transactions of the Lancashire and Cheshire Historical Society* 18 (1966).

Discussion

JESUS DE ALBA MARTINEZ: Do you think that the curriculum information of Horrocks includes astronomy as a kind of Quadrivian curriculum?

ALLAN CHAPMAN: Yes, that's perfectly right. Certainly when you'd been to Cambridge, or Oxford, or any other major European University at that period, or even in the Grammar School, you would have studied the Quadrivian, including astronomy, geometry and simple calculation. This would have given you a basic familiarity with the laws of proportion, such as would have been taught in the ancient Roman schools, and which came to be enshrined in the curricula of Europe's medieval Universities. Horrocks would have picked up quite a lot of information about the classical universe, especially of Ptolemy (probably including Johannes Sacrobosco's *De Sphaera Mundi*, c. 1240) as part of the Quadrivian. In Cambridge, however, there would probably have been no formal teaching of the *new* astronomy. Indeed, Horrocks pretty well tells us that he was taught no new astronomy in Cambridge! Therefore, you are quite correct, Sir, for like Kepler, like Gassendi, like all astronomers in Europe, Horrocks would have had a Quadrivian training.

DAVID SELLERS: The well-known translation of the *"Venus in Sole Visa"* by Whatton describes Venus as coming into the image that was projected by Horrocks from the top right and that description has been fatefully reproduced in the Ford Madox Brown painting at Manchester Town Hall and the paintings by Lavender and Eyre Crowe of Horrocks observing, but is it not the case that a Galilean telescope would have inverted the image, but not reversed it, and that would actually mean that the stained-glass window at Hoole Church, although fanciful, is nevertheless the right way round?

ALLAN CHAPMAN: You're absolutely right there, Sir, it was! We don't know exactly whether he was using a Galilean or a Keplerian. I suspect a Galilean although a few years later Gascoigne in Leeds is certainly observing with a Keplerian. I suspect the telescope he would have bought for half a crown would have been a general telescope for looking at horse-riding and things of this sort, which of course would have given you the right-way-up image, so I suspect he was working with a Galilean, although there's no proof; he doesn't say it.

DAVID HUGHES: I love the first painting [by JW Lavender]. What was special about 1903? Why did he happen to paint it then?

ALLAN CHAPMAN: I don't know why. It's quite remarkable – between 1879, I think, when the Manchester corporation commissioned the Ford Madox Brown, till 1903, with the 1891 Eyre Crowe which was shown at the Royal Academy, there seems to have been a sort of burst of some fascination in Jeremiah and his friends. I suppose one could say too that the great interest in the area of Much Hoole, and the Reverend Robert Brickel, the windows, the sundial ... I think it probably had something to do with a sense of England's glory, England's greatness, and of course the movement of naturalistic history painting. What you do have in British art in the late 19th century is a looking for great scenes to paint, such as, let's say, "The Death of General Gordon at Khartoum", and "Captain Cook's discovery of Australasia". This becomes a sort of tradition in English art in the late 19th century; very naturalistic, photographic in its accuracy. I suspect our friends Horrocks and Crabtree fell into that tradition as great Englishmen.

DAVID HUGHES: Is it true that the Ford Madox Brown painting was exhibited with the other two paintings at Southport in 1903 for the British Association for the Advancement of Science?

ALLAN CHAPMAN: I don't know, David.

DAVID HUGHES: I think the three were exhibited together at that time and it was certainly at Southport, but there are very few details about it. I thought you might know.

ALLAN CHAPMAN: I didn't know that, so I'll see if I can confirm it, because, although they couldn't take one from the Manchester Town Hall wall, there was a canvas one in the Manchester art gallery; so yes, it would have been very easy to put them all together in Southport.

ROBERT VAN GENT: You mentioned this problem of how the manuscript came into various hands through Christiaan Huygens. It was in 1661 he got a copy of the manuscript. He knew he wanted to publish about it, so he just simply passed on Horrocks' notes.

ALLAN CHAPMAN: Do you know that for certain?

ROBERT VAN GENT: Yes, you can find details in the correspondence.

ALLAN CHAPMAN: Is there correspondence on it?

ROBERT VAN GENT: Yes, there is correspondence. I can give you details.

ALLAN CHAPMAN: Thank you. I do appreciate that. I did suspect that because John Evelyn mentions being at an early gathering of the Royal Society in 1661 where King Charles II was present. They looked at the Moon and he mentions Monsieur Huygens being present. So, Huygens was clearly in England.

ROBERT VAN GENT: He was also saying that there would be a Mercury transit, so he was going to watch the transit.

ALLAN CHAPMAN: This is what historical research is all about: Material comes out, and I thank you for that.

JAMIE MATTHEWS: Is there any evidence of scepticism amongst the rational community when these results came out? After all, they couldn't be verified, since it was a once in a lifetime event. And how soon was it that any of the predictions that arose from Horrocks' and Crabtree's observations could be verified?

ALLAN CHAPMAN: Well, of course they wouldn't have been able to see another transit till 1761. Yet nobody denied or doubted the truth of Horrocks' and Crabtree's claims for, after all, the major tables did predict an inferior conjunction of Venus for 23 or 24 November 1639, and all placed that conjunction very close to the solar disk, so there was no reason to doubt the authenticity of the observation. When Christiaan Huygens obtained a manuscript copy of Horrocks' 'Venus' in 1661, he regarded it as a work of great significance, and Huygens was instrumental in securing its publication by Johannes Hevelius in Dantzig soon afterwards. Now, had Horrocks, or one of his posthumous English admirers, been able to secure a publication from some printer in Paternoster Row

in London, it could have come years before Hevelius' edition in 1662. It is also clear by
the 1660s the early Royal Society Fellows were viewing Horrocks' as a figure of truly
international standing, and Horrocks' Emmanuel College contemporary, the Reverend
John Wallis, F.R.S (now Savilian Professor of Geometry at Oxford) took on the task of
editing Horrocks' and Crabtree's surviving correspondence for international publication
in 1672–3.

NICK KOLLERSTROM: In the Ford Madox Brown picture you have the image just projected up onto a wall, and you said that Horrocks and Crabtree wouldn't have used this
method – there would've been some sort of bar connecting the telescope with the screen.
Would you like to comment on it?

ALLAN CHAPMAN: You're absolutely right, Nick: You couldn't have madde a reliable
observation without both the telescope and the projection screen being fixed to a common
axis – such as a wooden bar. With such an arrangement, one could gently nudge the
whole optical system to make it track the sun. Over the years I have tried to replicate
the observing techniques of Horrocks and his contemporaries. I have made cross-staffs,
tried to measure solar and lunar diameters with holes – or "foramen" – devices, and made
solar projection systems to view sunspots. And I can assure you, you cannot keep a solar
image within a circle drawn on a piece of paper, unless that paper, and the telescope, are
all part of one optical axis. So I feel that Ford Madox Brown was wrong when he showed
Crabtree observing by simply *aiming* the solar image across the room.

STEVE DICK: Would Horrocks have know of John Wilkins, Bishop John Wilkins, 1638
"The Discovery of a World in the Moon"?

ALLAN CHAPMAN: He could have done. Wilkins is also one of my own great heroes, not
to mention that he was Warden of my college. Horrocks could have read Wilkins' "Discovery", but he never mentions it. Yet surprisingly, on his circa 1635 list of astronomers
works he knew, there are no Englishmen, and the only English scientific writer whom
Horrocks subsequently mentioned, if my memory serves me right, was the navigation
book author Edward Wright. In 1638, moreover, Wilkins was only 24, and not much
older than Horrocks himself. So I am afraid that there is no record of Horrocks being
acquainted with Wilkins' work.

STEVE DICK: Why would Horrocks not have told everybody else to look for the transit?

ALLAN CHAPMAN: Well he did! He told not only Crabtree, but also wrote to his brother
in Liverpool, and got probably Crabtree to write to Samuel Foster in London, which of
course he mentioned, "Can you tell Mr Foster?" Which tends to indicate that Crabtree
had a London correspondence with Gresham College which I'd love to know more about.
Clearly Foster and Crabtree knew each other. But there simply was only a month before
the event. Horrocks was not in a position to send this to printing presses, and there were
no international journals in those days, so he depended on letters to friends and I think
he's very lucky, considering communication delays, that Crabtree was able to secure a
co-observation.

Transits of Venus: New Views of the Solar System and Galaxy
Proceedings IAU Colloquium No. 196, 2004
D.W. Kurtz, ed.

© 2004 International Astronomical Union
doi:10.1017/S1743921305001237

Jeremiah Horrocks's Lancashire

John K. Walton

Department of History, University of Central Lancashire, Preston PR1 2HE, UK
(email: jkwalton@uclan.ac.uk)

Abstract. This paper sets Jeremiah Horrocks and Much Hoole in the context of Lancashire society on the eve of the English Civil War. It focuses on the complexities of what it was to be a "Puritan" in an environment where religious labels and conflicts mattered a great deal; it examines the economic circumstances of county and locality at the time, pointing out the extent to which (despite widespread and deep poverty) the county's merchants were looking outwards to London, northern Europe and beyond; and it emphasizes that even in the apparently remote and rustic location of Much Hoole it was possible for Horrocks to sustain a scientific correspondence and to keep in touch with, and make his contribution to, developments on a much wider stage.

This contribution is intended to complement Allan Chapman's plenary lecture by developing the local and Lancashire dimension to Jeremiah Horrocks's astronomical activities at Much Hoole, less than three years before the outbreak of the great Civil War in 1642, which culminated in the execution of Charles I in 1649 and ushered in Britain's only period of republican government, which lasted until the Restoration in 1660 (see Chapman 1994, to which this paper runs in parallel). Some of the earliest skirmishes of that long and highly significant conflict took place in Lancashire, and the county was unusual in the extent to which it combined a serious, doctrinaire and influential Puritan presence in and around Manchester, in the area where the region's famous textile industries were already beginning to develop, with an area in the west (the coastal plains to the west and south-west of Preston, in the Hundreds of West Derby and Amounderness) where Roman Catholicism survived strongly, especially among the landed gentry (Walker 1939; Haigh 1975).

The religious dimension of the Civil War is often highlighted in the literature, with the Puritans as the staunchest supporters of Parliament pitted against the Catholics as the most determined (though sometimes problematic) allies of the King (Broxap 1910). It is therefore highly relevant to begin by looking at the question of Horrocks's religious affiliation, especially as the apparently straightforward label of "Puritan" that was fixed to him by Victorian biographers and commentators has been hard to shake off. As an illustration, the *Guardian* of 5 June 2004 referred to Horrocks simply as a "20-year-old Puritan" (with the implication that he really was a very serious young man) even after interviewing Chapman himself, who is very sceptical about the application of this label to the first recorder of the Transit of Venus (Marston 2004, pp. 18–21, is more supportive of the idea of Horrocks as Puritan, though aware of the contemporary flexibility of the term). Placing Horrocks in the context of place and time lends further weight to Chapman's scepticism, and enables us to take it further than the very limited direct biographical evidence allows.

In the first place, we should emphasize the elasticity with which the term "Puritan" might be applied in the Lancashire of the 1630s. The core concept does involve a strong, serious Protestantism that rejected ceremonial and fripperies in divine service, sometimes to the extent of iconoclasm. It prioritised careful listening to assiduously prepared sermons over mere attendance at ceremonial services, and at its core was the believers' commitment to exploring their relationship with God and finding out whether they were to be saved or damned according to the divine plan. At an extreme it could embrace Calvinist doctrines of predestination, but their concentration on the fate of the individual soul did not prevent strong Puritans from aspiring to be their brothers' and sisters' keepers, with a commitment to suppressing excessive merriment and immoral behaviour which might involve (for example) the suppression of alehouses, dancing and popular festivities (Wrightson 1974; Poole 2002). Prioritising the individual conscience sometimes also led to challenges to established hierarchies when they seemed to behave in ungodly ways, and Lancashire not only contained Puritan supporters of Parliament against the ungodly Crown, but also opponents of the authority of bishops, so that the area around Manchester in the mid-seventeenth century saw the formation of an alternative Presbyterian church organisation in which ministers were elected by godly members of the congregation rather than being nominated by bishops, colleges, wealthy laymen or the Crown (Halley 1869, Vol. 1, Chapter 4). This embrace of a very limited form of church democracy should be set alongside the well-established (but much debated) association of Puritanism with the development of trade and the "rise of capitalism", and with relative freedom of secular scientific enquiry (as opposed to dangerous pursuits like astrology). It might be tempting to see this as particularly important in the case of Horrocks (Walton 1987, Chapter 3; Richardson 1969). But the actual use of the label "Puritan" in pre-Civil War Lancashire, especially as a term of abuse deployed by opponents and enemies, went far beyond this core stereotype. It could be applied to anyone who seemed to take religion seriously, read the Bible and attended sermons, even if they also drank, hunted and played bowls, as at least one "Puritan" cleric did. John Angier, a Puritan minister on any definition, was capable of decrying a perceived culture of overwork as denying his parishioners the necessary time to attend to religious duties, an inversion of the orthodoxy about the Puritan work ethic. Lay people with little or no formal schooling could be quite capable of discussing biblical texts and precepts in alehouses, in Lancashire villages, although we do not know whether this applied to Much Hoole (Richardson 1969, p. 84; Heywood 1937; Sachse 1938).

As Chapman makes clear, we know little about Horrocks's actual religious affiliations, but it seems quite possible that contemporaries, especially in Much Hoole, might have regarded him as a Puritan, at least on the less exacting of these definitions. He was, after all, a serious scholar and gave his mind to his books and apparatus, in a place where such concerns would normally be non-existent. We should remember that to be at all convincing as a Puritan it was necessary to be literate to be able to access and interpret the Bible, even though there are reports of prodigious feats of memory as regards sermons and Biblical passages in what was still a predominantly oral culture (for an example see Richardson 1969, p. 102). Reading was taught before writing in what schools there were, which were attended by only about 10% of children (on one definition); but it was a minority accomplishment at this time, almost vanishingly so among the lower orders: 64% of a sample of tradesmen, 86% of husbandmen, 94% of servants and labourers, and 98% of women in a Lancashire sample were unable to sign their name, and therefore, in the overwhelming majority of cases, to have the cultural capital for religious reading, speculation and discussion (Rogers 1975b; Walton 1987, Chapter 3).

As we might expect under these conditions, and in a firmly hierarchical society, active Puritanism was most visible among the gentry and the emergent early capitalists who organised local manufacture and linked Lancashire up with external markets, especially in London, Ireland and northern Europe; but it had ramifications among the 'middling sort of people' of craftsmen, artisans and yeoman farmers who, according to some interpretations, made up the backbone of Parliamentary support in the Civil War, and a leavening of genuine interest among some of the lower orders (Manning 1976). As we shall see, the Stones brothers, who built Carr House in Much Hoole, where Horrocks observed the transit, for their locally-based farmer and landowner brother, might seem to fit squarely into this pattern, as a London haberdasher and an Amsterdam merchant; but they defy simplistic assumptions by showing no direct evidence for or against Puritan leanings of their own (Marston 2004, pp. 13, 20-1). Most such people emerged from the upper yeomanry or the younger sons of the gentry, a level of social distinction that required far less wealth to sustain it in Lancashire than was the case further south, and a level not far removed from Horrocks's own social origins, despite the romantic assumptions about his alleged poverty that are current in the older literature (Blackwood 1978; Marston 2004, pp. 7 and 15, n. 24).

The limited evidence available on Horrocks's life suggests that he would have come into contact with plenty of Puritan influences. Toxteth, near Liverpool, where he was born in 1619, had originally been settled by Puritans from the Bolton area, which chimes in exactly with the available evidence on Horrocks's family, including the unsubstantiated suggestion of a Bolton connection (Richardson 1969, p. 96, refers to Alexander Horrocks, who shared a Cambridge education with Jeremiah, as 'the puritan pastor of Deane', near Bolton, in the 1620s and 1630s; but no connection can be substantiated); and Liverpool Corporation was early in endowing Puritan "lectureships" to provide an island of active Protestantism in the surrounding sea of West Lancashire Catholicism (Walton 1987, Chapter 3). The family connection with watch-making, part of an emerging complex of craft and metalworking industries in the Liverpool area at this time, which was to make its own contribution to the early Industrial Revolution more than a century later, also fits the pattern of Puritan engagement with trade beyond the locality and access to external ideas; and the fact that Horrocks's family had access to sufficient resources to send him to Cambridge for three years suggests a comfortable background among the upper levels of the 'middling sort' (Marston 2004, pp. 7-8; Langton 1979). We have no direct evidence of the sort of 'sponsored mobility' initiated by a patron that often propelled promising boys from the local grammar schools to Oxford or Cambridge at this time, although it would be tempting to assume that the Alexander Horrocks who attended an unspecified college at Cambridge University in the generation before Jeremiah might have set an extended family precedent and otherwise smoothed the path in some way; and the lack of external patronage for Horrocks's astronomical work would tend to support the assumption that there was enough money in the family to support him in his scholarly pursuits, no doubt with the goal of an eventual career in the Church (Richardson 1969; Stone 1966). Once at Cambridge he would certainly have encountered current Puritan views at close hand, not least in his own college, Emmanuel (Richardson 1969). And in Much Hoole, on his return to his native county, he would probably have stood out very distinctively from the society around him, unless there were popular Puritans and 'sermon-gadders' in the hamlet who have remained hidden from the historian's view.

We do not know much about Much Hoole in the first half of the seventeenth century. Although it was situated just outside the Catholic strongholds of Warrington Deanery and West Derby Hundred, it was part of that Lancashire that the Puritans of the Manchester area regarded as a 'dark corner of the land', where what strong formal religious influences there were would be at least as much Catholic as Protestant, and where older beliefs survived alongside a Christianity that might receive little doctrinal understanding or assent among the poor (Haigh 1975; Richardson 1969; Poole 2002). This was, notoriously, a county of sprawling parishes, ill-endowed chapels and poorly-paid, sometimes incompetent incumbents, although the standard had been raised considerably since the late sixteenth century (Haigh 1975; Richardson 1969). Moreover, the poor in the

Lancashire of the second quarter of the seventeenth century really were poor: the memory of the famine years of 1623-4 and of the epidemics (apparently including bubonic plague) of 1630 would still be present, and there were more to come in the troubled 1640s. Famine had threatened in the Rochdale area as recently as 1638 (Rogers 1975a; Appleby 1978; Walton 1987, Chapter 2). Much Hoole was on the edge of the Catholic heartland of barely accessible mosses and marshland that was south-west Lancashire, around Croston and Martin Mere, where flooding regularly made the churches inaccessible (Walton 1987, Chapter 2; Walker 1939). It was on an old route to a crossing point on the River Ribble, and the county town of Preston was accessible, as was Liverpool to the south-west; but it was unusual among Lancashire villages in having no resident gentry at all, like neighbouring Little Hoole, Longton and Tarleton, and even the neighbouring towns were of no great size (Blackwood 1978). On the best recent estimates, Preston in 1664 had a population of 1890, making it the third largest town in Lancashire south of the banks of the River Ribble, while Liverpool's great days were still a century in the future. Its main trade was still coastal and with Ireland, and a population of 1273 left it seventh in the same league table. Preston at least drew in the Lancashire gentry for Assizes and Quarter Sessions, and was the market centre for a wide area, at which opinions could be exchanged and business transacted beyond the bounds of the locality; but the Much Hoole village community has the hallmarks of a small and isolated place on any definition (Stobart 2003; Walton 2000, pp. 123-4).

Lancashire in the second quarter of the seventeenth century was, indeed, an impoverished county, although its position near the bottom of national league tables of wealth based on taxation returns needs to be qualified by awareness that it was probably easier to evade tax or get away with artificially low property valuations in areas like this, where the weakness of central government was indicated by (among other things) the very extent of Roman Catholic survival in face of ostensibly penal discrimination (Walton 1987, Chapters 1-2). Within the broader setting, the area around Much Hoole was particularly undeveloped, lying as it did at a distance from the emergent textile industries of the Manchester district or the coal mines of south-west and south central Lancashire (Langton 1979). In a context of substantial population expansion across the county between 1563 and 1664, Leyland Deanery, in which Much Hoole was situated, saw growth of only 14% during this period, compared with 89% for Manchester Deanery, the main centre of the emergent textile industries, and a surprising 121% for the bucolic and peripheral Amounderness Deanery, north of the River Ribble and mainly west of Preston (Phillips & Smith 1994, p. 9). Much of the population growth elsewhere in the county merely deepened poverty and vulnerability to crisis by putting pressure on limited resources; but the contrast between the Much Hoole area and the neighbouring deaneries suggests a failure to share in developments that were occurring nearby.

Horrocks's horizons, however, were not bounded by his locality. Not only had he been brought up near the emergent (though still tiny and isolated) port of Liverpool, he had spent the years between 1632 and 1635 far from his roots, at a Cambridge college that was noted for its Puritan leanings, and he was capable of sustaining correspondence with like-minded people at a distance, even from this apparently isolated corner of Lancashire (Walton 1987, Chapters 2-3; Marston 2004, pp. 8-9). If he was indeed employed by the Stones family at Carr House, as Chapman surmises, he was in touch with the wider currents of economic growth that are associated with Lancashire's links with London and overseas trade: the currents along which new Protestant ideas also flowed, nourishing Lancashire Puritanism; and Horrocks's contacts with William Crabtree of Salford bring him into a similar orbit (Willan 1980). As Chapman points out, the Stones family endowed a font and provided sacramental plate for Much Hoole church, which shows that they were not Puritans on the narrow core definition cited above, although Marston argues that the items in question were sufficiently plain and lacking in ostentation for this not to rule out adhesion to a version of Puritanism, or at least to preclude any accusations of High Church leanings in those who commissioned them (Marston 2004, p. 21). The descriptions of 'Thomas Stones of London, haberdasher, and Andrew Stones of Amsterdam, merchant', who paid for the construction of Carr House for their brother, reveal direct contacts between this part of Lancashire and the burgeoning economy of northern Europe, forming part of a wider pattern whereby Lancashire merchants who had made good in the metropolis and beyond endowed schools, churches and lectureships in their home county in this period (Jordan 1962). Horrocks's Salford correspondent and fellow astronomer, William Crabtree, was also part of this culture as a cloth merchant with metropolitan and international contacts. A lot of this activity, but not all of it, had Puritan associations, and some of its scientific dimensions crossed the religious divide between Protestant and Catholic which in other circumstances, and only three years later, might be regarded as the defining ingredient in Civil War allegiances in the county. According to Charles Webster there was also a Carr House connection with the antiquarian Christopher Towneley, a junior member of the Catholic family of Towneley Hall near Burnley who had astronomical interests. Towneley seems to have facilitated correspondence between Horrocks, Crabtree and three other young northern astronomers across Lancashire and Yorkshire (Webster 1966). Awareness of such networks across the county and beyond makes Much Hoole seem less of a bucolic backwater, and brings out the extent to which a leavening of Lancashire society was responding and contributing to the intellectual ferment of the outside world on the eve of the Scientific Revolution. One of the morals of this story is that you patronise the provinces at your peril.

This paper has demonstrated that we can add extra dimensions to the Horrocks story by placing it in the context of locality and county, and of the religious, cultural and economic currents, complexities and conflicts of the time. It also shows that, despite its bucolic setting and apparent isolation from the main currents of European thought, Much Hoole was not an outrageously improbable place from which to observe the Transit of Venus in 1639.

References

Appleby, A.B. 1978 *Famine in Tudor and Stuart England* (Liverpool: Liverpool University Press)

Blackwood, B.G. 1978, *The Lancashire Gentry and the Great Rebellion* (Manchester: Chetham Society)

Broxap, E. 1910 *The Great Civil War in Lancashire* (Manchester: Manchester University Press)

Chapman, A. 1994, *Jeremiah Horrocks and Much Hoole*

Haigh, C. 1975, *Reformation and Resistance in Tudor Lancashire* (Cambridge: Cambridge University Press)

Halley, R. 1869, *Lancashire: its Puritanism and Nonconformity* (Manchester: Tubbs and Brook, 2 vols.)

Heywood, O. 1937, (ed. E. Broxap), *Oliver Heywood's Life of John Angier of Denton* (Manchester: Chetham Society)

Jordan, W.K. 1962, *The Social Institutions of Lancashire: a Study of the Changing Patterns of Aspiration in Lancashire 1480-1660* (Manchester: Chetham Society)

Langton, J. 1979 *Geographical Change and the Industrial Revolution* (Cambridge: Cambridge University Press)

Manning, B. 1976 *The English People and the English Revolution* (London: Heinemann Educational)

Marston, P. 2004 *Jeremiah Horrocks: Young Genius and First Venus Transit Observer* (Preston: University of Central Lancashire)

Phillips, C.B., & J. Smith 1994 *Lancashire and Cheshire since AD 1540* (London: Longman,)

Poole, R. (ed.) 2002 *The Lancashire Witches: histories and stories* (Manchester: Manchester University Press)

Richardson, R.C. 1969 *Puritanism in North-West England* (Manchester: Manchester University Press)

Rogers, C.D. 1975a *The Lancashire Famine of 1623* (Manchester: Manchester University Extra-Mural Department)

Rogers, C.D. 1975b 'The Development of a Teaching Profession in England, 1574-1700, with special reference to Lancashire', Ph.D. thesis, University of Manchester

Sachse, W.L. (ed.) 1938 *The Diary of Roger Lowe* (London: Longman)

Stobart, J. 2003 *The First Industrial Region: North-West England 1700-60* (Manchester: Manchester University Press)

Stone, L. 1966 'Social Mobility in England 1500-1700', *Past and Present* 37

Walker, F.X. 1939 *Historical Geography of Southwest Lancashire before the Industrial Revolution* (Manchester: Chetham Society)

Walton, J.K. 1987 *Lancashire: a Social History 1558-1939* (Manchester: Manchester University Press)

Walton, J.K. 2000 'North', in P. Clark (ed.), *The Cambridge Urban History of Britain*, Vol. 2 (Cambridge: Cambridge University Press), pp. 111-31

Webster, C. 1966 'Richard Towneley (1629-1707), the Towneley group and seventeenth-century science', *Transactions of the Historic Society of Lancashire and Cheshire* 118, pp. 51-76

Willan, T.S. 1980 *Elizabethan Manchester* (Manchester: Chetham Society)

Wrightson, K. 1974 'The Puritan Reformation of Manners, with Special Reference to Lancashire and Essex', unpublished Ph.D. thesis, University of Cambridge

Discussion

STEVE DICK: Speaking of economic development, is there any relationship between our Jeremiah Horrocks, and the famous but much later Horrocks cotton industry in this area, which I was reading about yesterday in the museum.

JOHN WALTON: My answer is that I don't know, but it would be interesting to find out.

ALLAN CHAPMAN: His name, Horrocks, might actually be a modified version of the name for some kind of ... I think it was a kind of ox indigenous in medieval Lancashire; the Orrock was some kind of big brute this might be the origin of "Horrocks" ...

JOHN WALTON: He probably was 19 stone and 6-feet tall. The Victorians sometimes spelled his name with an "x" instead of "cks"; I am not sure what the origin of that was.

Robert van Gent and Nick Kollerstrom

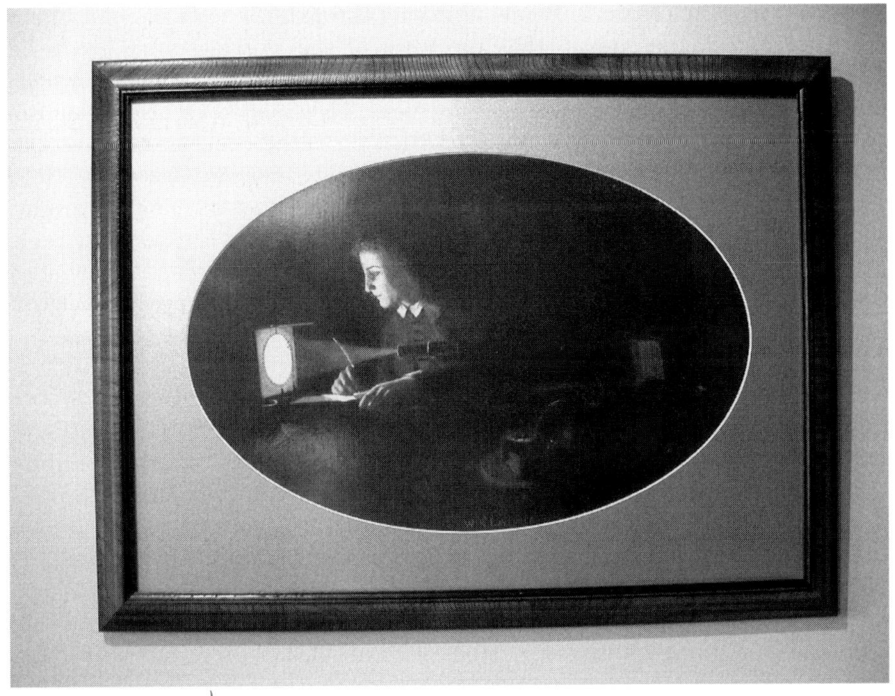

Jeremiah Horrocks as imagined by J.W. Lavender in 1903. This copy hangs in Carr House; the original is in Astley Hall Museum and Art Gallery, Chorley, Lancashire.

Transits of Venus: New Views of the Solar System and Galaxy
Proceedings IAU Colloquium No. 196, 2004
D.W. Kurtz, ed.

© 2004 International Astronomical Union
doi:10.1017/S1743921305001249

William Crabtree's Venus transit observation

Nicholas Kollerstrom

Science and Technology Studies Department, University College London,
Gower Street, London WC1 E6BT, UK
email: nk@astro3.demon.co.uk

Abstract. The close collaboration between the two North-country astronomers Jeremiah Horrocks and William Crabtree gave them special insight into the new astronomy published by the recently-deceased Kepler, whereby Horrocks became the only person to apprehend that the Rudolphine tables were in fact predicting a Venus transit in 1639. This paper focuses especially upon William Crabtree's role and contribution. A comparison is made with an earlier, unsuccessful endeavour by these two concerning a possible transit of Mercury. Much of the record of their work was lost during the civil war. Finally, thanks to Christiaan Huygens, Horrock's manuscript was published by Johannes Hevelius in Danzig, in 1662.

1. A Mercurial error

The North-countrymen gathered round Jeremiah Horrocks comprised the beginning of the tradition of English astronomy.[1] There had earlier been individuals like Thomas Hariott, who had scrutinised the Moon with his telescope, but he started no tradition, by way of others following his work. It was primarily through the work of John Flamsteed, Britain's first Astronomer Royal, that the work of this north-country group became known in Britain, as he learnt his astronomy from these people: Richard Towneley and William Gascoigne, William Crabtree and Horrocks. They drew their inspiration from the new astronomy of Kepler and combined this with a very practical approach to apparatus. According to a letter of 1640, Crabtree visited Yorkshire and saw Gascoigne's telescope 'amplified and adorned with new inventions of his own.' After describing the invention to the young Horrocks, he reported back to Gascoigne: 'My friend Mr Horrocks professeth that little touch I gave him of your inventions hath ravished his mind quite from itself and left him in exstasie between admiration and amazement.' For the period that concerns us, however, that micrometer device had not as yet been fitted to a telescope.

Horrocks revered Kepler as 'the most learned astronomer who had ever lived,' expressing the view that: 'His death was an event that must ever have happened too soon; the science of astronomy received the lamentable intelligence whilst left in the hands of a few trifling professors who had kept themselves concealed like owls until the brightness of his sun has set'[2] (Kepler died in 1630, and Horrocks composed this in 1640). His *Venus in Sole Visa* kept breaking into verse: 'Who, mighty shade, shall sing thy praises?' he asked, although he felt at liberty to amend Kepler's newly-published Rudolphine Tables. He and his colleagues were effectively the nucleus of Keplerian astronomy in England.[3]

Two things had to come into the world for transits across the Sun to be discerned: the telescope, and Kepler's astronomy. The elliptical orbits of the latter enabled the planetary

---

[1] Wilbur Applebaum, "Keplerian Astronomy after Kepler", *History of Science*, 34 (1996), 451-504.

[2] A. Whatton, *Memoir of the Life and labours of . . . Horrocks* 1859 (with *Venus in sole Visa*, pp. 109–202), p. 177.

[3] I. Bernard Cohen, *The Newtonian Revolution* 1980, CUP, p. 229.

nodal axes to be located in a definite manner, with their ascending and descending nodes both opposite each other and with the Sun in the middle: an instance of what Kepler meant by constructing a 'physical' astronomy. The dramatic success of Pierre Gassendi in Paris, seeing the transit of Mercury in 1631 as predicted in the Rudolphine Tables, vindicated the new Keplerian approach. The mathematics of elliptical motion was harder than that for circles, but this first-ever successful solar transit prediction convinced the cognoscenti that the effort was worthwhile. The extreme smallness of Mercury seen by Gassendi suggested that Kepler's 'solar parallax' – the angle subtended by the Earth's semi-diameter from the Sun – was too large.[4] Kepler's Rudolphine Tables also predicted a Venus transit on 6 December 1631 (New Style = Gregorian Calendar), a thing equally unseen hitherto. It actually turned up on 7 December – not that this mattered, because it happened before sunrise in most of Europe.[5]

William Crabtree and Jeremiah Horrocks met in 1636, and an intense correspondence came to pass between them, only some of which survives – involving the devising of a new lunar theory and the Jupiter-Saturn inequality, which do not here concern us. Crabtree described Horrocks as 'my friend and second self'[6] and posterity came to hear of Horrock's seminal work very much via Crabtree. On 29 September 1638, the 20-year-old Horrocks wrote to his 28-year-old friend Crabtree about a likely forthcoming transit of Mercury. Kepler's tables had not predicted it, giving Mercury's latitude during its forthcoming solar conjunction as nearly 18 minutes of arc. That was too wide for a transit (the Sun's angular radius we reconstruct as having then been $16' 8''$), whereas other tables, Horrocks explained, were claiming it would be no more than $13'$ or $14'$. This inferior conjunction would happen on 21 October 1638, he wrote (Old Style = Julian Calendar), explaining that for this possible transit of Mercury with the Sun, he intended to attach his telescope to an 'oblong stick, carrying a plane surface at right angles to itself, on which to receive the Sun's image.' He drew a circle on the sheet, with numerical markings.[7] He described his procedure in the same way in his *Venus in sole visa*. No such transit took place, with Mercury passing over the Sun at $20'$ latitude, well outside the limit for a transit.[8] Nothing comparable to the great achievement of Pierre Gassendi in Paris in 1631 took place; however, this was an important dry-run for this North-country duo, in preparation for their rendezvous with destiny over the Venus transit of 1639 – likewise not predicted by Kepler.

By way of comparison, earlier in the century Galileo had been corresponding with a colleague about how the errors of Mercury in current tables could go up to ten degrees.[9] Owen Gingerich has confirmed by his reanalysis of tables of this period that such errors then existed. Mercury had far larger errors than any other planet.[10] Galileo's positions for Mercury were often out by two or three degrees. Such errors of pre-Kepler tables

---

⁴ Horrocks assumed that the sizes of Mercury, Venus and Earth were in proportion to their distances from the Sun, which turned out to be quite a happy conjecture: Curtis Wilson, "Horrocks, Harmonies and the Exactitude of Kepler's Third Law", 1978, in Wilson, *Astronomy from Kepler to Newton* Variorum, London, 1989, p. 249.

⁵ Its last stage was visible from most of Italy and south-east Europe.

⁶ Letter, Crabtree to Gascoigne, 7 Aug 1640 in Whatton, *op. cit.* (2), pp. 53–55, 54.

⁷ Horrocks, "The Transit of Venus across the Sun," p. 315 in *Opera Posthuma*, Ed. J. Wallis, 1673.

⁸ Kindly ascertained by Bernard Yallop.

⁹ Letter, Ottavio Brenzoni to Galileo, Padua, 19 December 1605 (No. 130, Favaro's *Opere* of Galileo).

¹⁰ Owen Gingerich, *The Eye of Heaven, Ptolemy, Copernicus, Kepler* 1993 AIP, pp. 233, 385.

remind us of what Gingerich ascertained, that the Rudolphine Tables improved planetary positions by an entire order of magnitude.[11]

2. Omitted by Kepler

Kepler spotted the 1631 Venus transit while preparing the ephemeris for that year; however, he prepared these only up to 1636 and never got round to preparing one for 1639 – and, in consequence, failed to notice that his own Rudolphine Tables predicted a second Venus transit.

Histories hitherto have strangely averred that these tables erred in failing to predict this[12] – e.g., Dreyer: 'the Rudolphine tables threw Venus quite off the Sun's disc';[13] or, more recently, 'Kepler's tables ... predicted that Venus would just miss crossing in front of the solar disc'.[14] One author has tried to argue that Kepler's unduly large solar parallax caused him to miss the event.[15] In fact, the Rudolphine Tables specified a celestial latitude of $-7'$ $45''$ for this Sun-Venus conjunction,[16] as would have produced a transit (it was actually at $-8'$ $51''$, so they erred by an arcminute), and they predicted the event nine hours early because their Venus longitude erred by half a degree. Only one person in the world appreciated the significance of this! In Horrocks' words, 'The more accurate calculations of Rudolphi very much confirmed my expectations; and I rejoiced exceedingly in the prospect of seeing Venus'.[17] He requested that Crabtree look out for it.[18]

3. 'Rapt in contemplation'

It dawned upon Horrocks mere weeks before the event, in October 1639, that the transit was likely.[19] This short notice was his excuse for not having informed his friends: most of whom, he explained, were 'preferring their hawks and hounds, to say no worse.' He anticipated the event with 'much anxiety', because it boded bad weather: Jupiter and Mercury would be crowding together with Venus and the Sun, and from his experience he knew that this threatened bad weather.[20] Especially Mercury's conjunction with the Sun 'is invariably attended with storm and tempest.'[21]

William Crabtree's attic window of his thatched cottage at Broughton must have had a clear view of the setting Sun. The Venus transit lasted five or six hours, barely forty minutes of which were visible in the UK. It would have been fully visible in America

---

[11] Ibid, p. 386.

[12] R. Proctor, *Transits of Venus* 1878, p. 12: According to his [Kepler's] calculations, 'Venus passed not near enough for a transit to occur.'

[13] J. Dreyer, *A History of Astronomy from Thales to Kepler* Dover, 1906, NY 1953, fn. p. 420.

[14] D. Sellers, *The Transit of Venus*, Leeds 2001, p. 80

[15] Eli Maor, *June 8 2004 Venus in Transit* PUP 2000: Kepler's tables showed that Venus would pass 'just below the sun's disc,' and his 'correction applied to a topocentric observer' had caused his error, pp. 32–3.

[16] Whatton (ref. 2) p. 180. Owen Gingerich recently (HASTRO-L, 8 July 2004) reconstructed the Rudolphine Tables as a computer program, and found that they did indicate a Venus transit. Horrocks' estimate of the Venus latitude was $-8'$ $31''$ (Whatton p. 185).

[17] Whatton (ref. 2) p. 111.

[18] He also wrote to his brother Jonas, and asked Crabtree to inform Samuel Foster in London.

[19] Allan Chapman, *Three North-Country Astronomers* 1982 (40 pp) p. 13.

[20] Mercury, Venus, Jupiter and the Sun were all, during the transit, within five degrees of longitude.

[21] Whatton (ref. 2), p. 113.

as Horrocks noted – had anyone there been capable of realising this. Its first contact started with the Sun a mere five degrees above the horizon. It was in conjunction with the descending node of Venus, giving the date of 4 December for the event (New Style), in contrast with the present ascending node Venus transit on 8 June (the Venus node axis being stationary). Not a single observatory existed in England or France, as these two prepared to view the event.

"Rap't in contemplation he stood, motionless, scarce trusting his senses, through excess of joy." This was the description given by Jeremiah Horrocks of his colleague William Crabtree, while seeing Venus move across the face of the Sun in 1639.[22] The historian sees this ecstasy as the human experience of innovation: the first-ever experience, of Venus crossing the solar disc. For comparison we could describe Galileo's awe during that week when he first beheld the moons of Jupiter in 1610 – his Introduction to the *Sidereus Nuncius* described how the 'Maker of Stars' then spoke to him, concerning Jove and Jupiter, and his destiny. Or, the condition of Kepler when his third law of planetary motion dawned upon to him in 1618: 'it is my pleasure to yield to the inspired frenzy...',[23] he wrote. The personal experience of innovation is here recorded, at times when the world changes.

Horrocks' account rather gives the impression that William Crabtree's moment of joy caused him to miss his chance – it had been cloudy during most of the transit, and then clouds parted 'shortly before sunset':

> 'In a little while, the clouds again obscured the face of the Sun, so that he could observe nothing more than that Venus was certainly on the disc at the time.'

Afterwards, he made 'so rapid a sketch' of Venus as it had passed across the sun's disc. One finds this account written up in the histories. Though admitting that Crabtree was 'a person who has few superiors in mathematical learning,' the young Horrocks seems to have wished to depict his colleague as one who rather missed the time available to map the transit accurately, and so had to sketch it from memory. Did Crabtree really miss his life's biggest opportunity, leaving behind a mere sketch from memory?

Let us here recall that Crabtree had transcribed Kepler's Rudolphine Tables into a more convenient decimal notation rather than fractions[24] which gives us a hint of his careful, methodical manner. The estimates then made of Venus' angular size may be compared with its true value:

Horrocks	$1'\ 12''$
Crabtree	$1'\ 3''$[25]
Actual size, then	$1'\ 3''$[26]

It can never appear larger than $1'\ 6''$. Crabtree's measurement was thus within a second of arc. He appears as the first astronomer to take a significant measurement having second-of-arc accuracy: for that hazy sunset, projected as a screen image of something like 20 cm across, was that mere fabulous good fortune?

---

[22] Whatton (ref. 2), p. 129.

[23] Johannes Kepler, *Harmonices Mundi* 1618, trans. by E. Aiton, A. Duncan, J. Field, *The Harmony of the World* Amer. Philosophical Soc. 1997, p. 391

[24] Whatton (ref. 2); also Allan Chapman, *William Crabtree: Manchester's first Mathematician* Man. Stat. Soc. 1995.

[25] Whatton (ref. 2) p. 130.

[26] As noted by the present writer: N. Kollerstrom, 'Crabtree's Venus transit measurement,' *Quarterly Journal of the RAS*, 1991, 32, 51

4. Hevelius' incredulity

The coronation of the Merry Monarch Charles II fell on 23 April 1661, the day of a Mercury transit across the Sun (3 May, New Style). The Dutchman Christiaan Huygens attended the event, during which he got to hear about a certain manuscript of Jeremiah Horrocks, together with some fragments of correspondence with Crabtree. There was, he was informed by Sir Robert Moray, no prospect of them being published in England, and might he find a publisher?[27] He knew the eminent Polish astronomer Johannes Hevelius, and that he was observing this Mercury transit, so he decided to take the manuscript copy he was offered, with the intention of passing it on to the latter.[28] And so it was that a manuscript composed in the North of England in 1640 *Venus in Sole Visa* was first published in Poland in Danzig, where Hevelius' observatory was located (nowadays called Gdansk), in 1662 – as a postscript to the latter's report on the Mercury transit.[29]

It was a century of dizzy change, which began with Kepler having all the stars of heaven within an 'eighth sphere" a mere two German miles wide – and ended with Flamsteed estimating the size of Sirius in an infinite universe. These North-country astronomers were the first ever to apprehend the real immensity of the solar system, and they did so by keying into the notion of solar parallax: in simple language, that meant how very small Venus and Mercury looked as they journeyed across the Sun. Horrocks estimated the Sun's distance as 'at least 15,000 semi-diameters of the Earth'[30] which is two-thirds of the correct value.

Kepler's third law gave the relative sizes of the planetary orbits but not their actual size, and in fact Kepler's solar system was too small by a factor of seven. Solar parallax was the key parameter indicating how far away the Sun was, and thus how large the solar system really was. In 1662, when Hevelius published Horrocks' manuscripts, he made one correction: in place of Horrock's value for the 'solar parallax' of 14 seconds of arc he made this 41 seconds of arc[31] – he was unable to credit so drastic an enlargement of the solar system.[32] Or, was this a mere slip of the pen on his part?[33] Tycho Brahe's value for comparison had been two minutes of arc. Flamsteed in 1672 visited Richard Towneley at Burnley, and together they made a solar parallax measure of just $10''$, by using a Mars-against-the-stars method with an eyepiece micrometer (the modern value is $9''$).[34] Flamsteed reported this in his first letter to the Royal Society.[35]

Sir Isaac Newton was not willing to believe this, and his first edition of his *Principia* used Hevelius' value: it cited the solar parallax as $40''$.[36] His second edition adjusted this

[27] On 11 April, Huygens had dinner at Gresham College with Robert Moray, Lord William Brouncker and others, when the publication of the Horrocks papers could have been discussed (private communication by R.H. van Gent).

[28] Letters of Huygens to Hevelius (Nos 892 and 917), 21 September and 9 November 1661, in Christiaan Huygens *Oeuvres Complètes* The Hague 1890, vol. 3, pp. 334-335 and 385-386. This correspondence indicates that Moray had earlier received the manuscript from Paul Neile.

[29] Frances Wilmoth, Ed., *Flamsteed's Stars* 1997, Boydell, p. 56.

[30] Whatton (ref. 2), p. 151.

[31] J. Hevelius, *Mercurius in Sole Visus*, Danzig 1662, Note to Ch. 6, p. 124.

[32] Peter Aughton, *The Transit of Venus, The Brief, Brilliant Life of Jeremiah Horrocks*, London 2004, p. 127.

[33] A year later, in a letter to Huygens, he discussed the Crabtree and Horrocks measurements, in connection with Mercury and Venus' magnitudes: Letters of Huygens (ref. 28), vol. 4 (No. 1099), pp. 308-310, 19 Feb 1663.

[34] Letter of Flamsteed to Oldenburg, 16 November 1672, in: *The Correspondence of John Flamsteed, First Astronomer Royal*, Ed. Forbes, Murdin and Willmoth, vol. I, 1995, p. 185.

[35] J. Birks, *John Flamsteed, the First Astronomer Royal*, 1999, p. 73.

[36] Whatton (ref. 2), p. 83.

to 10″– giving no credit to Flamsteed, from whom, we may assume, he obtained this key parameter.[37] Newton's view of these estimates (there having been no other Venus transit since) was written up in his *System of the World*:

> 'Venus appeared to Mr Crabtrie only 1′ 3″; to Horrox but 1′ 12″; though by the mensurations of Hewelcke and Huygens without the sun's disc, it ought to have been seen at least 1′ 24″'[38]

Based on his understanding of solar system dimensions, he is judging his fellow-countrymen of having taken too small measurements of Venus' size. He preferred Huygens' estimate.[39] It thereby became a little-appreciated fact that, in the year 1639, a British astronomer's moment of joyful rapture did not prevent him from taking a planetary measurement accurate to a single second of arc.

Acknowledgements

I'm grateful to Wilbur Applebaum, Robert van Gent, David Sellers and Bernard Yallop for substantial assistance.

Discussion

STEVE DICK: You say Newton accepted the 40″ value, he knew of the other values but for some reason he chose that one. Do you know any more details on that?

NICHOLAS KOLLERSTROM: Yes, let's just quote from the third edition – Newton's final opinion in the last year of his life. Venus appeared to Mr Crabtree only 1′ 3″; to Horrocks 1′ 12″, though by the mensurations of Hewelcke and Huygens without the sun's disk, it ought to have been seen at least 1′ 24″. So he's giving more weight to these European astronomers.

WALTER BRISKEN: It seems that if they had telescopes at that time they could have made direct visual estimates of the size of these planets. Was that not possible for some reason? Why did they rely on the transit to determine the diameters?

NICHOLAS KOLLERSTROM: Well, I think a practical astronomer should comment on this, but it was just apparently seeing the black shade against the sun that gave them the ability to tell its physical size, and I think you would need much more detailed telescopic apparatus for estimating just by looking at a planet. I don't think they had that apparatus at the time.

ALLAN CHAPMAN: Just to take up on the point which you had made, Nick, about measuring very, very small objects: of course, before the development of the micrometer there was no way in which you could make an in-telescope measurement with great accuracy, and of course various people were experimenting with this; Huygens is doing it, as Cassini is doing it in France, and so on, and of course Gascoigne, who was part of the total Northern group, actually succeeds in doing it with a pair of pointers in 1640. But that wasn't widely known until the 1660s and Robert Hooke, for instance, in the

---

[37] Corroborated by the 9″.5 value for solar parallax obtained by Cassini (using the 1672 observations of Mars by Richter), which he published in 1684: A. van Helden, *Measuring the Universe*, 1985, p. 154.

[38] Isaac Newton, *The Principia*, Motte translation, Vol II, UCLA Press 1963, p. 565 (His *System of the World* was published posthumously and its date of composition is uncertain).

[39] For Huygens' obtaining of this value, see Letters of Huygens (ref. 28), 4: letter to Hevelius, 25 July 1662 (No. 1037), pp. 181 382.

comet of 1664, tries to measure quite successfully the diameter of the nucleus of the comet by watching the comet set beyond a distant building on his skyline, where he mentions on the top of this building there was a weathercock. He notices the comet set behind the weathercock and he timed the periods, where different parts of the cometary nucleus were actually obscured by the bar of the weathercock. Then he measured subsequently the diameter of the weathercock, triangulated the distance of it from his observing post in Gresham College, so he was actually using the weathercock as a marker bar, and from this drew the conclusion that the comet's nucleus, I think, was 10″ and its general coma about 50″. But before the invention of the micrometer in the 1660s, there was no way of measuring a planet or a very, very small size in the heavens, and hence using the sun's disc was the ideal way of doing this because you could divide it into a fraction of 30″ or 31″.

MYLES STANDISH: This reminds me of what people speculated with Galileo. He put on the side of his telescope something that looked like a tennis racket with cross-hatches – and he would move it back and forth until Jupiter just exactly fit one of the squares. Then he could count off the distances to the different satellites by measuring them with squares. The square effectively was a measurement of the relative diameter of Jupiter, so it's a kind of a cross-staff method.

A window in Carr House, Much Hoole, Lancashire

Transits of Venus: New Views of the Solar System and Galaxy
Proceedings IAU Colloquium No. 196, 2004
D.W. Kurtz, ed.

© 2004 International Astronomical Union
doi:10.1017/S1743921305001250

Venus transits – A French view

Suzanne Débarbat

Observatoire de Paris, SYRTE/UMR 8630, Paris, France

Abstract. After a careful study of Mars observations obtained by Tycho Brahé (1546-1601), Kepler (1571-1630) discovered the now-called Kepler's third law. In 1627 he published his famous *Tabulae Rudolphinae*, a homage to his protector Rudolph II (1552-1612), tables (Kepler 1609, 1627) from which he predicted Mercury and Venus transits over the Sun. In 1629 Kepler published his *Admonitio ad Astronomos...* Advertisement to Astronomers (Kepler 1630), *Avertissement aux Astronomes* in French *Au sujet de phénomènes rares et étonnants de l'an 1631: l'incursion de Vénus et de Mercure sur le Soleil*. This was the beginning of the interest of French astronomers, among many others, in such transits, mostly for Venus, the subject of this paper in which dates are given in the Gregorian calendar.

1. The 1631 and 1639 Venus transits

The predicted Mercury transit being the first to occur, Gassend (1592-1655), frequently known under the form Gassendi, then a professor in the south of France, later to be a professor to the *Collège Royal* in Paris, decided to observe it. Being a friend of Peiresc (1580-1637) he was introduced to astronomy and to Kepler's elliptic orbits. Confident in the prediction, he began (being in Paris at that time) with Mercury, on November 7, and succeeded with a solar image having a diameter of 24 cm. Gassend made a careful study of his data; his conclusions were favourable to elliptic orbits and Kepler's tables, but he considered that Kepler was in error for the conjunction.

Gassend (1632) decided to observe the Venus transit, on 1631 December 6 as predicted, despite the fact that Kepler had written that it could not be seen from Europe. From a map, established by Jean Meeus, it can be seen that the transit could be observed only from Asia. From better ephemerides it was seen, during the XVIIIth century and later, that the conjunction occurred only on December 7, and that it was close to the limb of the Sun.

In the "Advertisement to Astronomers" the next predicted Venus transit could occur only in 1761, the 1639 one being only a close approach. With no success in 1631, the French astronomers went into other fields. Gassend and Peiresc began observations of the satellites of Jupiter discovered by Galileo Galilei (1564-1642), having in view predictions for improving the determination of longitudes.

Meanwhile, following a revision of Kepler's Tables, young Horrocks (1618/19-1641) and his friend Crabtree (1610-1644) were the only ones in Great Britain, and in the world, to observe the 1639 December 4 transit of Venus (Horrocks 1672). From the corresponding Meeus map, conditions were favourable, even with a more-or-less cloudy day, as it was.

This success caught the attention of many Europeans. The Scottish Gregory (1638-1675), so well known by his reflector, mentioned in his book published in 1663 the capability, from any Mercury or Venus transits, to determine the solar parallax. Having in mind the 1677 Mercury transit, astronomers had to wait.

In France, the *Observatoire Royal*, created in 1667 by Louis XIV, was under construction. The following year the astronomers suggested the invitation of Cassini (1625-1712), this astronomer and professor in Bologna having published his high-level tables for the prediction of the eclipses of the Galilean satellites of Jupiter. In 1669/70, Picard (1620-1682) measured a terrestrial arc along the Paris Observatory meridian line using a portable quadrant, a zenith sector and a level equipped with refractors and with the micrometer he had devised with Auzout (1622-1691).

Having deduced from these measurements the dimension of the Earth, the astronomers could think about the solar parallax. This was made using Mars observations from Cayenne (Guyana) by the French astronomer Richer (1630-1696) and from Paris by Cassini. While Horrocks, from the diameter of Venus, had multiplied by a factor of 4 the Sun-Earth distance established by Copernicus (1473-1543), the new measurements brought this to a factor of 20 in 1672.

One year before the 1677 Mercury transit the Danish astronomer Roemer (1644-1710), then in the Paris Observatory, employing the eclipses of the satellite Io observed from 1666 to 1676, discovered the fact that light has a finite speed. The same year, 1676, began the mapping of the west French coast, by Picard and La Hire (1640-1718), using as a reference the Paris Observatory meridian.

Such were the works performed by the French astronomers when, as we say in French, "Halley vint". In England the first Astronomer Royal, Flamsteed (1646-1719) was installed in the Royal Observatory (nowadays the Royal Observatory Greenwich), created in 1675. A young assistant, Halley (1656-1642), was sent to the southern hemisphere to improve cartography and navigation (Cook 1998; Halley 1981). When on the Island Saint-Helena he observed on 1677 November 7 a Mercury transit, with a 24-foot refractor. Having a good pendulum clock, equipped, as made by Huygens (1629-1695) in the Low-Countries, Halley could obtain the duration of the complete phenomenon. This brought him to consider the following year that Venus could be of better use to determine the solar parallax (Toulmonde 2001). But, according to Kepler, the next such transit was to occur in 1761, giving plenty of time to the astronomers.

2. The 1761 and 1769 Venus transits

The interest of astronomers (Van Helden 1985) came through the various studies made by Halley (from 1720 the second Astronomer Royal). His important publication *Methodus singularis...*, published in 1716 in the Philosophical Transactions, was translated into English by Ferguson (1710-1776), in his book "Astronomy" published one year before the 1761 transit. The first translation of Halley's paper into French was recently published in the magazine *l'Astronomie*, in a translation made by Bonche & Toulmonde (2001). The French complete title given is *Méthode singulière pour déterminer la parallaxe du Soleil ou sa distance à la Terre par les observations de Vénus dans le Soleil, proposée devant la Royal Society par Edmond Halley, Docteur dans les Deux Droits, secrétaire de la Société.*

However, 1761 was far from 1716; the astronomers had a time to wait, but Halley's method was given. While the Greenwich astronomers had duties regarding improvements for astronomy and navigation, in France they were busy with the consequences of Newton's *Principia* and with a general map of France. In Greenwich, Bradley (1693-1762) deduced, from his positional observations, nutation and aberration, while his colleagues were working on the lunar distances method for navigation, and the horologer Harrison (1693-1776) on his marine clocks.

In France, the Cassinis, Lacaille (1713-1762) and others established the reference meridian line from Dunkirk to the Pyrenees. In 1750 Lacaille (ms: "Avis aux astronomes") departed to the Cape of Good Hope with several purposes, among them the parallax of Mars and of the Moon; for the success of this work, he requested his colleagues from many countries to perform similar observations. In a note to this call, Lacaille spoke about Halley's proposal and, mostly, about the error he had given which appeared to him rather small, given the difficulties in the observations.

Having observed Mercury transits in 1723, 1736, 1743 and 1753, the astronomers were still waiting for the 1761 Venus transit. But they had also made studies about the best possible method for the determination of the solar parallax from Venus (Woolf 1959; Dumont 2001). Halley had assumed that the accuracy for the timings of contacts could be less than one second of time, asking his colleagues to observe on a more-or-less international level. But, Halley having died in 1742, Delisle (1688-1768), who had met him in 1724 during his voyage in England, recalled Halley's idea. He also suggested, instead of using the duration of the transit from two stations, to observe the timing of the same contact from such places.

In April 1760, Delisle gave to the *Académie des sciences* a "Mémoire" with a more complete study than his previous papers published in 1723 and 1743 (Delisle 1743). He also made a "Mappemonde" showing the best observing places. Despite the fact that he made copies of the mémoire and that 200 maps were sent around the world, none of the maps can be found in France. The best known 1761 map is the one published by Ferguson "A Plain Method" in 1760. Sent well in advance to his correspondents, Delisle's map and mémoire reached their destinations and almost every one agreed to do his best.

The British and the French appeared as the most concerned by the phenomenon. In France it was observed by Maraldi (1700-1788) at the Paris Observatory, Lalande from the *Palais du Luxembourg*, Jeaurat (1724-1803) at the *Observatoire de l'Ecole Militaire*, Messier (1730-1817) from the *Observatoire de la Marine*, then situated in what is now named *Hôtel de Cluny* and *Musée du Moyen-Age* in the *Quartier Latin*. Lemonnier (1715-1799) and La Condamine (1701-1774) were with Louis XV who requested astronomers to see the phenomenon.

Many others in France made similar observations (Mém. Acad. Sci. 1763), while four teams were sent abroad: Le Gentil de la Galaisière (1725-1792) was at sea on June 6, trying to observe from the boat; Chappe d'Auteroche (1728-1769), upon invitation from the empress Elisabeth II (1709-1762) of Russia, was in Siberia, in Tobolsk, observing the transit after the first two contacts (Chappe d'Auteroche 1763); Pingré (1711-1796) went to the "Ile Rodrigue" close to "Isle de France", nowadays "Ile Maurice" or Mauritius Island and could observe the whole transit (Pingré 1763); Cassini III (1714-1784) was in Europe, in Vienna (Débarbat & Dumont 1996), to observe with Father Liesganig (1719-1799) and the Archduke Joseph (1741-1790).

¿From 62 stations, the French Academy of sciences received about 120 reports giving values of the solar parallax lying between 8.''6 and 10.''6. With such a difference, far more that thought by Halley, the astronomers were not satisfied. A better international organisation appeared to be needed with an improved location of the stations. Delisle decided, in 1763, to leave the *Hôtel de Cluny* to Messier. Lalande (1762) had already given over the call to his colleagues, establishing a 1769 map and delivering it to the *Académie* in May 1760. It was published in 1762 (in the 1757 issue) together with his mémoire; there is apparently no written proof that Mme Lepaute (1723-1788) made the 1769 map, while she made so many calculations clearly mentioned by Lalande.

After only 8 years, most of the 1761 observers were ready for 1769. As examples, Messier was at the *Collège Louis-le-Grand* in Paris, Maraldi and Cassini at the Observatory, while Le Gentil, Chappe d'Auteroche (1772) and Pingré (1767) were abroad: Le Gentil (1778), who had decided to stay in the Indian Ocean, receiving orders from France, was in Pondicherry in due time, but the sky had some clouds and, back in France, he wrote a book concerning his long voyage and his scientific works between 1760 and 1771; Chappe d'Auteroche went to California and was able to observe the complete transit before his death on August 1, but his results were not lost due to his collaborators; Pingré was the third astronomer, being a specialist for checking at sea chronometers made by Berthoud (1727-1807); after his 1761 trip for the transit, he made four more voyages and was in Saint-Domingue in 1769, being able to catch the Venus transit at its beginning.

From 63 stations the French *Académie des sciences* received about 150 reports, mostly through Lalande in a similar way as Delisle for the 1761 transit. The results lie between $8.''43$ and $8.''80$. Lalande, in his careful study, reduced the solar parallax to values between $8.''55$ and $8.''63$. The result being not so precise as expected, many astronomers made deep studies of the 1761 and 1769 transits. As examples: Delambre (1749-1822) with $8.''55$; Laplace (1749-1827) with $8.''81$; Encke (1791-1865), in Germany, with $8.''57$. The improvement is far less than expected and other methods began to appear for the determination of the solar parallax.

Nevertheless, astronomers from all over the world, having assumed for this parallax values between $8.''5$ and $8.''8$, and having by mid-XIXth century better instruments, including photographic techniques, they began to think about the next Venus transit. They took into account the fact that more precise local values of the coordinates of the stations were needed to improve the accuracy of the results. With their modern instruments, they could obtain them accurately for longitude and latitude.

3. The 1874 and 1882 Venus transits

The *Académie des sciences* began to think, in 1870, of a national commission for the 1874 transit, and called several academicians to be its members such as Laugier (1812-1872) and Delaunay (1816-1872) from Paris Observatory. But they died in 1872. Puiseux (1820-1883) was the secretary of the commission, while Le Verrier (1811-1877) and others were among its members. In 1873 Dumas (1800-1884) was called to be its president and Fizeau (1819-1896), a physicist of a very high level reputation, was in charge of the photographic aspects. He installed in the Luxembourg garden all that was needed for the photographers with an equatorial refractor. Another one was set up on the upper platform of the Luxembourg palace while an artificial Venus transit was taking place on the roof of the Paris Observatory, not far away.

All the stations were to be equipped with two equatorial refractors together with instruments for hydrography, geophysics and meteorology. Two stations were chosen in the southern hemisphere: *Ile Campbell* with Bouquet de La Grye (1827-1909) and *Ile Saint-Paul* with Mouchez (1821-1892) who gave details in "Passage de Venus ... 1874" (Mouchez, no date). Two were chosen in the northern hemisphere: Peking with Fleuriais (1840-1895), a marine officer, and Japan with Janssen (1824-1907). Two more stations, not so well equipped, were decided later: Saigon with Héraud, an engineer and hydrographer, and Nouméa with André (1856-1927).

Most of the expeditions departed in July 1874, the helpers having left France in April with the equipment, while the scientists were with the scientific instruments which number was more than 30. The equipment was more or less the same for five stations. Janssen – under the direct leadership of the *Académie* – decided to experiment a new instrument he called a revolver, later a *revolver photographique* with *daguerréotype*. His first model was built, in 1873, by Deschiens. But Janssen was not satisfied; he requested the Rediers, in 1874, to built a new one. This last original model is in the collections of the Paris Observatory and in exhibition on the occasion of the 2004 June 8 transit of Venus. A paper was published by F. Launay (2001) in the May issue of *l'Astronomie*, and an extended one, in English, is under final preparation. A third model was made by the Rediers and given, after his death, by his daughter Antoinette, to the *Conservatoire des Arts et Métiers*. There are experimental plates in the Paris Observatory but, unfortunately none of the real plates has been found, up to now.

Having been informed about the new apparatus Airy (1801-1892), the Astronomer Royal at that time, asked Dallmeyer, in England, to build some Janssen's revolvers, while De La Rue (1815-1889) decided to build others. One is preserved in the Sydney Powerhouse Museum. In total, according to Launay, 9 Janssen's revolvers were in use in the world in 1874. This employment is responsible for the development of Marey's work (Marey, 1853-1904) followed by the films and the cinema. Tisserand (1848-1896) was, with Janssen, in charge of the determination of the coordinates of the station and responsible for the chronometers (Saito & Shinozawa 1973).

After the campaign the derived results gave to the solar parallax values such as $8.''80 \pm 0.''06$ and $8.''85 \pm 0.''06$, while the astronomers had estimated the error at only $\pm 0.''01$. After the 1761 and 1769 campaigns and the careful studies made later, in various countries, the most probable value was given between $8.''5$ and $8.''8$. For Lalande, as already mentioned, it was between $8.''55$ and $8.''63$. Not so different from what was given one century later for 1874. This explains why, at national and international levels, astronomers wanted to increase cooperation for the following transit to occur in 1882 and visible from Europe, but only at its beginning.

In view of this 1882 event, the *Académie des sciences* decided to do its best to increase the number of observing sites and the international cooperation. Under the name *Conférence internationale du passage de Vénus* (1881), the meeting took place in 1881 October 5 at the Paris Observatory the director of which, after the death of Le Verrier who died in 1877, was Admiral Mouchez.

Among the French attendees were most of those having been in the 1874 expeditions such as Tisserand, Bouquet de la Grye, d'Abbadie, Fleuriais, Hatt, Le Clerc, Perrier, Perrotin. Were also present from other countries and from observatories, Foerster (Berlin), Weiss (Vienna), Hirsch (Neuchatel), Liais (Rio de Janeiro), Puzajon (San Fernando), Stone (Oxford) and Van de Sande Bakhuysen (Leiden). There were also present delegates from several countries, other than those already mentioned: de Azcarate (Spain), Broch (Norway), Govi (Italy), Mansila (Argentina), Moesta (Chili), Oudemans (Low Countries), Pechele (Denmark) and Viegas (Portugal).

The president was Dumas proposed by Italy as the president of the similar French commission. The two vice-presidents were Foerster and Weiss, the two secretaries were Hirsch and Tisserand. Great Britain, who had organized, in 1881, a committee for the 1882 Venus transit, under the chairmanship of Stokes, had sent to Paris one of its members, Stone; he informed the members that the British committee had decided to provide 16 teams located in 6 different spots.

At the Paris international meeting many questions, already raised at various national levels, were discussed, mostly based on the 1874 transit and on new experiments made on Mercury. Comparisons of success or difficulties were also under discussion; decisions taken at national levels were presented. They mostly concerned photography, including British, German, French comments. Proposals for the national stations were given by the delegates of each country. Several sub-commissions were created for the places of stations, the instruments and the observational methods, together with the organization of reductions of the collected data.

On the last day of the meeting, October 13, several proposals were adopted by the members of the commission, including a text about the instructions for the observations of contacts. These proposals were sent to Governments which were to be free to follow or not to follow the content. A Table of the result, concerning the stations and their locations, was established and the complete report printed and sent all over the world.

For France several missions were sent: Cuba, under d'Abbadie (1810-1897) (whose private Observatory in Abbadia, close to Hendaye in the far south of France, has to be visited) with Callandreau (1852-1904) from Paris Observatory, the marine officer Chapuis (1843-1889) for photography and de La Baume Pluvinel (1860-1938) in an independent station; Martinique under Tisserand with Bigourdan (1851-1932) from Paris Observatory and, among the instruments, a heliometer; Florida under colonel Perrier (1833-1888); in Mexico Bouquet de La Grye and Héraud who were in one of the 1874 mission; Chili, a marine officer de Bernardières. Argentina received in Santa Cruz Fleuriais (1840-1895), in Chili went Hatt (1840-1915), in Rio Negro Perrotin (1845-1904) also with a heliometer; Argentina installed in Bragado a station to which the marine officer Perrin (1852-1926) was added. Being in Cape Horn, the marine officer Courcelle-Seneuil was asked to attempt the observation on December 6 (Passage de Vénus ... 1883).

The disappointment was great in all countries (Sellers 2001), the results being not so accurate as the astronomers thought. From the observations Newcomb (1835-1909), using the 1761 and 1769 transits, had obtained for the solar parallax, $8''.79 \pm 0''.05$. Adding the 1874 and the 1882 Venus transits he gave $8''.79 \pm 0''.02$ in 1891. Not yet $\pm 0''.01$ as obtained by Gill (1843-1914), with the planet Mars at the 1877 opposition, with a concluded value $8''.78 \pm 0''.01$.

In 1896, during the meeting of the *Conférence internationale des étoiles fondamentales*, held at the Paris Observatory, it was decided to fix the solar parallax at a more or less mean value issued from all the previous results. Included were old French data for 1761 and 1769, and the recent ones from 1874 and 1882, derived from all recent campaigns, together with other methods. The adopted value, to make the astronomical constants uniform, to be used in the ephemerides, was fixed at $8''.80$.

The discovery of the small planet (433) by Witt (1866-1946) in 1898, named Eros, showed that this body could approach the Earth at only 0.15 astronomical units, instead of about 0.37 for Mars and 0.26 for Venus. An international campaign was launched for the close 1900/1901 approach with a meeting in Paris. Loewy (1833-1907), then director of the Paris Observatory, was in charge of the compilation of the data to derive the solar parallax, but he died in 1907 before the end, and the report was finalized by the British astronomer Hinks (1873-1945) and printed (Hinks 1901) under the title: *Rapport sur la détermination de la parallaxe solaire au moyen des observations de la planète Eros, faites dans les années 1900 et 1901*.

But this is no longer the story of the Venus transits.

4. Final remarks

To conclude some French reliefs, left for the people of any time, are mentioned:

- A medal made after the 1874 transit and offered to all participants of the *Conférence internationale du passage de Vénus* held in Paris in 1881.

- In some places there are symbolic pieces related to the XIXth century transits. Two examples: in Saint-Paul with a special stone and in the surroundings of Nagasaki, the pyramid erected more than one century ago and still there.

- A ceiling in an octagon room on the first floor of the Perrault building of the Paris Observatory installed in 1886. Looking at it carefully are seen the Sun, Venus, a small Cupidon with an arc and an arrow, *amours* or angels observing, submitting elements to Urania or ready to announce, with a horn, the phenomenon. Also seen are, in a golden medal, Halley, as a homage to his call to observe Venus transits, Delisle in a silver one

and, facing them, Le Verrier who had made a study of about 9 000 values of the solar
diameter. A description of this ceiling follows in the next section in French.

Figure 1. The painting by Dupain (1847–1933) that decorates the ceiling of the octagonal west
rotunda of the Perrault Building of the Paris Observatory showing an allegory of the passage of
Venus over the sun. Copyright Paris Observatory.

5. Allégorie du passage de Vénus devant le Soleil

La peinture, œuvre de Dupain (1847-1933), orne le plafond de la Rotonde Ouest, de
forme octogonale, du Bâtiment Perrault de l'Observatoire de Paris. L'Amiral Mouchez,
qui en est alors directeur, souhaite créer un décor dans cette partie du rez-de-chaussée
du bâtiment oú il a fait installer, à partir de 1879, certains des instruments les plus
prestigieux des collections. D'une part, il fait réaliser les portraits des astronomes qui se
sont trouvés, au cours du temps, à la tête de l'Etablissement. D'autre part, il envisage
un plafond décoratif, représentatif de travaux qui y sont menés. Il envisage, dès 1880,
une allégorie du passage de Vénus sur le Soleil. Il a lui-même observé un tel passage en
1874 et un autre doit se produire en 1882.

En 1884, Dupain essaye sur place son esquisse du Passage de Vénus dans cette salle,
la peinture, de grande dimension, devant y prendre place en 1885. En fait elle ne le sera
qu'en octobre 1886, ayant - dans l'intervalle - figuré dans une exposition. Le Soleil, sur
son char attelé de chevaux bondissants, est entouré de vapeurs et de nuées rougeoyantes.
Vénus, nue et la tête surmontée d'une étoile, est entrainée dans une belle envolée par
deux oiseaux blancs tenant en leurs becs les fils qui l'emportent. Accompagné d'un petit

Cupidon portant arc et flèche, cet ensemble baigne dans des nuages clairs d'une grande légèreté.

Au bord de l'allégorie, au pied de ces personnages, Dupain a placé, au centre assise sur un pilier, Uranie la tête couronnée de lauriers et d'étoiles. Sa main gauche s'appuie sur un globe terrestre oú se devinent les continents. Sa main droite tient du matériel d'écriture au-dessus d'un grand registre ouvert, que lui présente un autre angelot. Deux autres volumineux et beaux ouvrages, représentés plus à gauche, sur lesquels s'appuie un nouvel angelot soutenu par un nuage blanc et vaporeux, contiennent certainement les données de prédiction des éléments du passage de Vénus. Dans sa main gauche il tient un document, noué d'un ruban, tandis que, de sa main droite, il manipule la vis d'entrainement de l'oculaire d'une petite lunette, à monture équatoriale, avec laquelle il observe Vénus.

A droite de cette partie de l'allégorie un troisième angelot, aux belles ailes déployées, tient dans sa main gauche une trompette de la renommée en prévision, sans doute, de l'annonce de l'instant précis du phénomène. Son bras droit s'appuie sur une médaille d'or, qu'il couvre de lauriers, représentant Halley. A droite Delisle figure en médaille d'argent et, un peu plus à droite, leur faisant face, c'est Le Verrier, lequel a mené une étude portant sur plus de 9 000 valeurs du diamètre du Soleil, que l'on reconnaît. Un bandeau "Passage de Vénus" lui fait suite avec une partie d'ancre de marine débordant d'un soubassement en pierre blanche.

References

Bonche E., Toulmonde M., La Méthode de Halley, in *l'Astronomie*, vol. 118, p. 314-318, 2001.

Chappe d'Auteroche J., extract of a travel in Sibérie, pour l'Observation de Vénus sur le disque du Soleil, faite à Tobolsk le 6 juin 1761, *Mémoires de l'Acad. Sci.*, p. 337-377, 1751, publ. 1763.

Chappe d'Auteroche J., Voyage en Californie..., written and published by M. de Cassini fils, Paris, Jombert, 1772.

Cook A., Edmond Halley - Charting the Heavens and the Seas, *Clarendon Press*, Oxford, 1998.

Conférence internationale du passage de Vénus, *Procès-verbaux*, Imprimerie Nationale, 1881.

Débarbat S. et Dumont S., Deux pélerins de la cartographie scientifique en Europe centrale et orientale, in *XYZ*, n 67, p. 70-76, 1996.

Delisle J.-N., Extrait d'une lettre de M.Delisle, écrite de Petersbourg le 24 Août 1743, et adressée à M. Cassini Servant de supplément au Mémoire de M. Delisle de 1723, p. 105, pour trouver la parallaxe de Mercure dans le disque de cet Astre, Histoire et Mémoires de l'Académie des sciences, partie Mémoires, p. 419-428. et planche XI, 1743.

Dumont S., Les expéditions du XVIIIe siècle pour les passages de Vénus, in *l'Astronomie*, vol. 118, p. 290-295, 2001.

Gassendi P., Mercurius, in *Sole visus et Venus invisa anno 1631*, Paris, 1632.

Halley E., The three Voyages of Edmond Halley, in *the Paramore 1698-1701*, Ed. N. Thrower, The Hakluyt Society Publ., Londres, 1981.

Hinks A., Rapport sur la détermination de la parallaxe solaire au moyen des observations de la planète Eros, faites dans les années 1900 et 1901, Carte du Ciel I, sn, sd.

Horrocks J., Excepta ex Epitolis Jerem. Horroccii ad Gul. Crabtrium..., Londres, œuvre posthume, 1672, part Observationes Célestes HabitæToxyethæ, propè Liverpoliam ANGLI, a JEREMIA HORROCCIO, several editions.

Kepler J., Astronomia nova..., Pragae, 1609.

Kepler J., Tabulæ Rudolphinæ..., Ulm, 1627.

Kepler J., Admonitio ad astronomos..., Leipzig 1629, édition consultée Frankfurt, 1630.

Lacaille N., Avis aux astronomes...à l'occasion des observations...hémisphère austral, Imprimé de 4 pages, Manuscrits A 6.9 de la Bibliothèque de l'Observatoire de Paris.

Lalande J., Mémoire sur les passages de Vénus... 1761 et 1769...., Académie des sciences - Mémoires, p. 232-250, 1757, publ. 1762.

Launay F., Le revolver photographique de Jules Janssen, in *l'Astronomie*, vol. 118, p. 306-308, 2001.

Le Gentil de La Galaisière G., Voyage dans les mers de l'Inde..., Paris, 1778.

Mémoires de l'Académie des sciences, volume de 1751 publ. 1763: numerous observations; among them Lemonnier (with Louis XV, Château de Saint-Hubert), Maraldi (Observatoire de Paris), Lacaille (Conflans-sous-Carrières), Lalande (Palais du Luxembourg), ...

Mouchez E., Passage de Vénus sur le Soleil du 9 décembre 1874, nl, nd.

Passage de Vénus sur le Soleil du 6 décembre 1882 - Rapports préliminaires, 1883.

Pingré A., Observation du passage de Vénus devant le disque du Soleil le 6 juin 1761, faite à Rodrigue dans la mer des Indes, Mémoires de l'Acad. Sci., p. 87, suivi de Remarques de Lemonnier, p. 88-89 et 105-111, de Lalande, p. 93-95 et 111-112, puis Pingré J., Observations astronomiques pour la détermination de la parallaxe du Soleil, faites en l'isle Rodrigue, p. 413-486, Vol 1761, publ. 1763.

Pingré A., Mémoire sur le choix... passage de Vénus du 3 juin 1769..., Lu à l'Académie des sciences en décembre 1766, en janvier et en février 1767. Cavelier Libraire, 1767.

Saito K. et Shinozawa S., On the Observations of the Transit of Venus over the Sun with Particular Emphasis on the december 9, 1874 Event observed in Japan (en japonais), Tokyo Astronomical Observatory Ed., 1973.

Sellers D., The transit of Venus, Maga Velda Press, 2001.

Van Helden A., Measuring the Universe. Cosmic Dimensions from Aristarchus to Halley, Chicago and London, 1985.

Toulmonde M., La parallaxe du Soleil, in *l'Astronomie*, vol. 118, p. 274-289, 2001.

Woolf H., The Transits of Venus - A story of eighteenth-century science, *Princeton University Press*, 1959.

Discussion

DAVID SELLERS: I have seen papers by Delisle from the 1720s and the 1750s right up to the eve of the 1761 transit explaining how to calculate the solar parallax from the transit, but when you read them they all seem to be Halley's method, which was based on measuring the duration. Do you know when Delisle actually came up with the method that bears his name, which only requires the measurement of the time?

SUZANNE DÉBARBAT: For Halley it is the timing between the contacts observed from two different stations, while for Delisle it is the measurement of the same contact – the timing – from different places. He began to think of that after he had visited Halley in 1724. Delisle went to England – as far as I remember he met also Newton – and he met Halley, and they discussed it. He began to think of the subject in 1723 before meeting Halley. He departed from France in 1723 for Russia, and during the time he was in Russia he thought also on the matter and he published a paper in 1743 – the final paper is the one I have mentioned. So you see I have a feeling that he was thinking of the different possibilities to determine the solar parallax from different places during the time he was in Russia. Was it exactly what you wanted to know?

DAVID SELLERS: Yes, although I thought the 1743 paper was still explaining more or less Halley's method, rather than his own. I was just wondering how late it was that Delisle actually came up with that unique method of his.

SUZANNE DÉBARBAT: I don't know more.

DAVID HUGHES: The Delisle method is of course only half as good as the Halley method, because the inaccuracy of the Delisle method is ...

Suzanne Débarbat: I don't know. I'm not sure that the accuracy is so much less; do you think so?

David Hughes: Oh, yes.

Suzanne Débarbat: You have tried some calculation? Done an experiment?

David Hughes: It's exactly half as ...

Suzanne Débarbat: But Halley's method is not so accurate as he thought.

David Hughes: Halley's method depended entirely on this timing and of course ...

Suzanne Débarbat: Oh, yes!

David Hughes: Well, what fascinated me about Halley was he thought that the clocks would be good enough to give an accuracy of 1 second, and then of course the black drop effect ...

Suzanne Débarbat: I have heard in some place half a second about Halley. Did you find that? I've seen half a second for the timing of the contacts.

David Hughes: Well, if his accuracy was half a second he would have got the astronomical unit to one part in a thousand; and as it was an accuracy, I thought, of just 1 second that's were his one part in 500 came from.

Steve Dick: Can you say any more about the French photographic method in 1870. I believe it was the same as the American: the long focus photographic method.

Suzanne Débarbat: I don't remember for the photographic plate; they had a 6-ft and an 8-ft, but I cannot remember which one was for the contact and which one was for the photographic plate. It is said in the paper itself which one was used for the photographic plate, but it could be only 6 or 8.

Steve Dick: The Americans used a 40-foot photo-heliograph to get the scale ...the long focus.

Suzanne Débarbat viewing the transit; Nicole Capitaine and June Kurtz

Wayne Orchiston at the window of Carr House from which it is thought that Jeremiah Horrocks observed the 1639 transit.

Transits of Venus: New Views of the Solar System and Galaxy
Proceedings IAU Colloquium No. 196, 2004
D.W. Kurtz, ed.
© 2004 International Astronomical Union
doi:10.1017/S1743921305001262

James Cook's 1769 transit of Venus expedition to Tahiti

Wayne Orchiston

Anglo-Australian Observatory & Australia Telescope National Facility,
PO Box 296, Epping, NSW 2121, Australia
wo@aaoeapp.aao.gov.au

Abstract. After the failure of the 1761 transit to provide a reliable value for the astronomical unit, the focus shifted to the 1769 event, and Britain mounted an ambitious program, with overseas observing parties dispatched to North Cape (Norway), Hudson Bay (Canada) and newly-discovered Tahiti in the Pacific. Lieutenant James Cook was in charge of the Tahitian expedition, ably assisted by fellow-astronomer, Charles Green, and they were supplied by the Royal Society and the Royal Observatory at Greenwich with telescopes and other scientific instruments. The main observing site was set up at Fort Venus, and supplementary transit stations were established on Irioa Island (Moorea) and Taaupiri Island (off the east coast of Tahiti). June 3 was warm and clear, and all observers successfully recorded the transit, but on the journey home 'the curse of the transit' prevailed and more than half of them fell ill and died. Back in England, Cook wrote up the transit observations for the *Philosophical Transactions of the Royal Society*, but for some inexplicable reason only used data obtained at Fort Venus. It was left to Oxford astronomer, Professor Thomas Hornsby, to derive a meaningful figure for the solar parallax, and he utilized the Tahitian data and observations made at four other sites to arrive at a figure of 8."78. But discordant results obtained by other researchers fuelled controversy over the effectiveness of transits of Venus as a valid means of determining the astronomical unit. In fact, the solar parallax obtained by Hornsby was remarkably similar to the currently-accepted value of 8."794148, thereby discrediting Beaglehole's oft-quoted claim that the Tahitian observations were a failure. Although more than a dozen men were involved in the Tahitian transit program, most of their records have been lost, and remarkably few of the instruments they employed can now be identified. Yet for those of us with Pacific affiliations, Cook's first voyage to the South Seas occupies a special place in transit of Venus history.

---

1. Introduction

During the eighteenth and nineteenth centuries transits of Venus were viewed as ideal tools with which to attack that fundamental yardstick of solar system astronomy: the value of the solar parallax (P) and hence the astronomical unit. Although the transit of 1761 marked the first great scientific project undertaken on an international scale (Woolf 1959), results were inconclusive, with values of P ranging from 8."28 to 10."6. Understandably, the focus then shifted to the up-coming 1769 transit.

2. The preparations

During the 1761 transit, Britain and Prussia were at war with Austria, France, Russia, Saxony and Sweden, but by 1769 the Seven Years' War was a distant memory. So, what better was of 'competing for peace' than through scientific supremacy? Thomas Hornsby, Professor of Astronomy at Oxford University, alluded to this when he wrote: "It behoves us therefore to profit as much as possible by the favourable situation of Venus in 1769, when we may be assured the several Powers of Europe will again contend which of

them shall be most instrumental in contributing to the solution of this grand problem."
(Hornsby 1765: 343). In Britain, a Transit of Venus Committee was formed by the Royal
Society in 1767, featuring (amongst others) the Astronomer Royal, Nevil Maskelyne,
and Dr John Bevis, James Ferguson and James Short, all leading British astronomers of
the day. After referring to Hornsby's paper (1765: 339), they reviewed possible overseas
observing sites, settling on North Cape (Norway), Hudson Bay (Canada) and a suitable
Pacific site. Fortuitously, Samuel Wallis returned from a voyage of exploration in May
1768 and announced the discovery of King George's Island – now known as Tahiti – at
precisely the desired Pacific location! Moreover, the climate was admirable, the locals
were friendly, and there was a suitable harbour at Port Royal (Matavai Bay).

The Transit of Venus Committee adopted all three localities, and successfully peti-
tioned King George III for £4,000 to finance the Pacific expedition (see Banks n.d.: 512-
513). They also requested a suitable vessel, and the Admiralty spent £2,307.5s.6d pur-
chasing the *Earl of Pembroke* (Deptford Yard Officers 1768a), and a further £2,293.17s.7d
modifying it for the voyage (Deptford Yard Officers 1768b). Renamed *Endeavour*, this
370 ton ex-Whitby collier proved cramped quarters for nigh on 100 men, comprising
officers, marines and able-bodied seamen from the Royal Navy, and non-naval personnel
referred to collectively as 'supernumeraries'. Most of the latter were members of Banks'
scientific retinue, but a notable exception was one of the two official astronomers on the
voyage, Charles Green.

Green (see Orchiston 1998b: 19-20) was eminently qualified for the task at hand, as he
had participated in the Royal Observatory's 1761 transit of Venus programme. Born in
Yorkshire in 1734, he served as a teacher before finding employment at the Observatory,
but when Maskelyne became Astronomer Royal in 1765 and eventually appointed his
own assistant, Green found himself without a job. His solution was to serve as purser on
R.N. Aurora, before accepting the *Endeavour* appointment. Beaglehole (1968: ccxlii) has
characterized Green as "... a highly conscientious as well as sprightly person."

The second astronomer on the *Endeavour* was none other than James Cook, who had
dual responsibilities – and salaries – as commander of the vessel and astronomer (e g
see Beaglehole 1968: cv; cxxvi, Banks n.d.: 513-514). Badger (1970: 30) claims that we
are fully justified in regarding Cook as a 'scientist', in that he "... was a scrupulously
careful observer ... [and] His attitude to measurement and inquiry would do credit to
any conventionally-trained scientist." Six years Green's senior, Cook also was born in
Yorkshire, and while still a teenager found service on North Sea coastal colliers. Eager to
learn and improve, his training in seamanship received a distinct fillip in 1755 when he
joined the Royal Navy as an able bodied seaman, and from 1757 to 1767 he learnt and
practiced the principles of maritime astronomy, hydrographic surveying and cartography
whilst serving on the American coast (Ritchie 1978; Skelton 1954). During this period,
he also observed the 5 August 1766 total solar eclipse, a short report on which (see
Cook 1767) was published in one of the most prestigious scientific journals of the day,
the *Philosophical Transactions of the Royal Society* (of London). So when the Admiralty
was reviewing possible commanders for the *Endeavour*, Cook was already known for his
prowess in maritime and non-maritime astronomy, even though there were other officers
who far out-ranked him (Beaglehole 1974). At the time he held the rank of master and
commanded a naval schooner, the *Grenville*, but on 25 May 1768 he received a letter
from the Admiralty advising that "... we have appointed you first Lieutenant of His
Majesty's Bark the Endeavour now in Deptford, and intend that you shall command her
during her present intended Voyage ..." (Admiralty 1768).

The Royal Society, with assistance from Nevil Maskelyne, was eager to provide Cook
and Green with the best possible astronomical instruments (Howse 1979; Howse &

Figure 1. Gregorian reflector 1900-136, by Short, with object-glass micrometer (Courtesy: Science Museum, Science and Society Picture Library).

Hutchinson 1969; Orchiston 1998b: 36-58), namely: "2 Reflecting telescopes of two feet focus ... [with] 2 Wooden stands ... An astronomical quadrant of one foot radius, made by Mr Bird ... An astronomical Clock [by Shelton] and Alar[u]m Clock. A brass Hadley's sextant ... of Mr Ramsden. A Journeyman Clock bespoke of Mr Shelton. 2 Thermometers of Mr Bird. 1 Stand for Bird's quadrant." (Beaglehole 1968: cxliii). The two telescopes were Gregorian reflectors made by the celebrated Scottish instrument-maker, James Short (Green & Cook 1771: 398), and one of these was furnished with a Dollond object-glass micrometer (see Fig. 1). The Navy Board also provided a Gregorian telescope that Cook had used previously on the *Grenville*, and before the departure of the *Endeavour* he had a micrometer made for it, convinced that this accessory "... will be of great service in the observation of the Transit Vinus [*sic*] ..." (Cook 1768). This telescope had a focal length of 18 inches (Green & Cook 1771: 416), and we know from later Cook voyage documentation that it was made by Watkins (Beaglehole 1967, I: 243). In addition, Dr Daniel Solander, one of Banks' party, had access to a Gregorian telescope of 36 inches focal length (Green & Cook 1771: 412). Apart from these four Gregorian reflectors there must have been others, but no documentation of these has survived. The only other scientific instrument of astronomical relevance was a watch, which Maskelyne loaned to Cook (Howse & Hutchinson 1969). The clocks were there to provide a local time service during shore-based stopovers; the quadrant and sextant allowed observations vital in the determination of latitude and longitude, and maintaining a local time-service; and the telescopes were primarily supplied for the all-important observations of the transit.

3. The voyage and the transit

On 26 August 1768 the *Endeavour* sailed from Plymouth on an astronomical extravaganza (Betts 1993; Orchiston 2004), and one of the most expensive and ambitious expeditions ever undertaken by Mother England (Orchiston 1998a). To all intents and purposes this was a scientific voyage: at issue was the most pressing problem in world astronomy, and at stake was British pride and prestige.

After rounding Cape Horn, the *Endeavour* penetrated the Pacific, and on 13 April 1769 anchored in Matavai Bay on the northern coast of Tahiti more than seven weeks

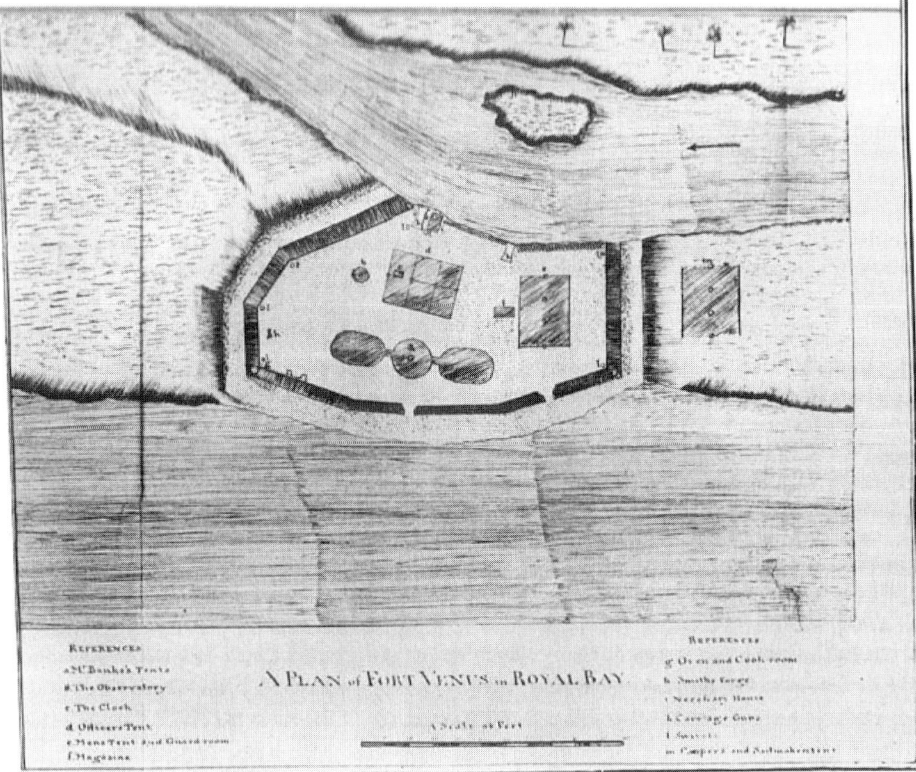

Figure 2. Plan of Fort Venus, Matavai Bay (after Parkinson 1784: Plate IV), showing the location of the circular tent observatory which housed the quadrant and journeyman clock, and the large rectangular tent with the astronomical clock.

before the scheduled transit. Nearby Point Venus proved an ideal observatory site: it was easily accessed from the ship, and lay on a narrow strip of land that could readily be fortified (for it was vital that there were no local interruptions during observation of the transit). Thus, Fort Venus came into existence, replete with an assortment of tents for men and instruments (Fig. 2). Green & Cook (1771: 398) report that near one of the tents "... stood the observatory, in which were set up the journeyman clock and astronomical quadrant: this last, made by Mr. Bird, of one foot radius, stood upon the head of a large cask fixed firm in the ground ..." This quote reveals the way in which the quadrant and the telescopes were used: empty casks originally used to transport food or water were sunk into the sand and filled with sand for ballast. In this way they provided sturdy and reliable foundations for the delicate scientific instruments.

A key factor that would later influence the success of the entire expedition was accurate determination of the latitude and longitude of Fort Venus, and Green made this one of his priorities. To determine the latitude of the site he used quadrant observations of meridian zenith distances of the Sun obtained between May 6 and June 27 and meridian zenith distances of fifteen bright stars observed between June 21 and July 4, deriving a mean value of 17° 29′ 15″ S (Green & Cook 1771: 405-406). For longitude, he measured 'lunars' with the quadrant or Ramsden sextant on sixteen evenings between April 30 and June 30, arriving at a figure 149° 36′ 38″ W of Greenwich. In close accord was the value of 149° 32′ 30″ W, obtained by Green and Cook from telescopic observations of Jovian

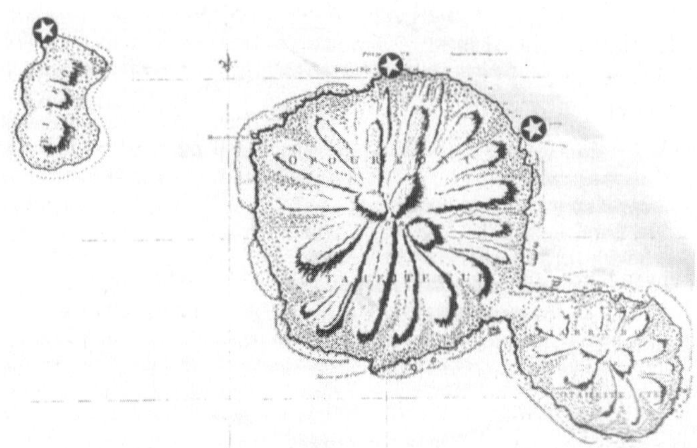

Figure 3. The stars indicate the locations of the three Tahitian transit of Venus observing sites. From west to east they are: Irioa Island, Fort Venus and Taaupiri Island.

satellite eclipses made on seven different nights between June 4 and July 6 (Green & Cook 1771: 407-409).

As the date of the transit neared, Cook observed that the Fort Venus region was not always blessed with clear skies, so he decided to follow the advice proffered him before the voyage by the President of the Royal Society (see Douglas n.d.: 516) and establish two ancillary observing stations (see Fig. 3). One of these was sited on Irioa Island just off the north-western tip of adjacent Moorea, and comprised no more than "... a Coral rock about 150 yards from the shore ... it was about 80 yards long and 60 broad and had in the middle of it a bed of white sand large enough for our tents ..." (Beaglehole 1963, I: 284). The other supplementary observing station was on Taaupiri Island, an islet off the shore of Tahiti to the east-southeast of Fort Venus (see Beaglehole 1968: 97n). Fortunately there were enough instruments to accommodate three observing stations (Cook 1770), while Green had made sure that there would be suitable observers by providing lessons in nautical and transit astronomy during the voyage out from England (Beaglehole 1968: cxxxiv).

June 3, the day of the long-awaited event, loomed warm and clear throughout Tahiti, which boded well for the three teams of observers. At Fort Venus, Cook and Green used the two Short telescopes (Green & Cook 1771: 398), while Solander carried out independent observations with a much larger, more powerful Gregorian reflector (Beaglehole 1968: 98). A Swede by birth and a disciple of Linneaus, Dr Daniel Solander (1733–1782) was Banks' botanical mainstay on the voyage (Beaglehole 1968: cxxxiv-cxxxvi). A fourth member of the Fort Venus team was Robert Molyneux (1747–1771), a Master on the *Endeavour* with a particular aptitude for marine surveying and cartography (see Beaglehole 1968: cxxxii). He, too, was furnished with a telescope (Molyneux 1769: 559), and although no description of this instrument has survived it was probably the Watkins reflector that had kept Cook company on American coast. We know that this Gregorian reflector was at Fort Venus during the transit, and that Cook made use of it between second and third contact (Green & Cook 1771: 416). Perhaps Molyneux had recourse to it during the ingress and egress phases. As the principal observing station, it is reasonable to presume that the astronomical clock remained at Fort Venus for the transit (although this is nowhere stated in the literature). John Satterley, the *Endeavour*'s ever-popular carpenter, was assigned responsibility for the clock and the thermometer (Molyneux 1769:

559). He, Green and Molyneux were all destined to die in early 1771, while the *Endeavour* was *en route* to England, following the disastrous Batavia and Cape Town stop-overs.

The Taaupiri observing team was made up of Zachary Hicks, Charles Clerke, Richard Pickersgill and Patrick Saunders (ibid.). As Second Lieutenant on the *Endeavour*, Hicks (1739–1771) was Cook's second-in-command, while Third Lieutenant Clerke (1743–1779) ranked fourth in the hierarchy. Clerke had a special affinity for astronomy, and was to take over Green's responsibilities after the latter's untimely decease during the voyage home. Assisting Hicks and Clerke were Pickersgill (1747–1779), like Molyneux a Master with a talent for surveying and cartography, and Saunders, a little-known and quarrelsome Midshipman. Nowhere is there a listing of the specific instruments assigned to Taaupiri Island, but Molyneux (op. cit.) states that Green provided both ancillary transit parties "... with Telescopes & every thing necessary ...". Precisely how many telescopes were supplied is not stated, but long after the transit Cook (1771a: 694) identified all four members of the transit party as observers. Local time, meanwhile, would have been provided by either the journeyman clock or the alarum clock.

The Irioa Island observing party also comprised four members: John Gore, Jonathan Monkhouse, William Brougham Monkhouse and Herman Spöring (Molyneux 1769: 559). American-born Third Lieutenant John Gore (~1730–1790) was an "... invaluable practical man ..." (Beaglehole 1968: 595) who ranked third in the officer's hierarchy, while Jonathan Monkhouse was a Midshipman, and his older brother, William Brougham Monkhouse, was the *Endeavour*'s surgeon. Both Monkhouses were to die on the voyage home, William in November 1770 and Jonathan in February of the following year. The fourth member of the observing party was Swedish-born Spöring (~1733–1771), who had trained in medicine and then worked as a watch-maker, but came on the voyage, with Banks, as an artist and draftsman. He also was to die on the way home. On the day of the transit Banks identified the two observers as Jonathan Monkhouse and Gore (Beaglehole 1963, 1: 284), whereas Cook says "... I sent Lieutenant Gore in the Long-boat to York Island [Moorea] with Dr Monkhouse and Mr Sporing to observe the transit of Venus, Mr Green having furnished them with Instruments for that purpose." (Beaglehole 1968: 97). To confuse matters further, long after the transit Cook (1771a: 694) identified Spöring and Jonathan Monkhouse as the two observers. Nor do we know the number of telescopes involved, or their appearance. In his journal entry of 2 June, Banks describes how: "Before night our observatory was in order, *telescopes all set up and tried* &c. and we went to rest anxious for the events of tomorrow." (Beaglehole 1963, I: 284, my italics). This indicates that there were two or three telescopes, and the preparations mentioned by Banks strongly suggest that Gregorian reflectors rather than portable marine telescopes were involved. Nor is there any information about the time-keeper used, but this would have been either the journeyman clock or the alarum clock. Finally, we should note that although Joseph Banks accompanied the Irioa Island party to Moorea, he took no part in the transit observations (Beaglehole 1963, I: 284-285).

We know from various letters penned by Cook (e.g. Cook 1770) that all three parties successfully observed the transit, which lasted for a little over six hours, and that Fort Venus fully justified its name in that "... no Indians was allow'd to come near us that nothing might disturb the Observation." (Molyneux 1769: 559). The main focus of all three observing parties was the ingress and egress contacts, but where telescopes were furnished with micrometers, observers also were able to measure the diameter of Venus while superimposed on the Sun's disk. Cook reported from Fort Venus: "This day prov'd as favourable to our purpose as we could wish, not a Clowd [*sic*] was to be seen the whole day and the Air was perfectly clear, so that we had every advantage we could desire in Observing the whole of the passage of the Planet Venus over the Sun's disk: we very distinctly saw an Atmosphere or dusky shade round the body of the Planet which very much disturbed the times of the Contacts particularly the two internal ones [see Fig. 4]. Dr Solander observed as well as Mr Green and my self, and we differ'd from one another in observeing [*sic*] the times of the Contacts much more than could be expected. Mr Green's Telescope and mine were of the same Magifying [*sic*] power but that of the D^r was greater than ours." (Beaglehole 1968: 97-98). We know that Green & Cook (1771: 410) used a magnification of 140×. In the course of the transit Green and Cook also measured the diameter of Venus. They used one of the Short telescopes with the Dollond object-glass micrometer, deriving mean values of 54.″97 and 54.″77 respectively. On three different occasions Cook also measured the planet's diameter with a telescope of 18 inches focal length (probably the Watkins reflector) and a Dollond micrometer, obtaining mean values of 56.″8, 56.″28 and 56.″02 (Green & Cook 1771: 412-418). He also used the Short reflector and Dollond micrometer to take a series of measurements of the separation between Venus's limb and the limb of the Sun, and to measure the diameter of the Sun (Green & Cook 1771: 413-415).

With the transit behind him Cook was now free to open the sealed orders from the Admiralty, and these revealed that this scientific venture would now become a voyage of exploration, in search of the mooted southern continent, *Terra Australis Incognita*. In fact this quest proved singularly unsuccessful, and after re-discovering and then circum-navigating New Zealand and charting much of the east coast of Australia, Cook made his way home to England via epidemic-ridden Batavia (present-day Jakarta) and Cape Town. In the process, seven of the thirteen individuals directly involved in the transit of Venus observations perished, while an eighth (the troublesome Patrick Saunders) deserted while the *Endeavour* was anchored at Batavia. Given this mortality rate, fatalists could be excused for thinking that the Tahitian transit carried a curse of Tutankhamen-like proportions!

4. From transit to parallax

As soon as the transit was over, all observers handed over their records to Green for safe-keeping (Cook 1771a: 692), and when the *Endeavour* reached Batavia he sent copies of them to the Royal Society, enclosed in a packet that Cook sent to the Admiralty via a Dutch ship (Cook 1771b: 694). The *Endeavour* quit Batavia on 26 December 1770 and it was only when Green died of dysentery on 29 January 1771 that Cook discovered the shocking state of the astronomical records: "... my first care was, to preserve, for the perusal of the Royal Society, all his papers that contained any Astronomical observations of what nature soever; many of which I had never seen before; and I found far from having been kept in that clear order their importance seemed to require ..." (ibid.; cf. Wales 1788: i.). Unfortunately, Green never briefed Cook on precisely which records were sent to the Royal Society, so Cook "... caused copies of all those that had any relation to the Transit of Venus, and for fear of any accident happening to us, I put them on board His Majesty's Ship Portland addressed to Mr Maskelyne ..." (Cook 1771b: 695). This occurred on 10 May 1771, when the *Endeavour* and *Portland* were anchored at Ascension Island (Beaglehole 1968: 469). Cook (1771a: 693-694) reveals that the letter to Maskelyne contained the following enclosures:

"1. Observations of equal altitudes of the un for the time; and observed altitudes or Zenith distances of the sun and stars for the latitude.
2. Observations of Jupiters Satellites for the longitude; and of the times of the contacts of the limbs of the sun and Venus observed by Mr Green.
3. Lunar observations of the Moons distances from the sun and fixed stars for the

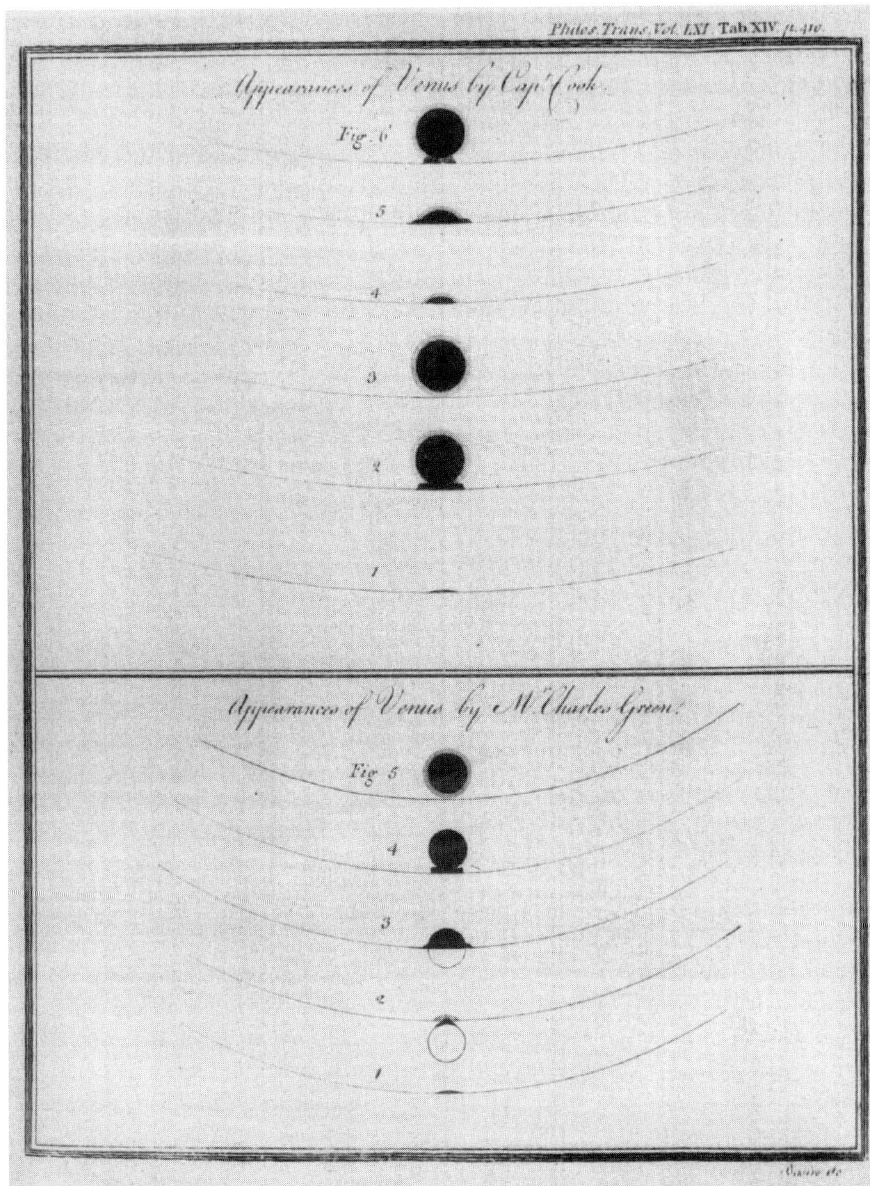

Figure 4. Drawings by Cook and Green illustrating the transit contacts, as seen at Fort Venus (after Green & Cook 1771: facing page 410).

longitude; and Capt Cook's observations of the times of the contacts of the limbs of the sun and Venus.

4. Mr Green's observations of the diameters of the Sun and Venus; nearest approach of their centers; difference of Declination, distances of their limbs in a direction parallel to the Equator; all observed with Dollond's Micrometer.

5. Observations of the transit of Venus made at York Island [=Moorea] by Mr Monkhouse and Mr Sporing; and Dr Solander's observations of the two external and first internal contacts of Venus at Georges Island.

6. Observations of the transit of Venus at Morton's Island [=Taaupiri Island] by Lieu^t Hicks, Clerk, Saunders, and Pickersgill."

Cook also explains to Maskelyne why these documents were sent to him rather than the Royal Society: "If I recollect right Mr Green has made some mistake in the observations he sent home, of the beginning and end of the Transit, as it was by him observed; at least I do not find the true times that he observed the different contacts faithfully entered in any of his books or papers; on the contrary I find them put down in two places, and different from each other, and neither the one nor the other are precisely the same as they were observed; the alterations that have been made will appear from the inclosed papers, and from them you will be able to judge how far it was reasonable to make such alterations, and this is the reason why I wish you to have the perusal of these papers before they are laid before the Royal Society ..." (Cook 1771a: 693).

Subsequently, Cook had yet another copy made "... of all or most of the observations relating to the Transit, that I know to be authentic, made by Mr Green, my self and others ..." (Cook 1771b: 695), and on 11 July 1771, as the *Endeavour* reached England, he sent these and all of Green's papers to the Secretary of the Royal Society (ibid.).

Once he was safely settled in England, Cook faced the non-trivial task of trying to make sense of these confusing observations, and it is much to his credit that he subsequently was able to publish a report in the *Philosophical Transactions of the Royal Society*, with Green serving as a deserving posthumous co-author (Green & Cook 1771). "Observations made, by appointment of the Royal Society, at King George's Island in the South Seas" appeared in 1771, spans 25 journal pages, and provides readers with details of the transit (including contact drawings by Cook and Green), lists of observations made for time-keeping purposes and in order to determine the latitude and longitude of Fort Venus, and includes some magnetic and tidal records. The transit itself occupies approximately half of the paper, but only details the observations made by Cook, Green and Solander – notwithstanding the aforementioned problems associated with Green's contact timings. Perhaps this is why, at the bottom of the very last page of the paper, Cook acknowledges Maskelyne's assistance in preparing the final manuscript. Surprisingly, none of the contact timings made at Irioa Island or Taaupiri Island is included, and indeed the sole mention of these two observing stations is almost an aside: "Some of the other gentlemen, who were sent to observe at different places, saw at the ingress and egress the same phenomenon as we did; though much less distinct, which no doubt was owing to their telescopes being of less magnifying power ..." (Green & Cook 1771: 411). Although at least three copies of the Irioa Island and Taaupiri Island observations were at one time or another sent to the Royal Society (one set via Maskelyne), and all of the original transit papers were also lodged there, none of these records has survived (G. Richardson, pers. comm., 2000), and searches by the author at other 'obvious' repositories such as the RGO Archives at Cambridge and the National Maritime Museum have also proved fruitless. Thus, we cannot examine the original records to determine why Cook chose not to include contact times from the ancillary observing stations. Nor can we see what observations – if any – Molyneux contributed from Fort Venus, and how Green's various timings listed in the original records compare and contrast with those in the published paper.

While Cook was working on the Royal Society paper, the talented Thomas Hornsby was analysing the British transit observations with the aim of coming up with a mean-ingful figure for the solar parallax. After reviewing the available readings he decided to employ data provided by the British Tahitian and Hudson Bay missions, Chappe's French expedition to Baja California, Rumovsky's Russian site at Kola and Hell's Danish station at Vardö. Critical to the success of this exercise were reliable latitudes and longitudes for the different observing stations and consistent second and third contact timings, but even this selective dataset presented Hornsby with problems. For instance, the times that Cook, Green and Solander registered for these two contacts varied by as much as thirteen

Table 1. Contact Timings at Tahiti (Local time).

Contact	Cook	Green	Solander
1	7h 21m 25s	7h 21m 20s	7h 21m 46s
2	7h 38m 55s	7h 38m 55s	7h 39m 08s
3	13h 09m 56s	13h 09m 46s	—
4	13h 27m 45s	13h 27m 57s	13h 27m 56s

seconds, and similar discrepancies also characterized the first and fourth contacts. Table 1 lists the various contact times, and this compilation shows that Cook and Green were sometimes in accord and Solander was the 'odd man out', while at other times Green and Solander agreed and Cook's was the dissident reading. In other words there was no consistent pattern, so it was not possible to derive correction factors for the different observers.

Although we now know that such discrepancies are to be expected and that the variations documented in Tahiti were by no means excessive, at the time Cook was just a little disappointed (Beaglehole 1968: 98) – even if the reasons for them were only too obvious. In his Royal Society paper he wrote how it was "... very difficult to judge precisely of the times that the internal contacts of the body of Venus happened, by reason of the darkness of the penumbra at the Sun's limb, it being there nearly, if not quite, as dark as the planet. At this time a faint light, much weaker than the rest of the penumbra, appeared to converge towards the point of contact, but did not quite reach it ... in like manner at the egress the thread of light was not broke off or diminished at once, but gradually, with the same uncertainty: the time noted was when the thread of light was wholly broke by the penumbra." (Green & Cook 1771: 411). What Cook was alluding to was the notorious and long-misunderstood 'black drop effect' (see Schaefer 2001; Pasachoff et al., these proceedings), but this did not prevent Hornsby (1771) from coming up with a solar parallax value of 8.″78 (cf. Howse & Murray 1997), a figure that is remarkably similar to the 8.″80 obtained by Pingré and the currently-accepted value of 8.″794148.

Had Hornsby's and Pingré's parallax values been the only ones obtained from the 1769 transit then the Earth-Sun distance would have been beyond dispute, but unfortunately Hell came up with a figure of 8.″70, while Euler obtained 8.″63, Lalande 8.″55–8.″63 and Planmann 8.″43 (see Woolf 1959: 190-191). The problem was that the difference between 8.″80 and 8.″43 equated to a difference in the astronomical unit of $\sim 6.57 \times 10^6$ km, which was totally unacceptable. Was a value for P in the 'high suite' of Hornsby and Pingré correct, or in the 'low suite' represented by Euler, Lalande and Planmann; and what of Hell's intermediate figure? Hopefully, the next pair of transits, in 1874 and 1882, would resolve this issue.

In fact both nineteenth century transits came out in favour of the 'high suite' of figures, as did other non-transit means of determining P, and Newcomb's reworking of all four transit results and review of the entire system of constants between 1891 and 1895 showed conclusively that a value of $P = 8.″794$ was in order (see Dick et al. 1998). Hornsby and Pingré had been right all along ... and with the benefit of hindsight we must applaud the following remarks, made by Hornsby back in 1771: "... from the observations made in distant parts by the astronomers of different nations, and especially from those made under the patronage and direction of the [Royal] Society, the learned of the present time may congratulate themselves on obtaining as accurate a determination of the Sun's distance, as perhaps the nature of the subject will admit." (Hornsby 1771: 574).

5. Concluding remarks

Lieutenant James Cook's first expedition to the South Seas was primarily undertaken as a voyage of science, and it was only after the transit of Venus was successfully observed that it became a voyage of exploration. Cook and Green were the 'official' astronomers on the *Endeavour* and were supplied by the Royal Society and the Royal Observatory at Greenwich with the best scientific instruments available, but Cook, Solander and some of the officers also brought along their own instruments. Three different Tahitian observing stations were established and all observers successfully recorded the transit, but there is confusion about how many active observers there were at the different stations, and the telescopes and time-keepers they used.

Moreover, when Cook came to write up the official account of the transit he only drew on observations made by himself, Green and Solander at Fort Venus, and for some reason ignored the dedicated observers based at Irioa Island and Taaupiri Island. To compound the mystery, the currently whereabouts of their original observations, and those made by Green and Solander, is unknown. Skelton (1954: 116) has noted that many of the coastal charts prepared during the voyage have disappeared, and it would seem that the transit records followed a similar path.

Most of the astronomical instruments associated with the Tahitian transit observations have also suffered a similar fate, and the current whereabouts of very few of them is known. While two Short telescopes with indisputable transit affiliations are in the Science Museum, London, there is no proof that either of these is the one that accompanied Cook to Tahiti (see Howse 1979; Orchiston 1998b). The current whereabouts of the Watkins reflector is unknown, but the telescope used by Solander at Fort Venus is probably the Heath and Wing reflector now housed in The Museum of New Zealand Te Papa Tongarewa, in Wellington (see Orchiston 1999). The situation regarding the Tahitian Bird quadrant is little better: admittedly, there are three different 12-inch Bird quadrants in the Science Museum, London, but there is no proof that any of these was on the *Endeavour*. The saga of the Cook voyage time-keepers is only marginally better: neither First Voyage journeyman clock nor alarum clock has survived (Howse & Hutchinson 1969), but Howse & Murray (1997) believe that the Shelton astronomical clock used at Fort Venus is now in the National Museum of Scotland, in Edinburgh.

While Cook's First Voyage was an outstanding success in terms of anthropology, botany, geography and zoology, and Cook acquired an instant reputation as a successful explorer, the astronomical outcome of the voyage was in dispute. Although Hornsby used the Tahitian data (amongst others) to derive a value for the solar parallax of 8.″78, which is remarkably similar to the currently-accepted figure, others came up with results that ultimately raised doubts about the validity of these transits as a means of determining the astronomical unit. Thus, Beaglehole (1968: cvi; cxliv) has made the oft-quoted claim that the Tahitian observations were a failure, but we concur with Waldersee (1969) and strongly dispute this. While he is a revered figure and is undoubtedly the world's leading Cook authority, Beaglehole was no astronomer, and did not have access all the evidence. As Sir Richard Woolley (1970: 135) has so aptly reminded us, nineteenth century astronomers using data from the 1761 and 1769 transits "... knew the distance to the Sun much better than present-day astronomers know the distance to the centre of the galaxy."

Acknowledgements

I am grateful to the following for their assistance: Ms Gudrun Richardson (The Royal Society), Peter Hingley (RAS Library), Adam Perkins (RGO Archives, Cambridge), Dr

Stuart Ryder (Anglo-Australian Observatory) and staff at the Mitchell Library (Sydney), British Library (London), National Maritime Museum (London) and the Science Museum (London). I also wish to thank the Science Museum (London) for permission to publish Fig. 1, the Donovan Trust (Sydney) for helping fund a research trip to England, and the IAU for awarding a Travel Grant that made it possible for me to attend the Preston meeting where an earlier version of this paper was given.

References

Admiralty 1768. Letter to Lieut. James Cook, dated 25 May. Original in the Public Record Office, London (Adm 2/94). Abstract in Beaglehole, 1968: 609.

Badger, G.M. 1970. Cook the scientist. In Badger, G.M. (ed.). *Captain Cook Navigator and Scientist*. Canberra, Australian National University Press. Pp. 30-49.

Banks n.d., Transactions of the Royal Society relative to the sending out people to Observe the transit of Venus in 1769. Original in the National Library of Australia, Canberra. Reprinted in Beaglehole, 1968: 511-514.

Beaglehole, J.C. (ed.) 1963. *The* Endeavour *Journal of Joseph Banks 1768-1771*. Two volumes. Second Edition. Sydney, Public Library of New South Wales in association with Angus and Robertson.

Beaglehole, J.C. (ed.) 1967. *The Journals of Captain James Cook On His Voyages of Discovery. III. The Voyage of the Resolution and Discovery 1776-1780*. Two volumes. Cambridge, Cambridge University Press.

Beaglehole, J.C. (ed.) 1968. *The Journals of Captain James Cook On His Voyages of Discovery. I. The Voyage of the Endeavour 1768-1771*. Cambridge, Cambridge University Press.

Beaglehole, J.C. 1974. *The Life of Captain James Cook*. London, Hakluyt Society.

Betts, J. 1993. The eighteenth-century transits of Venus, the voyages of Captain James Cook and the early development of the marine chronometer. *Antiquarian Horology*, 21: 660-669.

Cook, J. 1767. An observation of an eclipse of the Sun at the island of New-found-land, August 5, 1766. *Philosophical Transactions of the Royal Society*, 57: 215-216.

Cook, J. 1768. Letter to the Admiralty Secretary, dated 27 July. Original in the Public Record Office (Adm 1/1609). Reprinted in Beaglehole, 1968: 621.

Cook, J. 1770. Letter to the Admiralty Secretary, dated 23 October. Original in the Mitchell Library. Reprinted in Beaglehole, 1968: 499-501.

Cook, J. 1771a. Letter to Nevil Maskelyne, dated 9 May. In the Royal Society Council Minutes of 11 July 1771. Reprinted in Beaglehole, 1968: 692-694.

Cook, J. 1771b. Letter to the Secretary of the Royal Society, dated 11 July. In the Royal Society Council Minutes of 25 July 1771. Reprinted in Beaglehole, 1968: 694-695.

Deptford Yard Officers 1768a. Letter to the Navy Board, dated 27 March. Original in the Public Record Office, London (Adm 106/3315). Abstract in Beaglehole, 1968: 606.

Deptford Yard Officers 1768b. Letter to the Navy Board, dated 7 October. Original in the Public Record Office, London (Adm 106/3315). Abstract in Beaglehole, 1968: 624.

Dick, S.J., Orchiston, W., & Love, T. 1998. Simon Newcomb, William Harkness and the nineteenth century American transit of Venus expeditions. *Journal for the History of Astronomy*, 29: 221-255.

Douglas, J. n.d., *Hints* offered to the consideration of Captain Cooke, Mr Bankes, Doctor Solander, and the other Gentlemen who go upon the Expedition on Board the

Endeavour. Original in the National Library of Australia, Canberra. Reprinted in Beaglehole, 1968: 514-516.

Green, C., & Cook, J. 1771. Observations made, by appointment of the Royal Society, at King George's Island in the South Seas. *Philosophical Transactions of the Royal Society*, 61: 397-421.

Hornsby, T. 1765. On the transit of Venus in 1769. *Philosophical Transactions of the Royal Society*, 55: 326-344.

Hornsby, T. 1771. The quantity of the Sun's parallax. *Philosophical Transactions of the Royal Society*, 61: 574-579.

Howse, D. 1979. The principal scientific instruments taken on Captain Cook's voyages of exploration, 1776-80. *Mariner's Mirror*, 65: 119-135.

Howse, D., & Hutchinson, B. 1969. *The Clocks and Watches of Captain James Cook 1769-1969*. London, The Antiquarian Horological Society.

Howse, D., & Murray, A. 1997. Lieutenant Cook and the transit of Venus, 1769. *Astronomy & Geophysics*, 38(4): 27-30.

Molyneux, R. 1769. Remarks in Port Royal Bay in King George the thirds Island. Original in the Public Record Office (Adm 55/39). Reprinted in Beaglehole, 1968: 551-564.

Orchiston, W. 1998a. From the South Seas to the Sun. The astronomy of Cook's voyages. In Lincoln, M. (ed.). *Science and Exploration in the Pacific. European Voyages to the Southern Oceans in the 18th Century*. Woodbridge, Boydell Press, in association with the National Maritime Museum. Pp. 55-72.

Orchiston, W. 1998b. *Nautical Astronomy in New Zealand. The Voyages of James Cook*. Wellington, Carter Observatory.

Orchiston, W. 1999. Cook, Banks and the Gregorian telescope in The Museum of New Zealand Te Papa Tongarewa. *Journal of the Antique Telescope Society*, 18: 4-9.

Orchiston, W. 2004. The astronomical results of Cook's voyages. In Robson, J. (ed.). *Captain Cook Encyclopaedia*. London, Chatham Publishing, Pp. 31-35.

Parkinson, S. 1784. *A Journey of a Voyage to the South Seas, in His Majesty's Ship, The Endeavour*. London, printed for Stanfield Parkinson.

Ritchie, G.S. 1978. Captain Cook's influence on hydrographic surveying. *Pacific Studies*, 1: 78-95.

Schaefer, B.E. 2001. The transit of Venus and the notorious Black Drop Effect. *Journal for the History of Astronomy*, 32: 325-336.

Skelton, R.A. 1954. Captain James Cook as a hydrographer. *Mariner's Mirror*, 40: 92-119.

Waldersee, J. 1969. Sic transit: Cook's observations in Tahiti, 3 June 1769. *Journal of the Royal Australian Historical Society*, 55: 113-123.

Wales, W. 1788. *Astronomical Observations, Made in the Voyages which were Undertaken by Order of His Present Majesty, for Making Discoveries in the Southern Hemisphere* ... London, Elmsly.

Woolley, Sir Richard 1970. The significance of the transit of Venus. In Badger, G.M. (ed.). *Captain Cook Navigator and Scientist*. Canberra, Australian National University Press. Pp. 118-135.

Woolf, H. 1959. *The Transits of Venus: A Study in the Organisation and Practice of Eighteenth-Century Science*. Princeton, Princeton University Press.

Discussion

STEVE DICK: I guess the question is whether they got it right by accident or ...?

WAYNE ORCHISTON: Well we don't know that; it's easy to be wise after the event.

JACQUELINE MITTON: Is it known whether observations were made by projection or otherwise? I was looking quite recently at a book which suggested there was a phrase like "all eyes to the telescope". You just mentioned the possible use of filters with a sextant or quadrant, but as far as I am aware no safe filters had been devised at that time. So what were they up to? Were they risking their eyesight, or did they project, or …?

WAYNE ORCHISTON: That's an excellent question. It's one that has occurred to me, too, Jacqueline, and I've not been able to find any documentation. This is one of the problems with going through this literature. There are descriptions of what was done, but not the actual precise detail of the observing techniques. Given that sextants were used for solar observations, obviously there were filters that were suitable. I had presumed that they would have adapted those for the Gregorian telescopes, and one would assume that they would have made plenty of test observations with these sorts of instruments here in Britain before they left on the expedition. But I would really like to get some documentation of the evidence for that, so I don't know. Thank you for raising it though.

NICK KOLLERSTROM: You described the expedition with the distinguished astronomers on board. Would you confirm that they used the lunar method for longitude in their voyage. This is one of the few main journeys when it was the main method of finding longitude, and it continued to be used as a backup after the clocks – chronometers – came in. Was this a very difficult thing to do?

WAYNE ORCHISTON: For the first voyage, of course, there were no chronometers on the *Endeavour*. They only came with the second and third voyages, so the lunar method of determining longitude was the standard practice that was used for navigation on the first voyage. Part of the training that Green was required to give any of the lower officers – and particularly able bodied seamen who wanted to move up through the hierarchies (as Cook himself had done) – was to teach them the basic principles of lunars, so they were learning nautical astronomy as well as transit astronomy during the voyage. And you will also recall, of course, the longitude prize – there was a prize of up to £30 000, if I remember rightly – to provide accurate chronometers that could be used on round the world voyages. One of those who was on the committee to make decisions about the allocation of their funding was Maskelyne, the Astronomer Royal, who, of course, was pushing the lunars method and hoping that he would get that prize for the Royal Observatory. So there were actually vested interests – very interesting politics – to look at the first voyage versus the second and third, in relation to navigation in terms of lunars and the different chronometers.

DAVID SELLERS: Just a quick point on the use of filters: there is a paper by Maskelyne – I think it's in relation to the 1769 transit – where he said that his preference is to use smoked glass, as opposed to red glass, which he doesn't like. I'm not sure what the red glass consisted of.

WAYNE ORCHISTON: Thank you, David, I appreciate that.

ALLAN CHAPMAN: Back on the filter subject: in the Museum of the History of Science, for instance, where there are several Short telescopes surviving with their cases and all of their micrometers, and everything, they all used to carry standard filters as normal, just to screw them in – often in deep red and often deep green. These, of course, were not for the transit as such, but were just for solar observing. I think we are now very much more concerned about eyesight than they were. William Lassell – in 1851 when he went to observe the eclipse of the Sun at Norway with a 3 3/4-inch Merz refractor – describes seeing the Sun come out of eclipse, and he describes how it started shattering his eye-piece filters, and he kept several of them in his waistcoat pocket, and every time a filter shattered he'd screw a new one in; that's how they operated in those days.

WAYNE ORCHISTON: Yes, thank you. Certainly there's a lot of evidence that there were filters issued with most of these Short telescopes, and the comment you make, Allan, about the later times is well documented; a lot of observers with the 1874 transit had problems with filters.

MARILYN HEAD: I have been always intrigued by the secret instructions given to Cook, especially as they effected us [New Zealanders] ... and you say that the expedition was primarily a scientific one, not a political one?

WAYNE ORCHISTON: Yes, various people have put forward that latter view, but in the public eyes when this voyage was announced, when it was publicised, and also in the eyes of the Royal Society, of Royal Observatory, and certainly of the Navy, this was a scientific expedition. It was only after the science had been achieved, if in fact it was possible, to address a second issue. Of course, with the benefit of hindsight, and also looking at the *raison d'être* of the second and the third voyages, it becomes much easier to interpret the second part of that first voyage as a voyage of discovery and exploration – which indeed it was – but that was not the primary function, and that was not the reason that the funding was allocated by King George III.

WALTER BRISKEN: We have been talking mostly about the scale of the solar system, but when did we become aware of the eccentricity of the Earth's orbit around the Sun and how does that relate to the early measurements of AU?

WAYNE ORCHISTON: I'm not really the person to answer that. There are a whole range of people in this audience who would be much more competent to. I'm sure Alan Chapman or Dave Hughes ...

ALLAN CHAPMAN: I missed the exact question, I'm sorry.

WALTER BRISKEN : What effect does the eccentricity of the Earth's orbit have on these measurements?

WAYNE ORCHISTON: ... and who was the first one to detect that?

ALLAN CHAPMAN: Who was ...? I'm afraid I don't know that.

STEVE DICK: Myles do you have anything to add?

WAYNE ORCHISTON: ... but who was the first person to actually publish on the eccentricity? Do you know?

MYLES STANDISH: ... Greek astronomers, collectively.

STEVE DICK: Yes, it's an interesting question: what's the magnitude of the effect of the eccentricity. Maybe we'll mull that over overnight.

Transit of Venus: New Views of the Solar System and Galaxy
Proceedings IAU Colloquium No. 196, 2005
D.W. Kurtz, ed.
© 2005 International Astronomical Union
doi:10.1017/S1743921305001274

Observations of the 1761 and 1769 transits of Venus from Batavia (Dutch East Indies)

Robert H. van Gent

Institute for History and Foundations of Mathematics and the Natural Sciences,
Utrecht University, The Netherlands
email: r.h.vangent@astro.uu.nl

Abstract. This paper discusses the observations of the 1761 and 1769 transits of Venus by the Dutch-German clergyman Johan Maurits Mohr (1716–1775) from Batavia (Dutch East Indies). We will investigate how Mohr became interested in observing this phenomenon and how he made the necessary preparations. Finally, the fate of his observatory and his instruments will be discussed.

1. Introduction

In 1761 and 1769, the Dutch-German clergyman Johan Maurits Mohr (1716–1775) made successful observations of the transits of Venus from Batavia (Dutch East Indies). Earlier research, notably by van der Bilt (1940) and Zuidervaart (1999), has already supplied a very detailed picture of Mohr's scientific activities in Batavia, but until now this has only been available in Dutch. More recent research by Zuidervaart & van Gent 2004 has uncovered more information on Mohr's instruments and on his motivation for observing the transits of 1761 and 1769.

2. Mohr's education and early years in Batavia

Johan Maurits Mohr was born in 1716 in Eppingen, Germany, and enrolled at the University of Groningen as a student in theology in August 1733. Three years later he completed his studies with a public disputation on the role of visions in the Bible.

After successfully passing an examination by the Amsterdam board of the Dutch United East India Company (Verenigde Oostindische Compagnie [VOC]) in December 1736, Mohr was appointed as minister for the Dutch East India Church in Batavia (present-day Jakarta in Indonesia) with a monthly allowance of 90 Dutch guilders. In the following year, Mohr set sail for Batavia on the Dutch East Indiaman *Oostrust* and, after a brief call at the Dutch provisioning station at the Cape of Good Hope, arrived at Batavia in October 1737.

After a brief service as a vicar in Batavia, Mohr was commissioned to lead the Portuguese community at the Dutch settlement at Galle in Ceylon. However, he requested permission to stay in Batavia to perfect his knowledge of the Portuguese language and, after his marriage with Johanna Cornelia van der Sluys, he was appointed in February 1739 to lead the Portuguese community in Batavia instead.

In 1743 Mohr was appointed as rector of the newly founded Theological Seminary of Batavia on account of being a "man of learning and of good qualities who is dearly beloved in the Indies". So, already after a few years in Batavia he had acquired the reputation of a scholar. During his tenure as rector, Mohr published several theological works, including translations of the Bible into the Portuguese and Malay languages.

Mohr resigned as rector in February 1753. Shortly before, in April 1752, Mohr had married again (his first wife died in September 1750). His new wife was Anna Elisabeth van 't Hoff, the Asian-raised daughter and principal heiress of Jan van 't Hoff, a very wealthy VOC official. Anna Elisabeth also had money of her own as she had been married twice before to wealthy men.

Up to the late 1750s, we have no evidence of Mohr's interests in astronomy and the related sciences but that was to change as the 1761 transit of Venus approached.

3. The Venus transit of 1761

In May 1760, the French astronomer Joseph-Nicholas Delisle wrote to the Dutch astronomer Dirk Klinkenberg, inquiring on behalf of the *Académie des Sciences* whether the Dutch authorities would be able to assist in transporting and stationing a French astronomer in Batavia to observe the transit of Venus. However, when Delisle subsequently learned that the Royal Society of London was planning to send the astronomers Charles Mason and Jeremiah Dixon to observe the transit from Sumatra, he decided that a French observer would not be necessary in the Dutch East Indies. On the other hand, as Delisle later wrote to Klinkenberg, it would be a pity if the transit went unobserved at Batavia and he urged Klinkenberg to contact the VOC authorities at Batavia so that perhaps a local observer could make measurements with the modest instruments that happened to be available there.

The transit of Venus of was indeed observed from Batavia on 6 June 1761, and not by a professional astronomer. It was in fact observed by Gerrit de Haan, the head of the department of mapmakers in Batavia, who was assisted by Pieter Jan Soele, a skipper employed by the VOC.

Their observations were made from a beach, just outside of Batavia, on the estate "Kliphoff" owned by Johan Maurits Mohr, who apparently also supplied a few of the used instruments. The measurements were made with the help of two Gregorian reflectors (with focal lengths of 18 and 27 inches), an octant made by the London instrument maker George Adams and some accurate pocket watches. Both the beginning and the end of the transit were successfully observed and Mohr also made measurements of the progress of Venus across the Sun's disk during the transit. Although the observations were mostly made by de Haan and Soele, it was Mohr who prepared a detailed account (in Dutch) that was sent to the *Hollandsche Maatschappij der Weetenschappen* in Haarlem and which was published in their *Verhandelingen* (Mohr 1763).

4. Preparing for the Venus transit of 1769

Although the next transit (of 1769) would only be partially visible from Batavia, Mohr decided that it should be observed as completely as possible with the best instruments that were available. With the death of his father-in-law Jan van 't Hoff in April 1763, Mohr's wife had inherited a large fortune and with this money they had been able to acquire a large plot of land to the south of Batavia along the Molenvlietse dike on which a luxurious building was erected with an observatory in the uppermost storey.

The construction of the 80-foot high edifice started in 1765 and it was completed in 1768. According to contemporary records, the costs of Mohr's observatory had been estimated at 80 000 "ryksdaalders" (200 000 Dutch florins) and it outranked nearby "Buitenzorg" (present-day "Bogor"), the residence of the governor-general of Batavia, in magnificence.

Figure 1. View of Mohr's observatory with a Chinese temple in the foreground. Drawing in ink made around 1770 by the Danish artist Johannes Rach. (National Library, Jakarta)

In the mean time, Mohr had sent letters to the Leiden professor of mathematics and philosophy Johan Lulofs, commissioning him to select and purchase astronomical and mathematical instruments and books for equipping his observatory. According to the VOC archives, shipments were sent to Batavia in 1765 and 1767 containing various (unspecified) astronomical instruments.

Fortunately, we have a few depictions of Mohr's observatory (see Figs 1 and 2) and descriptions by foreign visitors such as from the French naturalist Louis-Antoine de Bougainville when he made a call at Batavia in October 1768 during a 28-month journey around the world:

"We could never be tired with walking in the environs of Batavia. Every European, though he be used to live in the greatest capitals, must be struck with the magnificence of the country around it. This is adorned with houses and elegant gardens, which are kept in order, in that taste and with that neatness which is peculiarly observable in all the Dutch possessions. I can venture to assert that these environs surpass those of the greatest cities in France, and approach the magnificence of those of Paris. I ought not to omit mentioning a monument, which a private person has there erected to the Muses. Mr. Mohr, the first clergyman at Batavia, a man of immense riches, but more valuable on account of his knowledge and taste for the sciences, has built an observatory, in a garden belonging to one of his country-houses, which would be an ornament to any royal palace. This building, which is scarce completed, has cost prodigious sums. Its owner now does something still better, he makes observations in it. He has got the best instruments of all kinds from Europe, necessary for the nicest observations, and he is capable of making use of them. This astronomer, who is doubtless the richest of all the children of Urania, was charmed to see M. Verron. He desired he should pass the nights in his

Figure 2. Left: Front view (eastern face) of Mohr's observatory; Right: Side view (northern face) of Mohr's observatory. Architectural drawings by J. Clement (*Hollandsche Maatschappij der Wetenschappen*, Haarlem)

observatory; unluckily, not a single one has been favourable to their purposes. M. Mohr has observed the last transit of Venus [of 1761], and has communicated his observations to the society of Harlem; they will serve to determine the longitude of Batavia with precision."†

De Bougainville's favourable description of Batavia is noteworthy, as the Dutch settlement on Java was generally thought of as an insalubrious place and was much feared by seamen and visitors, who knew it as the "graveyard of the Europeans".

5. The Venus transit of 1769

On the morning of 4 June 1769, Mohr was ready to observe the long-awaited transit of Venus. Compared to 1761, he was now much better equipped and could make use of the facilities of his new observatory and far superior instruments. During the past year he had also made several observations to determine the precise geographical co-ordinates of his observatory. Nevertheless, despite his long and careful preparations, his observations of the transit were nearly a failure.

The transit started nearly four hours before sunrise at Batavia and during the first two hours the Sun was hidden by clouds. However, around 8 o'clock the sky cleared and

† L.-A. de Bougainville, *Voyage autour du monde, par la frégate du roi La Boudeuse, et la flûte L'Étoile; en 1766, 1767, 1768 & 1769* (Paris, 1771). Quoted from the English translation by John Reinhold Forster published in 1772 (pp. 425–426).

Mohr could observe Venus on the solar disk with his 3½-foot focal length, Dollond-made Gregorian reflector. When Venus was near to the solar limb he also made observations with a heliometer. Mohr timed his observations with an astronomical pendulum clock with a Graham-type escapement made by the London clockmaker John Shelton. A 2½-foot astronomical quadrant made by the London instrument maker John Bird served to determine the errors of Mohr's clock.

Mohr did not prepare a detailed report of his observations until the end of the year, as he also wanted to include his observations of the transit of Mercury across the Sun's disk on 10 November. A Dutch account of his observations of both transits was sent to the *Hollandsche Maatschappij der Weetenschappen* in Haarlem and were duly printed in the society's *Verhandelingen* of 1770 (Mohr 1770a, 1770b).

Mohr also prepared a Latin translation of his report which he delivered into the hands of the British explorer James Cook when his ship the *Endeavour* made a call at Batavia in December 1770 for repairs and victualling. On his arrival in England in July 1771, Cook sent Mohr's report with his own report on the transit of Venus to the Royal Society and both reports were published in the same year's issue of the *Philosophical Transactions* (Mohr 1771).

6. Mohr's last years

After the Venus transit of 1769, Mohr continued to improve his observatory and several more shipments of astronomical and meteorological instruments to Batavia are recorded in the VOC archives. Mohr also made daily observations of the weather and the compass variation at Batavia, but these observations were never published.

Mohr's last-known publication dates from 1773 and gives a vivid description of the volcanic eruption and subsequent collapse of the nearly 3000-m mountain Gunung Papandajan, located some 165 km southeast of Batavia, that had occurred in August 1772 (Mohr 1773).

Mohr died in October 1775, only a few years before the founding of the first scientific society in the Dutch East Indies, the *"Bataviaasch Genootschap van Kunsten en Wetenschappen"* (Batavian Society of Arts and Sciences).

7. The fate of Mohr's observatory

In 1780, only five years after the death of its founder, Mohr's observatory was severely damaged by a heavy earthquake. Mohr's widow, Anna Elisabeth van 't Hoff, passed away in May 1782 and shortly afterwards permission was granted to the future owner of Mohr's estate to "modify, lower or demolish" the building. However, this was not done until several decades later as the building was used until 1809 to house company clerks and for some time afterwards it was used as army barracks.

Evidence of the rapid speed in which the observatory was falling into ruin is found in the correspondence of the young Francis Beaufort, who at a later date would become famous as an admiral and hydrographer. Signed on as a 15-year old captain's servant he visited Batavia in the summer of 1789 and described the abandoned observatory built by a certain 'M. de More' in a letter to his parents as follows:

> "Such an Observatory I never saw. There is but one room in the house calculated to observe in, which is at the top of the house, near one hundred feet from the ground, affording only the most awkward viewing and so built that if you just stamp in any one part of the house it shakes the Observation room. Indeed we find that a person walking about on the stairs or in any [...] room of the house shakes

the Horizon and makes the objects turn about in the Equatorial Telescopes [...] In
short, I never saw or heard of such a place in my life."†

Exactly when Mohr's observatory was completely demolished in not known, but in
1844 only its foundation stones were reported to be visible.

8. The fate of Mohr's instruments

For a long time it was believed that Mohr's astronomical and meteorological instruments had all been lost. Recent work, however, has revealed that some of his astronomical
instruments have in fact survived and are preserved in various Dutch museum collections.

Mohr's astronomical and meteorological instruments were bought from his widow in
1776 by the clergyman Johannes Hooijman, who would become one of the most active
members of the subsequently founded *Bataviaasch Genootschap*. However, Hooijman
rapidly realized that many of Mohr's instruments had suffered from the hot and humid
climate of Batavia and were unusable. As there were no good instrument makers in
Batavia, he had no other option but to ship them back to Amsterdam for repairs.

For as yet unknown reasons, the instruments were not repaired and for many years
they lay forgotten in the attic of an Amsterdam storehouse. When they resurfaced again
in 1789, their origin was largely forgotten, and they were offered to the newly founded
scientific society, *Felix Meritis*, for their observatory. From then on, the whereabouts of
several of Mohr's instruments can be followed up to the present day (for more details,
see Zuidervaart & van Gent 2004).

References

Mohr, J.M. 1763 *"Waarneminge over den schijnbaaren loop van Venus over de Zonneschijf,
in zijn begin, midden en einde, den 6 Juni 1761"*, Verhandelingen uitgegeeven door de
Hollandsche Maatschappye der Weetenschappen te Haarlem 7, part I, 380–391.

Mohr, J.M. 1770a *"Waarneeming van Venus, by haaren uitgang van de Zonne-schyf, gedaan den
4. Juny 1769, te Batavia op het Observatorium (liggende op 6 gr 12 min Zuider br.)"*, Verhandelingen uitgegeeven door de Hollandsche Maatschappye der Weetenschappen te Haarlem 12, 123–130.

Mohr, J.M. 1770b *"Waarneeming van Mercurius by zynen uitgang van de Zonne-schyf, gedaan
den 10. Nov. 1769, te Batavia op het Observatorium"*, Verhandelingen uitgegeeven door de
Hollandsche Maatschappye der Weetenschappen te Haarlem 12, 131–134.

Mohr, J.M. 1771 *"Transitus Veneris & Mercurii in eorum Exituè Disco Solis, 4to Mensis Junii
& 10mo Novembris, 1769, observata"*, Philosophical Transactions of the Royal Society of
London 61, 433–436.

Mohr, J.M. 1773 *"Bericht nopens het springen en instorten van een brandenden Zwavel-
Berg, met het droevig gevolg van dien, op het eiland Java, in de Maand Augustus 1772
voorgevallen: Met bygevoegde aanmerkingen nopens dit verschynsel"*, Verhandelingen uitgegeeven door de Hollandsche Maatschappye der Weetenschappen te Haarlem 14, part II,
82–96.

van der Bilt, J. 1940 *Venus tegen de Zonneschijf 1761; 1769: Een bladzijde uit de geschiedenis
der Nederlandsche sterrenkunde.* J.B. Wolters, Groningen.

Zuidervaart, H.J. 1999 *Van 'Konstgenoten' en Hemelse Fenomenen: Nederlandse Sterrenkunde
in de Achttiende Eeuw*, chapters 16–17. Erasmus Publishing, Rotterdam.

Zuidervaart, H.J. & van Gent, R.H. 2004 "A Bare Outpost of Learned European Culture on
the Edge of the Jungles of Java: Johan Maurits Mohr (1716–1775) and the Emergence of
Instrumental and Institutional Science in Dutch Colonial Indonesia", *Isis: An International
Review devoted to the History of Science and its Cultural Influences* 95, 1–33.

† Quoted from A. Friendly, *Beaufort of the Admiralty: The Life of Sir Francis Beaufort* 1977,
31.

Discussion

FRITZ BENEDICT: How is it the Dutch remained healthy and every time an Englishman stopped at Batavia he dropped dead?

ROBERT VAN GENT: Well, not all the Dutch were healthy – you either acquired a resistance or you died young there. That also applied for the Dutch.

WAYNE ORCHISTON: Very, very interesting paper, and it's nice to get a different perspective other than just the Cook or the American versions of transits. My question though is: What would be the outcomes of the analysis of those observations in 1761 and '69 in terms of values for the solar parallax?

ROBERT VAN GENT: Well, the first ones – the 1761 observations – were not used because they were published in the Dutch language, so outside the Netherlands they were unknown. Furthermore, van der Bilt – the Dutch astronomer who researched these observations in the 1940s – already came to the conclusion that the observations of the 1761 transit had been flawed in some way because the times and the solar altitudes didn't match, so there was something definitely wrong with those observations. The other observations which Mohr made himself in '69 appeared to be alright, but in the subsequent analysis Encke did not use them, but there was also, of course, the problem of the longitude of Batavia. Now Mohr did determine his longitude and he did publish it in the *Philisophical Transactions*, and even Cook reports that the longitude determination by Mohr was fairly useful for him on the last leg of his voyage, because he could correct his maps. But for some reason I think people still considered that Mohr's determination of longitude was not precise enough, and so that's why Mohr's observations were rather neglected afterwards.

NICK KOLLERSTROM: Just about the fate of Captain Cook: Did you say he or his crew perished from malnutrition?

ROBERT VAN GENT: No, it was dysentery – not all of his crew, but it was quite a number of his crew.

NICK KOLLERSTROM: I'd heard that it was on Hawaii he was stabbed.

ROBERT VAN GENT: That's a later voyage. This is Cook's first voyage.

MARY BRÜCK: I'd like to thank you very much for a fascinating talk, because it's such an interesting story from the human, as well as the astronomical point of view, on how somebody could do all this on his own on the other side of the world, when it took a whole year to get your instruments ordered and delivered. But it was very useful to have plenty of money, and also he may not have been the only one who wedded a wealthy wife in order to help his career, but that's another story!

[general laughter]

Transits of Venus: New Views of the Solar System and Galaxy
Proceedings IAU Colloquium No. 196, 2004
D.W. Kurtz, ed.

© 2004 International Astronomical Union
doi:10.1017/S1743921305001286

The 1761 transit of Venus dispute between Audiffredi and Pingré

Luisa Pigatto

INAF – Astronomical Observatory of Padova, Vicolo dell'Osservatorio, 5, Padova, Italy
email: pigatto@pd.astro.it

Abstract. On 6 June 1761 the Dominican Giovanni Battista Audiffredi observed the transit of Venus in his little observatory at the Monastery of Santa Maria sopra Minerva in Rome. Soon after, he published an anonymous short report in Italian, and in the first months of 1762 he published a complete Latin essay about his transit observations. Late in 1762, and in 1765, the French abbé Alexandre-Gui Pingré, who had observed the transit at the Rodriguez Isle, to the south of the equator, presented to the French Royal Academy of Sciences the results of the solar parallax determination derived from comparison of observations made in different geographic places. He had excluded the Roman data because – he said – of the lack of a fundamental quantity, the longitude of the Monastery, concluding that the Roman observations were imperfect. In order to defend his scientific reputation, Audiffredi published two Latin essays concerning the solar parallax determination, the *Investigatio parallaxis solaris* in 1765, and the *De Solis parallaxis Commentarius* in 1766.

1. Introduction

In the 18th century, Italy, fragmented into a number of more or less small States, didn't own a national Observatory like those of Paris (1667), Greenwich (1675), Berlin (1705), St. Petersburg (1725) and Vienna (1731), equipped with modern instruments which could co-ordinate astronomers in view of important astronomical events. In the first half of the century the only public Observatory, founded in 1712 and operative since 1726, was that of Bologna (Baiada *et al.* 1995).

Nonetheless, since the telescope was invented many scholars and amateurs of astronomy observed the sky with long and even better telescopes. Places of observations were rooms in private houses, monasteries, loggias in the aristocratic palaces, terraces and gardens. Though only on 23 December 1757 a papal Bull cancelled the prohibition against writings in favour of Copernicanism, the scientific climate in 18th-century Rome was very vivid, especially in the world of religious orders. The Religious devoted themselves to the study of modern science mainly to verify its principles with those of the Catholic Faith, but also often due to an authentic passion for science, in particular astronomy. For these reasons the Religious' scientific research was of very high quality. As an example, we mention the famous Jesuit Ruggiero Boscovich who made astronomical observations from the terrace of the "Collegio Romano", the "Minimi" Fathers François Jacquier and Thomas Le Seur, who observed at the Monastery of "Trinità dei Monti". In 1739 – 1742, Jacquier and Le Seur edited the famous commentary to Newton's *Principia* which introduced Newtonianism to the learned society of the Roman papacy. Important astronomical events were carefully observed, and the results soon published. So it was for the transit of Mercury of 6 May 1753 and that of 7 November 1756. The phenomenon gave the astronomers the

Table 1. Table 1. Observers, places and instruments for the 1761 transit of Venus observed in Italy (1 Parisian foot = 32.4833 cm; 1 Roman palm = 22.3322 cm; 1 Neapolitan palm = 26.4 cm)

OBSERVER	PLACE	INSTRUMENTS
Audiffredi	Rome	9 Roman-palms Refractor
	Monastery of S. Maria sopra Minerva	9 Roman-palms Divini Refractor
Asclepi *et al.*	Rome	5 Roman-palms quadrant
	Collegio Romano	12 Roman-palms Divini Refractor
		8 Roman-palms Refractor
Spagnius		20 Roman-palms Huygens Refractor
Jacquier	Rome	not mentioned
Le Seur	Monastery of S. Trinità dei Monti	
Beccaria	Turin	3 Parisian-feet Quadrant
Canonica	Small tower at Beccaria's home	
Revelli		40 Parisian-feet Refractor
Poleni	Padova - his home	clouded sky
Zanotti	Bologna	2 1/2 Parisian-feet Quadrant
Frisi	Public Observatory	10 Parisian-feet Refractor
Canterzani		22 Parisian-feet Refractor
Marini		6 Parisian-feet Refractor
Matteucci		8 Parisian-feet Refractor
Casali		
Ximenes	Florence	4 Parisian-feet Newtonian reflector
	Specola at Collegio	
	S. Giovanni Evangelista	
Carcani	Naples	5 1/2 Neapolitan-palms Quadrant
	Royal College	23 1/2 Neapolitan-palms Refractor

opportunity to better define the elements of the planet's orbit and, at the same time, allowed them to practice for observing the very expected transit of Venus of 1761.

1.1. *The 1761 transit observed in Italy*

In western Europe, the 1761 transit of Venus should have been visible before and during the egress phase in the morning. Father Niccolò Maria Carcani of the "Scuole Pie", professor of mathematics at the Royal College of Naples, received a long letter from the famous French astronomer Joseph-Nicolas Delisle concerning the best methods to observe the transit. Following Delisle's suggestion, Carcani published the letter in the Italian journal "Novelle Letterarie" edited in Florence in order to make Delisle's advice known to Italian astronomers. Delisle suggested two different methods to observe the transit of Venus: the first method consisted of observing the differences between the transit timing of the planet's centre and that of the solar limbs seen through the horizontal and vertical wires in a quadrant. The second method consisted of observing the differences in right ascensions and declinations between the planet's centre and the solar limbs, using an equatorially mounted telescope with a good micrometer equipped with wires perpendicular and parallel to the diurnal motion of the Sun and Venus. The transit was observed, weather permitting, all around Italy, as shown in Table 1. Observations were soon published in pamphlets and in magazines (Asclepi (1761); Beccaria (1761); Poleni (1761); Zanotti (1762); Ximenes (1761); Carcani (1761)).

2. Audiffredi's astronomical observations

Giovanni Battista Audiffredi (1714-1794) (Fig. 1) was a Dominican who was appointed librarian in 1749 at the "Biblioteca Casanatense", the famous Library founded by the

Figure 1. Oil portrait of Giovanni Battista Audiffredi.

cardinal Girolamo Casanate (1620-1700) in Rome that was inaugurated in the large complex of the Dominican Monastery of Santa Maria sopra Minerva in 1701 (Fioravanti 1994). Dominicans were known as the faithful guardians of the Catholic orthodoxy. So, it is not surprising that the Dominican Monastery was the seat of the Roman Inquisition during the 16th and 17th centuries. In the very same place Galileo abjured the Copernican system.

It was in the period of his new duties as Librarian that Audiffredi had a small observatory built at the top of the Monastery of Santa Maria sopra Minerva. He was very fond of astronomy since his youth, so "put astronomers' books aside", he decided he had to observe the sky directly himself (Audiffredi 1754). In 1751, as soon as he got the instruments to observe every kind of celestial phenomenon, he started to draw the meridian line with the method of the correspondent heights of the Sun, so that the line lay exactly in the meridian plane. The meridian line, engraved "over a prismatic piece of marble placed on a quite solid ground", fixed "with strong lime and by means of an iron web connected to the wall", was so precise that "the time didn't deviate even from a second". The gnomon was about 14 Parisian feet high; its vertex was "the centre of a hole in a brass plate about two lines thick, sized as a truncated cone" (Fioravanti 1994).

With his meridian line Audiffredi determined the Pole altitude, i.e. the latitude of his observatory ($41°\ 54'\ 33''$). In the years 1751, 1752, 1753 he observed eclipses of Jupiter's satellites, lunar occultation of stars, and the transit of Mercury of 6 May 1753. Comparing his results with those of Paris and Bologna observatories, he realised that some mistakes were evident in the meridian difference between Rome – St. Peter Church and Paris – Observatory. From his computations he concluded that the "Minerva" Monastery was to the east of Paris Observatory by a value that couldn't be greater than $40^{\rm m}\ 20^{\rm s}$ (Audiffredi 1754).

2.1. *Transitus Veneris*

On 6 June 1761 Audiffredi observed the transit in his little observatory. On 13 June he published the true times of the third and fourth contacts of Venus with the solar limb in the *Diario ordinario Romano*; he also published a short note of his observations in an anonymous six-page booklet in Italian, and the same anonymous report appeared in the *Novelle Letterarie* of 1761.

Before he published his results, Audiffredi got in touch with the Religious, who had observed the phenomenon in Rome, in order to verify the reliability of his data. He contacted Jacquier and Le Seur, the Jesuit Benvento, the Fathers of the "Scuole Pie" who had observed the transit at the "Collegio Calasanzio", wrote to the Jesuit Leonardo Ximenes in Florence and to Eustachio Zanotti in Bologna. The Dominican was enthusiastic for the excellent agreement of the internal contact time recorded by him with those of the other observers, taking into account the difference of longitude. This time was one of the quantities which could be used to calculate the solar parallax.

So, in the first months of 1762 Audiffredi published a complete Latin essay, *Transitus Veneris ante Solem* (Fig. 2) with all his observations. As in a clear and detailed handbook, the Dominican described all the operations he had performed before, during and after the transit. The first operation was to control the delay of the pendulum clock with regard to the midday of the meridian line, in order to reduce the observations to the true time correspondent to the hour-angle of Sun and Venus. He arranged a 9-Roman-palms telescope (2 meters) with a micrometer "formed of four very thin wires crossing at 45 degrees" to observe the planet during its transit over the Sun, and a 19-Roman-palms telescope (4.24 meters), with the objective lens by the "famous Eustachio Divini", to observe the planet's egress.

He observed 33 positions of Venus over the solar disc. For each of them, he recorded the contact time of the western and eastern limb of the Sun with the vertical wire of the micrometer, the contact time of the western and eastern limb of Venus with the first oblique wire, with the vertical wire and with the second oblique wire respectively (see Fig. 3).

In order to reduce his observations, he chose the solar diameter of $31'\ 37''$, made a mathematical interpolation of the planet's path, reduced the positions of Venus and Sun to the true values and determined the minimum distance from their centre by trigonometric method as shown in his drawing (Fig. 4). At the end, he presented the elements of Venus's orbit he had calculated from his observations, which he communicated to his Italian colleagues, intending to calculate the Sun's parallax as soon as he could get data from other European observers.

Figure 2. The frontispiece of the *Venus Transit* of 1762.

3. The affair Audiffredi – Pingré: an 'anonymous' dispute

In late 1762, the French Abbé Alexandre–Gui Pingré (1711-1796) gave a detailed
account of the Sun's parallax determination to the members of the French Royal
Academy. In his long memoir, he analysed the Venus transit observations that French
and foreign astronomers had performed in different geographical places, and gave the

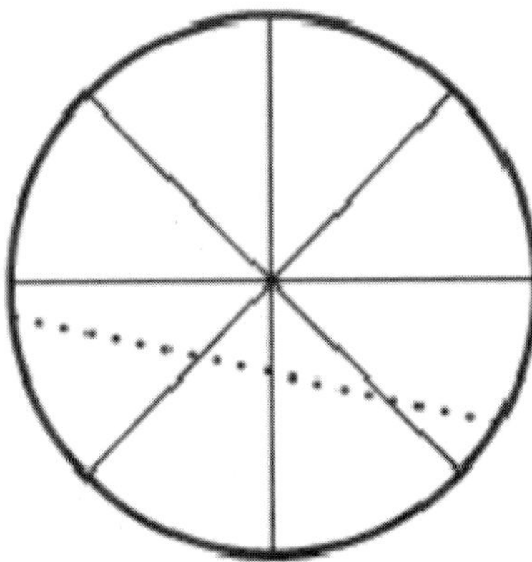

Figure 3. Scheme of the four wires micrometer. The circle represents the telescope field, dots the Venus path on it.

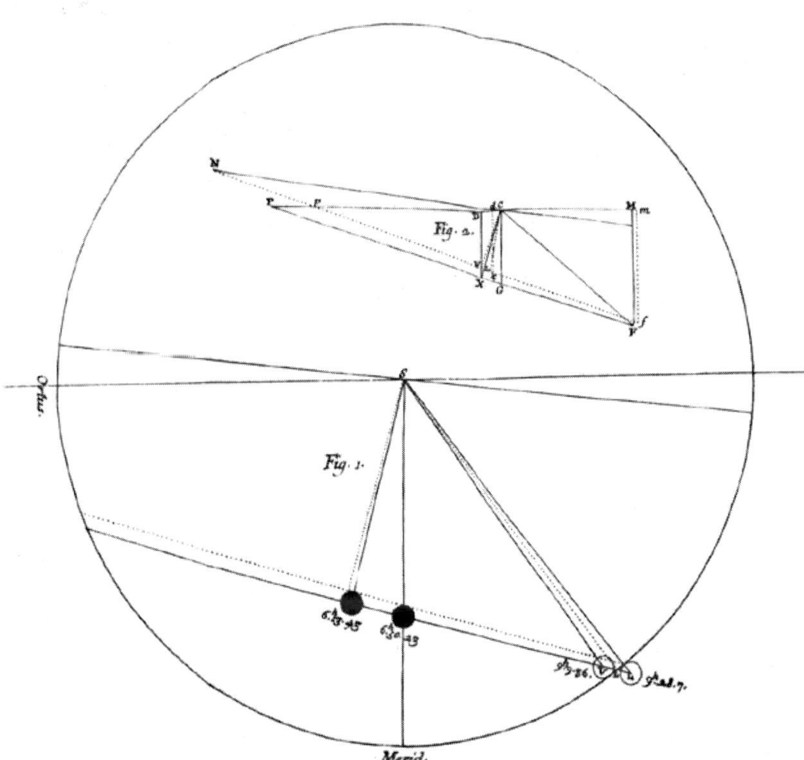

Figure 4. The drawing by Audiffredi shows the trigonometric method to determine the minimum distance between the Sun's and Venus's centres. The circle represents the solar disc.

results of his computation. For this purpose, he used three different methods: the first, by comparing the duration of the whole path of Venus over the solar disc; the second, comparing the minimum distance between the centre of Venus and Sun; the third, comparing the time of the last internal contact of Venus with the solar limb. This third method required a "very precise knowledge of the longitude of the locations where the end of the transit was observed", while it was sufficient to know the latitude within few minutes (Pingré 1763a).

Pingré stated that he couldn't take into account some listed observations because of the lack of the value of longitude. As far as Rome was concerned, he underlined that the *Connoissance des Temps* (the French ephemerides) gave the value of the St. Peter Church: "Is it there, or on the same meridian – he asked himself – that the transit of Venus was observed by an anonymous who gives the exit time at 21h 09m 36s? They say the place was S. Maria sopra Minerva". The abbé concluded: "In order to use the Roman observations it is necessary to know the precise difference in latitude, and especially in longitude, between the two places". This comment was not offensive at all, but it demonstrated that the only source of information was the anonymous paper Audiffredi had published in the *Novelle Letterarie* where the requested quantities were lacking, but where it was said: "The meridian line has been carefully examined in this same month of June during the solar eclipse, and has been found very precise" (Audiffredi 1761). It seems surprising that Pingré didn't see the following and most exhaustive work by Audiffredi, *Transitus Veneris*, before deriving his conclusion about the Sun's parallax. As a matter of fact, this book was published in the first months of 1762, long before Pingré could be able to come back at home from the Rodriguez Isle, where he went to observe the transit. In the very same book he should be able to find the requested value of longitude together with the name of the author, which was mentioned in the "permission to publish" at the end of the volume. Besides, not even the observations performed by Jesuits at the "Collegio Romano" were taken into account, even if the Jesuit Asclepi published them in the *De Veneris per Solem Transitu* (Asclepi 1761).

We assume that Audiffredi was offended by Pingré's remarks because the abbé demonstrated he had never considered, or never seen, the astronomical observations the Dominican started publishing since 1753, where it was possible to find any information concerning the small observatory at the top of the "Minerva" Monastery. So, the misunderstanding was born for many scholars' bad habit – also a habit today – not to read carefully the other authors' works or, worse, to ignore them. It was improbable that Audiffredi's book did not arrive at the prestigious *Académie de France*, since the Dominican, just because of his position as a "Prefect" of the most important library in Rome, certainly did take care to make new books known, in particular his books.

However, Audiffredi got a last nasty surprise when Pingré gave a second account about the Sun's parallax computation in 1765. This paper concerned a very unpleasant dispute between James Short (1717–1768) and Pingré. As a matter of fact, the French astronomer accused his English colleague to have tried "to overthrow the building" he had started to build, giving a parallax value of 8.″56, well lower than the mean value of 10.″60 calculated by Pingré himself (Pingré 1768). Short, starting from the first French memoir, wrote that Pingré seemed "to think there must be some mistake in the observation of Mr. Mason at the Cape of Good Hope, particularly with regard to the difference of longitude between Mr. Mason's observatory and Paris, because by comparing the observation of Mr. Mason at the Cape with the European observations, he finds the parallax of the Sun, from thence resulting, to be between 8″ and 9″, consequently differing from the determination by the observation at Rodrigues when compared with the same places" (Short 1763).

The dispute between the two astronomers was openly discussed on the pages of the two prestigious journals – the *Mémoires* and the *Philosophical Transactions* – while we can define that one with Audiffredi a hidden dispute with only one competitor. In this second memoir, Pingré mentioned, among others, the Roman observations only to demonstrate that Short's computations were wrong. In a few words, Pingré stated that the longitude determined by Mason at the Cape of Good Hope, at the time of the transit of Venus, was wrong, so his observations were useless. We note that European astronomers were greatly awaiting Mason's and Pingré's observations because they were the only ones made to the south of the equator, at the maximum distance in latitude with respect to the northern places, so with a maximum effect of Venus parallax.

In order to defend his own work, Pingré reviewed the observations used by Short (1764), criticising the values of longitudes the English astronomer had adopted for some places. On page 15 of his memoir the French abbé wrote: "The observation in Rome is made by an anonymous person, and it is impossible to evaluate its precision degree. On the contrary, I assume that this observation is imperfect, for the following reason: the longitude in Rome established by the author of the Memoir [Short], is the very same of St. Peter Church, but the Anonymous observed at the Monastery of Santa Maria sopra Minerva, which is 4^s $1/2$ or 5^s to east of St. Peter. The author makes only two comparisons with the Roman observation, and this is a very wise precaution. As a matter of fact, more comparisons could demonstrate the defect of the anonymous observation in a very clear way" (Pingré 1768). So – Pingré concluded – this observation had had to be suppressed.

This was too much for Audiffredi, even if a small footnote, added evidently at the last minute when it was impossible to change the text for printing reason, said: "In the memoir quoted in the previous footnote [Pingré, 1763a], I fixed a St. Peter longitude different from that we can find in the *Connoissance des Temps*, with which the Roman observation can be accepted; besides, the author became well known, the observation is no more anonymous, it is by a very intelligent astronomer, Father Audiffredi". Perhaps Pingré had seen the *Investigatio parallaxis solaris* Audiffredi published in 1765 to give all the details of his observations and the value he had calculated for the Sun's parallax.

We can assume that the Dominican felt highly humiliated before the members of the French Royal Academy. His revenge arrived soon after, and was much more subtle than the bibliographic ignorance and the inaccuracy of the French abbé were rough. First of all, Audiffredi wrote a letter to the perpetual secretary of the Academy, Jean-Paul Grandjean de Fouchy (1707-1788), who read it before the academicians, so "honouring him". Perhaps this was the only way to be taken into consideration by French astronomers. The correspondence between the two stimulated Audiffredi to publish, in 1766, a second Latin work about the parallax determination, *De Solis parallaxi Commentarius* (Fig. 5), dedicated to the same De Fouchy. In the preface of his book Audiffredi wrote that he had decided "not to be more involved in astronomers' disputes". He had got "the absolute peace of mind about the solar parallax", however maintaining his own opinion about his observations, independently from anyone who supported them.

3.1. *The Commentarius*

The *Commentarius* followed the same plan that the *Investigatio* of 1765, as we can see comparing the titles of the four chapters. First of all Audiffredi illustrated how he was able to determine in a very precise way the 'true' difference of longitude between Rome–Minerva and Paris–Observatory, comparing astronomical observations performed by Roman and Parisian astronomers during the Sun's eclipse of 13 June 1760, the Moon's eclipse of 28 March 1755 and the transit of Mercury over the Sun of 6 May 1753 (Ch. 1); then, he re-established the difference of longitude between Paris and Stockholm starting from the difference between Rome and Stockholm (Ch. 2); he demonstrated the trigonometric method to calculate the parallax using the planet's internal contact, and how to reduce the observation to the Earth's centre (true value), given a predefined Sun's parallax (Ch. 3); finally, he determined single values of the Sun's parallax comparing the Roman observations with those of other geographic places (Ch. 4). This last chapter is divided into two sections: the first shows the calculations concerning geographical places

Figure 5. The frontispiece of the *Commentarius* of 1766.

with well-known longitude, the second deals with the two places – the Cape of Good Hope and the Rodriguez Isle – where the English astronomer Mason and Pingré, respectively, observed the transit and at the same time determined the longitude. This chapter put an end to this peculiar dispute five years long. Audiffredi concluded that the value of the Sun's parallax derived comparing his data with Mason's at the Cape was congruent,

Table 2. The Sun's parallax calculated by the contact method; mean value 9.″26 (Audiffredi 1766).

Observation places	Time difference (calculated)	Time difference (observed)	Parallax
Cape - Stockholm	$9^{\mathrm{m}}\ 56^{\mathrm{s}}\!58$	$9^{\mathrm{m}}\ 12^{\mathrm{s}}\!00$	9.″25
Cape - Greenwich	$8^{\mathrm{m}}\ 16^{\mathrm{s}}\!77$	$7^{\mathrm{m}}\ 39^{\mathrm{s}}\!50$	9.″25
Cape - Paris	$8^{\mathrm{m}}\ 37^{\mathrm{s}}\!79$	$7^{\mathrm{m}}\ 59^{\mathrm{s}}\!00$	9.″25
Cape - Bologna	$8^{\mathrm{m}}\ 37^{\mathrm{s}}\!79$	$7^{\mathrm{m}}\ 59^{\mathrm{s}}\!00$	9.″25
Cape - Rome	$7^{\mathrm{m}}\ 27^{\mathrm{s}}\!62$	$6^{\mathrm{m}}\ 54^{\mathrm{s}}\!00$	9.″25
Stockholm - Rome	$2^{\mathrm{m}}\ 28^{\mathrm{s}}\!96$	$2^{\mathrm{m}}\ 18^{\mathrm{s}}\!00$	9.″26
Stockholm - Bologna	$2^{\mathrm{m}}\ 08^{\mathrm{s}}\!18$	$1^{\mathrm{m}}\ 58^{\mathrm{s}}\!66$	9.″26
Stockholm - Paris	$1^{\mathrm{m}}\ 39^{\mathrm{s}}\!81$	$1^{\mathrm{m}}\ 32^{\mathrm{s}}\!50$	9.″26
Stockholm - Greenwich	$1^{\mathrm{m}}\ 18^{\mathrm{s}}\!79$	$1^{\mathrm{m}}\ 13^{\mathrm{s}}\!00$	9.″26
Greenwich - Rome	$1^{\mathrm{m}}\ 10^{\mathrm{s}}\!17$	$1^{\mathrm{m}}\ 05^{\mathrm{s}}\!00$	9.″26
Greenwich - Bologna	$0^{\mathrm{m}}\ 49^{\mathrm{s}}\!39$	$0^{\mathrm{m}}\ 45^{\mathrm{s}}\!66$	9.″24
Greenwich - Paris	$0^{\mathrm{m}}\ 21^{\mathrm{s}}\!02$	$0^{\mathrm{m}}\ 19^{\mathrm{s}}\!50$	9.″28
Paris - Rome	$0^{\mathrm{m}}\ 49^{\mathrm{s}}\!15$	$0^{\mathrm{m}}\ 45^{\mathrm{s}}\!50$	9.″26
Paris - Bologna	$0^{\mathrm{m}}\ 28^{\mathrm{s}}\!37$	$0^{\mathrm{m}}\ 26^{\mathrm{s}}\!16$	9.″22
Rome - Bologna	$0^{\mathrm{m}}\ 20^{\mathrm{s}}\!78$	$0^{\mathrm{m}}\ 19^{\mathrm{s}}\!34$	9.″31

while the same comparison with Pingré's at the Rodriguez Isle gave a value of 11″, very far from the truth. He demonstrated that Pingré's data gave a mistake of 5.″87 compared with Mason's, so Pingré's observations were useless. Table 2 summarises Audiffredi's 'scientific revenge'. He presented it with the following words: "To imitate in part Pingré, I submit to the reader's eyes a not inelegant table, which shows parallaxes calculated by comparison of the observations made in places with known position" (Audiffredi 1766). Obviously, the data from Rodriguez Isle are omitted.

4. Conclusions

In this dispute Audiffredi was the winner. He was able to demonstrate his high-level scientific preparation, his astronomical works were appreciated by the most important European astronomers, and his observations were no more criticised. May we give any extenuation to Pingré? I think so. As he told in his first memoir, Rodriguez Isle, a small volcanic isle almost desert in the Indian Ocean and French colony at that time, proved to be an uncomfortable place for the transit observation. Pingré had had to install the instruments without any technical support, the hut for the instruments was insufficient to shelter them from the wind and dust and to protect them from animals and children (Pingré 1763a). The worst came when the English invaded the isle during the Seven-years war, at the end of June, soon after the transit observation. Pingré was despairing of coming back to France and presenting in time an account of his observations to the European scientific community (Pingré 1768). For this reason he compiled a first report and tried to send a copy to London by the "same who had occupied the isle" (Pingré 1768). A very short note arrived to the French Academy and was communicated on February 1762 (Pingré 1763b). On March 1762 Pingré was in Lisbon on the way back to home, just 9 months after the Venus transit. We can assume that Pingré, for the unpleasant circumstances and the hurry to communicate his results, had made some mistakes in reducing and copying his data to send to Europe. In addition, when he arrived in Paris, it was late, thus for his calculations he took into account only the observations published in the most important European journals.

Finally, it is notable that the scientific dispute concerned just a crucial problem of that time, i.e. the precise determination of longitude. The methods used by astronomers – eclipses, occultations, lunar distances – were insufficient both for the oceanic trips and for a very precise determination of the Sun's parallax. As everybody knows, the longitude problem at sea was solved by John Harrison's timekeeper in the second half of the 18th century, after the numerous attempts of obstructionism by the Astronomer Royal Nevil Maskelyne (Howse 1997; Betts 1997).

Acknowledgements

I would like to acknowledge the Biblioteca Casanatense, Rome, for the permission to publish Audiffredi's portrait and the frontispiece of the *Commentarius*, the Biblioteca Universitaria of Padova for the frontispiece and table of the *Transitus Veneris*.

References

Asclepi, G. M. 1761 *De Veneris per Solem transitu, exercitatio astronomica habita in collegio Romano Soc. Iesu a Patribus ejusdem Societatis anno 1761*, Roma, typis Io. Generosi Salomoni.

Audiffredi, G.B. 1754 *Phaenomena Caelestia Observata Romae*, Roma, Ex Typographia Generosi Salmoni.

Audiffredi, G.B. 1761 *Novelle Letterarie pubblicate in Firenze*, vol. 22, col. 484–491, 500–503, 514–520.

Audiffredi, G.B. 1762 *Transitus Veneris ante Solem observati Romae Apud PP. S. Mariae super Minervam 6 Junji 1761. Expositio historico- astronomica. Accedit descriptio aurei nummi CN. Domitii Ahenobardi 1761*, Roma, Ex Typographia Generosi Salmoni.

Audiffredi, G.B. 1765 *Investigatio parallaxis solaris ex selectis aliquot observationibus transitus Veneris ante Solem qui accidit die 6. Junii 1761 collatis cum Romana observatione habita apud PP. S. Mariae super Minervam.*, Roma, ex Typographia Hermathenaea.

Audiffredi, G.B. 1766 *De Solis parallaxi ad V. Cl. Grandjean de Fouchy Acad. Scient. Paris. a Secret. Commentarius*, Roma, ex Typographia Hermathenaea.

Baiada, E., Bónoli, F., Braccesi, A. 1995 *Museo della Specola*, Bologna, University Press.

Beccaria, G.B. 1761 *Novelle Letterarie pubblicate in Firenze*, vol. 22, col. 471–475.

Betts, J. 1997 *John Harrison*, Greenwich, The National Maritime Museum.

Carcani, N.M. 1761 *Novelle Letterarie pubblicate in Firenze*, vol. 22, col. 696–702, 714–720.

Fioravanti, R. 1994 In *Giovanni Battista Audiffredi*, (ed. Angela Adriana Cavarra), Roma, 73–93.

Howse, D. 1997 *Greenwich Time and Longitude*, London, Philip Wilson Publishers.

Pingré, A.-G. 1763a *Histoire de l'Académie Royal des Sciences, Anné 1761. Avec les Mémoires des Mathématique et de Physique, pour la même année*, Paris, Mém., 413–486.

Pingré, A.-G. 1763b *Histoire de l'Académie Royal des Sciences, Année 1761. Avec les Mémoires des Mathématique et de Physique, pour la même année*, Paris, Mém., 87.

Pingré, A.-G. 1768 *Histoire de l'Académie Royal des Sciences, Année 1761. Avec les Mémoires des Mathématique et de Physique, pour la même année*, Paris, Mém., 1-34.

Poleni, G. 1761 *Ad Gabrielem Manfredium ... Epistola in qua agitur de Veneris inter Solem et Tellurem transitu anno 1761*, Padova.

Short, J. 1763 *Phil. Trans. R. Soc.*, London, vol. 52, 611–628.

Short, J. 1764 *Phil. Trans. R. Soc.*, London, vol. 53, 300–345.

Ximenes, L. 1761 *Osservazione del passaggio di Venere sotto il Disco Solare accaduto la mattina del dí 6 Giugno 1761, all'Osservatorio di S. Giovanni Evangelista, e ridotta al tempo del Meridiano Fiorentino*, Firenze, Stamperia Imperiale.

Zanotti, E. 1762 *De Veneris ac Solis congressu. Observatio habita in astronomica specula Bononiensi Die 6 Junii 1761*, Bologna, typis Laelii a Vulpe.

Discussion

WAYNE ORCHISTON: Thank you, Luisa, for your paper. Two quick questions: firstly, does either of the telescopes that Audiffredi used still survive and, if so, are they preserved in museums?

LUISA PIGATTO: No, they all disappeared. Perhaps there are some instruments in some depositories, because Audiffredi had a high position – he was a prefect of the Library. There is a museum in Rome, the "Astronomical Copernican Museum", where there is a Divini's telescope like that used by Audiffredi, but I don't know where this instrument comes from. We have to remember that in 1870 the new Italian State took over all of the buildings, instruments, libraries, and monasteries of the papal states, and then many instruments were perhaps lost.

WAYNE ORCHISTON: My second question, which hopefully needs a shorter answer, is once Pingré became aware of the fact there wasn't a problem with the longitude of Audiffredi's observatory, did he incorporate Audiffredi's data in his final reduction to come up with the figure of 8″80 for the solar parallax, or did he still not use Audiffredi's data in his final analysis?

LUISA PIGATTO: Pingré didn't take into account Audiffredi's data at all.

JESUS DE ALBA: When you mentioned that a Dominican friar is involved: I remembered that the Copernicus' *De revolutionibus* (first edition) copy that is in Guadalajara belongs to the Dominican Monastery. It has a hand-written name, Cardinal of Capua [Nicholas Schönberg].

LUISA PIGATTO: Since Dominicans compiled the index of forbidden books, the same books were in their library; it could be that a Dominican picked some books from it.

MARY BRUCK: I would like to say thank you very much for that Luisa. I didn't know about this observatory in the Church of Santa Marie sopra Minerva. And it's just fascinating because we always associate that Church with the Galileo trial, and I think most of us didn't realise that they did astronomy there. Now, I would just like to ask you a question: Were they connected with the Collegio Romano? They were the predecessors of the present Vatican, were the Jesuits. Did these Dominicans and Jesuits share their information, or did they work together in any way?

LUISA PIGATTO: It doesn't seem they collaborated. For example, for this transit, Jesuits published their own observations, but they didn't mention Audiffredi's observations.

MARY BRUCK: It's strange.

LUISA PIGATTO: I think that the relationship with the Jesuits was friendly, but at the same time they were in competition.

MARY BRUCK: About the meridian: you would have thought that the other place would have had their own meridian, but they didn't?

LUISA PIGATTO: All the places devoted to astronomical observations had their meridian line in order to determine the true time.

MARY BRUCK: It's just all very interesting. I would also to like to ask when did that observatory cease to exist, when did they stop doing the astronomy there? Was there any successor of Audiffredi?

LUISA PIGATTO: Audiffredi's observatory at Santa Maria sopra Minerva was the Dominican's private observatory; it stopped working when Audiffredi, in his old age, stopped observing.

Luisa Pigatto, Jon Ponsonby and Suzanne Débarbat

John Butler and Mary Brück

Transits of Venus: New Views of the Solar System and Galaxy
Proceedings IAU Colloquium No. 196, 2004
D.W. Kurtz, ed.

© 2004 International Astronomical Union
doi:10.1017/S1743921305001298

Observations of planetary transits made in Ireland in the 18th Century and the development of astronomy in Ireland

C. J. Butler

Armagh Observatory, College Hill, Armagh BT61 9DG, Northern Ireland

Abstract. We review the small number of known observations of planetary transits made in Ireland in the 18th century with particular reference to the 1769 observations of Venus by Charles Mason. Though inconclusive, there is evidence to suggest that planetary transits were instrumental in the foundation of at least one of the principal observatories in Ireland. In addition, we note the close personal involvement and the contributions of Nevil Maskelyne, the prime mover of the UK 1769 Transit observations, in the design and equipment of these observatories.

1. Introduction

It is difficult now to ascertain just how much astronomical activity there was in Ireland in the first half of the 18th century. There was only one university on the island, namely Trinity College or the University of Dublin, founded in 1592 and, whilst undoubtedly some astronomy was taught there by the Professors of Natural Philosophy and Mathematics, there were no public observatories and therefore few opportunities to contribute to the subject on a professional level. The fact that no observations from Ireland of the Transit of Venus in 1761 are listed by Woolf (1959), whereas there were eleven such observations from Great Britain, suggests that astronomical activity in Ireland at this time was at a very low level. In the second half of the 18th century, the situation changed dramatically with the establishment of the Royal Irish Academy, modelled on the Royal Society (McDowell 1985), and two professional observatories, one at Dunsink near Dublin (Wayman 1987) and the other in the ecclesiastical city of Armagh (Bennett 1990). What brought about this change and what part, if any, did planetary transits play?

2. Charles Mason's observations in 1769

The first planetary transit to be observed in Ireland that we have knowledge of was the Transit of Venus in 1769 which was observed by Charles Mason on behalf of the Royal Society. In the autumn of 1768, Mason had returned to Great Britain from the North American Colonies where he had been engaged for the previous four and a half years with his colleague Jeremiah Dixon in laying out the boundary between Pennsylvania and Maryland (the Mason-Dixon Line) and in measuring a meridian arc for the Royal Society. Shortly thereafter, Nevil Maskelyne, Astronomer Royal, asked for volunteers to take part in an expedition to northern Norway to observe the Transit of Venus in June 1769 (see Woolf 1959). These observations were intended to complement those, from the South Seas, by Green and Cook. Mason did not volunteer his services on this occasion, possibly because of the rather off-hand treatment he had received from the Royal Society at the beginning of a previous engagement to observe the transit of 1761 when he had to return to port after an attack from a French frigate. Maskelyne, aware of Mason's exceptional ability as a surveyor and astronomer, was anxious not to lose his expertise in this important project and suggested instead that he observe the transit from northwest Ireland. As Maskelyne emphasised in his outline of procedures for the 1769 transit in the Nautical Almanac, observations were required from a variety of locations, both to improve the accuracy of the determination of the Astronomical Unit and to overcome the vagaries of the weather. However, there was another reason for Maskelyne's choice of northwest Ireland, namely, that here at the time of first contact, Venus would have an altitude nearly twice it would have at Greenwich. Hornsby (1765) had warned, and Maskelyne was well aware, that at an altitude of 6° at Greenwich the times of first and second contact would be difficult to determine from a site so close to sea level. At 12° altitude, Mason would have a much better chance of making definitive observations from the northwest of Ireland.

Mason agreed to go and, by late January 1769, his terms had been agreed with the Council of the Royal Society, at 100 guineas for his maintenance and gratuity and 20 guineas for travel (Maskelyne 1769a). By the beginning of March, Mason was already on his way, first by road to Holyhead and then by the appropriately named vessel 'Venus' to Dublin (Mason 1770a). He set sail on 10 March but a storm during the night and following morning prevented the ship entering Dublin Bay and they were forced to round Howth Head and land about 20 miles north of Dublin. With customs formalities to attend to as well as the ongoing journey overland, he arrived in Strabane, County Tyrone on 19 March, nearly three weeks after leaving London. He immediately set about finding a suitable spot for his observatory.

Maskelyne had given clear instructions as to the type of site he was to select together with what soon turned out to be an optimistic schedule for the project:

> *You are expected by the Council of the Royal Society to proceed with the utmost dispatch to Londonderry or to any place towards the North West coast of Ireland lying between Londonderry and Galway Bay, which you shall find to be the most proper for the purpose of making astronomical observations. You will therefore chose a place free from fogs arising from bogs or lakes or very high land but a moderate elevation will be very advantageous. You will endeavour to settle at such a place by the 8th of March next, in order to be ready to observe the eclipses of Jupiter's satellites for two months before Jupiter's opposition to the Sun, and you will continue your observations here for the space of four months.*

By the 24 March, Mason had settled on the townland of Cavan†, near Lifford in County Donegal, within sight of the city of Londonderry. The next week was taken up with transporting his equipment to Cavan and setting up his observatory (with a movable top). The instruments he had been provided with, in common with the other expeditions of the Royal Society, included a 12-inch quadrant by Bird, a 2-foot Gregorian telescope (by Short or Bird) and a clock by Shelton (Howse 1989). On the journey to Ireland, the pendulum spring of his clock by Shelton had been broken in two places. This was replaced, but the readjustment meant that he was not able to make the gravity comparison with Greenwich at that time, as originally intended.

Observations started on 3 April 1769 and, with two months to go before the Transit, he should have had adequate time for many of the important preparatory observations. He proceeded to make observations of stars at equal altitudes for setting his clock and zenith distances for latitude. Though he made observations on 13 days in the remainder of April, bad weather set in and he was able to observe on only 9 days in May. However,

---

† The townland of Cavan should not be confused with the county town of Cavan some 60 miles to the south.

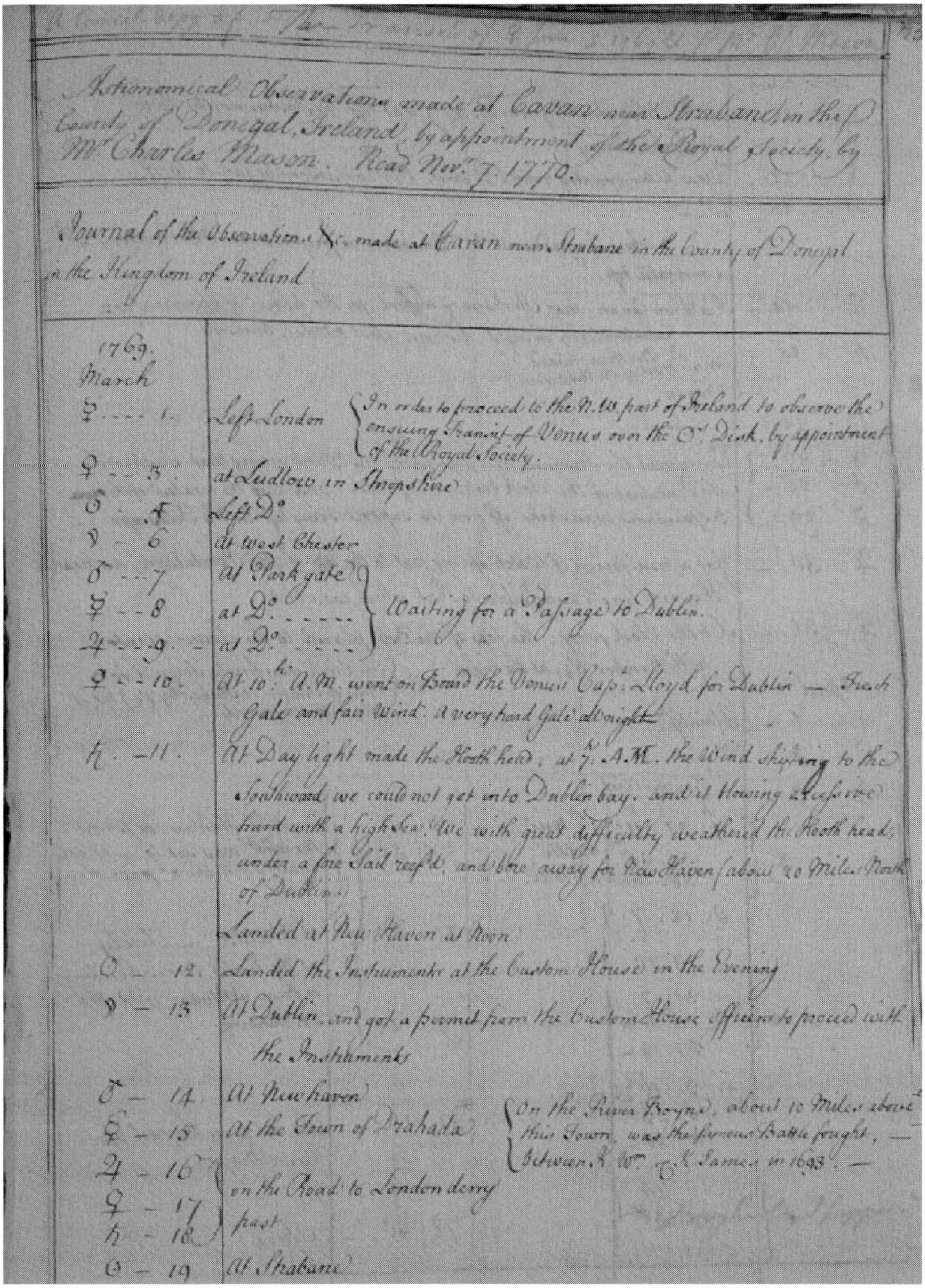

Figure 1. The first page of Charles Mason's Journal of Observations from Cavan, County Donegal. (courtesy of the Royal Astronomical Society)

on the crucial day of the Transit, it was clear enough for him to time the outer and inner contacts at ingress. In addition, like many other observers, he noted the black drop effect.

At internal contact, the air was much changed and the limb of Venus seemed to adhere to the Sun's limb by a protuberance that appeared like a dark shade: which seemed to prevent seeing the thread of light for about 40ˢ longer than expected...

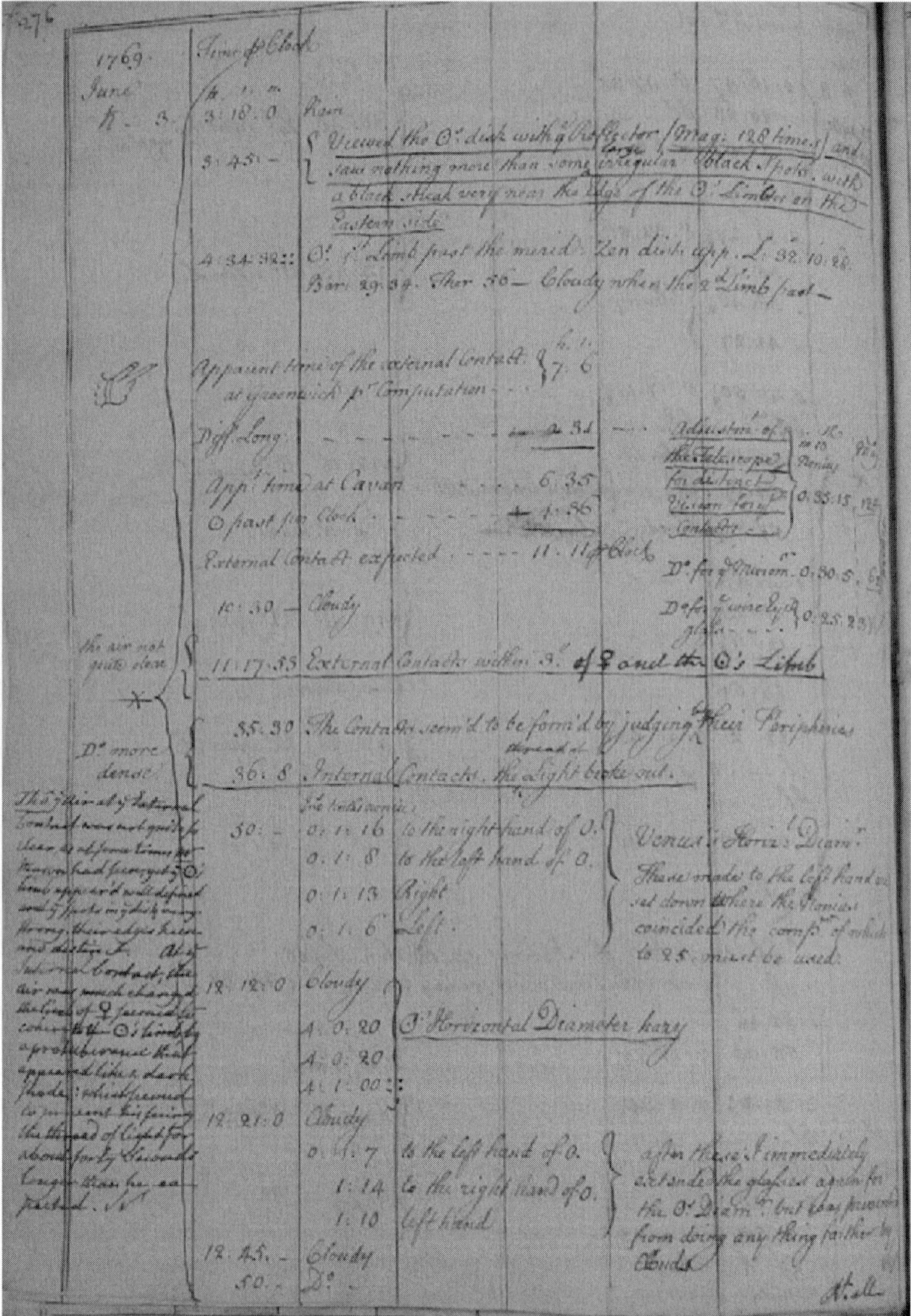

Figure 2. The page from Mason's journal showing his observations of the Transit of Venus. (courtesy of the Royal Astronomical Society)

Mason also remarked on a thin halo around Venus when seen against the Solar disk:

...there appeared a faint light shade (having a gentle fluctuating motion) round its periphery, and widest on that part farthest on the sun's disk ...very regular and well defined ...the whole shade was apparently of equal brightness...

Figure 3. A Quadrant of 12 inches radius by John Bird, made in 1768 for the Royal Society's Transit of Venus Observations the following year. (courtesy of the Royal Society)

Figure 4. An Astronomical Regulator by Shelton from the Kew Observatory, made for the Transit of Venus, 1769. (Armagh Observatory)

Though conditions were far from perfect, nonetheless, Mason had observed both contacts at ingress and when Maskelyne received word from him on 21 June, he was evidently very pleased. The only point of regret was that clouds and rain had prevented Mason from seeing the whole of the following solar eclipse, which, if he had observed it in its entirety, would have gone a long way to fixing his longitude. Though Mason had only been

Figure 5. The view from the townland of Cavan, Co Donegal, towards the southern horizon.
(photo by the author)

Figure 6. The entry in Mason's diary concerning his departure.
(courtesy of the Royal Astronomical Society)

contracted for four months, Maskelyne pleaded with him to stay on until his longitude was settled. He proposed that Mason observe fainter stars close to the Moon that could subsequently be observed at Greenwich. Maskelyne (1769b) wrote

> *You have been very diligent and taken a great deal of pains, a great part of which has been disappointed by cloudy weather, but go on and don't despair, for without more, what you have already done will go near being lost.*

Mason followed Maskelyne's suggestions and made additional observations of the occultation of stars by the Moon. But, time and again, when Mason tried to observe an occultation suggested by Maskelyne, he was clouded out, and when Mason succeeded, Maskelyne failed. Computed lunar positions, it seems, were insufficiently accurate at this time, and only observations on the same night at both Cavan and Greenwich would suffice for an accurate longitude.

At that time, Maskelyne knew nothing about how the observers sent to the North Cape and North America had fared and it would be two years before he heard details of the observations by Green and Cook in Tahiti. He was clearly delighted with Mason's observations but he emphasised that without an accurate longitude Mason's observations of ingress could not be combined with observations of egress by other observers in the northern hemisphere. The importance Maskelyne attached to Mason's observations intensified as he later learned of the failure through bad weather of Dixon and Bayley's observations from the North Cape. Meanwhile, to improve the accuracy of Mason's determination of longitude, Maskelyne had sent him from Greenwich a four-foot transit instrument. This instrument arrived at Cavan on 21 September and by 3 October Mason had it set up with meridian marks defined the following day. With an improvement in the weather, Mason made transit observations on 34 nights in October and November. In the end he had five successful observations of occultations in Cavan and thirteen observations of lunar transits for which Maskelyne had corresponding observations from Greenwich. With his longitude now defined he finally packed the instruments on 27/28 November and dispatched them to Greenwich via Dublin on 7 December.

Just how accurate Mason's surveying work was is illustrated by a table at the end of his journal containing 50 individual determinations of the latitude of his observatory at Cavan. The mean latitude from Mason's observations is $54° 51' 44.''8$ with a standard error of the mean of only $1.''8$, equivalent to 55 metres on the Earth's surface. The RMS deviation of a single determination of latitude was $13''$, equivalent to 400 m. When we consider that these deviations include errors in the tabulated stellar positions and reading error as well as errors in the division of the scale, we can appreciate not only Mason's expertise, but also the ingenuity of his friend, John Bird, who had manufactured his one-foot radius quadrant. Its probable that Mason's observations from Donegal represent one of the most precise bodies of astronomical data to have been collected in Ireland at that time.

With nearly nine months spent in Ireland on a project that he had expected to take, and was only contracted for, four months, Mason seemed to be in no hurry to depart as he didn't actually leave Cavan till 28 December. Arriving in Donaghadee on the County Down coast on 31 December he first attempted to cross the Irish Sea (with his horse) in a *small open boat*. However, bad weather as always kept him company, and about one third of the way over he was forced to return to the Irish coast where he was fortunate to regain land. The following day, he records that he made the journey in comfort in a house boat in four and a half hours. From the Scottish coast he continued his journey by road, finally reaching London at six o'clock in the evening of 13th January 1770.

Maskelyne (1770) gave an extended report to the Council of the Royal Society outlining the many observations made in Donegal by Charles Mason and their ultimate success in establishing the location of his observatory, which he wrote *is now settled with greater exactness than any other place where observations of the transit have been made.* In spite of the inclement weather and the many problems Mason had faced, he had eventually achieved all he had set out to do. It is worthwhile to ponder on why Mason had succeeded where some others had failed. The surviving letters from Maskelyne to Mason give the clue and demonstrate the importance to success of the iteration between the two correspondents. When Mason had so many difficulties determining his longitude, it was Maskelyne's encouragement, advice and help with additional observations from Greenwich that saved the day, for, in spite of the detailed instructions Maskelyne gave to each of his observers in advance, changes in the type of observations required, extra time and even additional equipment were necessary to complete the project. For Dixon and Bayley at the North Cape, no such iteration through correspondence with Maskelyne was possible, whereas with Mason, together with a bit of luck on the Transit day, it ensured eventual success.

3. Maskelyne, Hamilton and Ussher and the foundation of observatories in Ireland

Mason's observations were published in full in the Transactions of the Royal Society (Mason 1770b). Though this seems to have had little or no impact in Ireland, there is no doubt that the whole Transit of Venus exercise and Maskelyne's leading role in it, had conspicuously raised his profile, both amongst the fellowship of the Royal Society and beyond. With the publicity that surrounded Captain Cook's return and the major contributions his voyage had made to astronomy, botany, anthropology and navigation, Maskelyne was now a major player in British science. So when moves were first made in Ireland in the second half of the 18th century to promote astronomy, it was natural that Maskelyne should be consulted.

One of his earliest involvements was to host and tutor two Irish astronomers, namely Henry Ussher (1740-1790) and James Hamilton (1748-1815). Both were graduates of Trinity College, now in holy orders, and would in time become the first directors of the two public observatories at Dunsink and Armagh.

Hamilton had visited Maskelyne at Greenwich in the early 1780s where he had evidently developed his interest in the transits of the inner planets and their use in determining the dimensions of the solar system. He later wrote a comprehensive account listing values for the solar parallax from observations of the Transit of Venus made in 1761 and 1769 (Hamilton 1810). These he then combined with values for the Earth's radius from measurements of various meridian arcs to calculate the size of the astronomical unit. The value he derived for this was 95 million miles.

On 12 November 1782 Hamilton observed a Transit of Mercury from his house near Cookstown, County Tyrone. He wrote to Maskelyne with details of his work and Maskelyne promptly replied, congratulating him on his successful observations which, because of more favourable atmospheric conditions, were superior to those made at Greenwich. Maskelyne requested further details and subsequently decided to convey Hamilton's observations to the Royal Society for publication (Hamilton 1783). It was reported many years later by a clergyman in Armagh (Elliott 1882) that Richard Robinson, Archbishop of Armagh, was so impressed with Hamilton's work, or at least Maskelyne's recognition of it, that he decided to include an observatory in his plans for rebuilding the City of Armagh and to make Hamilton its first director. Hamilton, appointed in 1790 with a certificate of competence from the Astronomer Royal, lost no time in asking for Maskelyne's assistance in selecting instruments and clocks for the new observatory. Once again, Maskelyne, mindful of his colleague's needs, provided expert advice in abundance and took it upon himself to test the accuracy of several clocks at Greenwich before recommending them to Hamilton and the Archbishop (see Maskelyne 1790).

Likewise, in 1783, after Henry Ussher was appointed the first Andrews Professor of Astronomy in the University of Dublin, he went straight to London to order instruments from Ramsden and to meet with Maskelyne. Though Ussher had few scientific papers to his name and none, at that time, on astronomy, he was evidently well aware of the basic requirements of an observatory. There seems little doubt that he learned many of these from Maskelyne on his visits to Greenwich (Wayman 1987). In the event, Ussher, together with his architect, the little known Graham Moyers, created a landmark building in Dunsink Observatory. Indeed, Donnelly (1973) in her classic book on observatory architecture, considered it to be the prototype for later European and American observatories.

Figure 7. Nevil Maskelyne (1732-1811) the fifth Astronomer Royal, in 1785, by L.F.G. Van de Puyl (Courtesy of the Royal Society, photo by the author)

Though there were evident influences from Scandinavia, from Greenwich, and the Radcliffe Observatory in Oxford, it was Dunsink that clearly placed scientific practicality first and aesthetics second. Ussher wrote an excellent account of why Dunsink Observatory was built as it was, with great attention to the location of the Observatory and its observing rooms, the provision of stone pillar supports for instruments with foundations independent of the buildings in which they were housed, and the need for ventilation to reduce seeing problems close to and within the building (Ussher 1787). Though Ussher and Moyers were clearly extremely capable and practical men who together created a masterpiece, in all the above respects we see the guiding hand of the experienced astronomer, Nevil Maskelyne.

4. Astronomy in Ireland in the 19th Century and later

With two public observatories now established in Ireland, the way was open for promising young mathematicians to become engaged in astronomy. Edward Cooper of Markree, County Sligo and the third Earl of Rosse in Birr, County Offaly, became attracted to the subject and the stage was set for a rapid advancement of practical astronomy in Ireland with, in 1834, the largest refracting telescope in the world at Markree, and with the largest reflector in the world at Birr from 1845 till the end of the century. This was Irish astronomy's golden age, culminating with the discovery of spiral structure in galaxies,

Figure 8. A drawing of Henry Ussher (1740-1790) first Andrews Professor of Astronomy at Trinity College Dublin. (courtesy of Dunsink Observatory)

the compilation of the NGC Catalogue and the great advances made by the Grubb family in telescope manufacture and design.

The increase in astronomical activity in Ireland during the 19th century can be seen in the list of observatories that contributed observations of the Transit of Venus in 1882: Dunsink, Armagh, University College, Cork, Markree and Daramona. Clouds, however, were by this time already on the horizon as reforms in the land tenure system which started in the late 19th century led to the break up of the great landed estates which had financed most of the observatories. With their estates gone, the landed gentry were no longer in a position to support such an esoteric pursuit as astronomy and within a few decades Ireland had lost almost all of the impetus it had built up in the physical sciences in the previous century. It was not till the middle of the 20th century that Hermann Brück and Eric Lindsay, with help from the then Taoiseach and former mathematics teacher, Eamon De Valera, began a new renaissance in astronomy in Ireland.

Acknowledgements

Armagh Observatory is grant-aided by the Department of Culture, Arts and Leisure for Northern Ireland. The author would like to thank: Dr Peter Hingley of the Royal Astronomical Society, Dr Adam Perkins of Cambridge University Library and the staff of the Royal Society for providing access to archival material.

Figure 9. Armagh Observatory, founded in 1790 by Archbishop Richard Robinson. (photo by the author)

Figure 10. Dunsink Observatory, founded in 1785 by Trinity College, now a part of the Dublin Institute for Advanced Studies. (photo by the author)

References

Bennett, J.A. 1990 *Church, State and Astronomy in Ireland - 200 years of Armagh Observatory,* Armagh Observatory

Donnelly, M.C. 1973 *A Short History of Observatories,* University of Oregon Books

Elliott, Rvd. J. 1882. Newspaper report of discussion following a lecture by J.L.E. Dreyer, *18 December 1882 on the Transit of Venus*, Minutes Book for General Meetings of the Armagh Natural History and Philosophical Society, 1869-1892, Armagh County Museum.

Hamilton, J.A. 1783. Extract of a letter from the Rev. James Augustus Hamilton, M.A. to the Rev. Nevil Maskelyne, D.D., F.R.S. giving an account of his observations of the Transit of Mercury over the Sun, of Nov. 12, 1782, observed at Cookstown, near Dungannon, in Ireland, Phil. Trans. Royal Society, **73**, 453-455.

Hamilton, J.A. 1810 "An Essay on the Present State of Astronomical Certainty, with regard to the quantity of the Earth's magnitude, the distance of that planet from the Sun, and the absolute limit of the smallest possible interval from the Sun to any one of the fixed stars," Trans. Royal Irish Academy, **11**, 13-24.

Hornsby, T 1765 "On the Transit of Venus in 1769," Phil. Trans. Roy. Soc., **55**, 265-274.

Howse, D. 1989 "Nevil Maskelyne - the Seaman's Astronomer," Cambridge

Maskelyne, N. 1769a Letter to Charles Mason, dated 31 Jan. 1769, Archives of the Royal Greenwich Observatory, Cambridge University Library, Ref. 184.3

Maskelyne, N. 1769b Letter to Charles Mason, dated 24 June 1769, *ibid*, Ref. 184.9

Maskelyne, N. 1770 Manuscript; Council Minutes of the Royal Society, **Vol 6**, 67-70.

Maskelyne, N. 1790 Letter to J.A. Hamilton, dated 16 Feb. 1790, Armagh Observatory Archives M51.4, Butler, C.J. and Hoskin, M.A. 1987, J. Hist. Astron. **18**, 295-307.

Mason, C. 1770a *Astronomical Observations made at Cavan near Strabane in the County of Donegal, Ireland by appointment of the Royal Society*, Manuscript; Royal Astronomical Society, ADD MS 8.

Mason, C. 1770b *On the Transit of Venus, and other Astronomical Observations made at Cavan, near Strabane, in the County of Donegal, Ireland, by appointment of the Royal Society*, Phil. Trans. Royal Society, **60**, 80 & **60**, 454-496.

McDowell, R.B. 1985 *The Royal Irish Academy – a bicentennial history 1785-1985*, T. O'Raifeartaigh (ed), Royal Irish Academy, Dublin. 1-92.

Ussher, H. 1787 "Account of the Observatory belonging to Trinity College, Dublin", Trans. Royal Irish Academy, **1**, 3-21.

Wayman, P.A. 1987 *Dunsink Observatory, 1785-1985*, Royal Dublin Society.

Woolf, H. 1959 *The Transits of Venus – a study of Eighteenth-Century Science*, Princeton.

Discussion

ED BUDDING: John, what about the 13-inch Markree Refractor? Do you know what happened to it?

JOHN BUTLER: The lens still survived. The telescope was taken to Hong Kong and re-erected there by the Jesuits, but it was bombed in the air raids of the Second World War; I'm not sure who did the bombing, but anyway the telescope was destroyed. The lens survived and was taken to the Philippines, where it still is. It's in Manila, I think.

ED BUDDING: The observatory building is still there, isn't it? The base of the building is still in the grounds.

JOHN BUTLER: That photograph actually was taken about ten years ago, so that's all that is really left, just a few walls. The Castle is still there, but not the observatory.

MARY BRÜCK: I can add to that: last week they were selling the gate lodge of Markree.

JOHN BUTLER: The observatory was sold as well.

MARY BRÜCK: They are trying to get people to buy this and spend a million Pounds, or so, on account of it's astronomical associations.

SUZANNE DEBARBAT: Do you know how many small quadrants made by Bird are actually still surviving? Because you have shown one, so there could be very many.

JOHN BUTLER: The only one I know of is the one at the Royal Society, which is the one I showed a picture of. That's on display at the Royal Society at the moment actually.

WAYNE ORCHISTON: Given all the problems that Mason had on trying to establish his longitude, why didn't he observe Jovian satellites on the odd clear nights? They are frequent; they are listed in the nautical almanac – instead of waiting for lunar occultations.

JOHN BUTLER: He did observe some, but I think that Jupiter was in opposition, or going into opposition shortly after June, but he got one observation, I think.

MARY BRÜCK: Thank you, John. It's heartening to find that the observations of the Transit of Venus, even if they brought no astronomical results, did result in observatories being built in Ireland. Now there is such a great interest in this country [UK] in the Transit of Venus yesterday. Let's hope it has the same result this time.

John Butler and Simon Mitton

Transits of Venus: New Views of the Solar System and Galaxy
Proceedings IAU Colloquium No. 196, 2004
D.W. Kurtz, ed.

© 2004 International Astronomical Union
doi:10.1017/S1743921305001304

The American transit of Venus expeditions of 1874 and 1882

Steven J. Dick

NASA HQ, Code IQ, 300 E. St. SW, Washington, D.C., USA

Abstract. When in 1874 and 1882 Venus passed in front of the face of the Sun, most countries with a scientific reputation to keep or to gain made plans to observe the great event. The United States was no exception. The purpose was primarily to measure the solar parallax, and thereby determine the astronomical unit, the distance between the Earth and the Sun. With a $177 000 Congressional appropriation for the 1874 event, and $78 000 for 1882, the Americans sent out eight well-equipped expeditions for each transit. Under the U.S. Transit of Venus Commission, the responsibility fell to the U. S. Naval Observatory (Dick 2003). Relying heavily on photographic methods, the Americans returned 350 plates in 1874, and 1380 measurable plates in 1882. Simon Newcomb grew skeptical of the results, but in 1894 William Harkness produced a final value of the solar parallax, after adjustments with other constants, of $8.''809$, with a probable error of $0.''0059$, yielding an Earth-Sun distance of 92 797 000 miles, with a probable error of 59 700 miles. This was a significant improvement over previous estimates. How important were the transit of Venus observations? In the end it was Newcomb who had the final say, for it was his system of astronomical constants that was adopted internationally at a Paris conference in 1896. Ironically, just at this time other methods were proving more accurate than Venus transits. In determining a final value for the solar parallax from all methods, Newcomb gave all photographic observations of the 1874 and 1882 transit a weight of 2, compared to a weight of 40 for Pulkovo Observatory's determination of solar parallax from the constant of aberration. Thus the Venus transit observations played little role in the official value used for the astronomical unit in the 20th century.

1. Introduction

As the 19th century transits of Venus approached, the phenomenon had been observed only three times in recorded history: 1639, 1761 and 1769. By the 19th century the main interest in observing the transits of Venus, aside from their extreme rarity, was in determining the solar parallax, and thereby the scale of the solar system. At stake was reducing an uncertainty in the Earth-Sun distance by several million miles. Every difference of one hundredth of an arcsecond in the solar parallax, in other words, from $8.''79$ to $8.''80$, translated into approximately 100 000 miles. The methods had been worked out in the 18th century by Edmond Halley, Joseph-Nicolas D'Isle and others, and required observations from widely varying geographic locations. This inspired expeditions to remote locations around the world, including 8 American expeditions in 1874 and 8 more in 1882.

Aside from their disputed scientific results, these transit of Venus expeditions are of historical interest for the international disagreements over techniques and instruments, as an early example of international cooperation and rivalry in astronomy, and for their place in two broader historical trends: the determination of the fundamental astronomical constants, and the great scientific voyages of the nineteenth century. A great deal was at stake, not only for science but also for national interests. This is why in 1874 alone the

British would have 12 expeditions, the Russians 26, France and Germany 6 each, Italy 3 and Holland one. As the 19th century historian Agnes Clerke put it, when in 1874 Venus passed in front of the face of the Sun, "every country which had a reputation to keep or to gain for scientific zeal was forward to cooperate in the great cosmopolitan enterprise of the transit" (Clerke 1902). The United States was no exception; it had a growing reputation in science and was anxious to accelerate that growth. The 19th century transit of Venus pair offered a unique opportunity for the country to showcase its rising scientific talent – one that would not come again until the distant year in the 21st century: 2004.

2. Organizing in the United States

In the United States, Naval Observatory astronomer Simon Newcomb began the discussion of the transits of Venus with a paper in 1870, followed by a resolution before the National Academy of Sciences in April of that year (Newcomb 1870). At the recommendation of the National Academy, in 1871 Congress approved $2000 for "preparing instruments." The same bill also provided for Transit of Venus Commission, to be composed of the Superintendent and two professors of mathematics of the Navy attached to the Naval Observatory, the President of the National Academy of Sciences, and the Superintendent of the Coast Survey. In creating a government commission, the United States followed the lead of Germany (1869) and France (1870), while in Britain the responsibilities were shared by the Royal Observatory at Greenwich and the Royal Astronomical Society. The U. S. Transit of Venus Commission as originally constituted included Benjamin F. Sands, Benjamin Pierce, Joseph Henry (President of the National Academy at the time), Newcomb, and Naval Observatory astronomer William Harkness. Only Newcomb and Harkness would survive the full term of the Commission, which held its last meeting and effectively disbanded in 1891, after more than two decades. And of those two only Harkness believed to the end that the method produced a valuable result.

3. Instruments and methods

Among the first and most crucial decisions to be made by the American Commission, and by other countries, were the method of observing and the instruments required. The measurement of the relative position of the center of the planet and the center of the Sun was the method of choice, but there was nothing to lose in attempting to measure the exact moment when the planet came into contact with the limb of the Sun. Either method might be attempted visually or photographically, and with photography still in its infancy, especially as applied to astronomy, that decision between visual and photographic methods was by no means a foregone conclusion. But at the urging of Newcomb, from early on the U. S. Commission was drawn to the photographic method, although it hedged its bet by also making visual observations with small refractors.

The photographic method proposed by Newcomb was unique. Unlike most of the European participants, who also opted for the photographic method but devoted their attention to securing the best photographs, the American method proposed by Newcomb concentrated also on the problem of measuring the photograph. Because the measurements on the photograph were made in inches and fractions, Newcomb reasoned, and because the quantity to be determined was in minutes and seconds of arc, a precise knowledge of the scale factor was necessary in order to convert from linear to angular measurement. This conversion, according to Newcomb, was "the greatest difficulty which

Figure 1. The fixed horizontal telescope known as a photoheliograph, in which a weight-driven heliostat directs the Sun's rays through a lens, which focuses the image onto a photographic plate 38.5 feet away. The method was used by American and French observers. From Simon Newcomb, *Popular Astronomy* (Haughton and Mifflin: New York, 1878), p. 186

the photographic method offered." To meet it, Newcomb proposed an instrument conceived by Joseph Winlock and already in operation at Harvard College Observatory. This was a fixed horizontal telescope of nearly 40-foot focal length, through which sunlight was directed by a heliostat, a slowly turning mirror that kept the Sun's image stationary with respect to the telescope (Fig. 1). The lens and heliostat mirror – a piece of finely polished but unsilvered glass that reflected about 1/20th of the sunlight into the lens – were mounted on a four foot-high iron pier embedded in concrete. The lens, designed to give the best photographic image, formed that image four inches in diameter about 38.5 feet away on the photographic plate. The plate itself was held vertically on another iron stand next to a grid that was overlaid on each photograph for purposes of measurement. A special device consisting of 5-foot lengths of pipe was used to measure the distance from the lens to the plate within a hundredth of an inch – the crucial measurement on which the scale factor depended. To complete the setup, a transit instrument was to be used to align the system North-South, as well as for other purposes. This photographic method was to be used both for photographs of contact and of Venus as it moved across the face of Sun, with the latter believed to hold the most promise (Newcomb 1872).

At the request of Naval Observatory Superintendent Benjamin Sands, Congress appropriated $50 000 in June 1872 for the purchase and preparation of instruments, and another $100 000 in 1873 for the actual expeditions. The following year another $25 000 was appropriated to complete work and return parties home, for a grand total of $177 000 for the 1874 event alone, not including salaries and the use of Navy facilities and ships.

Figure 2. A 5-inch Alvan Clark refractor, one of eight manufactured for the American transit-of-Venus expeditions. Pictured here with Alfred Mikesell at the U. S. Naval Observatory in 1965.

This was a munificent sum indeed, one from which the country might well expect a decent scientific return.

With the first \$50 000 in hand, the Commission held its first meeting on July 22, 1872, and went to work in earnest on the matters of instruments, choice of stations and organization of the expeditions and their personnel. The Naval Observatory, with the majority of the members of the Transit of Venus Commission including its President (the Superintendent) and Secretary (Newcomb), was authorized to take charge of the details of the expeditions. Harkness drew up the specifications for most of the instruments, eight sets of which had to be manufactured, since the Commission had decided that the appropriation was enough to equip eight American parties. The instruments were constructed by a variety of makers. But for the most crucial of these instruments the Commission turned to Alvan Clark & Sons, then constructing the Observatory's Great 26-inch Refractor, used by Asaph Hall to discover the 2 moons of Mars in 1877. The firm not only made the 5-inch refractors for the visual observations (Fig. 2), but also the five-inch 40-foot photoheliograph lenses and the heliostat mirrors crucial for the photographic method, as well as the chronographs for precise registration of time. The polishing of the 7-inch mirrors, Newcomb recalled, was the most difficult part of the whole apparatus. The accuracy needed to be such that "if a straight edge laid upon the glass should touch at the edges, but be the hundred-thousandth of an inch above it at the centre, the reflector would be useless."

While the visual 5-inch refractors and the 40-foot photoheliographs were the most important equipment, a broken-tube transit instrument and sidereal clock were also crucial, as well as sidereal and mean time chronometers, and a chronograph for registering time. The transit instrument, designed by Harkness and built by Stackpole, was used not only for determining latitude, longitude and time, but also to insure that the central vertical line of the photographic plate holder could be very near the meridian. The clock used with the transit instrument, built by the Howard Clock Company of Boston, was designed for the rugged fieldwork and therefore not particularly elegant.

4. Stations and personnel

Once the methods were decided it was necessary to choose the stations. Although the entire transit would last about 4 hours, a very long time compared to the few minutes of totality for a solar eclipse, optimal weather was a prime consideration. To observe the parallax effect, both Northern and Southern Hemisphere stations were required, and after studying weather records it was decided to have three northern and five southern stations. In order to choose the stations, Newcomb began heavy correspondence with U. S. consulates and astronomers around the world. At the suggestion of Struve, Vladivostok was chosen as one northern station, with Nagasaki and Peking the other two. The southern stations chosen were Crozet Islands, Kerguelen Island, Hobart Town (Tasmania), Bluff Harbor (New Zealand), and the Chatham islands off New Zealand. In the end the Crozet Islands site was abandoned when the ship could not land due to severe weather, and two American stations would be located in Tasmania.

The personnel of each station consisted of one chief of party, one astronomer, one chief photographer and two assistant photographers, with a few parties having an additional astronomer and in one case, an instrument maker. The members of the parties were chosen with the greatest care, especially the chiefs and the all-important chief photographer. Two of the chiefs-of-party were from the Naval Observatory, two from the Coast Survey, one from the Army, one from the Navy and two from outside the government. The chief photographers were all professionals in photography, but the assistants were for the most part "young gentlemen of education, recent graduates of different colleges, who had been practiced in chemical and photographic manipulation."

5. Practice

For practicing visual observations, beginning in May 1873 an artificial Sun and Venus apparatus was mounted on a building near the War Department, about two thirds of a mile from the Observatory. Using the 9.6-inch refractor, a 5-inch telescope and a 4-inch comet seeker from the dome atop the Observatory, Newcomb, Harkness and Hall repeatedly observed the small black dot representing Venus impinging on the artificial Sun, which was a white circular disk. In the Spring of 1874 many of the participants gathered on the grounds of the Observatory to practice, with goal of improving the accuracy with which contacts could be observed (Fig. 3). At the same time, the photographic apparatus was set up and the photographic process rehearsed, with Henry Draper offering his services to the Commission for several weeks.

6. Results

A great deal could be said about the colorful details of the expeditions, the travel problems, the weather problems, the triumphs and heartbreak. But in terms of science, the most important detail is the result. Now began the saga of analyzing the observations. A total of some 350 plates were returned from the 1874 American expeditions. In October,

Figure 3. Spring, 1874 practice for transit of Venus on the grounds of the Naval Observatory. Standing at left nearest the photoheliograph is Admiral C. H. Davis, founder of the American Nautical Almanac Office and President of the Transit of Venus Commission at this time. Standing in front of him are Henry Draper and C. H. F. Peters (with hat); seated is Simon Newcomb. Asaph Hall is the tall man in front of the ladder, with hat; at far right (with stovepipe hat) is the Observatory's instrument maker, William F. Gardner. From a set of stereo views of Washington published by J. F. Jarvis, Washington, D.C.

1875 Newcomb still expressed optimism based on what he knew at that point about the observations. The optical observations of contacts made by the observers of all nations would, he believed, "by their combination give a value of the solar parallax of which the probable error will lie between $0.02''$ and $0.03''$." (Recall that means 200 000 to 300 000 miles). The American photographs alone, he further felt, "will give a result at least as accurate as this, and probably more so." However, Newcomb cautioned, "it is not to be disguised that there is a possibility of unforeseen perturbing causes being brought to light by a comparison of all the observations which will upset all our a priori estimates of probable error." (U. S. Naval Observatory 1875)

Newcomb's cautionary statement proved prophetic. Reporting on the results of the 1874 expeditions eight years later, Harkness recalled that after the parties returned, attention was first turned to the visual contact observations as the easiest to analyze, but "it was soon found that they were little better than those of the eighteenth century." Around the world the result was the same: The problem was that "the black drop, and the atmospheres of Venus and the Earth, had again produced a series of complicated phenomena, extending over many seconds of time, from among which it was extremely difficult to pick out the true contact. It was uncertain whether or not different observers had really recorded the same phase, and in every case that question had to be decided before the observations could be used. Thus it came about that within certain rather wide limits the resulting parallax was unavoidably dependent upon the judgment of the computer, and to that extent was mere guesswork." (Harkness 1883).

The photographic observations were thus all the more important, but here again disappointment was widespread. Harkness recalled that "it soon began to be whispered about that those taken by European astronomers were a failure." The official British report declared that "after laborious measures and calculations it was thought best to abstain from publishing the results of the photographic measures as comparable with those deduced from telescopic view." The problem was the Sun itself: "however well the sun's limb on the photograph appeared to the naked eye to be defined, yet on applying to it a microscope it became indistinct and untraceable, and when the sharp wire of the micrometer was placed on it, it entirely disappeared."

All hope focused on the American expeditions, which had returned with about 200 measurable plates in 1874 taken with the long-focus photoheliographs. In June 1875 the Commission charged Harkness with measuring these plates, and he devised a machine especially for this purpose. Harkness reported that 221 photographs yielded "excellent results" for the period between second and third contact when the planet was on the face of the Sun. However, those taken between first and second, and again between third and fourth contacts "proved of no value" because of the infamous black drop problem. In other words, while even long-focus photographic contact observations were no better than visual ones, there was reason for hope in obtaining results from the photographs of Venus fully upon the face of the Sun.

In the end, however, no result of the 1874 American transit of Venus expeditions was ever officially published, although in 1881 D. P. Todd (then an assistant in the Nautical Almanac Office) did publish a brief three-page article in which he determined a provisional value of $8.''883 \pm 0.''034$ (Todd 1881). By the eve of the 1882 transit the official American results remained uncertain, and it was a heated question whether parties should even be dispatched for the 1882 transit. There were eight more expeditions sent around the world.

To make a long story short, relying heavily on photographic methods, the Americans returned 350 plates in 1874, and 1380 measurable plates in 1882. And contrary to the common opinion today, a result was produced. Simon Newcomb grew skeptical of the results, but William Harkness (Fig. 4) produced a final value, after adjustments with other constants, of $8.''809$, with a probable error of $0.''0059$, yielding an Earth-Sun distance 0f 92 797 000 miles, with a probable error of 59 700 miles (Harkness 1891, 1894).

The publication of the official reports of the American Transit of Venus Commission, however, was another matter. Recalling the impressive tomes published by other nations, Harkness argued that the full publication of the American results was essential. His hope in this respect, however, was not realized. Aside from the *Papers Relating to the Transit of Venus*, and the *Instructions for Observing* it in 1874 and 1882, only Part I of a projected four parts of Observations was published. Part II, two volumes of reports of the eight 1874 parties and consisting of some description and much data, but no results, reached the page proof stage, and today exist as only a single copy of 564 pages in the Naval Observatory Library. Parts III and IV, which were supposed to be the results, were never published.

The reasons for this failure to publish again were bureaucratic. With the passing of the 1882 expeditions, the Transit of Venus Commission decided it would be best to combine the results of both transits into one report. By 1891, however, despite a recognition of the need "that the United States keep pace with other governments in publishing the results of its observations of these important transits," no further action was taken. The failure to publish an official American report, however, must be distinguished from the

Figure 4. William Harkness, who led the American efforts for the 1882 transit of Venus, and almost single-handedly achieved the final American result.

American result, which was not only obtained but published and discussed in the context of the other astronomical constants by Harkness in the *Washington Observations for 1885* (Harkness 1891).

7. Significance

How important were the transit of Venus observations? In answering this question we need to recall that the transits of Venus were only one method for determining the solar parallax. Ironically, just as the transit of Venus observations were producing an improved result, other methods became practical that gave even better results. This is true of asteroid parallaxes, but especially of the method involving the aberration of light. From the measurement of the aberration of light one can produce the light time; combined with newly accurate measurements of the speed of light, in which both A. A. Michelson and Newcomb were involved, an accurate distance can be determined. In the end it was Newcomb who had the final say in which methods were applied to the final solution for the astronomical unit, for it was his system of astronomical constants that was adopted internationally at a Paris conference in 1896. In determining a final value for the solar parallax from all methods, Newcomb gave all photographic observations of the 1874 and 1882 transit a weight of 2, compared to a weight of 40 for Pulkovo Observatory's determination of solar parallax from the constant of aberration (Table 1). Considering the probable errors, Newcomb's system and Harkness's system actually overlapped in

Table 1. Results of Solar Parallax Determinations, with Newcomb's Weights

	Solar Parallax arcseconds	Weight
From the mass of the Earth resulting from the secular variations of the orbits of the four inner planets	8.759 ± 0.010	9
From Gilly's observations of Mars at Ascension Island	8.780 ± 0.020	2
From Pulkovo determinations of the constant of aberration	8.793 ± 0.0046	40
From observations of contacts during the transits of Venus	8.794 ± 0.018	3
From the parallactic inequality of the Moon	8.794 ± 0.007	18
From determinations of the constant of aberration made elsewhere than at Pulkovo	8.806 ± 0.0056	28
From heliometer observations on the minor planets	8.807 ± 0.007	20
From the lunar equation in the motion of the Earth	8.825 ± 0.030	1
From measurements of the distance of Venus from the Sun's center during transits	8.857 ± 0.023	2

From Simon Newcomb, *The Elements of the Four Inner Planets and the Fundamental Constants of Astronomy* (Washington, 1895), 157. Solar parallax values are arranged in order of magnitude.

their values for solar parallax, and Newcomb came closest to overlapping the modern value of $8.''794146$ (Fig. 5).

By act of Congress dated July 26, 1886, the instruments and records of the Transit of Venus Commission were turned over to the Secretary of the Navy. In that year the instruments, valued at some \$30 000, were made the property of the Naval Observatory, and put into the hands of the Observatory's instrument maker. On April 25, 1891 the Commission held its last meeting, where Harkness reported on the status of the reductions. Along with the effective demise of the Commission, an interesting episode in astronomy passed into history. Today some of the instruments and records remain at the Naval Observatory, and memorial plaques still mark several of the sites of the observations.

The 19th century transits of Venus thus took their place in the long history of attempts to determine the astronomical unit, one of the fundamental constants of astronomy. Without vastly superior methods, Harkness cautioned in his 1894 paper, the value of the solar parallax was not likely to be improved from where it stood after the 19th century transits of Venus. Harkness did live to see the close passage of the minor planet Eros to Earth in 1900–1901, a method that gave a considerable improvement then, and again in 1930–1931.

Just before the 1882 observations, Harkness wrote the following poetic lines:

We are now on the eve of the second transit of a pair, after which there will be no other till the twenty-first century of our era has dawned upon the earth, and the June flowers are blooming in 2004. When the last transit season occurred the intellectual world was awakening from the slumber of ages, and that wondrous scientific activity which has led to our present advanced knowledge was just beginning. What will be the state of science when the next transit season arrives God only knows. Not even our children's children will live to take part in the astronomy of that day. As for ourselves, we have to do with the present ...

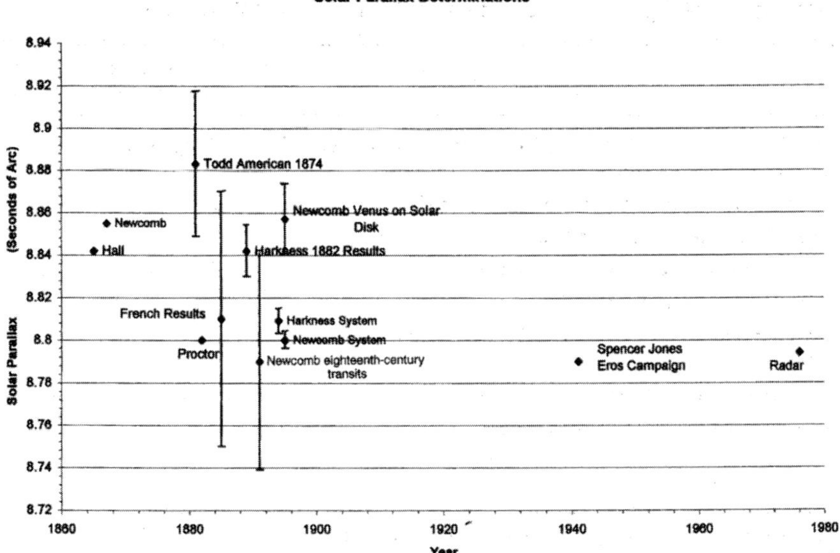

Figure 5. Selected solar parallax determinations, 1860–1976. Aside from the results of Hall and Newcomb in 1865 and 1867 from observations of the meridian parallax of Mars, all 19th century observations shown here were derived from the transits of Venus. Two 20th century results are shown for comparison; their error bars are so small they cannot be seen at this scale.

Harkness would surely marvel that radar methods have now determined the solar parallax to six significant figures, and that the mean distance to the Sun is known within a few meters. And he would surely marvel that when the flowers were blooming in June 2004, we would be observing transits of planets around other stars, with plans afoot to launch spacecraft just for that purpose. What the state of science will be with the next transit pair in 2117 and 2125, we cannot predict, any more than Harkness could have predicted orbiting spacecraft, nanotechnology, or electronic computers.

References

Clerke, A. 1902 *History of Astronomy during the Nineteenth Century*, chapter 6, A. and C. Black.

Dick, S.J. 2003 *Sky and Ocean Joined: The U. S. Naval Observatory, 1830-2000*, 238-273, Cambridge.

Harkness, W. 1883 Address by William Harkness, *Proceedings of the AAAS 31st Meeting, August 1882*, 77.

Harkness, W. 1891. The Solar Parallax and its Relation Constants in *Washington Observations for 1885*, Appendix III.

Harkness, W. 1894. On the Magnitude of the Solar System, *Astronomy and Astro-Physics*, **13**, 605-626.

Newcomb, S. 1870 On the mode of Observing the coming transits of Venus, *American Journal of Science*. **50**, 74-83.

Newcomb, S. 1872 On the application of photography to the observation of the transits of Venus In *Papers Relating to the Transit of Venus in 1874*, Part I, pp. 14-25. U. S. Transit of Venus Commission.

Todd, D. 1881. The Solar Parallax as derived from the American photographs of the transit of Venus, 1874, *American Journal of Science*, 21, 491-493.

U. S. Naval Observatory 1875 *Annual Report*, 80-81

Discussion

WAYNE ORCHISTON: Would you like to make any comment, Steve, about any surviving relics from the 1874 transit stations, in the light of the paper that's just been published?

STEVE DICK: Yes, there happens to be a nice paper in the *Journal of Astronomical History and Heritage* about Campbell Town. There are relics, but I think probably the best relics that still exist are at the Campbell Town, Tasmania, site. I don't recall what all is still there. There are foundations. Is there a pier there? I don't know any other site from the 19th-century expeditions of any country that still has that number of artefacts – unless somebody can correct me on that.

MARY BRÜCK: That wonderful table you showed at the end I think is most interesting. The astronomical one that has the most weight is the heliometer observations of minor planets. Now, the very first one made was done during the Venus transit in Mauritius by Gill, in 1874. When they went on that expedition Lord Lindsay had every possible type of equipment, but Gill knew that this new method was the thing, and he wasn't really worried about the Transit of Venus. He observed Juno, got a result, and published that result, and then, I think, he observed Mars two years later, and then they began wholesale observations of minor planets which turned out to be the better way. So I'm very glad to see that it got weight 20 as against weight 2 for the normal Transit of Venus observations.

Steve Dick and Mikhail Marov

Transits of Venus: New Views of the Solar System and Galaxy
Proceedings IAU Colloquium No. 196, 2004
D.W. Kurtz, ed.

© 2004 International Astronomical Union
doi:10.1017/S1743921305001316

The Mexican expedition to observe the 8 December 1874 transit of Venus in Japan

Christine Allen

Instituto de Astronomía, UNAM, México
email: chris@astroscu.unam.mx

Abstract. The voyage of the Mexican commission to observe the transit of Venus on 8 December 1874 in Japan is briefly recounted. The five-man expedition was led by Francisco Díaz Covarrubias. They succeeded in establishing two observing stations near Yokohama, one in Nogue-no-Yama and one on a hill called "The Bluff", and also in determining precise geographical positions for them. Clear skies allowed the observation of the transit at both stations. The results were presented in Paris in 1875, and published on the same year. They were meant as a contribution to be processed along with all other data obtained by different missions. The importance of the expedition for the development of early modern science in Mexico – particularly astronomy – is examined in the broad context of the social and political conditions then prevailing in the country. The relevance of the mission for the establishment of scientific, cultural and even commercial ties between Japan and Mexico is emphasized.

1. Introduction

The tradition of observing transits of Venus in Mexico seems to go back a long way. Recent research, for instance, has indicated that the Maya were able to observe transits with the naked eye and that they left some records of these observations in Mayapán, one of the last great Maya centers (Ruiz Gallut et al. 2001; Galindo-Trejo & Allen, these proceedings). In the so-called Fresco Hall, situated in front of the great pyramid in Mayapán, several frescoes are found, reasonably well preserved. The frescoes depict two personages flanking a Sun-circle. Most unusually, the Sun-circles contain within them reclining figures, as if they had been swallowed by the Sun. The figures may be associated with Kukulkan, one of the most important deities in the Maya culture and closely linked to Venus. Radio-carbon dating indicates that the paintings were done between 1200 and 1350 AD. During that interval of time four transits of Venus occurred. The transit of June 1, 1275 was visible at dusk from Mayapán, and it has been proposed as the one actually depicted in the frescoes (Ruiz Gallut et al. 2001; Galindo-Trejo & Allen, these proceedings). Confirming evidence for this identification can be found in the fact that the orientation of the building is such that the Sun-circles are illuminated grazingly by the Sun on dates centered on the summer solstice and separated by 73 days, one eighth of the synodic period of Venus.

Another historic observation occurred at the time of the 1769 transit (Cassini 1772). The French Academy of Sciences, seeking the best observing conditions, sent an expedition to San José del Cabo, Baja California. The expedition was headed by the Abbé Chappe d'Auteroche. On arrival at San José together with Spanish astronomers Vicente Doz and Salvador Medina, they contacted the Creole astronomers Joaquín Velázquez de León and José Antonio de Alzate, who also intended to observe the transit and had already established an observing station nearby. The weather was fine, and the transit was successfully recorded by all (Moreno 1986b).

But, most tragically, Chappe, Medina and a French technician died soon afterwards, victims of the yellow fever, an epidemic that caused more than 75% of the Baja California population to perish. However, the surviving astronomers returned safely with the precious data they had obtained during the transit, which were processed along with those of other expeditions (Díaz Covarrubias 1882). The observations of Velázquez et al. were published side by side with those of Chappe by the Académie Royale des Sciences de Paris. In this way, not only were the results of the expedition salvaged, but also the interest in astronomy in the New Spain, dormant for many years, was re-awakened. Velázquez, for instance, was able to conduct various geodetic studies, Alzate observed and studied the satellites of Jupiter, recorded several lunar occultations, and computed cometary orbits, notably that of the 1788 comet. The stage was set for the transit of the year 1874.

2. The 1874 transit

Curiously, interest in the upcoming transit started with a public lecture by Francisco Díaz Covarrubias at the meeting of the Humboldt Society, Mexico City in April 1874 (Díaz Covarrubias 1882). He explained in great detail the importance of the transit for the accurate determination of the distance to the Sun, and expressed the view that on the occasion of the next transit in early December, as many countries as possible should join forces to better determine this quantity. Those attending the lecture were clearly impressed. However, the situation in Mexico was by no means favorable to scientific endeavors. In fact, the young Republic had lost the war with the US some years previously, and also suffered the French intervention. The economic situation was chaotic, the country divided, civil strife rife...

The reforms of Juarez, continued by Sebastián Lerdo de Tejada, had met with considerable opposition from the church and the big land and mine owners. In the middle of this dramatic situation, several senators and cabinet members somehow managed to convince President Lerdo de Tejada that Mexico should begin to play a role in the international scientific community, and that the coming transit of Venus offered a great opportunity to put Mexico, so to say, on the international scientific map. After being assured of its feasibility, President de Tejada approved the project in early September, asked Díaz Covarrubias to form a commission and head the project, and provided the necessary funding.

In great haste, because time was short and the trip to the places that offered the best possibilities for successfully observing the transit was long and complicated, Díaz Covarrubias formed the Commission, selected and commandeered the necessary equipment and set forth to the Far East, where weather conditions were likely to be best.

The Mexican Commission (see Fig. 1) was composed of the following members:

- Francisco Díaz Covarrubias, Chief astronomer
- Francisco Jiménez, astronomer
- Manuel Fernández Leal, surveyor and analyst
- Agustín Barroso, photographer
- Francisco Bulnes, analyst and chronicler

Díaz Covarrubias found that several government agencies already had equipment suitable for the transit observations. So, from the Ministry of Mines and Development he obtained a zenith telescope, a barometer and a theodolite, from the National School of Engineering a second zenith telescope, and a precise chronometer, and from the Military Academy a refracting telescope and a second chronometer. To complete this equipment,

Figure 1. (left) Francisco Díaz Covarrubias. (right) The Mexican Commission. Back row: F. Jiménez, F. Díaz Covarrubias, F. Bulnes. Front row: A. Barroso, M. Fernández Leal.

Díaz Covarrubias took from his private observatory two altazimuth telescopes as well as various small instruments. With these instruments he planned on setting up two observing stations, in order to maximize the probability of fair weather at at least one of them.

3. The voyage

All the preparations for the expedition, as well as details of the voyage are described in great detail by Díaz Covarrubias in his book "Viaje de la Comisión Mexicana al Japón, 1874" (Díaz Covarrubias 1876). In this book he gives a detailed account not only of his observations, but also of the trip itself, of the land and people of Japan, of the history, economy and social customs of the Japanese, and even of the religious and sexual mores. It is written in a popular style, strongly reminiscent of Jules Verne.

For the trip to China (where he originally intended to go), he estimated that 55 days would be needed. The trip was by necessity quite indirect, since at that time of the year (just after the rainy season) the roads joining Mexico City with the Pacific coast were impassable, especially for an expedition carrying a considerable load of delicate instruments. The itinerary that Díaz Covarrubias planned is displayed in Table 1.

Table 1. Proposed itinerary of the voyage of the Mexican Commission

Mexico-New York (via Cuba and Philadelphia)	12 days
New York-San Francisco	8
San Francisco-Yokohama	25
Yokohama-Beijing	10
Total estimated	55 days

Preparations were completed in a short time, and on September 18 the Commission left Mexico City on their way to Veracruz. From Veracruz they sailed to Cuba, then north and upstream to Philadelphia, and onwards to New York. The trip from New York to San Francisco was by rail, on the newly completed Western Line. In San Francisco they stayed a few days, in order to establish some contacts with consular officers and to acquire photographic materials. For the long voyage to the East they boarded the steamer Vasco da Gama. The trip was not easy. They had stormy weather all along. Throughout the voyage, Díaz Covarrubias tried to get information on the places that offered the best weather conditions, and decided finally to set up his observing stations in Japan.

4. Arrival and setup

Díaz Covarrubias was already well known in Japan due to several techniques for determining geographical coordinates that had been developed and published by him, and translated into Japanese. As a matter of fact, on his departure, the Minister of Education Fugimaru Tanaka presented him with a copy of his booklet *"Nouvelle méthode pour determiner la latitude d'une station au moyen d'observations azimutales"* translated into Japanese and just published. The Commission arrived in Yokohama on November 9th. On their arrival, the Commission was received by Terashima Munemory, the Japanese Minister for Foreign Relations, who assured Díaz Covarrubias his full support for the project. Indeed, he offered to pay for the rent of the properties to be selected by the Commission for the erection of the observing stations, an offer that was courteously refused. On learning about the desirability of easy communication with other stations, he ordered a telegraph line to be constructed directly linking the stations with the telegraph office in the city of Yokohama. Also, all communications were to be made free of charges.

As it turned out, there was just time to obtain the necessary permissions, to determine the best locations and to start construction of two observing stations at "Nogue-No-Yama" and "The Bluff" (see Fig. 2):

Both stations were built by a Chinese worker, Mow Cheong, according to specifications by Díaz Covarrubias. By the end of November they were ready.

At that time, there were no diplomatic relations between Mexico and Japan, so that the formal introductions and the initial negotiations had to be conducted through the US Consul J.A. Bingham. Collaboration was always in the best terms, though. Díaz Covarrubias was even granted permission to hoist the Mexican flag at both stations, a most unusual concession in the absence of diplomatic relations.

Díaz Covarrubias established contact with the French and American Commissions in Nagasaki, under Jenssen and Davidson, respectively. At the request of the Japanese Government Díaz Covarrubias took on as trainee-assistants several gentlemen (from the Japanese Navy and the Education Ministry).

5. Preliminary observations

Díaz Covarrubias decided early on to set up two stations, but since they were by necessity close together (separated by just over 2 km) it was clear that his transit data would have to be processed together with those obtained by other missions. The method they chose to observe was the visual timing of the four contacts (to be used along with data from other missions for determining the solar parallax by Halley's method), but they also attempted to observe the position of Venus on the Sun's disk (to employ Delisle's method). For the latter procedure, it was imperative to know precisely the geographic

Figure 2. (left) The observing station at Nogue-No-Yama. (right) The observing station at "The Bluff".

coordinates of the stations, as well as to have accurately calibrated chronometers. They proceeded therefore to the determination of times, latitudes and longitudes (both absolute and differential). For the times, they observed meridian passages of a number of stars, for the latitudes they used four different methods (among them the so-called "Mexican method", developed by Díaz Covarrubias), and for the absolute longitudes they observed lunar culminations, occultations of stars and culminations of about 55 stars. They also exchanged a number of telegraph signals with Tisserand, and Davidson, of the French and American stations. The final results of these determinations, as given in the original publication, (Díaz Covarrubias 1875) are displayed in Fig. 3. The "Legación de Rusia" point was included at the request of O. W. Struve who was in Yokohama to observe the transit.

6. The December 8 transit of Venus

The morning of December 8, the transit day, appeared bright and clear in Yokohama, a great relief after several cloudy days. The transit was successfully recorded both at Nogue-No-Yama station, by Díaz Covarrubias himself, and at the Bluff, by Jiménez. The telescopes were fitted with projection screens, which enabled several people to observe the event. As a matter of fact, Díaz Covarrubias allowed the public to watch the transit, gave extensive explanations and answered many questions, only requesting absolute silence when the four times of contact were approaching. The Minister of Education was among those present at Nogue-No-Yama.

Unfortunately, Díaz Covarrubias had to share the disappointment of most of the missions, because the dreaded black drop effect was again present and spoiled the expected accuracy of the observations. His two stations were separated by close to 5 seconds in longitude. Since the transit occurred practically at the same time at both of them, in the absence of errors his measurements should have differed by exactly 5 seconds from

Nogue-no-yama	35° 26' 54"2
El Bluff	35 26 16.6
Legacion de Rusia	35 26 5.0
Torre del Matchi-guai-sho	35 26 47.9
Nogue-no-yama	— 9h 18m 33s.76
El Bluff	— 9 18 38.71
Legacion de Rusia	— 9 18 37.83
Torre del Matchi-guai-sho	— 9 18 37.22

Figure 3. Latitude (top) and longitude (bottom) determinations for the stations at Nogue-No-Yama and The Bluff, as well as for other significant points. The tables are taken from the original publication (Díaz Covarrubias 1875).

those of Jiménez. In fact, they differed by various amounts, up to 20 seconds. The measurements, as they appear in the original publication, are displayed in Fig. 4.

Even before the full reduction of his data, Díaz Covarrubias (1875) expressed doubts about the expected accuracy in the following words:

> "In spite of the great number of observations, we doubt very much that the 1874 transit will be able to furnish the correct second digit for the solar parallax."

However, he also held the optimistic view that by combining many observations from a large number of stations the errors would, to a certain extent, cancel out, and ultimately provide an improved value for the solar parallax.

The photographer of the Commission, A. Barroso, was able to secure some 14 photos of the transit. They were obtained with a very primitive camera, and were intended merely as a test for the application of the then very new photographic techniques to astronomical observations (see Fig. 5).

After the transit was over, Díaz Covarrubias proceeded to exchange further telegraphic signals with the French astronomers. During the trip back, via China and Europe, Díaz Covarrubias worked with Tisserand on the reductions of these determinations, and both were able to further refine their time and longitude measurements.

The results of the observations at both stations were reduced and prepared for publication by Díaz Covarrubias on his trip back to Mexico. Work was advanced to the point that, when Díaz Covarrubias arrived in Paris, the results were ready for publication (see Fig. 6). The Mexican Commission was the first to publish their results. As already mentioned, they were intended to be processed along with the results of other observers to establish the "best value" of the astronomical unit. To deal with a great number of results from different missions, G.B. Airy recommended that all data be processed in a prescribed way, so as to ensure uniformity among different groups, and only then submitted to him for final evaluation. The Mexican data were reprocessed according to Airy's recommendations, and duly submitted. They were received by the Greenwich Observatory on May 26, 1876.

7. Repercussions of the voyage – scientific and otherwise

President de Tejada was subjected to sharp criticism for funding such a "useless" project, at a time when the Nation was impoverished and had many other pressing needs.

Fases.	Horas medias de Nogue-no-yama.			
Primer contacto exterior.............	1874. —Diciembre 8 á	23ʰ 4ᵐ 7ˢ.0		
Primer contacto interior.............	»	» » 23 29 24.6		
Ruptura del ligamento...............	»	» » 23 30 25.6		
Formacion del ligamento.............	»	9 » 3 21 1.4		
Segundo contacto interior...........	»	» » 3 21 45.4		
Segundo contacto exterior...........	»	» » 3 47 55.5		

Fases.	Horas medias del Bluff.			
Primer contacto exterior.............	1874. —Diciembre 8 á	23ʰ 3ᵐ59ˢ.0		
Primer contacto interior.............	»	» » 23 29 50.0		
Ruptura del ligamento...............	»	» » 23 30 43.5		
Formacion del ligamento.............	»	9 » 3 21 20.9		
Segundo contacto interior...........	»	» » 3 21 50.9		
Segundo contacto exterior...........	»	» » 3 48 4.0		

Figure 4. Transit measurements obtained at Nogue-No-Yama by Díaz Covarrubias and at The Bluff by Jiménez. The table is taken from the original publication (Díaz Covarrubias 1875).

The trip of the Commission was perceived by some as a world tour with all expenses paid for some privileged individuals. Perhaps the haste of Díaz Covarrubias to publish his results was intended to silence such critics (and to prevent being accused of copying the results). But the success of the mission soon turned public opinion around, and Díaz Covarrubias was welcomed back enthusiastically (Moreno 1986b). Even the most severe critics soon afterwards supported the creation of an observatory for Mexico. In this manner, the voyage of the Commission paved the way to the foundation of the National Astronomical Observatory, created by presidential decree in 1876 and officially dedicated on May 5, 1878. The National Astronomical Observatory is the direct precursor of our present Institute of Astronomy.

A few years later, in 1887, astronomers from the National Observatory attended the Astrophotographic Congress in Paris, and were invited to participate in the Carte du Ciel project, one of the large international astronomical projects at the time. The next transit of Venus occurred in 1882 and was observable from Mexico. In fact, a French Commission, under Bouquet de la Grye, travelled to Puebla, and successfully observed the transit. The National Observatory, then at Chapultepec Castle, Mexico City, was already in operation, and had instruments suitable for observing the transit, among them a photoheliograph and a meridian telescope. Unfortunately, the skies were cloudy in Mexico City, thus preventing the observation of the transit there. But Mexican astronomers re-established contacts with their French colleagues, and these contacts ultimately lead to the invitation to the Carte du Ciel project (Moreno 1986a).

Going back to the 1874 transit, Díaz Covarrubias returned to Mexico most favorably impressed by the people and government of Japan. A firm basis for further collaborations with them was assured, leading in time to full diplomatic relations. Several circumstances were to facilitate the establishment of relations between Mexico and Japan. Let us recall that Mexico achieved independence from Spain in the early 19th century. After many conflicts and revolts, Mexico enjoyed a peaceful period during the presidency of Porfirio Díaz (1876-1880, 1884-1911), which led to relative prosperity, economic growth, and

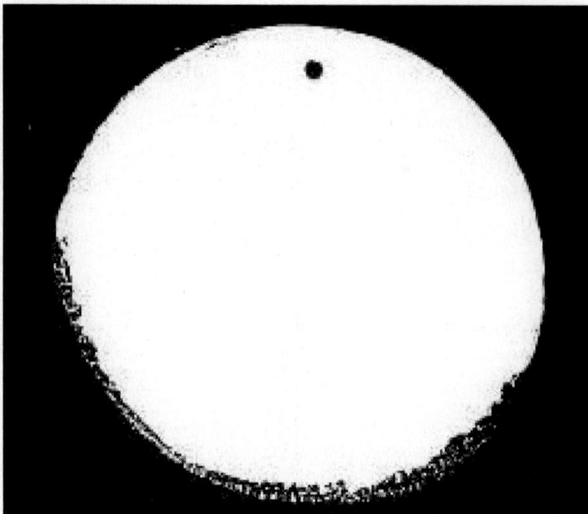

Figure 5. Photo of the transit obtained by Barroso.

international exchange. Due to its geographical location, its natural resources and a deliberate effort by the government to seek wider internationalism, Mexico offered advantages to Japan in its opening to the West during the Meiji Restoration (Alvarez 1996). After many centuries of being a closed society, Japan was seeking international contacts. Díaz Covarrubias realized that it would be of advantage to Mexico to look for a treaty for trading with Japan, and he was an active promoter of the establishment of full diplomatic relations. Díaz Covarrubias' book was helpful in providing much detailed information about Japan. The chronicler of the Commission, Francisco Bulnes, also published a book that enjoyed some success (Bulnes 1875).

In 1888 Ministers Matías Romero (Mexico) and Munemitu Mutsu (Japan) signed the Treaty of Friendship, Commerce and Navigation in Washington, D.C. This treaty was not only the first one signed by Japan with a Latin American country, but also the first written in completely equitable terms. It was also the first treaty signed by Mexico with an Asian nation. As a result of these negotiations, Mexico sent its first diplomatic mission with José Martín Rascón as chargé d'affaires, and Japan sent Tatemu Gozo as Consul in Mexico (Ota-Mishima 1982). Ironically, both the treaty and the transit are better remembered in Japan than in Mexico. So, for example, the municipal authorities in Yokohama rescued the masonry slab upon which one of Díaz Covarrubias' telescopes was erected and had it engraved as a plaque in remembrance of the Mexican expedition. It is now located in a Yokohama public park (see Fig. 7). Soon afterwards, the trade treaty led to the establishment of full diplomatic relations between Mexico and Japan.

In those times, Mexico was sparsely populated. The Mexican government saw immigration as a way to encourage growth. Japan already then had serious demographic problems, and therefore promoted the establishment of Japanese settlements in other countries. Díaz Covarrubias, having been very favorably impressed by the industriousness, the discipline, and the eagerness to learn shown by the Japanese people, vigorously promoted Japanese immigration to Mexico. In 1897 the first group of 36 Japanese immigrants arrived to Escuintla, Chiapas. Manuel Fernández Leal, a member of the original Commission, and later Minister of Mines and Industry, was directly responsible for this program. Later, other groups arrived to be employed at mines in the North or plantations in Yucatán. By 1941 a total of about 14500 Japanese immigrants had settled in Mexico.

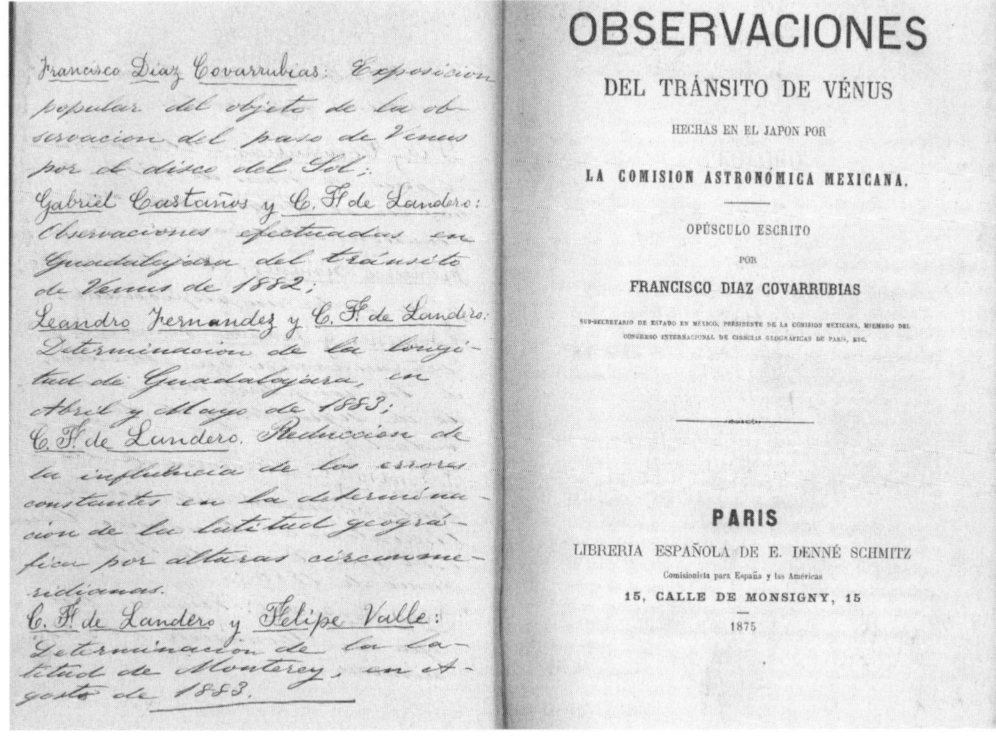

Figure 6. Original publication by Díaz Covarrubias 1875.

As Díaz Covarrubias had foreseen, they adapted very well, they prospered, and later generations were largely assimilated. In this way, the voyage of the Mexican Commission to observe the transit of Venus was important also for many non-astronomical reasons.

I will finish by quoting the words of farewell to the Commission of the Japanese Minister for Education (Díaz Covarrubias 1876):

> "Your presence in this country has been as pleasant as it has been useful, because you have given our youth instruction... Instead of bringing arms and destruction, you have brought to us the brotherhood of the sciences..."

And last, but not least, the transits won considerable public support for astronomy. Perhaps harking back to the old pre-Hispanic traditions that considered Venus an important deity, the transits caught the public's imagination. Just to mention an example: a pub in Mexico City opened with the very sober name *"El tránsito de Venus por el disco del Sol"* (alas, it no longer exists!). The success of the Mexican expedition became a source of national pride and contributed to provide a sense of cultural identity in Mexico.

Acknowledgements

I am indebted to Mr. I. Venteño, of the Autonomous University of Ciudad Juárez for providing a copy of the original Díaz Covarrubias (1875) publication. My sincere thanks to J. Galindo-Trejo and A. Poveda for useful discussions, to J. C. Yustis for obtaining the photographs and to P. Ronquillo for her help with the latex macros.

Figure 7. Masonry slab upon which one of Díaz Covarrubias' telescopes stood, engraved as a plaque remembering the Mexican expedition. This plaque is displayed in a Yokohama public park.

References

Alvarez J. R. 1996 *Enciclopedia de México*. 4467.

Bulnes F. 1875 *"Sobre el Hemisferio Norte once mil leguas. Impresiones de viaje a Cuba, los Estados Unidos, el Japon, China, Cochinchina, Egipto y Europa"*. Imprenta de la Revista Universal.

Cassini M. 1772 *Voyage en California pour l'Observation du Passage de Venus sur le Disque du Soleil, le 3 Juin 1769*. Paris, Antoine Jombert.

Díaz Covarrubias F. 1875 *"Observaciones del tránsito de Venus hechas en Japón por la Comisión Astronómica Mexicana"*. Paris: Libreria Española de E. Denne Schmitz.

Díaz Covarrubias F. 1876 *"Viaje de la Comisión Astronómica Mexicana al Japón para observar el tránsito de Venus por el disco del sol el 8 de Diciembre de 1874"*. México: Imprenta Poliglota de C. Ramiro y Ponce de León.

Díaz Covarrubias F. 1882 *"Exposición Popular del Objeto y la Utilidad de la Observación del Paso de Venus por el Disco del Sol"*. Guadalajara, México: Tipográfica de M. Perex Lete.

Moreno M. A. 1986a *"Historia de la Astronomía en México"* México, Fondo de Cultura Económica.

Moreno M. A. 1986b *"Odisea 1874, o el Primer Viaje Internacional de Científicos Mexicanos"* México, Fondo de Cultura Económica.

Ota-Mishima M. E. 1982 *"Siete Migraciones Japonesas en México"*. México, El Colegio de México.

Ruiz-Gallut M. E., Galindo-Trejo J., & Flores-Gutiérrez D. 2001 *"La Pintura Mural Prehispánica en México"*. Ed. B. de la Fuente, México, UNAM, p.265.

Discussion

STEVE DICK: Do you know any details about how that result was produced so fast from the 1874 transit? It was reduced at Greenwich – is that what you said?

CHRISTINE ALLEN: The preliminary data reduction was done along the way on the voyage back. There were data only for two stations and, of course, they couldn't give a value for the parallax – for the astronomical unit – but they did take into account things had to be taken into account – refraction, for example.

STEVE DICK: Why could they not give the value of the astronomical unit?

CHRISTINE ALLEN: Because the two stations were very close together. They couldn't possibly.

STEVE DICK: Was that data combined with others eventually?

CHRISTINE ALLEN: They were meant to be combined with other data. I don't know if that actually happened; I haven't been able to find out. I do know that the data did arrive at Greenwich observatory and that they were pre-processed by Covarrubias according to Airy's recommendations, but that's all I can say.

WAYNE ORCHISTON: A fascinating paper congratulations, I really enjoyed it.

CHRISTINE ALLEN: Thank you.

WAYNE ORCHISTON: What excited me particularly were the very early murals you presented there – these paintings dating between 1200 and 1230. I'm aware of the fact that Richard Stevenson has carried out a detailed analysis of all the early Asian records to try and determine whether there's any documentation of observations of transits in the Far East (as we call it) from here [UK] and by Chinese, Japanese, Koreans, and so on – knowing the dates of different transits – and he could find no documentation whatsoever, and so we all safely say the first transit was observed in 1639 from very near here [Much Hoole]. If, in fact, these claims that are made are correct, we are actually pushing back the boundaries for the earliest dates. So my question is this: when are these people who carried out this analysis going to publish in English so that the rest of the astronomical community can share this information, and also so people like Richard Stevenson can carry out an analysis and basically support them in this wonderful finding?

CHRISTINE ALLEN: The paper has actually been published in Spanish – I can give you the reference. I can give you a Xerox copy! I have it with me.

WAYNE ORCHISTON: . . . but in English?

CHRISTINE ALLEN: That's something I cannot say. There are three authors. One of them is a colleague of mine at the Institute, and the others are archaeologists and, I will try to convince them to present their work to the astronomical community.

[The following paper is a completely new discussion of this work that was researched, written and included in these proceedings as a result of this discussion.]

DAVID SELLERS: You said that the observation was by projection; did Covarrubias just time the internal contacts or did he take measurements during the progress, and if so how: using the projection method? Did they make accurate measurements of the position of Venus on the solar disc? Do you know?

CHRISTINE ALLEN: As far as I know, they didn't try to measure the position across the solar disc – just the times of the contacts.

RICHARD STROM: Just a follow-up to Wayne's comment: One might worry whether they didn't actually see sun spots; I wonder about what is known about Mayan observations that might have been related to sun spots. Do you know anything about that?

CHRISTINE ALLEN: Actually, one of the sun circles is covered with spots. This suggests that the Maya were familiar with sunspots and did not confuse them with Venus. Then, of course, you have this alignment of the building with a fraction of the synodic period of Venus, which again suggests an association with Venus itself – suggests, but cannot prove.

MARY BRÜCK: Thank you, Christine, for a charming story which has a happy ending; in fact, several happy endings.

Jeannie Allen and Christine Allen during the Transit

Steve Dick, Jim Message, Christine Allen and Jeannie Allen at Hoghton Tower

Transits of Venus: New Views of the Solar System and Galaxy
Proceedings IAU Colloquium No. 196, 2004
D.W. Kurtz, ed.   © 2004 International Astronomical Union
doi:10.1017/S1743921305001328

Maya observations of 13th century transits of Venus?

Jesús Galindo Trejo and Christine Allen

Instituto de Astronomía, UNAM, México, Apartado Postal 70-264, México, D.F. 04510,
email: galindo@astroscu.unam.mx; chris@astroscu.unam.mx

Abstract. With the advent of the 2004 transit of Venus, interest in historical observations of past transits has been rekindled. We present evidence suggesting that the Maya of the post-classic period actually observed at least one transit of Venus. The frescoes of Mayapan, which are proposed as a record of 12th- or 13th-century transits, are described and discussed in their astronomical context

---

1. Introduction

The city of Mayapan was the most important urban and military center of the Yucatan Peninsula during the postclassic period (1000-1519 AD). It is located at about 40 kilometers southeast of Merida, the present capital of the state of Yucatan. The first settlements date from the preclassic period (300 BC to 300 AD), but it was during the early postclassic that the city became prominent, and the Chichen-Itza heritage is readily apparent in its architecture. The main Mayapan temples resemble those of Chichen-Itza, although without the refinement of the latter. Mayapan shows a mixture of architectural elements characteristic of both the Maya and the peoples of Central Mexico. The city occupies an area of 4 square kilometers, and is surrounded by a wall that encompasses over 4000 structures. The central plaza includes buildings of civic, administrative and religious nature, as well as the living quarters of the governing classes (Peraza-Lope 1999; Milbrath and Peraza-Lope 2003).

According to Diego de Landa (1941), oral tradition indicates that Mayapan was founded during the second half of the 13th century by Kukulkan itself. This legendary personage, closely associated with Venus, governed the city during several years, returning later to Central Mexico. During his time at Mayapan the great pyramid bearing his name was erected, as well as a circular building with four entrances (Landa 1941). These buildings resemble respectively El Castillo and the Observatory at Chichen-Itza, but are smaller in size. Even the famous descending serpent formed by the Sun's rays at the time of the equinox on the stairway of El Castillo at Chichen-Itza has an analog in Mayapan, but it occurs during the winter solstice, when the shadow of the nine pyramid bodies is projected by the Sun on the balustrade (Arochi 1991). This is illustrated in Fig. 1. †

In the middle of the 15th century Mayapan was destroyed as a result of a civil war and its chieftains fled. In 1950, several structures and buildings at Mayapan were excavated and restored by the Carnegie Institution (see Jones 1952 for the designation of the Mayapan structures). But it was only in 1996 that a comprehensive project of excavation and stabilization was begun (Peraza-Lope 1999). This project resulted in the discovery of new frescoes and in the salvaging of previously reported ones. In this work, we analyze, from an archaeoastronomical point of view, the mural paintings in the so called Fresco Hall (structure Q.161 in the Carnegie scheme).

† All photos can be viewed in color in the electronic version of this paper

Figure 1. The City of Mayapan can be considered as a direct inheritor of Chichen-Itza, the great metropolis of the classic period. The pyramid of El Castillo in Mayapan also shows a hierophany at its north balustrade, but at sunset of the winter solstice. (Courtesy O. Casares Contreras).

Figure 2. The Fresco Hall in Mayapan is adjacent to the pyramid of El Castillo.

2. The murals in the Fresco Hall

The Fresco Hall is situated at the southern end of the Central Plaza of Mayapan, adjacent to the eastern wall of the Castillo (Fig. 2). It is a rectangular structure, with a small protrusion at one end. The painted wall, about 14-m long, lies at the center of the structure and is oriented in east-west direction. A shorter perpendicular extension (4-m long) is found at the easternmost end. A number of columns are still extant, which probably once supported a roof. The frescoes are still visible on the north and south surfaces of the central wall, as well as along the northern part of the slope of the adjoining Castillo.

The paintings are organized in horizontal rows. On both sides of the wall the themes are similar: we find a rectangular frame surrounded by a stripe and depicting two personages in profile, facing each other. At the center, between the two figures, a circular disk is located, inside of which another anthropomorphic figure is depicted in a descending position. In spite of the fact that the paintings are relatively well preserved, it is not easy to interpret them, since many details are blurred. In this paper, we propose an interpretation based on the available iconographic elements and aided by astronomical calculations that consider the orientation of the building that contains the murals.

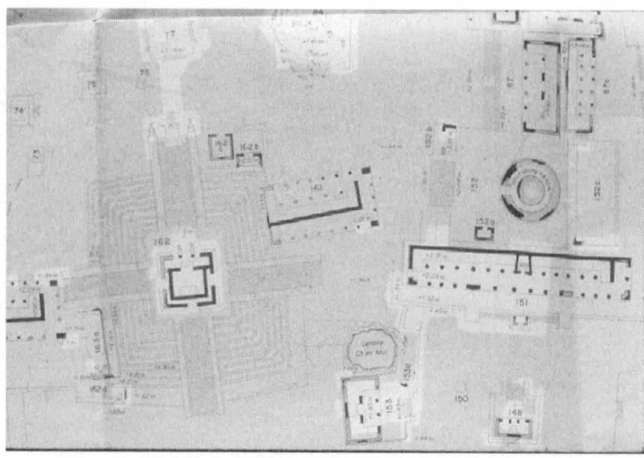

Figure 3. Map of Mayapan's Central Plaza. Buildings 152, 161 and 162 correspond to the Observatory, the Fresco Hall and the pyramid El Castillo, respectively (Pollock 1962). North is at the top.

A total of eight discs can still be recognized, but the surviving rectangular frames allow us to surmise that there were originally as many as 13. If indeed this was the original number of discs in the Fresco Hall, it would be very interesting, since the number 13 is extremely important in the Mesoamerican calendar. In fact, the ritual calendar, or Tzolkin, was organized in 20 periods of 13 days.

Basically, all 8 of the surviving representations are similar. In each, the pair of personages depicted in profile look towards the central circle. They seem to be walking towards it. They carry in their hands a long lance-like object, inclined upwards, that reaches the central image. This lance is painted in red and yellow, and touches what appears to be the head of a serpent in profile. The attire of the personages is quite rich: they wear sandals, earrings and elaborate headdresses. Their bodies and faces are painted in a very dark red, almost brown. Remarkably, their mouths are yellow, and so is a stripe running along their noses and foreheads. They also wear a kind of ornament under their chins, painted green and yellow.

In all cases the central element is painted yellow. The two personages in profile that we just described look towards it. This element is composed of concentric circles with four superimposed Sun rays, in the style of the postclassic representations found in Central Mexico. Inside the circles we find descending personages, with their legs flexed. They carry a kind of shield in one arm. Their faces are now quite blurred and do not allow a detailed description to be made. However, they all look different from one another, especially as regards the form and color of their faces (see Figs 4 and 5).

Despite the differences in time and place, the figures remind one of the well known mural at Teopancaxco, in Teotihuacan, where again two personages – probably priests, judging from their rich clothing – stand in front of each other. Between them a disk is painted, which occupies the central position in the whole mural. The disk has solar connotations, because in its interior is painted a glyph with a jaguar, a sign usually associated to the Sun – specifically to the setting Sun – in the Teotihuacan iconography (Ruiz-Gallut et al. 1996). Such solar discs are fairly common in Central Mexico, but we emphasize that apart from the ones in the Fresco Hall at Mayapan, none are known with personages painted inside them.

Radiocarbon dating has been carried out (Peraza-Lope, personal communication) and it indicates that the murals were painted between 1200 AD and 1350 AD. The style of the paintings of the Fresco Hall does not resemble the Maya style characteristic of that

Figure 4. Painted panel at the Fresco Hall in Mayapan. Two representations of the Lord of Night Yohualtecuhtli escort the solar disk within which a descending personage is depicted.

Figure 5. Detail of a painted panel at the Fresco Hall showing a descending personage, sumptuously dressed, in the interior of a solar disk. The pictorial design and the color palette used associate this mural to the cultural tradition of Central Mexico.

period. Rather, it is reminiscent of the images of the Borgia Codex, a pictorial document with calendric significance dating from approximately the same time as the murals, but showing clear influence of the cultures of Central Mexico (Seler 1906).

Another structure, called the Temple of the Painted Niches is located in the northern part of the Mayapan central plaza. Although its paintings have no obvious astronomical meaning, the building has been shown to have astronomical connotations (Ruiz-Gallut et al. 2001).

3. Archaeoastronomical analysis

3.1. *Measurements of the orientations*

The archaeoastronomical study of both murals required the measurement of the orientations of the Fresco Hall with respect to the celestial north. The geographic longitude of the site is $89°27'39''$, the latitude $20°37'44''$. We measured the azimuthal angle of both sides of the central wall containing the paintings, and the result was $88°56'$ for the north side and $88°16'$ for the south side. In both cases, the apparent horizon is defined by the upper part of the circular temple (henceforth the Observatory), located about 40 m to the east of the eastern flank of El Castillo. The measured height of this horizon is $7°10'$ and $9°30'$ for the north and south sides, respectively. At the time of the measurement (1998) the reconstruction work on this temple was not yet finished. At present, the upper part of the temple is almost twice as high. We considered two kinds of possible astronomical alignment for the Fresco Hall: with the Sun, along the central wall, and with stars, along the direction perpendicular to both sides of the wall. For our search, we have adopted the midpoint (1275 AD) of the interval during which the murals have been dated. Since the surrounding terrain is quite flat, we assumed an horizon height of only $1°30'$. We have also studied, for the same epoch, the direction perpendicular to the south side of the wall, with an azimuth of $178°16'$, assuming the same horizon height.

3.2. *Alignments with stars*

The pictorial designs on the mural can be recognized as having solar significance. However, since the alignments of both perpendicular directions do not point toward the Sun, we think their significance is rather of a ritual character, related to some solar concept derived from the Mesoamerican ideology.

The direction perpendicular to the northern side of the wall (with an azimuth of $358°56'$) points towards a region with few bright stars. However, it is a singular sky region: far from the Milky Way, it corresponds to the circumpolar region. In this region, the north celestial pole is surrounded by the Little Dipper and Draco. This suggests that the Maya wanted to point to the region of the sky where the Sun is never seen, the place of the dead which remains forever dark and serves as the great axis to the celestial sphere. It seems that we find here the concept of the nocturnal Sun, the Sun of the Underworld which is devoured by the Earth Monster at dusk, and which continues its labors until dawn. In this way, the cultural influence on the Maya of the peoples of Central Mexico would be apparent in the pictorial design representing the Lord of the Night, Yahualtecutli or Ahuau Ak'ab. In Central Mexico the nocturnal Sun (depicted as a sun-circle) is often escorted by personages having attributes of Yahualtecutli (similar to the pairs of figures in profile at Mayapan).

In contrast, the southern side of the wall points towards a part of the sky where the Milky Way is seen almost horizontally, with the constellations of Centaurus and the Southern Cross. The latter is particularly striking, because near upper culmination its vertical axis almost coincides with the direction of the southern side of the wall. This constellation, which describes a small arc over the southern horizon, would signal the Underworld regions, where the counterpart of the dark northern region is to be found. In the region far from the Milky Way the direction of the southern wall points toward the Large Magellanic Cloud, visible to the naked eye.

In view of the importance of the Observatory building (Fig. 6) regarding the Fresco Hall, we have also analyzed the direction in the sky towards which the central wall points. We have centered our search on the year 1275. In early May of that year, an observer placed at the central wall would have noted that the Jaws of the Sky Monster (which

Figure 6. The rounded building known as the Observatory at Mayapan. This structure is similar to El Caracol from Chichen-Itza, whose architectural elements show diverse astronomical alignments.

correspond to the Great Rift of the Milky Way) were emerging from over the Observatory (Freidel et al. 1993). In early August of the same year, a narrow portion of the Milky Way appeared over the Observatory, but there were also many bright stars nearby. In particular, the Orion constellation emerged precisely from behind the building. If we use for our search an apparent horizon of 17° to take into account the probable height of the original building, we find that three exceptionally bright stars appear to emerge from the top of the Observatory: Alpha Aql (Altair); Alpha Ori (Betelgeuse), and Alpha CMi (Procyon). The coincidence of the rising of these stars over the top of the Observatory, and in the direction determined by the central wall increases the astronomical significance of the alignment described earlier.

3.3. *Alignments with the Sun. The Venus connection*

We have also studied the possibility of an alignment with the Sun. The painted panels located at the side of El Castillo certainly suggest such an alignment, because the direction perpendicular to them points eastwards. A solar alignment would correspond to a grazing illumination of the paintings by the Sun. Even after its recent restoration, the Observatory does not appear to have the large dimensions mentioned by 19th century explorers (Marquina 1990). The northern part of the wall has a protrusion that would obstruct a grazing illumination by the Sun.

However, the southern side is clear, and allows the solar rays to penetrate when the Sun appears over the top of the circular building. With conditions as measured in 1998, we calculated the dates of grazing illumination of the southern part of the central wall to be April 2 and September 10. However, after the reconstruction of the observatory the dates were recalculated, and they turned out to be April 9 and September 2. The importance of these dates lies in the fact that sunrise on such days defines the division of the solar year of 365 days in a relationship of 2/3 with respect to the summer solstice. That is, counting from the first alignment, these dates divide the solar year in one period of 73 days before the solstice, one of 73 days after the solstice, and three periods of 73 days to complete the year.

Note that, according to the structure of the Mesoamerican calendar, in the course of 52 years of 365 days each, when the Tzolkin starts again simultaneously with the Haab, 73 periods of 260 days must pass. Thus, we are dealing with a calendric-astronomical alignment, like the ones that have been identified in several other Mesoamerican sites.

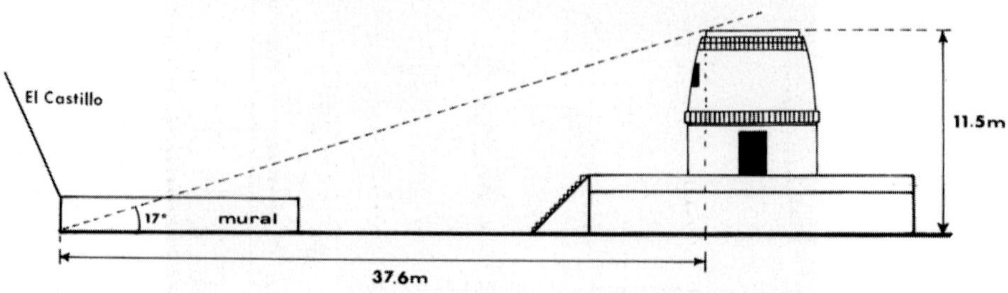

Figure 7. Reconstruction of the height of the Observatory from Mayapan in order to allow the grazing illumination of the mural painting at the Fresco Hall by solar rays. This event occurs on two dates which are important for the Mesoamerican calendrical system.

Most important for our purpose, however, this alignment is directly connected with the synodic period of Venus, 584 days, since the latter can be expressed as 8 periods of 73 days. This association with Venus appears probable when we remember that several circular buildings in Central Mexico, also dating from the postclassic epoch, have been associated with the planet Venus as a manifestation of the deities Quetzalcoatl and Xolotl, being the morning and evening star, respectively. Plausibly, the Maya at that time were assimilating the ideas related to a foreign deity, which they called Kukulkan. The angle necessary for the effect of grazing illumination to occur on the painted panels is about 17°, measured from the line joining the floor of the hall and the wall abutting the side of El Castillo (see Fig. 7). Taking into account the distance between that line and the circular building, that angle determines a height for that building of 11.5 m. The reconstruction of the building, considering data provided by 19th century explorers (Marquina 1990) gives a height of 10.65 m, a difference of less than 10 percent from the calculated value.

Attention should be called to several circular column bases in the Hall of Frescoes, which suggest the possibility that a roof might have existed. Such a roof would have obstructed the passage of the solar rays from the south side of the central wall. We estimate that the maximum column height to just allow the passage of the solar rays is 4.28 m. It is suggestive that the height of the wall at its westernmost extreme, where it meets El Castillo, is about 3 m. This renders plausible the idea that the columns too had this height, and allowed the free passage of the Sun's rays on the suggested days. This, in turn strengthens the importance of the suggested ritual and calendric interpretation, since the close relationship of the Sun and Venus is pointed out by the calendar and the orientation of the mural itself. Furthermore, it was possible to calibrate the synodic period of Venus by observing, every 365 days and three periods of 73 days, the rising of the solar disk over the top of the Observatory building. We recall that the Maya and other people of Central Mexico had a precise knowledge of the Venus cycle. The Dresden Codex, for instance records a synodic period of 584 days, as well as the dates of the four principal Venus stations : inferior and superior conjunction, and the dates of heliacal rising and setting (see Fig. 8).

The importance of Venus in Maya thought leads us to ask if the planet is depicted in the mural. As we know, during the classic period the Maya developed numerical tables to predict solar and lunar eclipses (Brickner & Brickner 1983, Martin 1993). At the time of both conjunctions, the planet is obliterated by the Sun's light, and these inferior conjunctions are not observable, unless they occur under special circumstances, as a

Figure 8. This stele was found in the Venus Platform at Chichen Itza and shows a remarkable characteristic of the motion of the planet. The hieroglyph over the bunch of reeds signifies "year". Eight small circles represent the number 8 and the vertical bar on the left, close to the Venus hieroglyph means the number 5. Further, the stele establishes that 5 synodic periods of Venus (5 × 584) equals 8 solar ones (8 × 365). Such an "equation" represents a remarkable example of the interest accorded to Venus by Maya priest-astronomers.

transit of Venus through the solar disc. Of course, a transit of Venus can normally be observed with the naked eye only when the Sun is close to the horizon, either at dusk or dawn (or through thin clouds). Then, due to the obscuring effect of the atmosphere, it is possible to recognize structures on the solar disc. It is, for instance, easy to distinguish sunspots, especially at times of maximum solar activity. For this reason we suggest that the descending personages depicted inside the sun-circles in the various panels of the mural might represent a Venus transit.

To explore such a possibility we have analyzed the transits of Venus that occurred between 1150 and 1400 AD (all dates are Gregorian) and were observable in Mayapan. During this period, four transits occurred: on November 30, 1153; June 1, 1275; May 30, 1283; and December 1, 1396. The last two events took place during the daytime, but the first two were observable at dusk. So, at 5:15 in the afternoon of November 30, 1153, when the solar limb was in contact with the horizon at an azimuth of 246°41′, Venus was observable on the solar disk, at about 1/4 of a solar radius from the limb. The times given correspond to the meridian 90° west of Greenwich. Similarly, at 6:31 in the afternoon of June 1, 1275, when the solar limb touched the horizon at an azimuth of 293°55′, Venus was located inside the solar disc, at about 1/3 solar radius from the limb.

Surely, the observation of the Sun with Venus inside its disc, both descending towards the horizon, corresponds quite closely to the depictions in the mural. The observations must have been carried out from a high site, likely from the top of the adjoining El Castillo, the highest structure in Mayapan. We think our proposed identification of the personage painted inside the Sun circles with Venus is plausible inasmuch as we are dealing with exceptional events that are nevertheless observable. The time of the observation could have been calculated, since the priest-astronomers carefully followed the trajectory of Venus for calendar purposes, and they would have noticed when an inferior conjunction was about to occur.

3.4. *Venus transits or sunspots?*

Although the proposed observations of Venus appear feasible, it is interesting to note that large sunspots could give a similar visual impression. However, the motion of such objects on the solar disk is very different. Whereas sunspots remain practically motionless during several hours, the changing position of Venus is easily recognizable even for naked eye observers.

For the proposed transits in 1153 and 1275 we have revised data on the solar activity obtained by different methods. The results are somewhat controversial. Schove (1955) uses historical records of aurorae and sunspots to estimate the solar maxima and minima, and obtains a strong maximum in 1151 and the next minimum in 1155. For the year 1276 he obtains a moderate maximum. However, results based on auroral activity often lead to ambiguities due to the confusion of aurorae with other phenomena like comets, noctilucent clouds, etc. To avoid such ambiguities, Wittman (1978) traced the solar cycles using sunspot observations alone. He determines two consecutive maxima for the years 1148.5 and 1161.2, so that the minimum would occur in 1154.9. No data are given near the 1275 event. On the other hand, Usoskin et al. (2004) have reconstructed sunspot numbers averaged over 11 years since 850 AD. They use physical models for the production of ^{10}Be in the Earth's atmosphere and data on the ^{10}Be concentration measured in ice cores from Antarctica and Greenland. They find that the level of solar activity for 1153 and 1275 is significantly lower than that found in earlier analyses. Therefore, we can assume that the transits of those years could have been observed without significant contamination by giant sunspots.

3.5. *Naked-eye visibility of a Venus transit*

The apparent angular diameter of Venus at inferior conjunction has a mean value of 61 arc seconds (Allen 1973). This angular size allows a comfortable visual perception of Venus as a little disk on the bright photospheric background. In fact, the minimum size of perceptible naked-eye dark structures on the solar disk is found to be 19.3 arc seconds (Keller & Friedli 1992).

Of course, the Maya could also have observed the transits of Venus by making use of a *camera obscura*, which could have been easily constructed from bark paper or vellum. The vertical shaft of the observatory at Xochicalco, which can be easily stopped to the required small aperture, suggests that the *camera obscura* effect could have been used there too.

3.6. *Further connections with Venus*

In spite of the blurring of the images inside the sun-circles which prevents us from appreciating details in them, we suggest that they represent a deity associated with Venus. This suggestion is supported by the motion of the planet relative to the Observatory. The planet rises from behind the Observatory to an altitude of 10° to 24° according to whether it is viewed from the slope of the Castillo, or from the eastern end of the Fresco Hall.

In Ruiz-Gallut et al. (2001) the apparent motion of Venus for one eight-year cycle was calculated. As is well known, the planet alternately ascends and descends, a motion that surely was observed by the Maya. Now, let us draw the silhouette of the Observatory

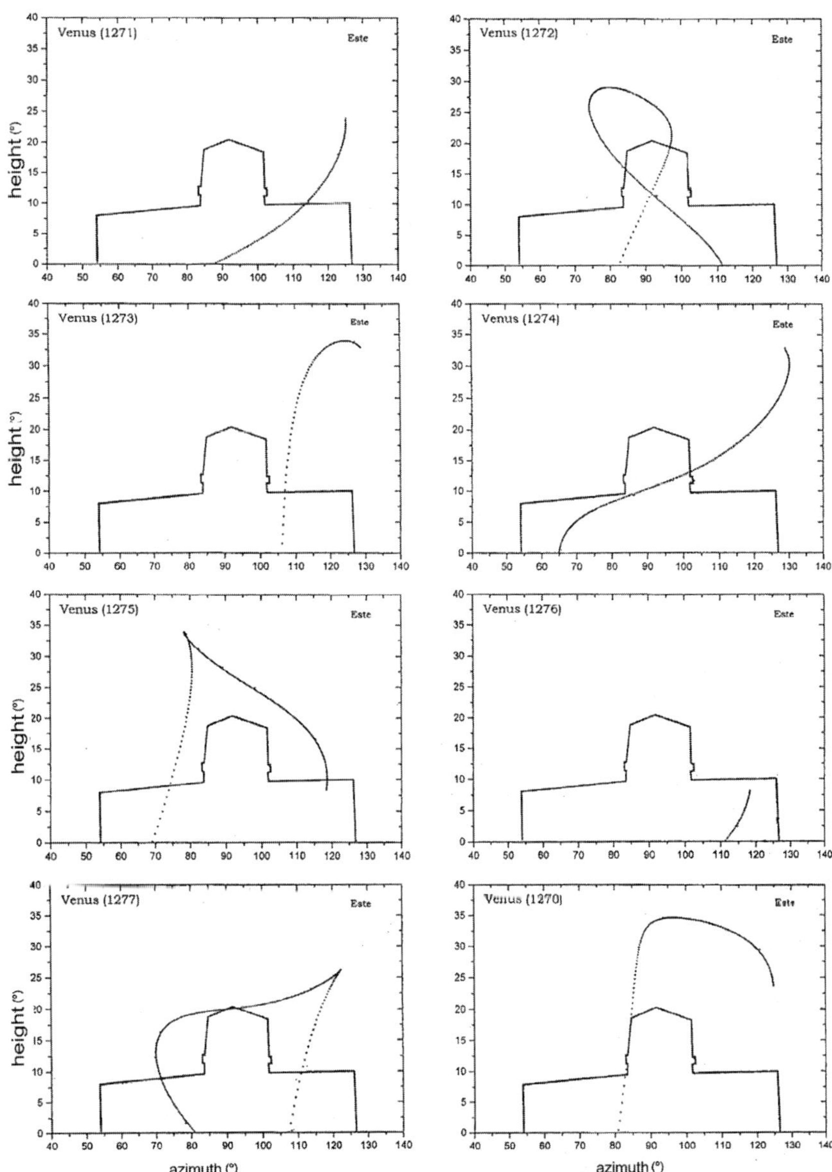

Figure 9. Apparent trajectories of Venus as Morning Star for the years 1271 to 1278 (Gregorian calendar) as registered by an observer situated at the east end of the central wall of the Fresco Hall at Mayapan. The apparent horizon formed by the outline of the Observatory is also depicted. The positions of Venus were calculated for the time of sunrise. Diagram adapted from Ruiz Gallut et al. (2001).

as seen from the easternmost end of the central wall and ask the question: When could Venus be observed emerging from behind this building? The apparent motion of the planet during the cycle 1271 to 1278 is drawn in Fig. 9 (adapted from Ruiz-Gallut et al. 2001). We can see, for example, in early 1271 the planet's day-to-day descent towards the southern side of the Observatory, behind which it disappeared on February 23. Later, in 1272, it rose from behind the building toward the end of April, and remained visible until early September. Venus reappeared as a morning star in early November 1273 and was visible until early April 1274. From mid-June 1275 until December 1275 the trajectory of Venus appeared to encircle the Observatory. A similar trajectory could be observed during January-August 1277. Note that during 1276 Venus was not observable as a morning star, since it did not reach sufficient height to appear above the building. Finally, in 1278 the planet emerged again from behind the top of the building and was observable as a morning star for the remainder of the year. The appearance of Venus as we have described is, of course, repeated every cycle of nearly eight years. Clearly there is a close relation of the planet with the Fresco Hall and the Observatory.

A very remarkable design calls for special attention. It is painted at the top of one of the rectangles that surround the sun-circle and the personages facing it. This design appears only on the southern side of the central wall, near its middle. It is a wavy line design, similar to a Yacametzli (a nasal ornament shaped like the Moon hieroglyph) but with two small circles underneath each end. On page 15 of the Madrid Codex we find a similar design, which bears the name of a personage and depicts an eye. Could the Mayapan design also represent an eye, perhaps indicating that observations were performed from that part of the wall?

It is worth noting that other parts of the southern wall, above the painted strips, also exhibit vestiges of designs, almost obliterated now. Nonetheless, one of these designs, shown to us by Peraza-Lope during a brief visit to the site in 2000, depicts a kind of eye with an oval ornament underneath, which resembles the representation used in Central Mexico for a star.

The Temple of the Painted Niches was also measured in order to determine its orientation with respect to the celestial North. We obtained an azimuth of $183°42'$ for its symmetry axis, and an apparent horizon height in that direction of $2°30'$. Once more, we find a southern orientation. If we search around 1275 AD we find an observational pattern similar to that found for the southern side of the central wall of the Fresco Hall. In this extremely southern region, the motion of the stars is in the form of short arcs. So, for several months the most striking constellation, the Southern Cross, points its vertical axis towards the horizon and beneath, thus signaling to the Underworld. It is not easy to find a correspondence between a particular design in the murals of this temple and an astronomical object. But we could, conversely, suggest an interpretation of the pictorial designs with some particularly striking celestial object in the direction considered. The open jaws of the serpent painted in Structure Q80 could be associated with the circular trajectories around the unseen (but clearly inferred) South Celestial Pole. The Nocturnal Sun could be associated with the personages brandishing the lances and accompanying the Sun-circles.

Recent archaeological explorations in Mayapan have revealed other mural paintings (Peraza-Lope, personal communication). Inside the Observatory, for instance, four painted niches were found, with very complicated designs. Their colors are bright and resemble those of the murals of the Fresco Hall. Several frames decorated with green feathers can be identified. There are even fragments of what could be sun-circles. It would not be surprising to find paintings with astronomical connotations, since the astronomical importance of the building is clearly established (Ruiz-Gallut et al. 2001).

4. The 2004 transit of Venus from Mayapan

In an attempt to recreate the Maya observations during the June 8, 2004 transit of Venus, one of us (J. G. T.) traveled to Mayapan. The completely flat terrain would have allowed a comfortable naked-eye observation from the main pyramid, with the light of the solar disk being naturally filtered by the atmosphere at dawn. The observational circumstances for the transit were not very favorable in Yucatan, because only the very final phase of the transit was to be visible for a few minutes. Unfortunately, a dense fog at sunrise made impossible the observation of the solar disk. This observation will be attempted again at the time of the next transit of Venus, in 2012.

However, the 2004 transit was observed from different parts of the world. It is interesting to examine images of the transit obtained from longitudes similar to that of Mayapan at the same phase of the event (see, for instance, Schaaf, 2004; http://www.vt-2004.org/photos and http://sunearth.gsfc.nasa.gov/sunearthday/2004/vt_gallery.htm). Images taken at sunrise without filters, reports of naked-eye observations of Venus on the solar disk and photos containing Venus in transit together with flying birds, as well as reports of comfortable naked-eye observations (Schaaf 2004) strengthen our suggestion that naked-eye observations by ancient observers of a Venus transit at dusk or dawn would indeed have been feasible.

5. Conclusions and summary

Despite the fact that the newly discovered paintings in the Fresco Hall cannot yet be fully interpreted, the study of the principal directions of the buildings has allowed us to highlight the importance of certain religious and calendric concepts that are related to astronomical events.

With regard to the images themselves, it is obvious that they include a mixture of stylistic elements and features related both to those found in Central Mexico and in Yucatan itself for the period we are studying. This makes it more difficult to present a unique interpretation, because cultural traits of both regions have to be taken into account.

It is worth mentioning that the circular temple, which we have called the Observatory, has been reconstructed in recent years, and mural paintings within it have been discovered. Ruiz-Gallut et al. (2001) showed that there is a close relationship between the Fresco Hall murals and those of the Observatory, and suggested that the building served as an actual astronomical observatory.

We finish up by summarizing the evidence which favors the interpretation of the Sun-circles in the Fresco Hall being a record of observations by the Maya of Venus transits. We also point out some problems and directions for future work.

1. The association of the Fresco Hall and the Observatory with Venus is supported by the following facts. (a) The orientation of the building is such that the Sun-circles are illuminated grazingly by the Sun on dates centered on the summer solstice and separated by 73 days, one eighth of the synodic period of Venus. (b) The relative orientation of the Observatory and the Hall of Frescoes is such that as viewed from the latter, the apparent trajectory of Venus appears to encircle the Observatory at the times near the transit, as it does every eight years.

2. The association of the Sun-circles (i.e, concentric circles usually painted red, with four protruding spikes) with the Sun is well documented in many instances. There is even a glyph representing the Sun which is a simplified, stylized version of the Sun-circles (see Fig. 10). However, the Sun-circles found elsewhere do not contain human figures inside them. The fact that all of the Sun-circles depicted in the Hall of Frescoes contain human figures in their interior is suggestive of some kind of eclipse or occultation of the disk of the Sun by a celestial body. Solar eclipses can be ruled out, since the Moon had a very clear and unique pictorial representation. Mercury is too small to be observed with the

Figure 10. Representation of two solar eclipses from page 71 of the Madrid Codex. Both central concentric hieroglyphs correspond to the Sun or Kin (in Maya language) which appear darkened and hang from the so called sky bands.

naked eye when it transits the Sun. A transit of Venus seems by far the most plausible of such events.

3. The figures inside the Sun circles cannot be clearly identified. They do, however, correspond to important priests or gods, and such important personages were often linked to celestial bodies. A mixture of styles is present in their attire, ranging from Central Mexico to Oaxaca and Yucatan. This is consistent with the idea that the Maya were in the process of adopting deities (like Kukulkan) imported from other cultures.

4. Certainly, a transit of Venus occurring at dawn or dusk is visible with the naked eye. The planet's apparent diameter makes it easily distinguishable on the solar disk.

5. The possibility that the Maya might have confused Venus with a sunspot can be ruled out for several reasons. First, the Maya continuously and carefully followed the trajectory of Venus, so that they would have been able to predict a transit, and be ready to observe the event as quite distinct from a sunspot. A sunspot, moreover, is visible over several days, whereas a transit lasts only a few hours, and this difference would not have escaped the keen Maya observers. Also, the fact that the time of the Venus transit corresponded to a period when the solar activity was not very intense should be considered.

Finally, we point out some problems that need future work to be clarified. The number of sun-circles with human figures inside is puzzling. Also puzzling is the fact that all human figures inside the Sun-circles seem to be different. There survive eight such figures, but there are vestiges of five more. The number 13 has intriguing calendric implications. But, why were so many figures painted if, at the most, four Venus transits could have been observed during the relevant period of time, and only two at dusk? Why are all the figures different? Could the Maya have known of previous Venus transits, perhaps observed at Chichen-Itza? Another important problem remaining is the establishment of a clear-cut identification of at least one of the personages with either an image connected with Venus, or with a deity or important person having Venusian connotations. The Mesoamerican gods usually show precise attributes in their garments and ornaments. Such an identification has not yet been found. We are in the process of examining several Maya codices to look for it. However, knowing the importance that Venus had for the Maya, it is conceivable that they would represent an extraordinary event like finding Venus inside the Sun's disk by depicting the Venus-God in different and unusual circumstances.

Acknowledgements

Our sincere thanks to R. Costero for a careful reading of the text and many suggestions, and to A. Poveda for interesting discussions.

References

Allen, C. W. 1973 *Astrophysical Quantities*. The Athlone Press, London.

Arochi, L. E. 1991 Concordancia Cronológica Arquitectónica entre Chichen Itzá y Mayapán. In *Arqueoastronomía y Etnoastronomía en Mesoamérica* (J. Broda et al. Eds, México, UNAM), 97-112.

Bricker, H.M. & Bricker, V.R. 1983 *Current Anthropology*, **Vol. 4**, No. 1, p. 1.

Freidel, D., Schele, L. & Parker, J. 1993 *Maya Cosmos: Three Thousand Years on the Schaman's Path*. New York.

Jones, M. R. 1952 Map of the Ruins of Mayapan, Yucatan, Mexico. *Current Reports No. 1, Department or Archaeology, Carnegie Institution of Washington*, Washington, D.C.

Keller, H.U. & Friedli, T.K. 1992 *Q. J. Roy. Astr. Soc.*, **33**, 83.

Landa, D. de 1941 *Landa's Relación de las Cosas de Yucatán.* (ed. A. M. Tozzer) Peabody Museum Papers No. 18. Harvard University Press.

Martin, F. 1993 *Latin American Antiquity* **4(1)**, 74.

Marquina, I. 1990 *Arquitectura Prehispánica, INAH*. México, 1006-1011.

Milbrath, S. & Peraza-Lope, C. 2003 *Ancient Mesoamerica*, **14**, 1.

Peraza-Lope, C. 1999 *Arqueología Mexicana*, **37**, 48.

Pollock, H. E. D. 1962 Introduction. In Mayapan, Yucatan, Mexico.*Carnegie Institution Publication No. 619, Carnegie Institution of Washington*. Washington, D.C., pp. 1-22.

Ruiz Gallut, M. E. et al. 1996 in *La Pintura Mural Prehispanica en Mexico*, ed. B. de la Fuente, Mexico, Inst. de Investigaciones Esteticas, UNAM, **Vol I(2)**, p. 343.

Ruiz Gallut, M. E., Galindo Trejo, J. & Flores Gutiérrez, D. 2001 Mayapán: de regions obscuras y deidades luminosas. Práctica astronómica en el Postclásico Maya. In *La Pintura Prehispánica en México, II Area Maya*. Tomo III, UNAM, México, 265-275.

Schaaf, F. 2004 *Sky and Telescope*. October, 85.

Schove, D. J. 1955 *J. Geophys. Res.* **60**, 127.

Seler, E. 1906 *Codex Borgia. Eine altmexikanische Bilderschrift der Bibliothek der Congregatio de Propaganda, Fide, Berlin*.

Usoskin, I. G., Mursula, K., Solanki, S., Schüssler & Alanko, K. 2004 *Astron. Astrophys.* **413**, 745.

Wittmann, A. 1978 *Astron. Astrophys.* **66**, 93.

Transits of Venus: New Views of the Solar System and Galaxy
Proceedings IAU Colloquium No. 196, 2004
D.W. Kurtz, ed.

© 2004 International Astronomical Union
doi:10.1017/S174392130500133X

Lord Lindsay's expedition to Mauritius in 1874

M. T. Brück

Institute for Astronomy, Royal Observatory, Blackford Hill, Edinburgh, EH9 3HJ, UK
email: mtb@roe.ac.uk

Abstract. An account is given of Lord Lindsay's lavishly equipped independent expedition to the Island of Mauritius in the Indian Ocean to observe the 1874 Transit of Venus. The expedition's secondary programme, the deriving of the solar parallax from observations of the minor planet Juno, is also described. This work proved a positive outcome of a generally disappointing event and brought about an important shift in the approach to the parallax challenge. The site on Mauritius where Lord Lindsay observed the Transit in 1874, now preserved as a National Monument, was the centre of celebrations during the Transit of 2004.

1. Dun Echt Observatory Transit plans

The Transit of Venus in 1874 was anticipated as "probably the most important astronomical event of the century" (1). Seventy-five observing stations were distributed worldwide, from France, Russia, Germany, the Netherlands, Great Britain, the United States, and their various dependent colonies. Of these, Britain sent eight teams of observers on expeditions which were centred on five locations – Egypt, the Hawaiian Islands, the Indian Ocean, New Zealand and the remote southern Kerguelen's Island – and were joined in collaboration by India, South Africa and Australia. The British effort was co-ordinated at the Royal Observatory at Greenwich, and masterminded with military efficiency by the Astronomer Royal Sir George Airy – "a generalship quite unparalleled in the annals of Science" (2).

Figure 1. (left) Lord Lindsay, 26th Earl of Crawford and Balcarres and (right) (Sir) David Gill, portrait by George Reid

One British expedition did not conform to the national plan – Lord Lindsay's privately funded expedition to Mauritius in the Indian Ocean(3). Lord (James Ludovic) Lindsay (later 26th Earl of Crawford) (Fig. 1), a talented amateur astronomer, engineer and photographer, founded an observatory on the family's country estate at Dun Echt, near Aberdeen, Scotland, which functioned successfully from 1872-92(4) and survived as a major component of the Royal Observatory Edinburgh(5). Lindsay's immediate ambition, when he began to make his plans in 1872, was to observe the approaching Transit of Venus, and preparations for the Transit went hand in hand with the furbishing of the new observatory. He recruited as his collaborator David Gill of Aberdeen (Fig. 1), a watchmaker by profession and a passionately enthusiastic and able amateur astronomer. Together they set about acquiring a variety of instruments, with particular reference to the needs of the Transit. In contrast to the official British expeditions, which were uniformly equipped with equatorial telescopes, Lindsay and Gill went their independent way. They decided that their principal instrument should be one which had been ruled out of the British plans – a heliometer, the choice of the official German and Russian expeditions. Lindsay sent Gill to St Petersburg to take advice from the great Otto Struve at the Pulkovo Observatory, after which they both attended a special Transit-planning meeting of the German Astronomische Gesellschaft, of which Struve was President, in Hamburg. There they met leading German and Russian astronomers, including Arthur Auwers of the Prussian Academy of Science, prime mover of the German Transit effort and Secretary of the German Transit Commission(6), who became Gill's close friend and later collaborator.

The Dun Echt heliometer (Fig. 2), constructed by the firm of Repsold of Hamburg, was of 4 inches (10 cm) diameter, and was to be handled specifically by Gill who familiarised himself thoroughly with its use at Dun Echt over many months. Preparations did not stop there. Lindsay also acquired a long focus (40 foot or 12.2 m) solar horizontal telescope, fed by a Foucault siderostat (Fig. 3), for photography of the Transit in the manner favoured by the American astronomers; and a 6-inch (15 cm) telescope equipped with a double-image micrometer for visual observations of the Transit as used by the official British expeditions. Together with a splendid portable transit theodolite by Troughton and Simms (Fig. 4), chronometers, and auxiliary apparatus, the Dun Echt expedition was, in the words of Agnes Clerke, "the epitome of modern resource and ingenuity"(7).

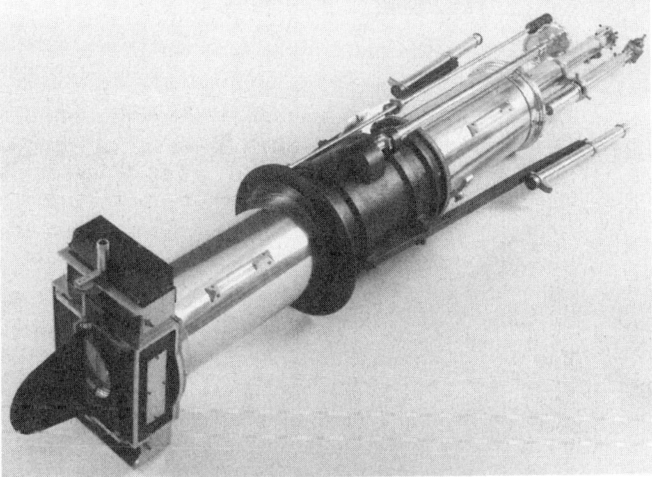

Figure 2. The Dun Echt heliometer (Royal Observatory Edinburgh)

Figure 3. The 40-foot solar telescope, with Ralph Copeland, at the eclipse expedition to Russia in 1887

Figure 4. Transit theodolite (Royal Museums of Scotland)

2. The Transit

Lord Lindsay travelled to Mauritius on his yacht, the *Venus*, with a crew of 22 men, accompanied by the astronomer Ralph Copeland, on leave from his post at Dunsink Observatory, Dublin, his photographer Henry Davis and the ship's surgeon. The "wearisome voyage" around the Cape took longer than anticipated due to unfavourable weather – almost four months, from July 9 to November 2. David Gill, bearing 50 chronometers, travelled separately by ship via the Suez canal, and reached Mauritius on 4 August, where he was welcomed by his countryman Charles Meldrum(8), Director of the Royal Alfred Observatory (a meteorological observatory) at Pamplemousse who also planned to observe the Transit. The observatory carpenter from Dun Echt, with domes and huts, travelled with Gill, and all building work was ready by the time Lord Lindsay and his companions sailed in. A resident of the island, Eduard de Chazal, offered the party a

Figure 5. The expedition's observatory at Belmont (Mauritius Transit of Venus 2004 Committee)

site on his estate at Belmont, sixteen miles from Port Louis (latitude 20° S), and put his house at the disposal of the astronomers for the duration of their operations (Fig. 5).

The crucial day of the Transit began cloudy, but cleared up sufficiently for a range of observations to be made with the heliometer, and a large number (270) of photographs of the sun to be obtained(9). (None of the photographs survives in the Dun Echt archives at the Royal Observatory Edinburgh). The material was sent back to Dun Echt for reduction, but never published: the overall results worldwide were disappointing.

3. Parallax of Juno

The Dun Echt astronomers, however, had another string to their bow. As well as observing the Transit, Gill planned a hitherto untried method to obtain the solar parallax – the measurement of the position of a minor planet (asteroid) relative to the stars in the evening and in the morning. The "diurnal" method of observing the parallax of a planet, and hence of the sun, had been suggested as far back as 1857 by G.B. Airy, but had not previously been put into practice. Instead of observing the planet from widely separated positions on the surface of the earth, the astronomer observes it from the same location after an interval of many hours, allowing the rotation of the earth to provide the baseline for parallax. This requires to be done when the planet is in or near opposition, the observations being made in the evening and again in the morning.

The idea of using a minor planet for parallax work originated with J.G.Galle of Berlin, who in 1872 pointed out that a minor planet, though more distant, had the advantage of a star-like appearance, allowing its position to be more precisely determined. He attempted it with the minor planet Flora at its opposition in October 1873, with the cooperation of a number of observatories, including Dun Echt, and corresponded with Gill on the subject(10). As Lord Lindsay explained: "The total parallactic displacement is very considerably less than in the case of an opposition of Mars, or a Transit of Venus, but we believed that the great accuracy attainable in the heliometric measurement of the distance of two minute points of light would more than compensate for this defect"(11). The minor planet chosen was Juno, which was in opposition one month before the Transit of Venus, on 5 November 1874. Lindsay and Gill prepared a list of comparison stars,

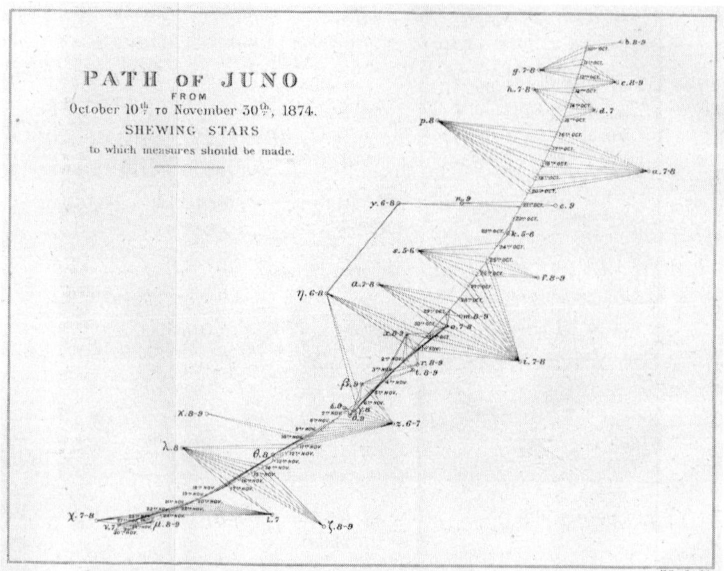

Figure 6. Path of Juno at opposition 1874 (*Monthly Notices of the Royal Astronomical Society* **43**, 1874)

against which Juno's position would be measured with the heliometer over a period before, during and after opposition (Fig. 6)(12). Gill was installed in Mauritius in good time, but the heliometer, which was transported by yacht, unfortunately arrived late, and Gill's observations began only on 12 November, continuing until 30 November, a total of fifteen nights. The span was enough, however, to establish the method and provide a result distinctly more reliable than the Transit of Venus could deliver. The procedure was also original, in that it combined for the first time the diurnal method with the use of a minor planet.

Gill's reductions, laid out in a 200 page paper, gave a value for the solar parallax of $8\rlap{.}''77 \pm 0\rlap{.}''04$ (probable error)(13). The experience gave him a "conviction that very powerfully influenced a considerable part of my future work"(14). Forgoing his salaried post at Dun Echt, he embarked on his next major and highly successful undertaking – the observation by the diurnal method of the opposition of Mars from the Island of Ascension in 1877, again using the heliometer, his favourite instrument. Thus, while the results of the actual Transit were universally disappointing, the Dun Echt expedition was by no means a wasted effort. Contrary to the pre-transit belief that "the coming observations are to serve astronomers until 2004"(15), the Juno series of observations, described in retrospect as "probably the most important result of all the many costly Transit of Venus expeditions"(16), initiated an improved way of finding the solar parallax(17). Nor did the instruments acquired for the Transit become redundant. Lord Lindsay's heliometer was used for Gill's observations of Mars, and the 40-foot solar telescope was employed by Copeland on three solar eclipse expeditions and for solar spectroscopy at Dun Echt and at the Royal Observatory Edinburgh.

4. Geodesy

The second substantial piece of work of the expedition was the determination of longitudes, not only at Mauritius but at various locations in the Indian Ocean. Telegraphic communication existed only at a few centres: astronomers relied on carrying the time with them by chronometer, and comparing chronometers with those at other stations. The longitude differences within Mauritius were established – between the Dun Echt site and those of Meldrum's observatory and of the German one at Solitude, a site south of Lindsay's where the observers M. Low and C.F. Pechüle also used a heliometer(18). For comparison with the British site at the neighbouring island of Rodriguez, the chronometers were transported by sea, a journey of 14 hours, and were also compared with those of the Dutch astronomers who had a station on the island of Reunion, and with the French station on St Paul's(19). Gill spent a month travelling through the Indian Ocean, linking also the Seychelles, and reaching Aden, Suez and and Alexandria, which were connected to Berlin by telegraph. In the last part of the chain, Gill had the cooperation of his German friend Auwers who was still at his own Transit station in Egypt. The exercise sparked Gill's abiding interest in geodesy.

5. Sequel

Over and above its scientific value, the Mauritius expedition was a springboard in the careers of Lord Lindsay's two colleagues. The parallax observation of Juno, repeated for Mars three years later, propelled Gill in 1879 to distinguished eminence as H.M. Astronomer at the Cape of Good Hope: the first to bring the good news of his appointment was Lord Crawford, "my former chief – an ever-true and loyal friend"(20). Ralph Copeland succeeded Gill at Dun Echt, and in due course, when Lord Crawford came to the rescue of Edinburgh's ailing Royal Observatory, became third Astronomer Royal for Scotland. "I cannot help feeling" said the present Earl of Crawford on the centenary of that event, "that he [his great grandfather] included in his gift his astronomer, his friend Dr Copeland"(21).

6. Mauritius revisited 2004

Before the Dun Echt astronomers dismantled their station on Mauritius in 1874, their host Monsieur de Chazal, made a generous gesture, which Lord Lindsay put on record in his Report of the expedition:

> I wish now to express my feelings of gratitude towards those who were so kind to me and mine during my three months' stay on their island. Towards Mons. de Chazal and his family I shall ever bear the warmest feelings of regard and sympathy. He most generously placed at my disposal, during my stay, the house and buildings of his estate at Belmont in the district of Poudre d'Or, in the north-east part of the island, about sixteen miles from Port Louis. When prostrated by fever, had I been his own son I could not have been cared for with more tenderness. Not content with thus ministering to the wants of the present, he executed a gift to the Government of the site upon which my observatory was placed, considering that it would be well, in the interests of science, that the actual spot on which the observations were taken should be preserved, and should not share the fate of the Base measured in 1753-54 by the Abbé La Caille, which unfortunately cannot now be identified(22). On this ground I claim for Mons. de Chazal the gratitude of future generations of astronomers and geodetecists.(23)

On de Chazal's death a few years later, Lord Lindsay, then President of the Royal Astronomical Society, paid further tribute to him at a meeting of the Society, as "a gentleman who, though not a Fellow of the Society, had done very much for her science, who had placed his observatory in the hands of Her Majesty's Government, and whose interest in the work was only equalled by his kindness during Lord Lindsay's illness."(24)

144   M. T. Brück

M. de Chazal's wish has been honoured by the Government of Mauritius. The site at
Belmont, where Lord Lindsay's observatory stood, has been cleared and its foundations
revealed. It is now designated a National Monument (Fig. 7), and a special committee
celebrating the 2004 Transit of Venus included in its festivities a commemoration of
that historic expedition and the naming of Lord Lindsay Avenue. Greetings were also
received from Dr. Adrian Russell and Prof. James Dunlop, respectively Director of the
UK Astronomy Technology Centre and Head of the Institute for Astronomy, University
of Edinburgh. These two institutions now occupy the site of the the Royal Observatory
Edinburgh. The Transit of 8 June 2004 was observed by thousands of schoolchildren and
adults on the actual site of the 1874 expedition. Among official guests taking part in that
happy event (Fig. 8) was the Honourable Alexander Lindsay, direct descendent of his
namesake of 1874.

Figure 7. Plaque on the walls of Lord Lindsay's observatory, now a National Monument

Figure 8. Celebration of the Transit of Venus 2004 on the 1874 site. From left to right: Cader
Kalla, Chairman of the Mauritius Museum Council; Motee Ramdass, Minister of Art and Cul-
ture; Dr Jhurry, a Representative of the District Council; Anil Gayan, Minister of Tourism and
Leisure and Alexander Lindsay

Acknowledgements

My warm thanks are due to Karen Moran, librarian at the Royal Observatory Edinburgh, and also Jason Cowan and Clive Davenhall for their skilled help in preparing the manuscript for publication. Marie-Jose Martial-Craig, secretary of the National Transit of Venus 2004 Committee, Government of Mauritius, has kindly provided the photographs and press reports of the 2004 celebrations. Her generous help and cooperation is quite particularly appreciated.

References

(1)   George Forbes 1874. *The Transit of Venus*. London: Macmillan and Co. Forbes, Professor of Natural Philosophy at the Andersonian University, Glasgow, was in charge of the British Transit observations in Hawaii.

(2)   George Forbes *ibid.*

(3)   A brief account was presented to the Archives Committee of the International Astronomical Union in Sydney 2003: M.T. Brück. 'Lord Lindsay's Transit of Venus expedition to Mauritius 1874.' *ICHA Newsletter* **5**, 24; *Journal of Astronomical History and Heritage* **6**, 37.

(4)   H.A. Brück 1992. 'Lord Crawford's Observatory at Dun Echt 1872-1892.' *Vistas in Astronomy* **35**, 81-138.

(5)   In 1888 Lord Crawford made a gift of the entire contents of his observatory, including its priceless Crawford Library, to the nation to furnish Edinburgh's new Royal Observatory which opened officially in 1896. H.A. Brück 1983. *The Story of Astronomy in Edinburgh*, Edinburgh; H.A. Brück 1992. *op. cit.* (4); M.S. Longair, H.A. Brück, 29th Earl of Crawford and M.T. Brück 1988. 'Centenary of the Royal Observatory Edinburgh 1888-1988.' *Vistas in Astronomy* **32**, 201-214.

(6)   Hilmar W. Duerbeck 2004. 'The German Transit of Venus expeditions of 1874 and 1882.' *Journal of Astronomical History and Heritage* **7**, 8-17.

(7)   Agnes M. Clerke 1902. *A History of Astronomy during the Nineteenth Century*, p 234. (Reprint edition, Decorah: Sattre Press 2003).

(8)   *Dictionary of National Biography*. (Compact Edition) 1975. Oxford University Press.

(9)   David Gavine 1982. *Astronomy in Scotland*. Ph.D thesis, Open University.

(10) J.G. Galle. Letter to Gill, addressed from Breslau, 1874 June 28. Dun Echt archives, Royal Observatory Edinburgh. Juno's distance from the Earth at opposition was .67 AU, compared with .52 for Mars.

(11) Lord Lindsay 1877. *Dun Echt Observatory Publications*, Volume 2, Introduction.

(12) Lord Lindsay and David Gill. 1874. *Monthly Notices of the Royal Astronomical Society* **34**, 279-300.

(13) Lord Lindsay and David Gill 1877. *Dun Echt Observatory Publications*, Volume 2, 1-212.

(14) David Gill 1913. *A History of the Cape Observatory*, H.M.S.O, p36.

(15) George Forbes *op. cit.* (1).

(16) Obituary of Lord Crawford. *Nature*, 1913 February 13.

(17) See Steven J. Dick 2004. 'The American Transit of Venus Expeditions of 1874 and 1882,' this conference.

(18) H. Duerbeck *op. cit.* (6), p11.

(19) Lord Crawford 1885. 'Mauritius Expedition 1874: Determinations of longitude and latitude.' *Dun Echt Observatory Publications*, Volume 3.

(20) David Gill 1913. *op. cit* (14), p39.

(21) 29th Earl of Crawford, reference (5).

(22) Brian Warner 2002. 'Lacaille 250 years on.' *Astronomy and Geophysics* **43**, pp2.25-2.26.

(23) Lord Lindsay 1877. *Dun Echt Observatory Publications*, Volume 2. Preface.

(24) Lord Lindsay 1879. *Observatory* **3**, no 25. p1.

Transits of Venus: New Views of the Solar System and Galaxy
Proceedings IAU Colloquium No. 196, 2004   © 2004 International Astronomical Union
D.W. Kurtz, ed.   doi:10.1017/S1743921305001341

Why did other European astronomers not see the December 1639 transit of Venus?

David W. Hughes

Department of Physics and Astronomy, The University, Sheffield, S3 7RH
email: d.hughes@sheffield.ac.uk

Abstract. The December 1639 transit of Venus was only seen and recorded in Much Hoole and Salford, Lancashire, England. It was visible, however, from all over Italy, France, Spain and Portugal. But no one was looking. This paper suggests reasons why.

1. Introduction

Imagine the scene. It is a typical cloudy day in late-autumnal Lancashire. In the mid-afternoon of Sunday, 4 December 1639 (Gregorian Calender, New Style = N.S.), equivalent to Sunday, 24 November 1639 (Julian Calendar, Old Style = O.S.) Jeremiah Horrocks (1618 – 1641) in Much Hoole, a village about 15 miles north of Liverpool, and William Crabtree (1610 – 1644) in Broughton near Manchester, 25 miles to the east of Horrocks' location, observe Venus starting to transit the solar disc. They were the only two on Earth who recorded observations of this rare event. Why? Considering that half our planet was pointing towards the Sun at the time (see Fig. 1), we can reasonably ask why about 50% of the population were ignoring what was going on. Let us consider a few possibilities.

(1) Only Horrocks and Crabtree knew about it. This is a reasonable supposition. Horrocks, a young Liverpudlian biblical clerk and children's tutor, with a fascination for astronomy in general and astronomical tables in particular, had only calculated that a transit would occur about a month before the event. In the Rudolphine Tables, of 1627, the German astronomer Johannes Kepler (1571 – 1630) had predicted that Venus would cross the Sun on 6 December 1631, and again in 1761. Unfortunately it was night-time in Europe at the time of the 1631 transit. After leaving Cambridge the young Horrocks amused himself by computing ephemeredes using Lansbergius' Tables. He soon realised that these tables were somewhat erroneous. However, using these tables, in conjunction with the Rudolphine Tables, enabled him not only to confirm Kepler's prediction that transits were spaced by about 120 years, but also to discover that they usually occurred in pairs, early December 1631 and 1639; early June 1761 and 1769; early December 1874 and 1882; early June 2004 and 2012, and so on.

In early November (N.S.) Horrocks predicted that in a mere four weeks there would be another Venus transit, on 4 December 1639. Four weeks was far to short an interval for news of this prediction to percolate even down to London, never mind across the Channel. And 1639 was before the foundation of the great scientific societies (Florence 1657, London 1662, Paris 1666). News travelled slowly. Scientific journals were unheard of. It was also at a time before the great observatories (Leiden 1632, Copenhagen 1637, Paris 1667, Greenwich 1675) produced a step function in the number of working astronomers.

Let us also speculate as to Horrocks' confidence in his prediction. Venus transits occur at inferior conjunction. Horrocks' prediction of the time of the 1639 Venus inferior

Table 1. Prominent scientists around the time of the 1639 transit.

Name	Birth/Death	Age at transit	Country or Place
Elias Ashmole	1617 – 1692	22	
Adrien Auzout	1622 – 1691	17	
Isaac Barrow	1630 – 1677	9	
Ismael Boulleau	1605 – 1694	34	France
Robert Boyle	1627 – 1691	12	
John Cutler	1608 – 1693	21	
Ren Descartes	1596 – 1650	43	Holland
Giovanni Domenico Cassini	1625 – 1712	12	Genoa
Johannes Fabricus	1587 –	52	
Galileo Galilei	1564 – 1642	75	Florence
Pierre Gassendi	1592 – 1655	47	Paris
William Gascoigne	1612? – 1644	27	London
James Gregory	1638 – 1675	1	
Johannes Hevelius	1611 – 1687	28	Gdansk
Robert Hooke	1635 – 1703	4	
Jeremiah Horrocks	1619 – 1641	20	Much Hoole
Christiaan Huygens	1629 – 1695	10	den Hague
Gottfried Kirch	1639 – 1710	0	
Germiniano Montanari	1633 – 1687	6	
Jonas Moore	1610 – 1679	29	London
Robert Moray	1608 – 1673	31	
Blaise Pascal	1623 – 1662	16	
Giovanni B. Riccioli	1598 – 1671	41	Bologna
Christoph Scheiner	1575 – 1650	64	Vienna / Nysa
Christopher Towneley	1604 – 1674	35	
Seth Ward	1617 – 1689	22	
Godefroy Wendelin	1580 – 1660	49	Bruxelles
John Wilkins	1614 – 1672	25	
Christopher Wren	1632 – 1723	7	

conjunction differed from those in other tables. Most gave the time as being hours sooner than he had calculated. Some even predicted that it would take place on the previous day (see Ferguson 1803). Horrocks not only observed the Sun on the Saturday, he also observed on Sunday from dawn, in case he might miss it. He might have thought, at the tender age of 20, that it was somewhat presumptuous to send his prediction to those more senior (and superior) members of the English intelligentsia with astronomical leanings, people with whom he was unacquainted. Sending letters in those days was not inexpensive and he might have thought it somewhat cheeky to write to the likes of Elias Ashmole (1617 – 1692), John Cutler (1608 – 1693), Jonas Moore (1610 – 1679), Robert Moray (1608 – 1679), Christopher Towneley (1604 – 1674), Seth Ward (1617 – 1789) and John Wilkins (1614 – 1672). It seems that the only letter he wrote on the subject was to his close friend William Crabtree.

(2) No one else was observing the Sun. Let us consider the European astronomical scene. Fig. 2 shows an expanded version of Fig. 1. In the region to the right of the bold line the Sun had set before transit started. Observers just to the left of the line would have been able to see the disc of Venus on the eastern limb of the Sun at sunset. Due to the absorption of sunlight in the Earth's atmosphere, and the size of the disc of Venus (just over one minute of arc in diameter), this could have been detected by the naked eye. The further one moves to the left of the bold line (i.e. westward) the longer the transiting Venus would have been in view. Horrocks, for example, first caught a glimpse

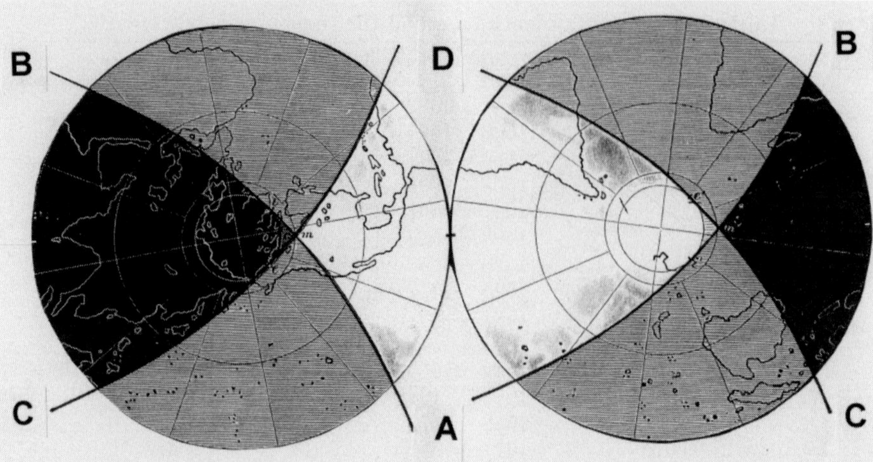

Figure 1. The visibility of the transit of December 1639. Everyone above the two lines marked AB would have been able to see the start of the transit. This region includes a considerable portion of Africa and western Europe and very nearly the whole of North and South America. Everyone below the lines marked CD could see the end of the transit. (This figure has been adapted from Plate III in *Transits of Venus* by R. A. Proctor, published by R. Worthington and Co., New York, 1875).

at 3:15 pm, and made other observations at 3:35 and 3:45. He noted that the Sun set at 3:50 (see, for example, Ferguson 1803, p. 497) this being about 5 minutes after the Sun had a solar zenith angle of 90°, due to atmospheric refraction). These times would have been 'Hoole' local time. The equation of time was about +3 minutes and it would take the Sun about 11.5 minutes to travel the 2.9° longitude west to Hoole, from the Greenwich meridian.

Fig. 2 shows that astronomers in places like Paris, Marseilles and Genoa would have had a similar amount of pre-sunset transit time as did Horrocks and Crabtree. Unfortunately, as one moves to the more favourable observing locations, such as Dublin, Barcelona, Madrid and Lisbon, astronomers become rather thin on the ground.

In the quest for 'other observations of the 1639 transit', one should really concentrate on the solar observers. This field was suffering from one of the typical astronomical lulls that often occur after a fast start. The work of Johannes Fabricius (1587 – ; the Frisian astronomer who wrote *De maculis in sole observatis* 1611), Galileo Galilei (1564 – 1642, who wrote three letters about sunspots in 1613; see *Istoria e Dimostrazioni intorno alle Macchie Solari*), John Greaves (1602 – 1652, Savilian Professor of Astronomy at Oxford University), Thomas Harriot (c1560 – 1621, England's first telescopic observer) and Christoph Scheiner (1573 – 1650, German professor at University of Ingolstadt) revolutionised solar astronomy. Sunspots were discovered to be actual 'blemishes' on the solar photosphere, even though Scheiner for many years thought that they were orbiting inner planets. Most early astronomers observed the Sun by projecting the solar image onto a white screen placed a convenient distance behind the eyepiece of their refracting telescopes. This was the approach used by Horrocks. Fig. 3 shows the Scheiner observing system. Interestingly, this image was shown to the artist Ford Madox Brown (1821 – 1893) when he was commissioned in 1880 by the city of Manchester to paint a mural of Crabtree observing the transit of 1639 (see Knobel 1903 and Wolf 1968), and he used it as a basis of his painting. Fig. 4 shows the helioscope used by Scheiner. A representation of this equatorial Keplerian telescope was incorrectly used in the Eyre Crowe

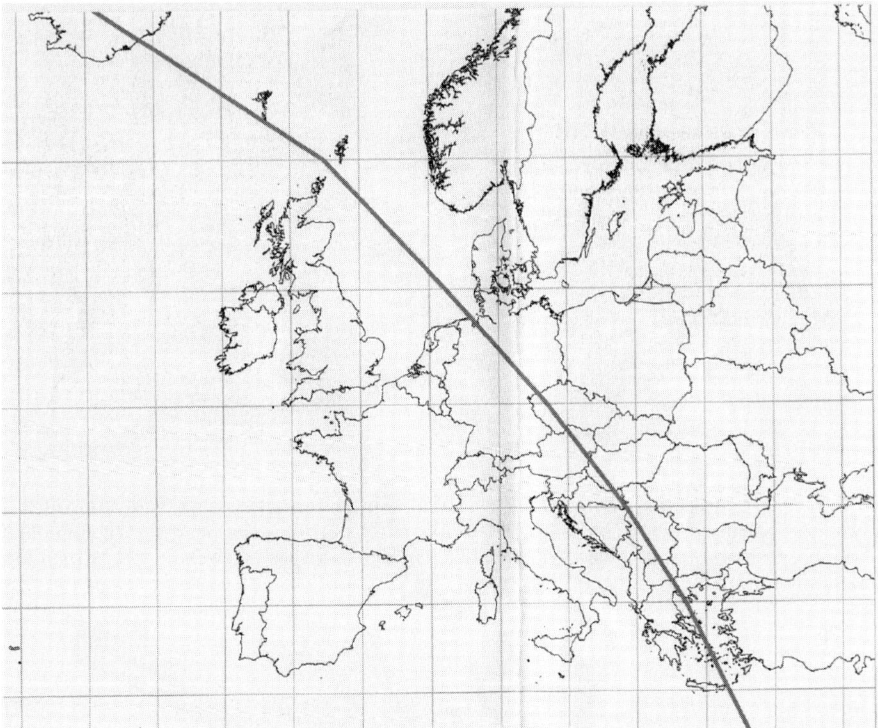

Figure 2. The European part of Fig. 1. The bold line across the diagram indicates the places where the solar altitude is $0°$ at the time of first contact (t_1). This line passes through the following points, (Latitude North; Longitude, degrees) (65; 19.0 W), (60; 2.5 W), (55; 7.0 E) (50; 13.7 E) (45; 19.0 E), (40; 23.3 E) (35; 26.9 E) (30; 30.1 E). These values were kindly calculated by David Sellers, using the algorithm given in Meeus (1989); see also http://www.phys.uu.nl/~vgent/venus/venus_text17.htm. Second contact (t_2) would occurs about 18.5 minutes after first contact. The horizontal altitude lines are spaced by $5°$, the $50°$ North line passing through the tip of Cornwall. The vertical longitude lines are spaced by $5°$. It takes the Sun 20 min to move $5°$ in longitude. Take the $40°$ latitude line as an example. Second contact would have just been visible in the heel of Italy at sunset, whereas observers in Mallorca could have seen Venus on the solar disc for an hour, in Madrid for 1 h 20 min, and in Lisbon for 1 h 50 min.

(1824 – 1910) painting of Horrocks observing the transit. Others (like Harriot) reduced the solar glare by observing near sunrise or sunset. Some also used tinted glass either in front of the telescopic objective or behind the eye lens. Many of them permanently damaged their eyesight.

Sunspot observations had been used to measure the spin of the sun and its rotation period of about 27 d (about a lunar month) about a fixed axis; this was of considerable fascination, especially as it is so much longer than the spin period of Earth. The next breakthrough, by Scheiner, was the discovery that the spin axis of the Sun was inclined at between $6°$ and $8°$ to the ecliptic. Most early observers commented on the fact that the spots were only to be found in a 'royal zone', a region that extended from the solar equator north and south to latitudes of about $30°$. Early solar science culminated with the publication of Scheiner's *Rosa Ursina, sive Sol ex admirando facularum et macularum suarum phenomeno varius* in 1630.

After this the discoveries dried up. Think how long it was before William Herschel (1738 – 1822) measured the ratio between umbral and photospheric radiation, Claude-Sevais

Figure 3. Father Christoph Scheiner (1575 – 1650) using a refracting telescope to project a solar image into an opaque screen. His assistant is apparently marking the positions of sunspots. This image was used as the basis of the Ford Madox Brown painting of William Crabtree observing the 1639 transit.

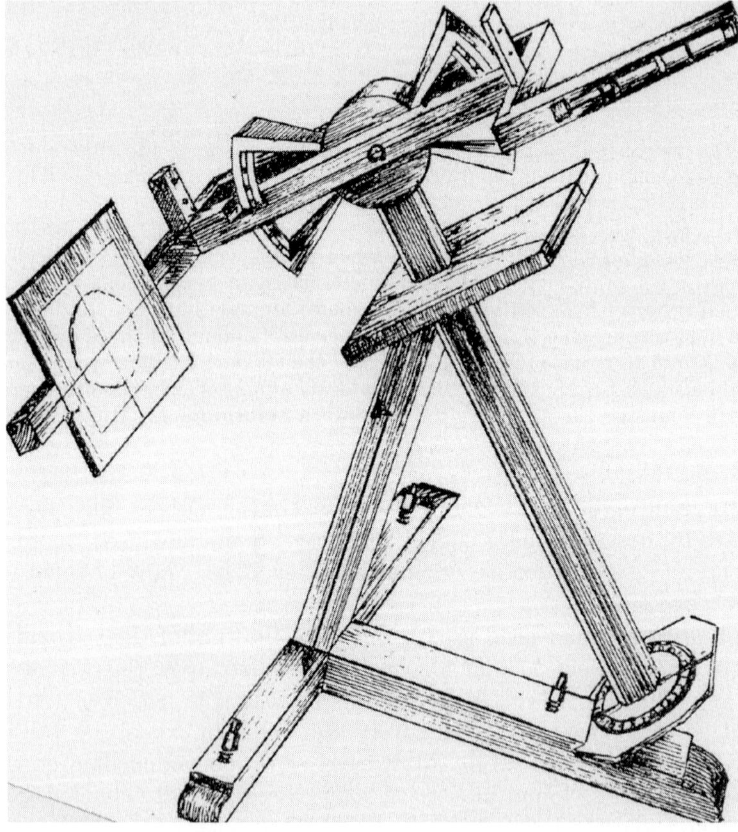

Figure 4. The Scheiner helioscope (see *Rosa Ursina* 1630) follows a principle introduced by Tycho Brahe and rotates about a polar axis. A reversed representation of this pioneering equatorial projection telescope was used in the Eyre Crowe (1824 – 1910) painting of Horrocks observing the transit, even though Horrocks probably used a system similar to that used by Crabtree.

Table 2. Key astronomical dates in the vicinity of the Horrocks Venus transit; 1600 – 1677

1600	Giordano Bruno burnt at stake for views on plurality
1603	Johann Bayer's Uranometria published
1604	super-nova in Ophiuchus, Kepler measures refraction corrections 1608
1608	Hans Lipersheim invents refracting telescope
1609	Kepler's 1st and 2nd laws published in Astronomia Nova
1609	Galileo uses refracting telescope, publishes Sidereus Nuncius, discovers the phases of Venus, sunspots, mountains on the Moon
1618	Kepler's 3rd law is published
1626	Wendelinus measures the rate of decrease of the obliquity of ecliptic
1627	publication of Kepler's Rudolphine tables
1630	Zucchi observes the belts of Jupiter
1631	Gassendi observes transit of Mercury and measures Mercury's diameter (this was predicted by Kepler)
1632	Galileo's Dialogues published
1632	Foundation of the observatory in Leiden, Holland
1633	Descartes' vortices theory is published
1635	Morin sees stars in day-time
1637	Foundation of first national observatory, Copenhagen, Denmark
1637	Horrocks notices inequality in mean motion of Jupiter & Saturn
1638	Phocylides Holwarda discovers first variable star, Mira Ceti
1638	Horrocks explains motion of lunar apsides
1639	Horrocks and Crabtree observe transit of Venus
1647	Hevelius publishes Selenographia and announces lunar libration in longitude
1651	Riccioli publishes Moon map
1651	Shakerley observes transit of Mercury
1654	Huygens gives details of Saturn's rings, and discovers Titan
1657	Foundation of Academia del Cimento, Florence
1658	Huygens makes first pendulum clock
1659	Huygens discovers markings on Mars
1661	Hevelius observes transit of Mercury
1662	Foundation of Royal Society
1663	James Gregory invents 'his' reflecting telescope
1664	Hook measures rotation period of Jupiter
1665	Giovanni D. Cassini publishes Jovian satellite tables
1665	Newton hit on head by apple at Woolsthorpe, eureka gravity moment
1666	G. D. Cassini discovers division in Saturn's rings, and measures rotation period of Mars
1666	Foundation of Academy of Sciences at Paris
1667	Foundation of Paris Observatory, Cassini in charge
1668	Hevelius publishes Cometographia
1669	Newton invents 'his' reflecting telescope
1671	Cassini discovers Iapetus, Richter measures non-sphericity of Earth
1672	Cassini discovers Rhea
1673	Flamsteed explains the equation of time
1675	Römer measures speed of light using Jovian satellites
1675	Royal Observatory, Greenwich, is founded
1677	Halley observes Mercury transit from St Helena and catalogues southern stars

Pouillet (1790 – 1868) measured the solar radiation flux, Heinrich Schwabe (1789 – 1875) discovered sunspot periodicity in the 1843 and Richard Carrington (1826 – 1875) discovered the differential rotation of the Sun in the 1860s. After the 1609 – 1630 excitement solar research went cool and solar observations probably became less and less frequent. The sun was regarded as 'old hat'. On that December day in 1639 maybe only Horrocks and Crabtree were bothering to observe. To quote Meadows (1970), 'The initial burst of enthusiasm for solar observation, which followed the introduction of the astronomical telescope at the beginning of the seventeenth century, waned fairly quickly.' We might go so far as to suggest that the thirty-year interval between the first astronomical use of the telescope and the Venus transit of 1639 was ten years too long. If only the transit had taken place ten years earlier, or the telescope had been invented ten years later, many of the solar observers mentioned above might have been hard at work, observing the Sun at the transit time.

Returning to Fig. 2, however, it is worth noting that Scheiner, living as he was in Vienna, can hardly be blamed. And not only had the sun set for Galileo, in Florence, he had unfortunately gone blind in 1637. Godfrey Wendelin (1580 – 1660) in Belgium was also in the 'dark' zone. The main surprise is that the French did not see it. Forbes (1975) divides the contemporary French astronomical effort into three. There was a group in the south around Aix-en-Provence and Marseilles, including Joseph Gaultier, and Father Pierre Gassendi who concentrated on trying to make new celestial discoveries with telescopes. There was a Parisian school, which included Father Marin Mersenne, Etienne Pascal, and Blaise Pascal, who concentrated on mathematics and theoretical optics problems. And there were the Jesuits who concentrated on teaching and the writing and publication of astronomical textbooks.

(3) It was Sunday afternoon. Let us be rather contentious. There are two other reasons why European astronomers might be excused from seeing the Venus transit (Carole Stott, private communication, 2004). It occurred on a Sunday, the Lord's Day, the day of rest. Maybe the astronomers were at home with their feet up, relaxing after a hearty Sunday lunch. Also, anyone who was logging the disc movement and evolution of sunspots would observe the Sun as early as possible in the day, and make their recordings. After taking the daily readings they would then relax. Afternoon observations were probably less common than morning ones.

2. Conclusions

So the timing of the invention of the telescope, the speedy measurement of the 'easy' solar parameters, the growing realisation that the 1627 Rudolphine Tables were inaccurate, and the timing of Horrock's tabular work, his shyness and his uncertainly about his results, all conspired to give England in general and Lancashire in particular an amazing astronomical first.

Acknowledgements

I would like to thank David Sellers for the preparation of Fig. 2 and Carole Stott for her encouragement and comments.

References

Ferguson, James 1803 *Astronomy Explained upon Sir Isaac Newton's Principles....11th Ed.* J. Johnson et al., London, p. 491.
Forbes, Eric G. 1975 *Greenwich Observatory, Volume 1: Origins and Early History (1675 – 1835* Taylor & Francis, London, p. 4.
Knobel, E. B. 1903, *The Observatory* 26, 424–425.
Meadows, A. J. 1970 *Early Solar Physics* Pergamon Press, Oxford, p. 3.
Meeus, Jean 1989 *Transits* Willmann-Bell, Inc., Richmond, Virginia.
Wolf, A. 1968 *A History of Science, Technology and Philosophy, Volume 1* George Allen & Unwin Ltd, London, p. 79.

David Hughes and Wayne Orchiston

David Sellers discussing Lord Lindsay with Mary Brück;
Takuji Tsujimoto and Naoteru Gouda

Transits of Venus: New Views of the Solar System and Galaxy
Proceedings IAU Colloquium No. 196, 2004
D.W. Kurtz, eds.

© 2004 International Astronomical Union
doi:10.1017/S1743921305001353

The Brazilian contribution to the observation of the transit of Venus

Ronaldo Rogério de Freitas Mourão

Museu de Astronomia e Ciências Afins do Rio de Janeiro and Universidade do Vale do Acaraú,
Sobral, Ceará, Brazil
email: mourao@ronaldomourao.com

Abstract. During the second half of the nineteenth century Brazilian astronomers participated in the observations of transit of Venus. In 1874 a Brazilian astronomer, Francisco Antônio de Almeida was sent by the Imperial Observatory to Nagasaki to use the "photographic revolver" invented by Jules Janssen. In 1882 three missions were sent by the Imperial Observatory (Brazil) to observe the transit in St Thomas (Antilles), Punta Arenas (Chile) and Olinda (Brazil). The value of the solar parallax obtained by the Brazilian Commission, led by Luís Cruls, was 8.″808, representing at that time one of the most precise values.

1. Introduction

The passage of planet Venus across the solar disk occurred twice in the 19th century, once in 1874 and again in 1882. During the occasion of the Venus transit on 9 December 1874 in Nagasaki, Japan, a young Brazilian astronomer, Francisco de Almeida participated in the French mission operating the French astronomer Jules Janssen's (1824–1907) astronomical revolver, considered the predecessor of the movie system.

During the transit of 6 December 1882, Brazil participated in the great first international enterprise of basic science, establishing three posts of observation out of Rio de

Figure 1. (Left panel) Francisco Antônio de Almeida. The Brazilian astronomer was commissioned and sent by Conde Prados, director of the Imperial Observatory of Rio de Janeiro, to study astronomy in France, when he was invited to participate in the French commission in Nagasaki, where he was responsible for the use of Jules Janssen's photographic revolver (1824–1907). With this device it was possible to obtain very close pictures that were produced on a type of film through which Venus was visualized in front of the Sun. (Right panel) Photographic revolver built by Rédier, used in Nagasaki in 1874. This invention is considered one of the prototype of the today's movies.

Figure 2. (Left panel) The Brazilian astronomer Julião of Oliveira Lacaille (1851–1926) who graduated in Engineering in France was the responsible for the observations of the passage of Venus by the solar disk in Olinda, 6 December 1882. (Right panel) Luiz Cruls, Chief of the Brazilian Mission, Brazilian astronomer of Belgian origin (1848–1908). Graduating in Engineering at the University of Gand, Cruls remained in the Belgian army up to 1874, when he asked for dismissal and came to Brazil. Invited and commissioned by Emperor Dom Pedro II, he was admitted at the Imperial Observatory under the administration of Emmanuel Liais for whom he substituted. He was the chief of the Brazilian mission that observed the transit of Venus, in Punta Arena, in 1882.

Janeiro: one in Olinda under the leadership of astronomer Julião de Oliveira Lacaille (1851–1926; Fig. 2 left), and the two others in St Thomas island in the West Indies, and in the city of Punta Arenas in Patagonia, respectively, under the command of hydrograph engineer Antonio Luis von Hoonholtz, a.k.a. Baron of Tefé (1837–1931), and of astronomer Luís Cruls (1848–1908; Fig. 2 right). In Pernambuco and in the West Indies, in spite of non-favourable weather conditions, there still was some success. The great success, however, was the mission to Patagonia, where Luís Cruls in Punta Arena observed all the phases of the phenomenon. As a result of that, several scientific communications with relative results to astronomy and to natural history accomplished by the Brazilian scientists were presented by Emperor Dom Pedro II to the Academy of Sciences at Paris.

The principal aim of the missions was to determine the distance from Earth to the Sun, one of the most fundamental concerns of astronomy. Besides being a unit of length in the solar system, the distance from Earth to the Sun is the base used to measure the distance to the stars. All the distances measured in the universe depend upon the determination of the ray of the orbit of planet Earth revolving around the Sun. To measure the distance of an object, be it terrestrial or celestial, we have to try to observe them from two different points, separated amongst themselves by a certain distance. The farther away the object to be observed, the greater the separation of the positions of observation needed. Thus, for the objects located within the solar system two distant points of observation from the surface of the Earth should be enough, whilst for a star, away from our solar system, it is necessary to use the diameter of the orbit that Earth describes around the Sun in one year.

Figure 3. (Left panel) The Commission of St Thomas, West Indies: interior of the east part of the central pavilion of the Brazilian Observatory. In this place the passage of the planets was observed by the meridian. From that observation rigorous determination of the local time was made, a fundamental factor to accurately measure the contact of Venus entering the disk of the Sun. (Right panel) The Commission of Punta Arenas, Chile: The Brazilian astronomer of Belgian origin Luiz Cruls was the chief of the Brazilian mission to Punta Arenas. The clear visibility of the Chilean sky allowed the whole phenomenon to be registered.

Up to the establishment of Kepler's three laws that rule the movements of the planets, especially the third law, the absolute distances of the heaven bodies in the solar system were completely erroneous (although the relative distances were well-known from the time of Copernicus). According to these laws, there is an exact relationship between the squares of the planets' orbital periods and the cubes of the radii of their orbits. Indeed, this law constitutes a link of connection of all bodies of the solar system amongst themselves. In fact, when determining the distance of any planet to Earth it is possible to find the distance of another planet from the sun due to their revolution periods around the Sun. In the observation of planet Mars in 1672, that the French astronomer of Italian origin, Giovanni Cassini (1625–1712), determined the distance of the Sun to be 140 million km, multiplying by 20 times the distance of the Sun to the Earth, consequently determining all the other measures of the solar system.

Taking this advantage from English astronomer James Gregory's (1638–1675) idea, astronomer Edmund Halley (1656–1742) suggested the observation of the transit of Venus across the solar disk as a method to best calculate the solar parallax. The passages or transits of Venus occurred when Earth, Venus and Sun are aligned in relation to a located observer in the terrestrial surface. At this time, Venus is projected as a black disk on the brilliant larger disk of the Sun. Halley suggested that the solar disk could be used as a background reference, making it easier to determine the movement of Venus. Thus, no device for measuring angle would be needed, because Halley suggested that the instant of contact should be measured of the black disk of Venus with the limb of the Sun. The use of this method justified at the time the rudimentary devices for measuring angles.

Transit of 1874

In 1874 several expeditions were sent to China, Japan, India, Egypt, Russia, Gulf of Bengal, Sumatra, Java, St Paul Island, Saigon and Australia, by France, Germany,

Figure 4. Francisco Antônio of d'Almeida, doctor of Physics and Mathematics who was sent by Conde Prados, director of the Imperial Observatory of Rio de Janeiro, to study astronomy in France. When participating in the French commission's transit of Venus in Nagasaki in 1874, Almeida was the responsible for the use of the photographic revolver created by the French astronomer Jules Janssen (1824–1907). This device is considered one of the precursors of the movies.

England and the USA to observe the phenomenon. In Nagasaki, as a member of the French mission, a young Brazilian astronomer, Francisco Antônio de Almeida (known in France as d'Almeida; Fig. 4), was responsible for the operation of the French astronomer Jules Janssen's (1824–1907) photographic revolver, a device considered as the precursor of cinema. The value found for the AU by several commissions was of 148 900 000 ± 2 000 000 km.

Transit of 1882

Immediately a campaign in the press and in the Parliament began against sending expeditions abroad. Whilst the Brazilian caricaturist, of Italian origin, Ângelo Agostini (1843–1910) made people laugh with his political cartoons against the Emperor and his idea of sending two missions abroad to observe Venus pass before the Sun. The senator Ferreira Viana (1834–1905), in the Parliament violently criticized the government's conjectures:

"Let's have this clear therefore, Minister of the Empire certainly that our astronomical missionaries will play the part of conductors of instruments for the wise men of France (laughs); on the other hand (the Parliament should admit the familiar sentence) they will make a big convescotte (laughs) in that we participated with the material part and French with the spiritual one, with the astronomical talents, with science, with the practical knowledge that they have acquired in similar observations and in the continuous study of the matter. Foreign missions. The mistake was reduced in half (1 000 000 km). Almost simultaneously, the determination began for the asteroids. The parallax of the Sun is an astronomical problem, that it is

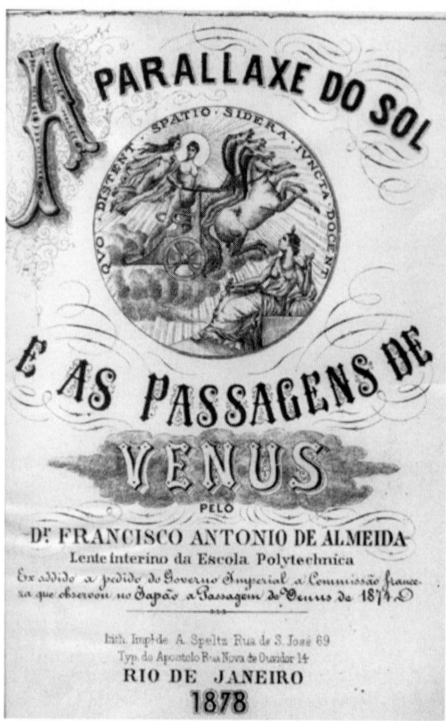

Figure 5. Frontispiece of the first book published in Brazil on the parallax of the Sun and the transit of Venus by the Brazilian astronomer Francisco Antônio of Almeida.

not vital; it doesn't have practical results to any human comfort or social interest, and it doesn't only depend upon the transit of Venus. Laplace already had made by the movement of the Moon; Priest Secchi refers to the process of speed [note of the author] of the light. There are more than ten or twelve processes to come to the foreseen result; and many authorities in the matter assured that none of them is evident and for sure. Consequently, it is a very problematic interest, even in the deep abstract root of science, however that we will put on an amount of money that has nothing of abstract, it is very positive (laughs)." (Annals of the Parliament, Section of March 22 of 1882, p.7).

Perhaps this was one of the most vivid debates ever to happen in Brazil about the use of basic science.

Actually, critics were justified in view of the determination of the solar parallax starting from the observation of the asteroids which had already begun to yield good and precise results compared to those obtained based in the observations of the passages of Venus. In fact, after having observed in 1874 on Mauritius Island, the Scottish astronomer Sir David Gill (1843–1914), disappointed with the method of the passage of Venus to determine the solar parallax, decided to use the planet Mars to settle the problem. For this purpose he decided to built an observatory on Ascension Island in the Atlantic ocean, where he accomplished a series of excellent photographic observations to measure the parallax of Mars in 1877 and, later, he used the same methods with recently discovered asteroids. As a result of those measures, the solar parallax was determined to be $8''.806$. For that reason Gill was not very enthusiastic to make any effort to observe the passage of Venus in 1882. In fact, in North America, in spite of the opposition of the president of the Congress (James Abram Garfield, 1831–1881 – later the president of the United States of America), a budget of about of US$175 000 was approved for the one mission

Figure 6. (Left panel) Frontispiece of Revista Ilustrada of 10 December 1880 by Ângelo Agostini (1843–1910) criticizing the budget expenditure directed to the observation of the transit of Venus by the solar disk. (Right panel) Frontispiece of Revista Ilustrada of 7 December 1882 by Ângelo Agostini (1843–1910) criticizing the difficulty in observing the transit in Rio de Janeiro, because of the foggy and rainy weather.

for the observation of the passage of Venus in 1874 in Nagasaki, in which the Canadian astronomer Simon Newcomb (1835–1909) participated. For the transit of 1882, the Congress of the USA approved a budget of US$85 000 without any political opposition, in spite of several astronomers' objection regarding the importance of the mission. Among them was Newcomb who had participated in the 1874 mission in Nagasaki. Once the resource was approved, he agreed to lead the mission to Wellington, South Africa, although he believed that the funds could have given better results with some other objectives. Several scientific bulletins with astronomy and natural history results were presented by Emperor Dom Pedro II to the Academy of Sciences at Paris for the Brazilian scientists.

The determination of the solar parallax by the Brazilian Commission, of $8.''808$ ($149\,400\,000 \pm 1\,000\,000$ km) represented at that time one of the most precise values. However, the error determined by Cruls is exactly the mentioned value in the speech made by Ferreira Viana. To have an idea, the accepted value in 1900 was of $8.''806 \pm 0.''004$ ($149\,000\,000 \pm 100\,000$ km), based on Eros' parallax and calculated by the English astronomer Arthur Robert Hinks (1873–1945).

Until the appearance of radar astronomy, Eros' positional observation during its closest approach to Earth provided the best means to determine the solar parallax. With this purpose, thousands of measures were accomplished during the favourable 1900–01 appearance and 1930–31. The discussion of all of observations from 1893 to 1966 allowed astronomer J.H. Lieske to conclude that the solar parallax was of $8.''79402 \pm 0.''00012$. This is almost exactly the length of astronomical unit, $149\,600\,000$ km, adopted by the International Astronomical Union with base upon the measures of the radar astronomy of

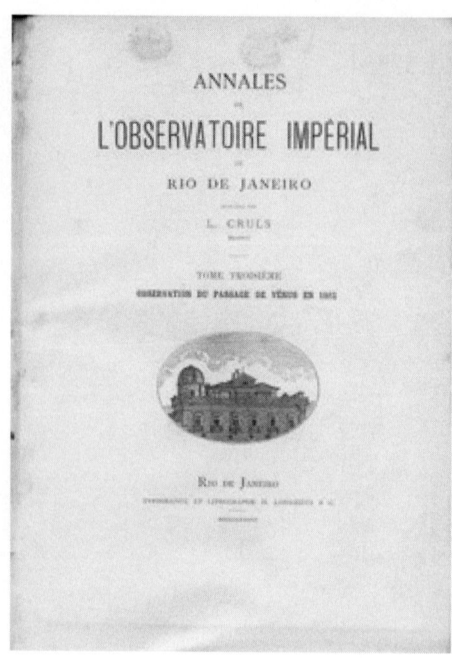

Figure 7. Frontispiece of the "Annals of the Imperial Observatory of Rio de Janeiro",
published by Luiz Cruls, exclusively dedicated to the observation of the transit of Venus in
1882.

the planetary distances. In 1974, the International Astronomical Union adopted the solar
parallax of $8.''794148 \pm 0.''00007$ ($149\,598\,870 \pm 1\,000$ km), the average value determined by
the USA, USSR and England through radar measurements at the beginning of the 1960s.
In 1976 the International Astronomical Union adopted the value $149\,597\,870 \pm 2$ km, obtained based on the light travel time by unit of distance.

In the 19th century Brazil honoured the commitments assumed in Paris, in 1881,
collaborating to determine this fundamental astronomical constant. It was a precious
opportunity for Brazilian astronomy that never again participated in international efforts,
since the observation of Eros in 1900–01 and 1930–31, to determine this measure with
greater precision.

The creation of a Museum for the Passage of Venus, in Olinda, to commemorate the
first participation of Brazil in a scientific enterprise of international repercussion is quite
justified in view of the fact that a resolution of the International Astronomical Union
decided that the sites where the passages of Venus were observed should be preserved,
as well as documents and instruments used during the events. In Brazil, we have a good
example of Sobral City Hall, in Ceará, that next to the commemorative monument of the
site for the observation 1919 eclipses, confirming the theory of the relativity, a Museum
of the Eclipse was built, providing a wide exhibition room and an enclosed observatory
to the building of the Museum, with the objective to publicize the event and to register
this important scientific effort.

Part 2

THE AU AND THE PC

Nicole Capitaine relaxes in Much Hoole churchyard after the transit

Myles Standish talking to Floor van Leeuwen

Transits of Venus: New Views of the Solar System and Galaxy
Proceedings IAU Colloquium No. 196, 2004
D.W. Kurtz, ed.

© 2004 International Astronomical Union
doi:10.1017/S1743921305001365

The Astronomical Unit now

E. M. Standish†

CalTech / Jet Propulsion Laboratory, JPL 301-150, Pasadena, CA 91109, USA
email: ems@smyles.jpl.nasa.gov

Abstract. The Astronomical Unit is one of the most basic units of astronomy: the scale of the solar system. Yet its long and colorful history is sprinkled liberally with incorrect descriptions and mis-quoted definitions – today as much as ever. Over the last half century, the accuracy of the au determinations has improved dramatically: optical (triangulation) methods have given way to modern electronic observations, high-speed computers, and dedicated efforts to improve planetary ephemerides. Typical uncertainties in the value of the au have decreased from many tens of thousands of kilometers to the present level of only a few meters. With the solar system providing a very clean, undisturbed dynamical model, the ephemerides have been used for a variety of exotic physical tests: alternative theories of gravitation, $d(G)/dt$, $d(\mathrm{au})/dt$, etc. In the beginning of this modern era, the author happened to be a witness to a couple of rather key events; more lately, a participant. A couple of these personal experiences are related.

1. Personal recollection

In the autumn of 1962 there was a discrepancy: the new determinations of the value of the au, made using radar measurements, did not agree with the classical determinations, made with optical (photographic) measurements of the asteroid 0433 Eros.

The very first Wednesday afternoon seminar that I attended as a graduate student at Yale University featured three speakers from the astronomy department: Dirk Brouwer described the optical determinations, James Douglas presented the radar values, and Ludwig Oster correctly argued why the effects of the solar corona upon the radar signal were not large enough to account for the discrepancy.

So, the matter was unresolved – an awakening in itself for me.

The next summer in the university's computer center I happened to speak to Brian Marsden: "How are things going?" His reply: "I'm trying to duplicate Rabe's solution, but I can't seem to reproduce his numbers."

Again by chance, in the following winter during a meeting at Yale organized by Brouwer, I happened to overhear Marsden telling Eugene Rabe: "I can't seem to reproduce your results." Rabe had not known of any problem and had no explanation.

The matter was virtually resolved over the next few years, as it was found that 1) the parameters involved in the optical solutions were highly correlated, 2) there were some errors in Rabe's partial derivatives, and 3) the sum-of-squares of the residuals had a very flat minimum. The thesis of J. H. Lieske (1968) provided the final assurance: he collected 8639 observations of Eros, covering the years 1893-1966, taken from 85 different observatories, based upon 106 different catalogues, each reduced to the FK4 stellar catalogue. Lieske adjusted the ephemerides to the observations using an integration of all 9 planets and Eros. The amended parameters were in accord with those from the radar determinations.

† Present address: JPL 301-150, Pasadena, CA, 91109, USA

I had been a non-participating observer of this little step of astronomical progress; it was later that I would take an active part.

It is worthy of note that the 1964 winter meeting started a tradition which was repeated for a number of years; eventually, the meetings, along with a nucleus of the participants, evolved into the American Astronomical Society's Division on Dynamical Astronomy.

2. What is the "au"?

There have been many incorrect descriptions and mis-quoted definitions of the astronomical unit. Even as recently as a month ago, there appeared in a popular magazine, the following:

> " ... table (page 68) says that the astronomical unit (a.u.) is 'based upon the mean Earth-Sun distance' – it's not the mean Earth-Sun distance itself. We now know enough about planetary motions to realize that Earth's average distance from the Sun is not fixed. It changes from one orbit to the next ..."

> " ... astronomers now treat the a.u. as a *defined* quantity rather than a measured one... the a.u. is the radius of an unperturbed circular orbit about the Sun with a period of 365.2568983 days (known as a Gaussian year). This works out to 149,597,870 kilometers ..."

Actually, the au is based upon the Gaussian constant, $k \equiv 0.01720209895$ (exact definition). This, in turn, is an old measurement of the Earth's mean motion.

In physics, one adopts units of length, mass, and time (cgs, e.g.); then, experiments provide the value of the gravitational constant, G. In astronomy, since a period or mean motion is much more easily measured than a distance in the solar system, the adopted units were chosen to be those of a solar mass, a mean solar day, and the gravitational constant $(= k^2)$. The au is then the unit of length which is consistent with the other three. As such, it is the result of a convention; it is not a defined quantity.

One equation which relates the au to the other units is Kepler's third law, $n^2 a^3 = k^2 M$. For a (massless) particle at 1 au from the sun in keplerian motion, we have $a = 1$ and $M = 1$, so that the mean motion is $n = k$. Thus, the period is simply $P = 2\pi/k = 365.2568983...$ days; this is the source of that (irrational) number in the "definition" quoted above.

Incidentally, even in keplerian motion, where "the Earth's average distance from the Sun [would be] fixed", neither the au nor the semi-major axis would be equal to the average distance. In keplerian motion, the mean distance is not the semi=major axis, a; instead, $< r > = a(1 + e^2/2)$. It *is* true, however, that $< 1/r > = 1/a$.

The last statement quoted above is the worst. Nothing "works out" to give the value of the au in kilometers. That number has been a holy grail for a number of centuries and has been the *raison d'etre* for the immense efforts put into the measuring of the transits of Venus, the conjunctions of Mars, the approaches of asteroids, and other endeavors.

As indicated above, the radar determinations of the value of the au expressed in kilometers are now more accurate than the optical determinations. However, both methods have a lot of similarities. In order to measure the au in kilometers, one measures some distance in the solar system, either by optical triangulation methods using earth distances as a baseline, or by recording the travel time of an electromagnetic signal along the path and converting that time into kilometers, given the speed of light. The value of that distance in kilometers is then compared with a corresponding value of the distance in au, obtained from an accurate planetary ephemeris.

The key issue is that accurate planetary ephemerides don't just happen to "work out" from a few equations. Modern ephemerides represent possibly the greatest dynamical system of all time: it has virtually no dust, no static electricity, no friction; it has been accurately measured over a substantial period of time; and it is now accurately modeled with high-speed computers and an understanding of gravitational physics. The modeling of the solar system dynamics is an extensive topic.

3. The creation of modern ephemerides

Back in the 1960's, it became apparent at JPL that the existing ephemerides were not accurate enough to support the increasingly stringent demands of spacecraft navigation. Those involved at that time made decisions that still to this day show a remarkable amount of foresight. The overall system continues to function as well as ever, even with many orders of magnitude increases in accuracy, both of the observational data to which the ephemerides are adjusted and of the resultant ephemerides themselves.

The ephemeris creation process may be looked upon as an attempt to play the part of Mother Nature: she has a set of laws to follow – the equations of motion; she is a very good numerical integrator of those equations; and she started at some time with a set of initial conditions – positions and velocities at some epoch, along with a number of associated constants, such as masses, etc.

The ephemeris creation process has the same three key ingredients:
   (*a*) the Equations of Motion, expressing gravitational physics,
   (*b*) an Integration Program, and
   (*c*) a set of Initial Conditions ($\mathbf{r}, \dot{\mathbf{r}}$, GM_i, etc.).

It is believed, in general, that a) the physics is well-known and b) the integration program has been adequately tested and been shown to be valid. The main key of the ongoing effort of ephemeris improvement, therefore, is c) the set of initial conditions.

The initial conditions are determined by a least-squares adjustment, fitting ephemerides to the set of observational data. As such, the accuracies of the ephemerides are a direct function of the accuracies of the observational data and their reductions.

The reductions follow basic mathematics and physics, for the most part, applying reference frame transformations, tracing the paths of electromagnetic signals, etc. In addition, there are certain adjustment factors, judiciously used where warranted (e.g., a transponder time delay, assumed to have not been previously calibrated.)

So, apart from the reduction procedures, it is the set of observational data which requires the most attention.

4. The observational data

The observational data are the products of many, many dedicated individuals, endowed with ingenuity and dedication: quadrants, telescopes, micrometers, photographic plates, meridian circles, traveling impersonal micrometers, cavity magnetrons, electronic computers, atomic clocks, radio telescopes, interplanetary spacecraft, VLBI, etc. And, the results of this advancing array of technology are reflected in the advance of accuracy in the observational ephemeris data.

References to the different sets of ephemeris data, as well as the data themselves may be found at the following website: http://ssd.jpl.nasa.gov/plan-eph-data/index.html

4.1. *Optical observations*

Until about 50 years ago, the only data used for ephemerides was optical – transit timings, photographs, micrometers, astrolabes. For these, a typical observational error has been on the order of one arcsecond [1″], improved over the years with significant effort, to about 0″.4 for the innermost four planets. This range corresponds to hundreds of kilometers. For the outer planets, the accuracies are now down to 0″.2, and in some cases, even 0″.1; here again, these correspond to hundreds, if not thousands, of kilometers.

4.2. *Mercury and Venus radar*

Starting in the 1950's, radar signals were bounced off the surface of the moon at first, then Venus, and then Mercury and Mars. The round-trip times of these signals immediately gave more that an order of magnitude increase in the accuracy of the observational data. Fig. 1 shows radar residuals of Mercury and Venus over the past decades. Even the earliest points show, for the most part, residuals below the 5-km level. And, much of the remaining scatter was caused by the variations in the planet's topography, typically variations of a couple of kilometers. Nowadays, for Venus, a topographic map, obtained from the Pioneer Venus Orbiter spacecraft, is used to eliminate most of that topography from the residuals; residuals are on the order of 1-2 km. For Mercury, an ellipsoidally-shaped planet is adjusted to the residuals.

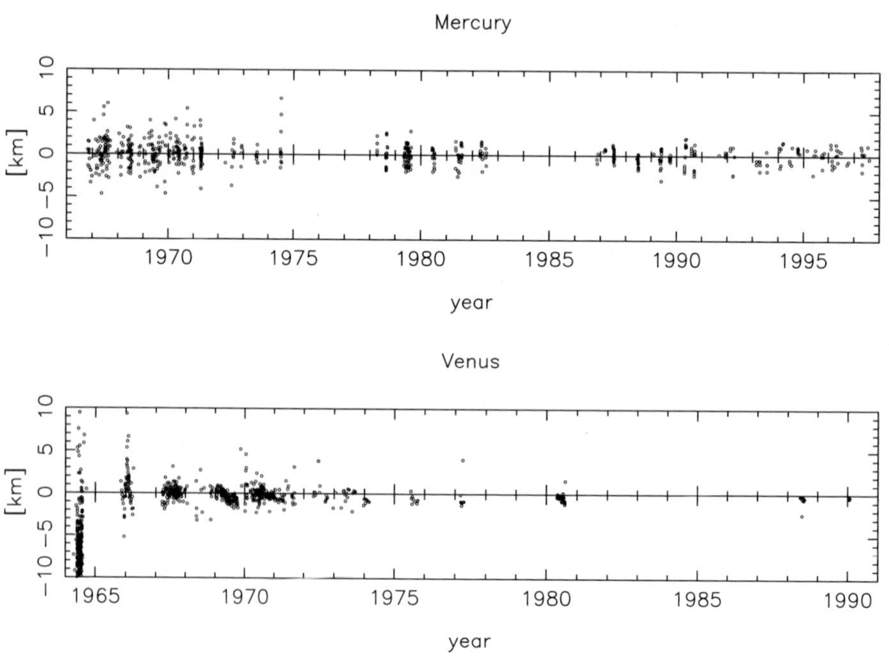

Figure 1. Radar ranging resiuduals of Mercury and Venus.

4.3. *Mars radar closure points*

On Mars, the topographical variations are much more severe. And, with the rapid rotations of that planet, the variations can be seen to change drastically within an observing run of only a few minutes. Fig. 2 shows residuals from two radar tracks on Mars, intentionately offset from each other by about 1.5 km in the vertical direction. They were taken on two different days, separated by over two years, and they have the property that the radar echoes of both days were reflected from spots on Mars with latitudes −18°.2 and with longitudes between 280° and 360°. Thus, the two tracks are composed of pairs of echoes where each member of the pair reflected off from the same spot on Mars;

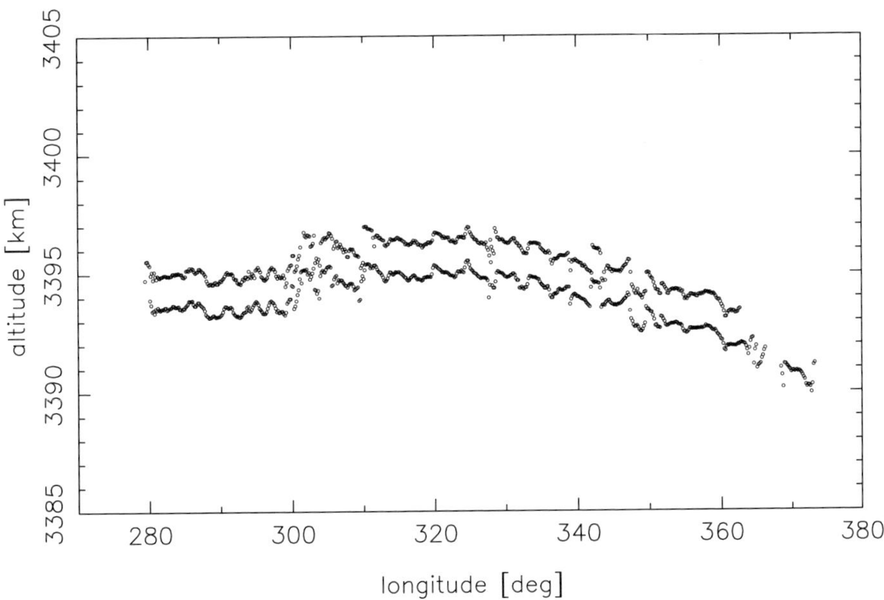

Figure 2. Two sets of Mars ranging residuals, 1971 and 1973, intentionally offset from each other by 1.5 km. On the two days shown, the radar bounced from the same locations on the surface of Mars. If the vertical offset were real, it would indicate an ephemeris drift between the two dates. The similarity of the two tracks shows how accurately one could determine such an offset – to an accuracy well below 100 m.

subtracting one from the other eliminates the topography and measures the ephemeris drift between the two days – determined with an accuracy well below 100 m.

Fig. 3 shows four such radar tracks, again intentionally offset in the vertical direction. These four were taken at differing latitudes, in order to show how the topography can vary as the latitude changes. The peaks of the tracks at longitude 120° are from the south flank of "Arsia Mons", the southernmost of the three prominent volcanoes, "Tharsis Montes".

4.4. *Mariner 9 range residuals*

In contrast to the radar-ranging which bounces off from the surface of a planet, the ranging data from a spacecraft are free from the variations in the planet's topography. Fig. 4 shows the ranging residuals from Mariner 9, in orbit around Mars, 1971–72. The Orbit Determination Program (ODP) at JPL was used to solve for the orbit of the spacecraft with respect to Mars, and the measurements were then reduced to the center of mass of the planet with an uncertainty significantly lower than the uncertainties in the ranging (timing) measurements themselves. For Mariner 9, since the frequency of the signal was relatively low, 2200 MHz, the free electrons in the solar corona contributed significantly to the delay of the signal, especially around the time of Mars' conjunction in late August, 1972. As a result, the points near the conjunction were severely downweighted. Models of the solar corona time-delay have been used with moderate success for removing the majority of the delay. However, it has been seen that the density in the corona can change significantly over the span of just an hour or so. Despite such problems, most of the Mariner 9 Mission yielded ranging measurements with uncertainties substantially below 100 m.

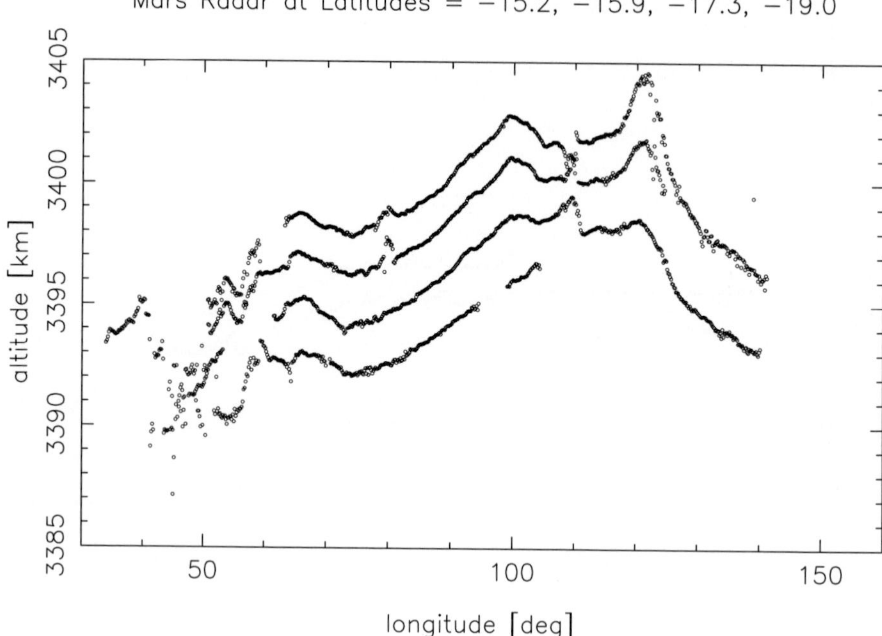

Figure 3. Four sets of Mars ranging residuals, intentionally offset from each other, each at a different latitude on Mars. Near longitude 120°, the tracks run over the south flank of "Arsia Mons", the southernmost of the three prominent volcanoes, "Tharsis Montes". The differences between each set show the variation in topography over the range in latitude.

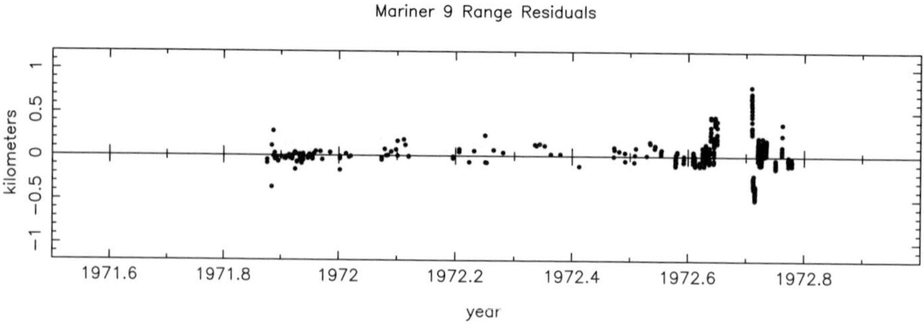

Figure 4. Ranging residuals from the Mariner 9 Spacecraft, 1971–72. Much of the scatter is less than 50 m until the times approaching conjunction when the signal passed through the noisy solar corona.

4.5. *Viking lander range and Doppler residuals*

The Viking Mission sent two spacecraft to Mars, each with an orbiter and a lander. The orbiters had dual frequency transponders; the landers had single frequency ones. Ranges in two frequencies allow the solar corona time-delay to be calibrated because of its dependency upon frequency. Thus, if the orbiter and lander ranges were taken close to each other in time, the delay, calibrated from the orbiter ranges, could be removed from lander ranges. The first plot of Fig. 5 shows the range residuals from the Viking

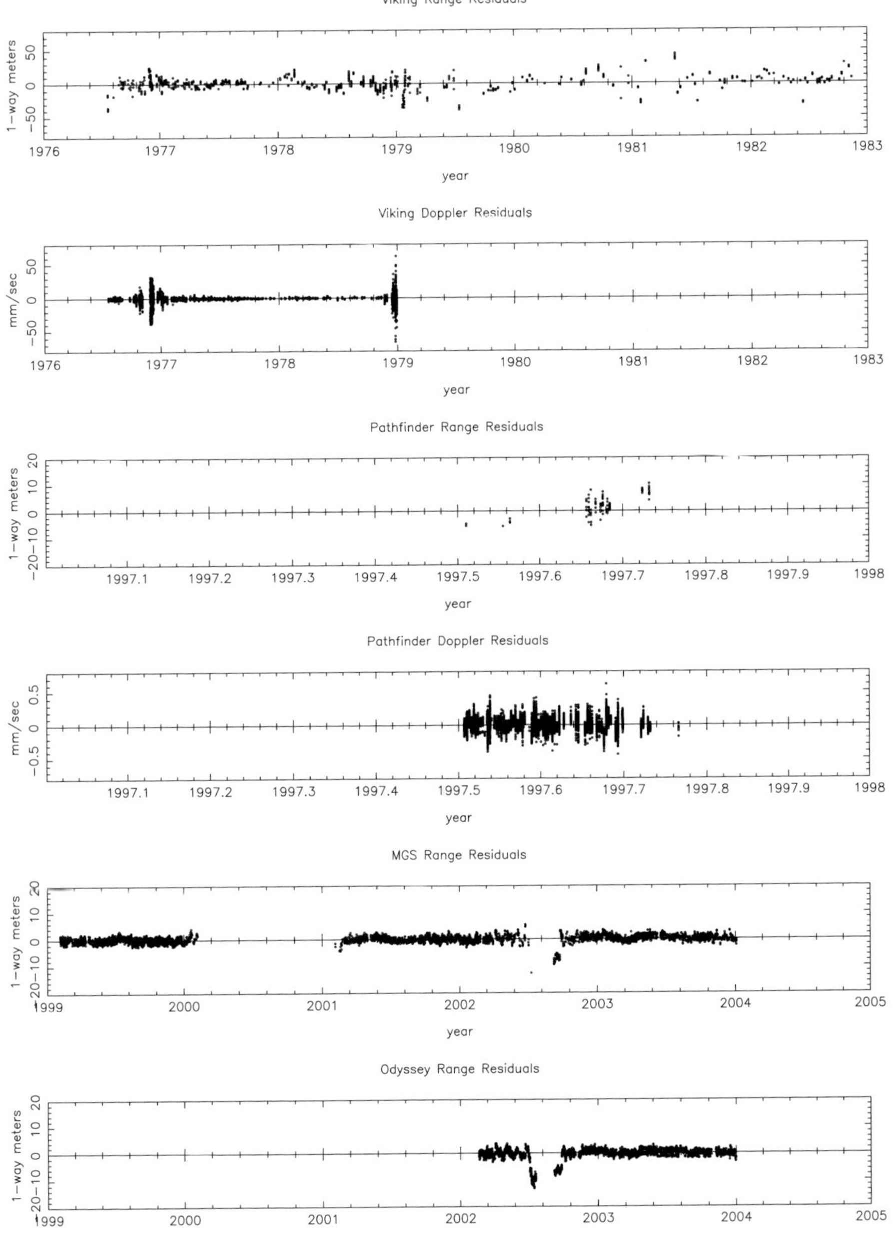

Figure 5. Spacecraft range and Doppler residuals. At the top are the Viking Lander range residuals, with an rms scatter of about 10 m away from the conjunctions. Secondly, Viking Lander Doppler; the noise around the conjunctions is evident. Third and fourth are the Pathfinder ranges and Doppler. The fifth and sixth plots show every tenth point of the over 230,000 range measurements from MGS and Odyssey, respectively. The scatter is about 1.3 m, with the exception of the conjunction in mid-August, 2002.

Landers; these were calibrated from the orbiters' dual frequency data while the orbiters were still active (1976-1980); after that time, a model similar to that used for Mariner 9 was applied. For modeling the lander ranges, one also needs to model the rotation of Mars, correcting the various relevant parameters, as well as determining the locations of the landers upon the surface. Various features of the rotation (precession, seasonal terms, etc.) may be estimated from such data – about 10-m uncertainty.

The second plot of Fig. 5 shows Doppler residuals from the Viking landers; in some sense, these data are redundant, given the existence of the range data.

4.6. *Pathfinder lander range and Doppler residuals*

The lander of the Pathfinder Mission produced a short set of range and Doppler measurements; they are given in the third and fourth plots of Fig. 5. The uncertainties of the ranges are seen to be only a few meters, though there are indications of unmodeled systematic errors, probably due to uncalibrated (and variable) electronic system delays.

4.7. *MGS and Odyssey range residuals*

The most accurate of the Mars data are the range residuals of the MGS and Odyssey orbiting spacecraft. The fifth and sixth plots of Fig. 5 show only every tenth point of the total data set; even so, over 23,000 points are shown. The scatter of these points, excluding those around conjunction in mid-2002, is only 1.3 m, excluding the points within six weeks of the Mars conjunction in mid-August, 2002.

4.8. Δ*VLBI residuals*

The final major set of observational data contains ΔVLBI observations of orbiting spacecraft with respect to the background radio sources, especially those of the (ICRF) International Celestial Reference Frame. Such observations give angular measurements in essentially one dimension: in the direction connecting the two participating radio antennas. Therefore, ΔVLBI observations between the Goldstone and Madrid complexes of the Deep Space Network provide determinations which are almost purely in right ascension, since the latitudes of the two sites are nearly equal. Observations between Goldstone and Canberra, on the other hand, are split about 50–50 between right ascension and declination. Figs 6 and 7 show the ΔVLBI observations presently being fit by the ephemerides of Venus and Mars, respectively. In each figure, two plots are given: the upper in milliarcseconds (mas); the lower in kilometers. The Venus points in Fig. 6, taken of the orbiting Magellan spacecraft, show a scatter of several mas. In contrast, the more recent observations in Fig. 7 show a scatter of less than a single milliarsecond or kilometer: a striking example of technological improvement.

5. The effects of the different types of data

The ephemerides of the four innermost planets and the moon are dominated by the data presented in the preceding section: ranging measurements and ΔVLBI. The ranging measurements, taken over various parts of the planets' orbits, provide all relative distances and angles between the Earth and Mercury, Venus, the Moon, and Mars, thus locking the whole system together. It is also true, though not readily envisioned, that the ranging measurements alone provide accurate mean motions of the planets with respect to inertial space. For a further discussion, see Williams & Standish (1989).

The orientation of the whole system, with respect to some external reference frame, is the only feature *not* provided by the ranging measurements. The orientation is provided by the ΔVLBI measurements which tie the system onto the background ICRF.

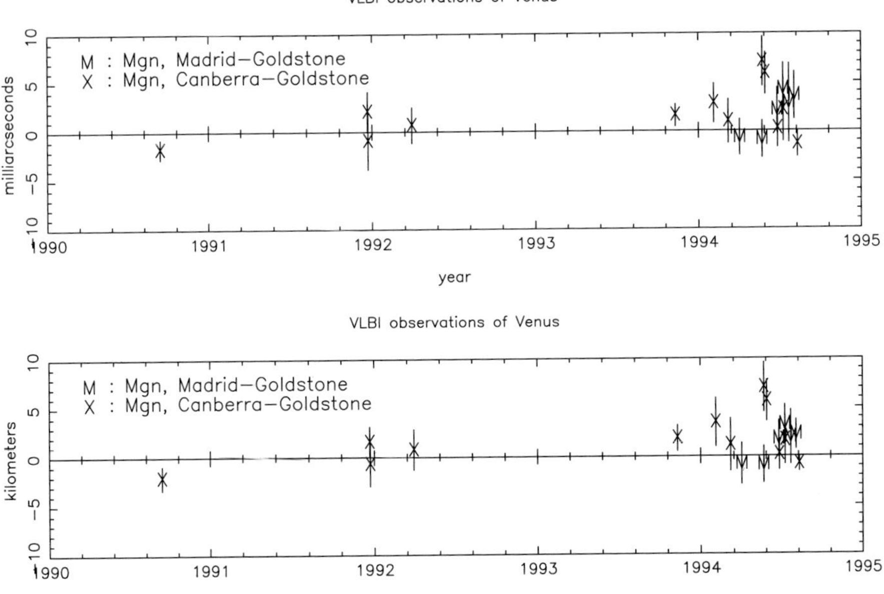

Figure 6. ΔVLBI residuals of Venus from the Magellan orbiting spacecraft, shown both in milliarcseconds and in kilometers. These observations are virtually one-dimensional. Those from Madrid–Goldstone are almost purely in right ascension; those from Canberra–Goldstone are about 50% each in r.a. and dec.

From the ranging and ΔVLBI, then, the relative positions are presently measured down near the 1-m level, and the mean motions are determined at the level of about 10 milliarcseconds/century. The orientation of the system onto the ICRF at the present time is accurate to a fraction of a milliarcsecond. As will be discussed in a later section, however, these accuracies deteriorate in time due to the perturbations of many asteroids whose masses are poorly-known at best, thereby rendering their perturbations not well-modeled.

For Saturn, Uranus, Neptune, and Pluto, there is basically only optical data: meridian circle timings, astrolabe timings, and photographic astrometry. A few various other data points exist, but these provide only momentary fixes in time – not enough for the determination of orbits with periods extending over decades. Jupiter is in between: a few ranges from former missions and some ΔVLBI points from the Galileo mission. So, in contrast to the inner planets, the ephemeris uncertainties for the five outermost planets remain above the 100-km level; substantially more for the outermost ones.

6. Testing with the ephemerides

It is no wonder that the inner solar system attracts those wishing to test various gravitational theories, asking if a modified set of equations of motion (i.e., an alternative theory of gravitation) can better fit the observational data. For instance, the PPN parameters of relativity, β and γ, are conventionally assumed to both be equal to unity; but, since they are programmed explicitly into the equations of motion, it is possible to solve for corrections to them, using the partial derivatives of the ephemeris coordinates with respect to those parameters: $\partial \mathbf{r}_i / \partial \beta$ and $\partial \mathbf{r}_i / \partial \gamma$. Any significant change to β or γ would then indicate a questioning of the original assumption that $\beta = 1$ and $\gamma = 1$.

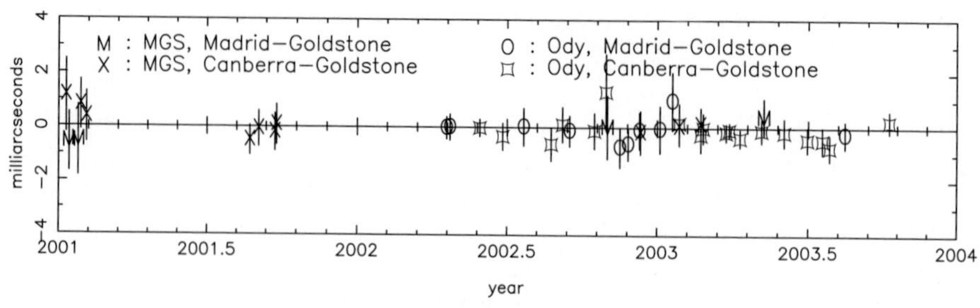

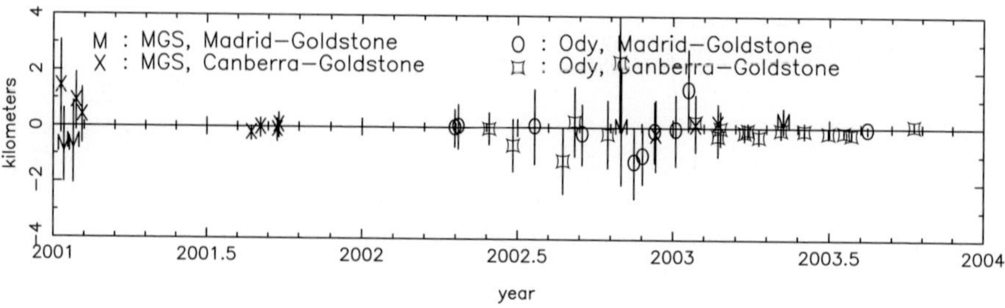

Figure 7. ΔVLBI residuals of Mars from the MGS and Odyssey orbiting spacecraft, shown both in milliarcseconds and in kilometers. These observations are virtually one-dimensional.

The following is a partial list of topics which have been tested, using the planetary and lunar ephemerides, requested by a number of theoretical researchers:

General Relativity : β, γ , J_2(sun)

Modified Newtonian Dynamics

New Weak Forces and Non-Newtonian Gravity

Equivalence Principle

Mach's Principle vs. Equivalence

Sun's gravitational-to-inertial mass ratio

$\dot{G}$

d(au)/dt

Behind all of this testing is the ongoing improvement of modern planetary and lunar ephemerides.

7. Asteroids

Even with such accurate observational data, the planetary motions show rather large uncertainties, for the planets are perturbed by the presence of many asteroids whose masses are quite poorly known. Furthermore, it's not possible to solve for the asteroid masses, other than for the biggest few, because there are too many of them for the data to support such an effort. As a result, the ephemerides of the inner planets, especially that of Mars, will deteriorate over time. Various experiments have shown that the ephemerides have uncertainties at the 1-km level over the span of the observations and growing at the rate of prehaps a km/decade outside that span.

A great deal of effort has been applied in order to represent the asteroid perturbations as well as possible. Studies of the estimations of masses of the most relevant 300 or so asteroids have been made by Fienga (2001) and by Krasinsky *et al.* (2001); Krasinsky *et al.* (2002) have also modeled a ring to represent the perturbations from the remaining thousands of small asteroids.

8. Values of the au

Table 1 is presented in order to emphasize the immense improvements in the determinations of the value of the au over the past century. The values given in the table are the differences from 149,597,870,691 m, the value used in the ephemerides of the late 1990's, JPL's DE403 and DE405 and IAA's EPM2000 (Pitjeva 2001). The optical determinations gave errors of many tens of thousands of kilometers, finally reduced to only 2500 km by the careful extensive work of Lieske.

Table 1. Deteminations of the au [TDB values]

		[meters]	
1931	Spencer Jones	+72,000,000	Eros
1941	Adams	-185,000,000	Stellar Doppler shifts
1942	Brouwer	-19,000,000	Lunar occultations
1950	Rabe	-73,000,000	Eros
1958	False Peak	-130,000,000	radar ghosts
	(me too)	-130,000,000	
1959	(me too)	-140,000,000	
1961	First Real	+1,000,000	radar
		-170,000	
		+2,000,000	
		-2,000,000	
		+130,000	
		$\rightarrow$ Venus Rotation	
		au [km]	
1964	IAU	+130,000	
1967	PEP	+760	ephemeris
1968	Lieske	+2,530,000	Eros
1969	DE69	+1350	
1976	DE96	+710	
1979	DE200	-31	Viking ('76-'79)
1995	DE403	0	Viking (-'82), LLR, radar
1997	DE405	0	
2000	EPM2000	0	
2004	DExxx	+7	MGS, Odyssey

$$d(\mathrm{au})/dt \; = \; +15 \; m/yr \; ?$$

The first radar reports were false identifications of signals, thought to be echoes from Venus. Right away, two more groups confirmed this wrong value. The first real radar measurements were in error by only $1000 - 2000$ km, soon to be reduced dramatically by increasing technology and the determination from MIT's PEP ephemeris. Measurements to spacecraft now have brought the value of the au to levels of only a few meters. The recent addition of the MGS and Odyssey ranges tends to indicate a value for the au which is a couple of meters shy of 149,597,870,700 m.

At the bottom of the table is the intriguing thought that maybe $d(\mathrm{au})/dt \neq 0$ (Krasinsky & Brumberg 2004). And, at the top is the statement that these are TDB values of the au, not TCB values, recently "recommended" by the IAU.

9. Conclusions

The au has a colorful history, full of mis-quotes, poor definitions, and varied measurements.

- The au is not defined; it is the result of a convention of units, one which has been used for many, many decades in astronomy.
- The determination of the value of the au in kilometers is the result of extensive ephemeris fitting to highly accurate measurements.
- The dominating error source for the four inner planet ephemerides is the perturbations from many asteroids whose masses are poorly known.
- There is still uncertainty in the value of the au: first, from the uncertainties in the ephemerides arising from the poorly-known asteroid masses, and secondly, from the possibility that the length of the au itself may not be constant over time.

Acknowledgements

The work described in this paper was carried out at the Jet Propulsion Laboratory, California Institute of Technology, under contract with the National Aeronautics and Space Administration.

References

Fienga, A.G. 2001 *private communication.*
Krasinsky, G.A. and Brumberg, V.A. 2004 *submitted to Cel. Mech. J.*
Krasinsky, G. A., Pitjeva, E. V., Vasilyev, M. V. & Yagudina, E. I. 2001 *Communication of IAA RAN* **139**, 1–43.
Krasinsky, G. A., Pitjeva, E. V., Vasilyev, M. V. & Yagudina, E. I. 2002 *Icarus* **158**, 98–105.
Lieske, J.H. 1968 *Astron. J.* **73**, 628–643.
Pitjeva, E. V. 2001 *Celest.Mech. & Dyn.Astr.* **80**, N 3/4, 249-271.
Williams, J.G. and Standish, E.M. 1989 in *Reference Frames* (J. Kovalevsky, I.I. Mueller and B. Kolaczek, eds.) Kluwer Academic Publishers, Dordrecht.

Discussion

DAVE MONET: There are several experiments such as PANSTARS, LSST, others that really think they are going to find all the asteroids down to a few hundred metres and hand that list to the community. Is that going to help at all?

MYLES STANDISH: The problem as much as anything is not where the asteroids are but their masses. So we need, first of all, an estimation of the diameters, volume - actually, when you get to the small guys I guess it's an estimation of the volume, and then you need the density. Right now for a number of the asteroids what we do is take the estimated volume or diameter and then assign a density according to the taxonomic class. And then we can actually solve for the density of overall taxonomic classes, but you cannot solve for many asteroids. I can try that, but I get a couple of negative masses, a couple densities which are 12; we tend not to believe those.

Nick Kollerstrom: You advised us that the mean distance in Keplerian orbit is not the semi-major axis. Did you tell us what it was?

Myles Standish: Yes. The mean distance – it was $a \times (1 + e^2)/2$; that little formula, is that what you meant? It's the semi major axis $\times (1 + e^2)/2$; that's the mean distance in a Keplerian orbit.

David Hughes: I mean, I love asteroids; what I really would like to know is the total mass of the asteroid belt. This perturbs the orbit of Mars; can you get a handle on this total mass using your work?

Myles Standish: Actually, following Krasinsky of St Petersburg, we have about 20 asteroids which we handle individually – the 20 most important affecting the orbits of Mars and the Earth. Then we have about 300 more which we put into the 3 taxonomic classes, as I've kind of alluded to, and then Krasinsky on top of all that puts a belt of asteroids out there and solves for the mass of the belt, so he has gotten somewhere around a sizable fraction of the mass of Ceres. I think about half the mass of Ceres for that belt. Do you recall Lena?

Elena Pitjeva: Mass of belt maybe less than half the mass of Ceres; about 50% of Ceres.

Myles Standish: 50% of Ceres – I thought so, yes. So that is an estimate, but you can change. So we have about 300 of the most important, and then another half of a Ceres mass up there. There is a paper by Krasinsky [et al. 2002]; I refer you to that.

John Ponsonby: I'm astonished that you didn't make any mention of pulsar timing as a contributory source of information about the AU. I would have thought this was a way of tying it in with the inertial frame, as well as getting precision measurements because they predict the time of arrival of single pulses to a few microseconds.

Myles Standish: But a few microseconds is 500 metres.

John Ponsonby: Sure, sure. It's not at a metre level, but somewhere along the line I would have thought it was a useful contribution.

Myles Standish: I have used them and looked at pulsar timings. Actually, the ranging swamps them out still, so they really don't contribute. One of the problems is, of course, you have to reduce them to the barycentre of the solar system; that is very poorly known because we don't know the masses of the planets as well as we could. As a matter of fact, there is a little story that when the people at MIT and we were reducing pulsar 19-whatever-it-is, we came up with a different period, a significantly different period, *monstrously different* – and the reason was we had different masses of Neptune in our ephemerides and Neptune puts in a 165-yr period into the barycentre, and with a little piece of that period it looks like a slope and gives you a different period. So, finding the barycentre is one problem, and then of course the σ on the observations is another.

Jacqueline Mitton: I notice that in conclusion you say that the astronomical unit is not defined. This is a real problem for people like me who write dictionaries where we are obliged to put in a definition, and also, because I often am writing for, or speaking to, children, or advising publishers of magazines, or indeed reading or doing similar to what *Sky & Telescope* did [incorrectly define the AU] which you criticised in a justifiable way. So the question is: okay, there is a technical basis for what we try to understand as the astronomical unit, but for the great public and children under 12 out there, have you come up with a form of words that you find acceptable?

MYLES STANDISH: [laughs] I guess it depends on whom it is acceptable to. I don't know . . . I put together the explanation that I showed to make it one or two sentences. Possibly, we could take the sentence that's in the explanatory supplement, and instead of using what they call the year, just use the mean motion, the effective average angular rate going around the sun, but then you have to say that it is also based on an old value, one that is over a century old. I don't know how old the Gaussian constant is . . . 150 years or something. But it is not easy; maybe we could work on it.

DENNIS MCCARTHY: Probably you have already solved for the rate of change of the astronomical unit, do you have a number?

MYLES STANDISH: Krasinsky & Bromberg have 15 metres per century in their paper and I am not sure whether that's cosmologically founded, or numerically founded, but I spoke to Dr Pitjeva just before my talk, because I knew you would probably ask me [laughter], and she gets about 5 metres per century

DENNIS MCCARTHY: But your number is from the solutions, from the ephemerides?

MYLES STANDISH: That's where her number comes from, yes. Now, there are other problems with ranging which I didn't want to admit to, but when you go into the electronics of a transponder there is a delay and, as a matter of fact, it's dependent on temperature and everything else. Some of these missions have been very carefully calibrated – the MGS, for instance. Viking was not so much, nor was Odyssey, and actually you put those Odyssey and MGS observations right next to each other, taken at the same time, and there's a couple of metres or so difference between them. So, when you have a solution that stretches from Viking in the 70s to now, then you have a 5-metre discrepancy, you can call it, kind of, an AU-dot. I prefer to call it biases in the equipment. So it's a trade off and we are not sure yet. Maybe with many, many more years of MGS alone we can go after that number.

JIM MESSAGE: Can I make one comment and ask one question. About the question of defined: of course there are two possible meanings of saying something is defined or not defined. You can define a concept in the sense of giving a meaning to it, or you can define it in the sense of saying this number is what it is, and I think that we have probably got a little bit of a confusion here, haven't we? I mean what you are saying is that the astronomical unit is not something where we state a number and say this is it. The number comes from the calculation from other assumptions that you have made. The criticism is not to say that the astronomical unit is not a perfectly well-defined concept, in the sense that as you have explained it follows from the equations that you are using. I was just wondering if there was a possible verbal confusion here.

MYLES STANDISH: Well, in the explanatory supplement, or in that quote, they use a term "based upon," and you could say based upon the definition of a numerical value for the mean motion of the Earth, and through Kepler's equation the AU follows. For somebody in school, . . .

JIM MESSAGE: Well, quite, there's your problem. So it's important to get the concept clear, if youre talking to people who haven't followed the story right through from the 1950s.

MYLES STANDISH: It took me three decades to figure this thing out!

[general laughter!]

JIM MESSAGE: I mean, I'm not there yet either, but my question is simply this, the 5 or 7 [metres] that is your present value for the difference of the astronomical unit from the received value: Is there an uncertainty associated with this? What is the σ?

MYLES STANDISH: There is a formal uncertainty, but I would not believe it. The uncertainty just comes from making many, many different solutions, by juggling the different parameters, and weighting the data sets differently, and seeing how much this number jumps around. I was so entranced with the value from 7 or 8 years ago – every time this thing came up plus or minus 6 or 7 [metres], I thought something's wrong. I kept trying to push it back. But now I think we have a real belief that maybe this is ...

JIM MESSAGE: But it's not 0? It's not 5 ± 5 [metres]?

MYLES STANDISH: No, I would say 7 ± 2. Now, I know that Dr Pitjeva has put out to another significant figure, so maybe that's significant. I don't know.

DON KURTZ: Myles, can you comment on sources of $\dot{a}$. Why?

MYLES STANDISH: Sources of $\dot{a}$? Why is it believed?

DON KURTZ: No, what's causing the astronomical unit to increase? What's the physical cause of the increase?

MYLES STANDISH: That's a cosmological argument I believe. There's a paper by Krasinsky & Bromberg from St Petersburg, and there is a fairly extensive argument by Bromberg, but I don't know it very well; I probably shouldn't comment.

DON KURTZ: Is there any contribution you could measure from either $\dot{M}$ from the sun or from accretion onto the sun, or is that just far too small for you still?

MYLES STANDISH: I've tried. We've put in $\dot{M}$ trying to model the course of hydrogen burning – the mass loss of the sun as it evolves – and to actually solve for anything past that, no. We can get some crazy answers if we do.

DON KURTZ: Something I would like to see (if I could talk you into producing it) would be a nice map of one orbit of the Earth, with the Earth and the Sun in a inertial frame, doing whatever they do in a year, exaggerating the scale to where you can see all these bumps from the perturbations from the asteroids, the planets ... some nice pretty picture to show us how non-Keplerian it really is.

MYLES STANDISH: It is certainly doable.

DON KURTZ: [in a tone of enticement] I'll put it in the proceedings if you do it.

[general laughter!]

MYLES STANDISH: Oh, boy! [wipes his brow]

[more laughter]

JAYMIE MATTHEWS: Apropos to the discussion of à, and you were talking about biases and the radar ranging measurements, a few years ago there were reports of anomalous timings from, I believe Pioneer and maybe Voyager as well, and I am just curious – I haven't heard much from those groups lately – I just wondered what your sense of what they were measuring was, and what's the consensus of your community on that?

MYLES STANDISH: Yes, I know very much of what you are talking about, and certainly there are some good people trying to figure out what this is all about. The only thing that scares me is how they handle the data right in the beginning, the raw data, because they certainly have looked into many, many different effects and not really been able to explain it. The problem ... the head guy, kind of would love to see some kind of very exotic explanation, new force or whatever, and I don't think that that has been really justified yet. So that's about all I should say on that, but handling the data is a real touchy issue here.

WALTER BRISKEN: Neptune was discovered by finding these anomalies in the ephemeris. Has there been any recent progress in trying to discover unknown bodies in our solar system this way?

MYLES STANDISH: Not in ten years or so, no. There was a lot of activity maybe 20 years ago – 15 years ago – but most of it was false because the wrong mass of Neptune was being used, and that affects the motion of Uranus. Once you put in the good mass of Neptune and adjust the orbit, the major part of the signature – almost the total signature – is gone. So, there is nothing that we see that really demands some kind of other explanation. Pioneer may be the one example, I don't know. Things *seem* to behave themselves.

MARILYN HEAD: As an amateur astronomer I was really fascinated. I do minor planet occultations, and this is the first time that I've heard that they are really useful, because what you are talking about – finding the diameter of asteroids, etc. – is useful, and it is important, in order to find out the mass, I assume. So it was just good to hear that, to know that what we are doing makes a contribution.

MYLES STANDISH: I'm sorry, the question is?

MARILYN HEAD: It was really just a comment. You know the minor planet occultations, were you trying to ...

MYLES STANDISH: Oh ... Oh! The occultations of asteroids?

MARILYN HEAD: Yes.

MYLES STANDISH: Or, the occultation of a star by an asteroid.

MARILYN HEAD: Yes.

MYLES STANDISH: Yes, certainly one of the major problems is the volume of an asteroid, because if you have a 20% error in the diameter, you have about a factor of two in volume (close to it) and that can make a big difference. So if the question is: "are these things useful," then, yes. Have you got any more of them?

[laughter!]

MARILYN HEAD: We try, we try. We get about 1 in 13, I think.

MIKHAIL MAROV: Could you tell us, based on the contemporary estimates for residuals in the motions of Neptune and Uranus, is it possible to estimate the total mass of the Edgeworth-Kuiper belt?

MYLES STANDISH: No, sorry. Especially because the optical observations are really the only measurements we have of Neptune and Uranus – and they exist back only until 1910, at best. It was then that there was a major improvement in observing technique, so the measurements before then are not as good. So, we don't even have a full period of Neptune; we have barely a period of Uranus. So the observation is just not strong enough. I can't put a number on what it would do to the observations, but I would be very, very surprised if we could pull out that signal.

NICOLE CAPITAINE: You provided a number in metres, but generally in astronomical tables what it is provided at first is the value of the astronomical constant in seconds, in time. So what is the first one? Is it the value in seconds, and then you derive the value in metres, or is it the ...

MYLES STANDISH: No, we actually solve for the value in kilometres and then derive the time. I think the reason could be very apparent. When we take a radar range, of course the measurement is in time. We use the given velocity of light to convert to kilometres, and then we have to convert to AUs to put into the ephemeris system, and that conversion is kilometres per AU. I never used the AU in seconds. It could be done. I mean, it cuts out the middle man, and I know that Irwin Shapiro one time made a comment that maybe we should run the whole ephemeris system using light time as the distance.

NICOLE CAPITAINE: Regarding the dot, the AU-dot, is it related to the timescale, is it in TCB or in TDB?

MYLES STANDISH: Either one. That won't change it ... well, it will change it in the eighth figure or something. [laughs] We're worried about the sign on the thing!

Transits of Venus: New Views of the Solar System and Galaxy
Proceedings IAU Colloquium No. 196, 2004
D.W. Kurtz, ed.
© 2004 International Astronomical Union
doi:10.1017/S1743921305001377

Precision time and the rotation of the Earth

Dennis D. McCarthy

U. S. Naval Observatory, Washington, DC 20392, USA
email: dmc@maia.usno.navy.mil

Abstract. Practical measurement of the passage of time requires the notion of a repeating phenomenon. The Earth's rotation has traditionally fulfilled this requirement. To cope with the impracticality of making precise measures of the Sun's hour angle or altitude, particularly in uncooperative weather conditions, various devices have been employed, but all have been calibrated with respect to astronomical phenomena related to the Earth's rotation, or to its orbital motion with respect to the Sun. Modern requirements for timing precision coupled with an increased understanding of the variability of the Earth's rotational speed are likely to bring about a change in the traditional relationship between precise timekeeping and astronomy. The historical background of this relationship and the current definitions are reviewed to show the development of timekeeping capabilities and the growing need for precise timekeeping. Possible future developments are outlined along with their advantages and disadvantages.

1. Introduction

Two components are involved in our measurement of time. First, we make use of a phenomenon perceived to be repeatable with sufficient regularity to serve our purpose. Second, we need to devise a convenient means to label the repetitions of this phenomenon. These elements correspond to the two properties of time measurement: interval and epoch. Historically we have used the repetitive cycles of celestial phenomena to serve as the basis for our time. The rising and setting of the Sun has provided the day; the phases of the Moon have provided the month and the position of the Sun with respect to the stars gives us the year.

These processes met historical timekeeping requirements with the exception of the times when the skies were not visible. For those times, various forms of clocks were created, and these were calibrated in terms of the traditional astronomical phenomena. When the need for extremely precise time arose, it became apparent that the Earth's rotation was not sufficient as the repeatable phenomenon. Even still we do maintain a relationship between clocks and astronomy. The background and issues involved are presented in detail in Nelson et al. (2001). This paper presents a summary of that paper and further explores the possibilities for the future.

2. Historical background

Time reckoned by the Earth's rotation has fulfilled the needs of society for time-keeping throughout history. The details of both astronomical time and "clock" time as implemented since the middle of the twentieth century are important in understanding the current timekeeping issues.

2.1. *Astronomical time*

Two primary methods of measuring time astronomically have been in common use for modern applications. The first is Universal Time, the time scale based on the rotation of the Earth on its axis. The second is characterized by a set of Dynamical Time Scales based on the revolution of the Earth in its orbit around the Sun.

2.1.1. *Universal time*

The form of Universal Time designated UT1 is the measure of astronomical time defined by the rotation of the Earth on its axis with respect to the Sun. It is nominally equivalent to mean solar time on the Greenwich meridian and reckoned from midnight. The mean solar day is traditionally described as the time interval between successive transits of the fictitious mean Sun over a given meridian. Historically, the unit of time, the mean solar second, was defined as $1/86\,400$ of a mean solar day (McCarthy 1991). In practice, Universal Time is determined, not by the meridian transit of the mean Sun, but by the observation of astronomical objects in conjunction with a conventional formula specifying UT1 in terms of Greenwich Mean Sidereal Time (GMST). The defining relation for UT1 with respect to the traditional optical reference system is given in Aoki et al. (1982). Beginning in January, 2003, a new defining expression has been implemented (Capitaine et al. 2000, 2003).

Apparent solar time, as determined by using a sundial or from measuring the altitude of the Sun, is the local time defined by the diurnal motion of the actual Sun. However, because the Earth's axis is not perpendicular to its orbital plane and because of its varying distance from the Sun, the time interval between successive meridian passages is not constant. This difference between mean and apparent solar time, called the equation of time, varies with the maximum being about 16 min. Until the early nineteenth century, apparent solar time was used as the argument for astronomical ephemerides. However, as clocks improved, apparent solar time was gradually replaced by mean solar time. Variations in the Earth's rate of rotation were clearly established in the 1930s when a seasonal variation was observed in comparisons of astronomical observations with accurate pendulum and quartz crystal clocks (Scheibe & Adelsberger 1936; Stoyko 1937). The time scale denoted UT2 was developed by adopting a conventional expression for the seasonal variation in the Earth's rotation, and this time scale was used briefly as a more uniform time scale determined from the Earth's rotation.

The sidereal day, which represents the Earth's period of rotation with respect to the stars is approximately 86 164.0905 mean solar seconds. Due to precession the sidereal day is about 0.0084 s shorter than the actual period of rotation in inertial space. Thus, the true rotational period of the Earth is approximately 86 164.0989 mean solar seconds. However, the mean solar day presently exceeds a day of exactly 86 400 SI (*The International System of Units (SI)* 1998) seconds by about a millisecond. Therefore, the Earth's period of rotation is currently about 86 164.0999 SI seconds.

The principal technique used to determine UT1 now is Very Long Baseline Interferometry (VLBI) observations of quasars. Although satellite laser ranging and tracking of GPS satellites provide estimates of UT1, because of the motion of satellite orbital nodes in space, VLBI provides the only rigorous determination of UT1.

2.1.2. *Dynamical time scales*

Ephemeris Time (ET) is a theoretically uniform time scale defined by the Newtonian laws of motion. When it became apparent that time based strictly on the Earth's rotation did not provide a uniform time scale, Clemence (1948) recommended a new standard of time based on the period of revolution of the Earth around the Sun, as represented by Newcomb's Tables of the Sun published in 1895. The measure of astronomical time defined in this way was given the name Ephemeris Time (ET). The working definition of Ephemeris Time was through Newcomb's formula for the geometric mean longitude of the Sun for the epoch of January 0, 1900, 12h UT (Newcomb 1895a),

$$L = 2794148.04 + 129602768.13T + 1.089T^2, \qquad (2.1)$$

where T is the time in Julian centuries of 36 525 days. The proposal was adopted by the International Astronomical Union in 1952 (*Trans. Int. Astron. Union, Vol. VIII* 1954). Using the value of the linear coefficient in Newcomb's formula, the tropical year of 1900 contains 31 556 925.9747 s, so the International Committee for Weights and Measures (CIPM) in 1956 defined the second of Ephemeris Time to be "the fraction 1/31 556 925.9747 of the tropical year for 1900 January 0 at 12 hours ephemeris time."

Even though it was defined in terms of the longitude of the Sun, ET was determined in practice by using conventional lunar ephemerides in comparing observations of the positions of the Moon with respect to stellar backgrounds. Thus, secondary time scales, denoted ET0, ET1, and ET2, were defined by the ephemeris actually used in making the estimation of ET (Guinot 1989).

The benefit of a more uniform time scale provided by ET, however, was complicated by the fact that it was not easily accessible. That fact, combined with the development of atomic time scales, meant that Ephemeris Time never became a significant practical time scale. However, ET did replace UT1 as the independent variable used in astronomical ephemerides beginning in 1960.

The concept of time scales based on the dynamics of the solar system was refined in 1976 when the International Astronomical Union (IAU) defined time-like arguments consistent with the general theory of relativity (*Trans. Int. Astron. Union, Vol. XVIB* 1977). This led to the development of Terrestrial Dynamical Time (TDT) and Barycentric Dynamical Time (TDB) (*Trans. Int. Astron. Union, Vol. XVIIB* 1980) that distinguish coordinate systems with origins at the center of the Earth and the center of the solar system, respectively.

In 1984 TDT replaced ET as the tabular argument of the fundamental geocentric ephemerides. It has an origin of 1 January 1977 0 h TAI (International Atomic Time), with a unit interval equal to the SI second, and maintains continuity with ET. In 1991 the IAU renamed TDT Terrestrial Time (TT). A practical realization of TT in terms of the atomic time scale, TAI, is (*Explanatory Supplement to the Astronomical Almanac, rev. ed.* 1992)

$$TT = TAI + 32.184\,\text{s}. \qquad (2.2)$$

The constant offset represents the difference between ET and UT1 at the defining epoch of TAI on 1 January 1958.

TDB was defined to be used as the time-like argument for ephemerides referred to the barycenter of the solar system. By adopting an appropriately chosen scaling factor, TDB varies from TT or TDT by only periodic variations, with amplitudes less than 0.002 s.

In 1991 the 21st IAU General Assembly introduced the general theory of relativity explicitly as the theoretical basis for the celestial reference frame and the form of the space-time metric to post-Newtonian order was specified. (*Trans. Int. Astron. Union, Vol. XXIB* 1992). At that time it also clarified the definition of Terrestrial Time and further refined the concept of dynamical time by defining two new time scales, Geocentric Coordinate Time (TCG) and Barycentric Coordinate Time (TCB) (Seidelmann & Fukushima 1992). The "coordinate" time scales TCG and TCB are complementary to the "dynamical" time scales TT (or TDT) and TDB. They differ in rate from TT and are related by four-dimensional space-time coordinate transformations (*IERS Conventions* 1996). These definitions were further clarified by resolutions adopted at the 24th IAU General Assembly in 2000 (*Trans. Int. Astron. Union, Vol. XXIVB*). The dynamical

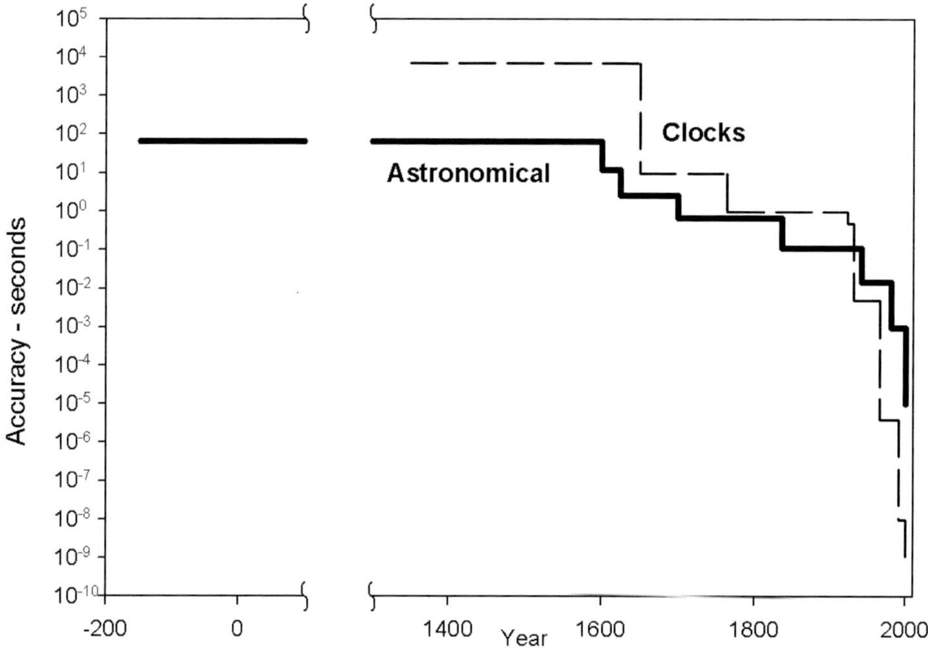

Figure 1. Development of timekeeping accuracy.

time scales are now used only for specialized studies and to develop astronomical ephemerides.

2.2. *"Non-astronomical" time*

To provide the means to measure the passage of time when it was not possible to use the Sun or stars, we have made use of various devices including clepsydrae, candles, and various forms of clocks. In the era of telescopic observations, pendulum clocks served as the standard means of keeping time until the introduction of modern electronics. Quartz crystal clocks were developed in the 1920s and 1930s (Gerber & Sykes 1966), and the first atomic clock was constructed in 1948 (*Natl. Bur. Stand. (U.S.) Tech. News Bull.* 1949) using the microwave absorption line of ammonia to stabilize a quartz oscillator. In 1955 Louis Essen and J. V. L. Parry (Essen & Parry 1955) constructed a practical caesium beam atomic clock and commercial caesium frequency standards appeared a year later. Norman Ramsey developed the hydrogen maser at Harvard University in 1960 (Goldenberg et al. 1960). Following the advent of atomic clocks, national timekeeping laboratories began to establish time scales based solely on atomic time, and these were coordinated by the *Bureau International de l'Heure* (BIH) (Guinot 2000). Some form of atomic time has been maintained continuously since 1955 (Guinot 1994/1995). The *Bureau International des Poids et Mesures* (BIPM) now has the responsibility for the maintenance of the standard international atomic time scale. Fig. 1 shows the development of the accuracy of astronomical and clock time scales with time.

2.2.1. *Atomic time*

Following the appearance of the first operational caesium beam frequency standard in 1955 at the National Physical Laboratory (NPL) in the United Kingdom (Essen & Parry 1957), the Royal Greenwich Observatory (RGO), U.S. Naval Observatory (USNO), and U. S. National Bureau of Standards (NBS) began to produce atomic time scales. The epochs of the USNO and NBS scales were made coincident and set equal to the observed UT2 on 1 January 1958 (Barnes et al. 1965). The details of the development of these scales into the current standard TAI (International Atomic Time) are contained in (Nelson et al. 2001).

2.2.2. *Atomic definition of the second*

Louis Essen and J. V. L. Parry of the NPL, in cooperation with William Markowitz and R. G. Hall at the USNO, determined the frequency of the NPL caesium standard with respect to the second of Ephemeris Time. Photographs of the Moon and surrounding stars were taken using the USNO dual-rate Moon camera over the period 1955.50 to 1958.25 to determine the Ephemeris Time from the position of the Moon at a known UT2. This information was used to calibrate the caesium beam atomic clock at NPL via simultaneous observations of the intervals between time pulses broadcast by radio stations WWV (then in Greenbelt, Md.) and GBR (Rugby, UK). The measured caesium frequency was $9\,192\,631\,770\,\mathrm{Hz}$ with a probable error of $20\,\mathrm{Hz}$ (Markowitz et al. 1958).

In October 1967 the atomic second was adopted as the fundamental unit of time in the International System of Units. It was defined as (*Metrologia* 1968) "the duration of $9\,192\,631\,770$ periods of the radiation corresponding to the transition between the two hyperfine levels of the ground state of the caesium 133 atom," thus making the second of atomic time equivalent to the second of Ephemeris Time in principle.

2.2.3. *Establishment of TAI*

The *Comité Consultatif pour la Définition de la Seconde* (CCDS) of the CIPM recommended guidelines for the establishment of International Atomic Time in 1970. It stated that "International Atomic Time (TAI) is the time reference coordinate established by the *Bureau International de l'Heure* on the basis of readings of atomic clocks operating in various establishments in accordance with the definition of the second, the unit of time of the International System of Units"(*Metrologia* 1971). The CCDS (BIPM *Com. Cons. Déf. Seconde* 1970) defined the origin so that TAI would be in approximate agreement with UT2 on 1 January 1958, 0 h UT2.

This definition was refined in 1980 to account for relativistic concerns with the statement, "TAI is a coordinate time scale defined in a geocentric reference frame with the SI second as realized on the rotating geoid as the scale unit" (*Metrologia* 1981). TAI, when formally adopted in 1971, was an extension of the BIH atomic time scale that had been continuous back to 1955. In 1988, responsibility for maintaining TAI was transferred from the BIH to the BIPM. Today approximately two hundred clocks maintained in fifty laboratories contribute to the formation of TAI.

3. Variations in the earth's rotational speed

Astronomical observations have shown that the Earth's rotation does not provide a uniform time scale. The secular deceleration as well as a wide spectrum of quasi-random and periodic fluctuations have been well documented (Lambeck 1980). The low-frequency variations in the Earth's rotation are demonstrated by the difference in time, ΔT, between the uniform scale of Ephemeris Time or Terrestrial Time and the variable scale

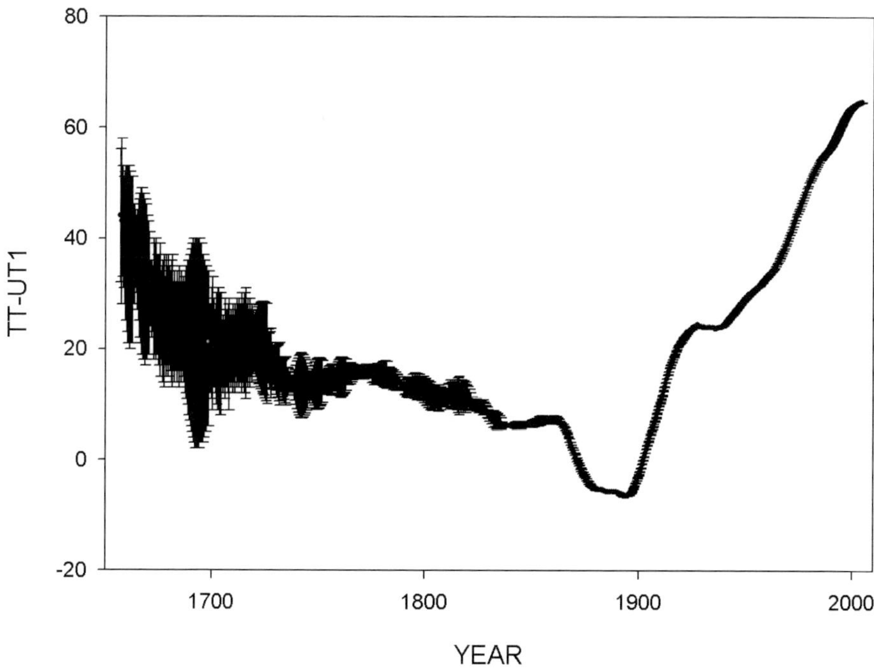

Figure 2. Variation of T since 1620 (in seconds).

of Universal Time. Values of ΔT have been summarized (Morrison & Stephenson 1986; McCarthy & Babcock 1986). Before 1955, the values are provided by observations based on observations of the Moon. After 1955, they are determined using atomic clocks. Fig. 2 shows the observations since 1620.

Stephenson (1997) shows that ΔT over the past 2700 years can be represented approximately by a parabola of the form

$$\Delta T = (31s/cy^2)(T - 1820)^2/(100)^2 - 20s, \tag{3.1}$$

where ΔT is expressed in seconds and T is the year. The derivative of ΔT,

$$LOD = (0.0017s/d/cy)(T - 1820)/100, \tag{3.2}$$

represents the excess length of day (the difference between the length of the astronomical day and 86 400 SI seconds) corresponding to the expression for ΔT above. According to this long-term trend, the rate of increase in the length of the day is about 1.7 ms per century. Fig. 3 illustrates observations of changes in the length of day during the era of telescopic observations, from 1620 onwards. Over this period, LOD has been increasing at about 1.4 ms per century (Morrison & Stephenson 1986). The actual value of the LOD departs from any long-term trend due to fluctuations on a time scale of decades. The epoch at which the mean solar day was exactly 86 400 SI seconds was approximately 1820. This is also the approximate mean epoch of the observations analyzed by Newcomb that resulted in the definition of the second of Ephemeris Time from which the SI second was derived (Newcomb 1895b). Fig. 4 shows more recent values of the excess length of day.

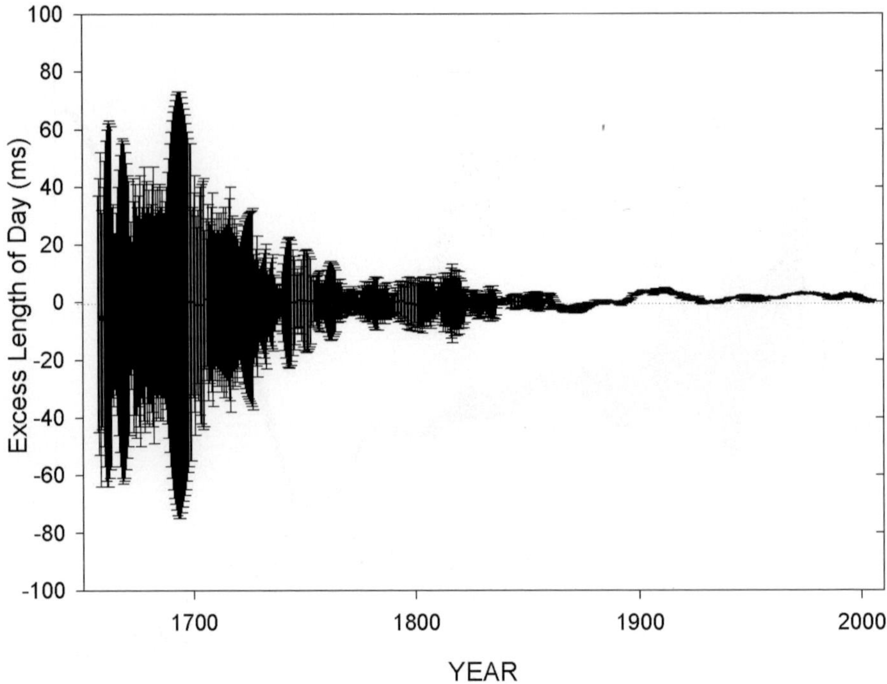

Figure 3. Variation of the excess length of day since 1620.

3.1. *Physical causes*

Three types of variation in the Earth's rotation have been identified: a steady deceleration, random fluctuations, and periodic changes. As early as 1695, Halley (1693, 1695) suspected an acceleration in the mean motion of the Moon from a study of ancient eclipses of the Sun. By the mid-eighteenth century, the lunar acceleration was fully established, and in 1754, Kant (1867) suggested that this acceleration might be an apparent phenomenon caused by a steady deceleration in the Earth's rotation due to tidal friction.

Jeffreys (1920, 1962) made the first quantitative estimate of global tidal friction in 1920. The rate of energy dissipation by tidal friction is now considered to correspond to a rate of increase in the length of day of 2.3 ms per century. To account for the observed deceleration, there must also be a component in the opposite direction of about 0.6 ms per century, which is possibly associated with changes in the Earth's shape caused by post-glacial rebound (Yoder et al. 1983) or with deep ocean dissipation (Egbert & Ray 2000).

Evidence for a long-term deceleration in the Earth's rotation, extending over millions of years, also exists in coral fossils that exhibit both daily and annual growth rings (Wells 1963). The evidence suggests that the rate of deceleration was substantially the same then as it is now (Runcorn 1966). Besides a steady decrease, the Earth's rotation is subject to frequent small, apparently random changes. The irregular changes in speed may be correlated with physical processes occurring on or within the Earth. They include decade fluctuations with characteristic periods of 5 – 10 years as well as variations, which occur at shorter time scales.

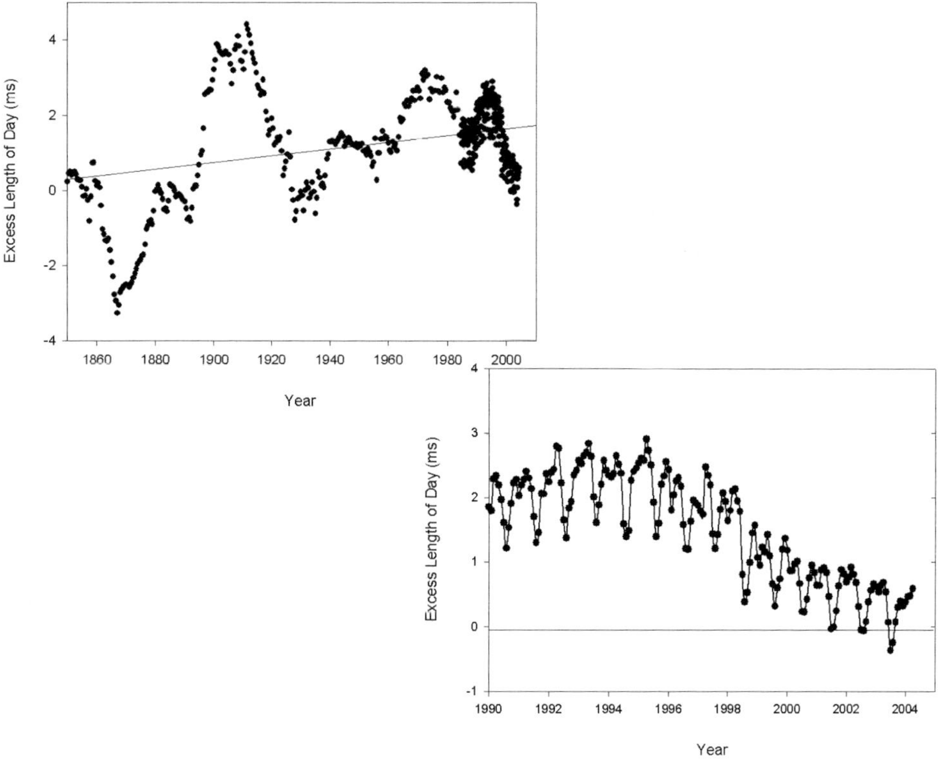

Figure 4. Variation of the excess length of day.

4. Coordinated Universal Time

The current definition of Coordinated Universal Time (UTC) was devised to satisfy the needs of two communities of users. Astronomers, geodesists and navigators would like a time connected with the angle of the Earth's rotation in space. Others, such as physicists and engineers, prefer a perfectly uniform time scale. Attempts to meet the needs of both communities led to the creation of UTC.

4.1. *History*

The term "Coordinated Universal Time" was introduced in the 1950s to designate a time scale in which adjustments to quartz crystal clocks were coordinated among participating laboratories in the USA and UK. The scale evolved over the years to the state where the BIH coordinated adjustments to an internationally accepted standard Coordinated Universal Time, designated UTC, that involved adjustments in both rate and epoch to stay in step with astronomical time.

The concept of the leap second, analogous to the leap day in the calendar, was proposed independently by Winkler (1968) and Essen (1968) at a meeting of the CIPM in 1968 (*Commission Préparatoire pour la Coordination Internationale des Échelles de Temps* 1968). It was proposed that integer steps of seconds replace the steps of 100 ms or 200 ms then being used. To meet the needs of navigators, it was suggested that coded information be incorporated in the radio time signals to indicate the difference between UTC and UT2.

The International Radio Consultative Committee (CCIR) formed an Interim Working Party to investigate requirements, submit proposals, and fix a date for the introduction of the new system. Specific proposals were made in Geneva in 1969, which were approved in January 1970. In its Recommendation 460 (International Radio Consultative Committee (CCIR) 1970a), the CCIR stated that (a) carrier frequencies and time intervals should be maintained constant and should correspond to the definition of the SI second; (b) step adjustments, when necessary, should be exactly 1 s to maintain approximate agreement with Universal Time (UT); and (c) standard signals should contain information on the difference between UTC and UT. The CCIR also decided to begin the new UTC system on 1 January 1972. At the IAU's 14th General Assembly in 1970, it was recommended that radio time signals should disseminate differences in the form of UT1 - UTC for navigational requirements.

Detailed instructions for the implementation of CCIR Recommendation 460 were drafted at a further meeting of Study Group 7 (International Radio Consultative Committee (CCIR) 1970b). The defining epoch of 1 January 1972, 0h 0m 0s UTC was set 10 s behind TAI, which was the approximate accumulated difference between TAI and UT1 since the inception of TAI in 1958, and a unique fraction of a second adjustment was applied so that UTC would differ from TAI by an integral number of seconds. The recommended maximum departure of UTC from UT1 was 0.7 s, and the term "leap second" was introduced. The correction DUT1 was introduced, having integral multiples of 0.1 s, to be embodied in the time signals such that, when added to UTC, they would yield a better approximation to UT1. In 1974, the CCIR increased the tolerance for UT1-UTC from 0.7 s to 0.9 s.

The present UTC system is defined by ITU-R (formerly CCIR) Recommendation ITU-R TF.460-5 (*ITU − R Recommendations*: *Time Signals and Frequency Standards Emissions* 1998): "UTC is the time scale maintained by the BIPM, with assistance from the IERS, which forms the basis of a coordinated dissemination of standard frequencies and time signals. It corresponds exactly in rate with TAI but differs from it by an integral number of seconds. The UTC scale is adjusted by the insertion or deletion of seconds (positive or negative leap seconds) to ensure approximate agreement with UT1." The interval between time signals of UTC is thus exactly equal to the SI second. A history of rate offsets and step adjustments in UTC is given at http://www.iers.org, and is shown in Fig. 5.

4.2. *Issues*

The primary reason for introducing the concept of the leap second was to meet the requirement of celestial navigation to keep the difference between astronomical time and atomic time small. However, this requirement has diminished because of the availability of electronic navigation systems, at the same time as the operational complexities of maintaining precise timekeeping systems without discontinuities have made the insertion of leap second adjustments increasingly difficult and costly. Recently created working groups of various international scientific organizations are now investigating the need to continue the leap second, with its technical inconveniences, or whether it would be better simply to discontinue the traditional relationship of timekeeping with the Earth's rotation.

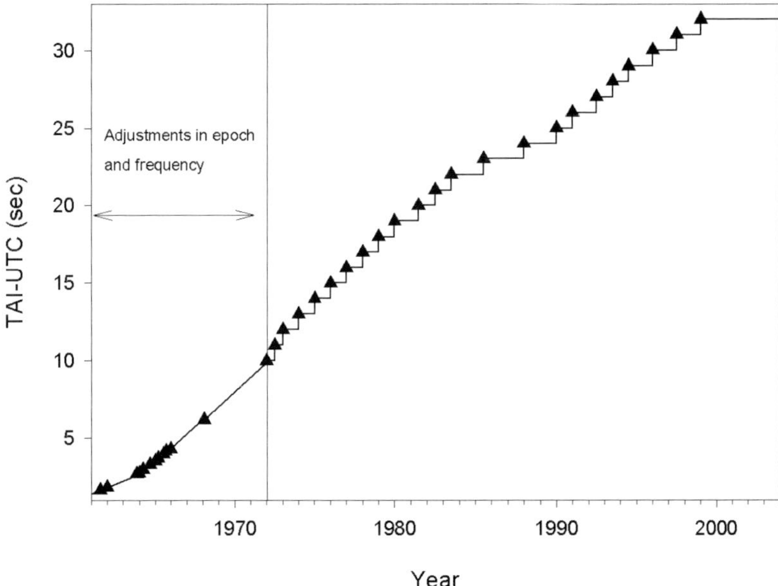

Figure 5. History of TAI-UTC showing the period of time when both rate and epoch were adjusted.

4.2.1. *Operational difficulties of preserving the leap second*

Modern commercial transport systems depend almost entirely on electronic navigation systems. It is likely that future systems will rely on this means of navigation even more, and it is clear that plans are currently being made to improve the satellite navigation systems to provide navigation accuracy, reliability, integrity and availability beyond current capabilities. Increasing worldwide reliance on satellite navigation for air transport is likely to demand systems free of unpredictable changes in timing epoch. Many telecommunications systems routinely rely on precise time synchronization. During the introduction of a leap second, communications can be lost until synchronization is re-established. However, only systems that depend specifically on time are affected by the introduction of leap seconds; systems depending on frequency have little or no sensitivity to epoch. In high speed communications that time stamp messages at the sub-second level, one second can be a significant length of time. Clocks normally count from 59s to 0s of the next minute. Leap seconds require a sequence of 59s, 60s, and then 0s of the next minute. Many computer systems have a problem introducing the second labeled "60". A similar concern is that when dating events using the Julian Day (JD) or Modified Julian Day (MJD) including fractions of a day, a positive leap second would create a situation where two events 1s apart can receive identical dates when those dates are expressed with a numerical precision equivalent to 1 s.

The possible introduction of one or two 61-s minutes per year into continuous site processes would directly affect synchronization if the leap seconds were not treated identically at the same instant at all cooperating sites. In some cases, the need to avoid disruptions has led to considerations of using nontraditional timekeeping systems, such as GPS Time or a time scale maintained by a private company. Continuing use of a non-uniform time scale including leap seconds in the face of these considerations could lead to the proliferation of independent uniform times adopted to be convenient for particular objectives.

4.2.2. *Operational difficulties of eliminating the leap second*

Some astronomers and satellite ground-station operators would prefer that leap seconds be retained. There is a significant amount of operational software at astronomical observatories and satellite ground stations that assumes implicitly that UT1 will always be within one second of the standard time. This assumption would no longer be true if leap seconds were eliminated. Fixing, testing and documenting this legacy code could be a major effort. It may also be difficult to change existing broadcast formats due to the prevalence of legacy hardware. Clocks exist that receive radio broadcast time signals to automatically display accurate time. These and similar devices might be affected adversely by a change in the broadcast format.

5. Global Navigation Satellite System (GNSS) time scales

Today, with GPS (*Navstar GPS Space Segment/Navigation User Interfaces* 2000) and GLONASS (*GLONASS Interface Control Document, Ver. 4.0* 1998), complemented by LORAN and other radio-navigation systems, celestial position determination is not as common as in the past. These systems and the augmentation systems they have fostered have been incorporated into various commercial applications. With extremely high accuracy and global coverage, satellite navigation systems have become a new public utility known by the general designation of Global Navigation Satellite Systems (GNSS).

5.1. *GPS*

The Global Positioning System (GPS), formally chartered in 1973, provides position with a precision on the order of a few meters or better and time with a precision on the order of 10 nanoseconds. Each satellite carries caesium or rubidium atomic clocks. The satellite and global tracking network atomic clocks are used to generate the time scale known as GPS Time, which is specified to be within one microsecond of UTC as maintained by USNO, modulo 1 second. The origin of GPS Time is midnight of 5/6 January 1980, so that TAI is ahead of GPS Time by 19 s, a constant value. As of 1 January 2004, GPS Time is ahead of UTC by 13 s. The algorithm defining the relationship between GPS Time and UTC includes a correction for leap seconds.

A new generation of GPS with enhanced capabilities, GPS III, is to be implemented in the future. The Earth-fixed orbit determination process for GPS, like virtually all other Earth-orbiting satellites, requires precise knowledge of UT1-UTC, and near real-time orbit determination must use predictions of this quantity. Today, these predictions are updated weekly (Bangert 1986). As GPS Time does not include leap seconds, the introduction of a leap second into UTC does not affect GPS users. The GPS operational control segment, however, must carefully account for the leap second step in UT1-UTC. Prior to a leap second event, two sets of Earth Orientation Parameters are provided to the GPS control segment. One is used up to the time a leap second is inserted and the second containing the new 1s step in UT1-UTC, is used after the leap second is inserted.

5.2. *GLONASS*

The Russian Global Navigation Satellite System (GLONASS) has many features in common with GPS (Daly 1991; Langley 1997). The system was fully deployed in early 1996 but currently there is not a full complement of operational satellites. The GLONASS design uses Moscow Time, (UTC + 3 h) as its time reference instead of its own internal time. Thus, users of this system are directly affected by leap seconds. During the process of resetting the time to account for a leap second, the system is unavailable for navigation service because the clocks are not synchronized.

5.3. *GALILEO*

The Galileo satellite navigation system is under development by the European Space Agency. The start of full operation is planned for 2008. As the internal time scale for Galileo will be continuous without steps, it has been expected that the Galileo system time will be a realization of TAI. However, in a letter to the CCTF disseminated at its April, 2004 meeting (Hahn 2004), the director of the ESA Galileo Project Office expressed the hope that any decision to discontinue leap seconds should be made in the near future. The Galileo system could then adopt a realization of UTC as its internal reference time scale without facing any discontinuities.

5.4. *Future satellite systems*

The growing utilization of satellite systems and their internal time scales may gradually become the primary source of time for many practical applications. Options for the future integration of the civil time scale and the GNSS timescales may be examined in the context of several desirable objectives (Nelson et al. 2004). These are:

(a) standardization of time scales;
(b) internal time scales without steps (no leap seconds);
(c) reduction in the number of time scales;
(d) provision for access to standard civil time.

6. Future time scale

There exists a variety of options for the future of UTC. Some of these options are identified and discussed below.

6.1. *Maintain the status quo*

The advantage of maintaining the present form of UTC is that established timekeeping practices will not require modification. On the other hand, if leap seconds were continued, the required number and frequency can only increase. By 2100 there would be a need for nearly two leap seconds per year. The current emerging problems and the resulting dissatisfaction with leap seconds will only continue to grow. The operational impact and associated cost of maintaining leap seconds in complex timekeeping systems must be considered in evaluating their continued use in the future.

6.2. *Increase the tolerance between UT1 and UTC*

An increment of several leap seconds could be inserted into UTC every few years , a "leap minute" in about fifty years, or a leap hour hundreds of years from now. The advantage of this approach is that it would be relatively easy to adopt.

6.3. *Periodic insertion of leap seconds*

A time step could be inserted into UTC at a well-defined interval, such as on 29 February every four years. The advantage is that the date would be predictable. However, the number of leap seconds would not be predictable and large time steps would still be required.

6.4. *Variable adjustments in frequency*

This alternative is similar to the original form of UTC that was abandoned. Introducing a variable atomic scale in step with solar time would cause significant disruptions to equipment and would not disseminate the unit of time, the SI second.

6.5. *Redefine the second*

This option would appear to be the most fundamental solution. However, it would be inconsistent with the usual practice in metrology, which is to adopt a new definition of a unit only when its realization under the old definition becomes the limiting source of experimental uncertainty and to maintain continuity between the old and new realizations. Changing the definition of the second to be closer to the current rotational second would alter the value of every physical measurement and render obsolete every instrument related to time. Moreover, the solution would be only temporary as the Earth continues to decelerate.

6.6. *Substitute TAI for UTC*

TAI is the fundamental atomic time scale from which other scales of uniform time are derived. TAI is related to UTC so that the increment to be applied to UTC to give TAI is equal to the total number of leap seconds plus 10 s. The advantage of TAI is that it is a continuous, atomic time scale without steps. However, TAI is currently not easily available to the precise time user and, as TAI is currently ahead of UTC by an offset of 32 s, a worldwide adjustment of clocks would be required if it were adopted as the scale of civil time. Promotion of two parallel time scales for civil timekeeping, one with leap seconds and one without would be potentially confusing. In addition, as UTC is recognized as the primary basis of civil time in resolutions of various international treaty and scientific organizations and by many conforming national legal codes, a worldwide change in the legal definition of time would be required if UTC were replaced by TAI.

6.7. *Discontinue leap seconds in UTC*

This option would permit continuity with the existing UTC time scale and would eliminate the need for future adjustments to complex timekeeping systems. If the current rate of deceleration of the Earth's rotation were to persist and no leap seconds were added, by 2050 the difference between UTC and UT1 would be about 1 min. By the end of the twenty-first century, the expected difference would be about 2.5 min (McCarthy & Klepczynski 1999). However, these differences are minor compared with the difference between apparent solar time and mean solar time (up to 16.5 min), mean solar time and clock time within a given time zone (nominally up to 30 min), or Daylight Saving Time and Standard Time (1 h). It is thus unlikely that the growing difference between clock time and levels of daylight would be noticeable for the foreseeable future. Therefore, the elimination of leap seconds would have no practical effect on the correspondence between civil time and solar time or on contemporary social conventions. The use of UTC without leap seconds would retain all the advantages of TAI. The transition to a continuous UTC system might be planned for a future date sufficiently far in advance that changes to existing hardware and software, where necessary, could be accommodated within the normal maintenance and replacement schedules.

7. Requirements of celestial navigation

There remains the need to meet the requirements of celestial navigation. Possible options suggested in Nelson et al. (2001) for addressing this need if the current UTC system were revised are considered. Additional alternatives may be identified as the issue is debated.

The IERS/USNO provides daily and weekly predictions of UT1-UTC in Bulletin A, available on the Internet at http://www.iers.org. The estimated accuracies are 0.0017 s at 10 days and 0.0039 s at 30 days. In addition, long-term projections might be included in the nautical almanacs with less precision. With the usual yearly schedule of publication, the extrapolation should not bring errors exceeding 1 s (leading to a position error of 0.5 km at most). Through both short-term and long-term UT1 predictions, it would be possible to complement the information to navigators by disseminating a correction to the argument of the ephemerides, as is done currently with DUT1.

8. Conclusions

Modern timekeeping conventions have loosened the strict relationship between astronomical time based on the Earth's variable rotation and a uniform time scale enabling modern navigation and communication technology. It is likely that this tie will be loosened further in the future as the need for precise timekeeping increases and the practical requirements for a solar-based time decrease. It is also likely that the astronomical determination of the Earth's rotation angle will improve in the future, enabling improvements in the near real time orbit determination of satellite systems. There is no technical reason to continue the relationship between the two. It is reasonable to treat the rotation of the Earth as an angle measured in a system of uniform time.

A suitable compromise for the future could involve the following:

(*a*) Increase the tolerance between UT1 and UTC to one hour, enabling the predicted date for the adjustment by one hour to be made many years in advance.

(*b*) Acknowledge that UT1 is essentially a measure of the Earth's rotation angle to be determined astronomically, and improve the accessibility of this information to users by electronic means.

It is critical that the international community decide the future of UTC soon. The allocation of resources to implement future systems is being planned now, and critical decisions may involve the nature of future timekeeping systems.

References

Aoki S., Guinot B., Kaplan G. H., Kinoshita H., McCarthy D. D., Seidelmann P. K. 1982 *Astron. Astrophys.* **105**, 359–361.

Bangert J. A. 1986 *Proc. 4th International Geodetic Symposium on Satellite Positioning* Austin, Tex.

Barnes J. A., Andrews D. H., Allan D. W. 1965 *IEEE Trans. Instrum. Meas.* **IM-14**, 228–232.

BIPM Com. Cons. Déf. Seconde 1970 **5**, 21-23; reprinted in *Time and Frequency: Theory and Fundamentals, Natl. Bur. Stand. (U.S.) Monograph 140* 1974 (ed. B. E. Blair), U.S. Govt. Printing Office, Washington, D.C., 19–22.

Capitaine N., Guinot B., McCarthy D. D. 2000 *Astron. Astrophys.* **355**, 398–405.

Capitaine N., Wallace P. T., McCarthy D. D. 2003 *Astron. Astrophys.* **406**, 1135–1149.

Clemence G. M. 1948 *Astron. J.* **53**, 169–179.

Commission Préparatoire pour la Coordination Internationale des Échelles de Temps 1968 *Proc.-Verb. Com. Int. Poids et Mesures* **36**, Annexe 1, 109–113; reprinted in *BIPM Com. Cons. Déf. Seconde* 1970, **5**, Annexe S 10, 121–125.

Daly P. 1991 *Acta Astronautica* **25**, 399–406.

Egbert G. D., Ray R. D. 2000 *Nature* **405**, 775–778.

Essen L. 1968 *Metrologia* **4**, 161–165.

Essen L., Parry J. V. L. 1955 *Nature* **176**, 280–282.

Essen L., Parry J. V. L. 1957 *Philos. Trans. R. Soc. London* **250**, 45–69.

Explanatory Supplement to the Astronomical Almanac rev. ed., (ed. P. K. Seidelmann), University Science Books, Mill Valley, Calif., 48.

Gerber E. A, Sykes R. A. 1966 *Proc. IEEE* 54, 103-116; reprinted in *Time and Frequency: Theory and Fundamentals, Natl. Bur. Stand. (U. S.) Monograph 140* 1974, (ed. B. E. Blair), U. S. Govt. Printing Office, Washington, D.C., 41–56.

GLONASS Interface Control Document, Ver. 4.0 1998 Coordination Scientific Information Center,Moscow.

Goldenberg H. M., Kleppner D., Ramsey N. F. 1960 *Phys. Rev. Lett.* **5**, 361–362.

Guinot B. 1989 "Atomic Time," in *Reference Frames for Astronomy and Geophysics* (ed. J. Kovalevsky, I. I. Mueller & B. Kolaczek), Kluwer, Boston.

Guinot B. 1994/1995 *Metrologia* **31**, 431–440.

Guinot B. 2000 , in *Polar Motion: Historical and Scientific Problems, IAU Colloquium 178* ASP Conference Series, Vol. 208 (ed. S. Dick, D. McCarthy & B. Luzum), Astron. Soc. Pacific, San Francisco, 175–184.

Hahn J. 2004 European Space Agency Galileo Project Office, Letter to the CCTF (March 26, 2004).

Halley E. 1693 *Philos. Trans. R. Soc. London* **17**, 913–921.

Halley E. 1695 *Philos. Trans. R. Soc. London* **19**, 160–175.

IERS Conventions (1996) (ed. D. D. McCarthy), International Earth Rotation Service Tech. Note 21, Observatoire de Paris, Paris, 84.

International Radio Consultative Committee (CCIR) 1970a , *XIIth Plenary Assembly CCIR* International Telecommunication Union, Geneva, 227; reprinted in *Time and Frequency: Theory and Fundamentals, Natl. Bur. Stand. (U. S.) Monograph 140* 1974 (ed. B. E. Blair), U. S. Govt. Printing Office, Washington, D.C., 31.

International Radio Consultative Committee (CCIR) 1970b , *XIIth Plenary Assembly CCIR* International Telecommunication Union, Geneva, 258a-258d; reprinted in *Time and Frequency: Theory and Fundamentals, Natl. Bur. Stand. (U. S.) Monograph 140* 1974 (ed. B. E. Blair), U.S. Govt. Printing Office, Washington, D.C., 32–35.

ITU-R Recommendations: Time Signals and Frequency Standards Emissions 1998, International Telecommunication Union, Radio-communication Bureau, Geneva,15.

Jeffreys H. 1920 *Philos. Trans. R. Soc. London* **A221**, 239–264.

Jeffreys H. 1962 *The Earth: Its Origin, History and Physical Constitution* 4th ed., Cambridge University Press, New York, 514.

Kant I. 1867 , in *Sämmtliche Werke* Leipzig, Vol. 1; Whether the Earth Has Undergone an Alteration of Its Axial Rotation, in *Kant's Cosmogony* 1968 (Translated by W. Hastie, (ed. W. Ley), Greenwood, New York, 157–165.

Lambeck K. 1980 *The Earth's Variable Rotation* Cambridge University Press, Cambridge.

Langley R. B. 1997 *GPS World* **8**, 46–51.

Markowitz W., Hall R. G., Essen L., Parry J. V. L. 1958 Phys. Rev. Lett. **1**, 105–107.

McCarthy D. D. 1991 *Proc. IEEE* **79**, 915-920.

McCarthy D. D., Babcock A. K. 1986 *Physics of the Earth and Planetary Interiors* **44**, 281–292.

McCarthy D. D., Klepczynski W. J. 1999 *GPS World* **10**, 50–57.

Metrologia 1968 **4**, 43.

Metrologia 1971 **7**, 43.

Metrologia 1981 **17**, 70.

Morrison L. V., Stephenson F. R. 1986 , in *Earth Rotation: Solved and Unsolved Problems* (ed. A. Cazenave), Reidel, Boston, 69-78.

Natl. Bur. Stand. (U. S.) Tech. News Bull. 1949, **33**, 17-24.

Navstar GPS Space Segment/Navigation User Interfaces, ICD-GPS-200C-004, 2000 ARINC Research Corporation, El Segundo, CA, 2000.

Nelson R. A., McCarthy D. D., Malys S., Levine S., Guinot B., Fliegel H. F., Beard R. L., Bartholomew T. R. 2001 , *Metrologia* **38**, 509–529.

Nelson R. A., McCarthy D. D., Gifford A., Bartholomew T. , *submitted to 2004 Proceedings of the European Frequency and Time Forum*

Newcomb S. 1895a *Astronomical Papers Prepared for the Use of the American Ephemeris and*

Nautical Almanac Vol. VI, Part I: Tables of the Sun, U.S. Govt. Printing Office, Washington, D.C., 9.

Newcomb S. 1895b *The Elements of the Four Inner Planets and the Fundamental Constants of Astronomy* U.S. Govt. Printing Office, Washington, D.C., 1895, Chap. 2.

Runcorn S. K. 1966 *Scientific American* **215** 26–33.

Schcibe A., Adelsberger U. 1936 *Phys. Zeitschrift* **37**, 38.

Seidelmann P. K., Fukushima T. 1992 *Astron. Astrophys.* **265**, 833-838.

Stephenson F. R. 1997 *Historical Eclipses and Earths Rotation* Cambridge University Press, 64.

Stoyko N. 1937 *C. R. Acad. Sci.* **205**, 79.

The International System of Units (SI), 7th ed. 1998 *Bureau International des Poids et Mesures*, Sèvres, 111-115.

Trans. Int. Astron. Union Vol. VIII 1954 Proc. 8th General Assembly, Rome, 1952 (ed. P. T. Oosterhoff), Cambridge University Press, New York, 66.

Trans. Int. Astron. Union Vol. XVI B 1977 Proc. 16th General Assembly, Grenoble, 1976 (ed. E. A. Muller & A. Jappel), Reidel, Dordrecht, 60.

Trans. Int. Astron. Union Vol. XVII B 1980 Proc. 17th General Assembly, Montreal, 1979 (ed. P. A. Wayman), Reidel, Dordrecht, 71.

Trans. Int. Astron. Union Vol. XXI B 1992 Proc. 21st General Assembly, Buenos Aires, 1991 (ed. J. Bergeron), Reidel, Dordrecht, 41-52.

Trans. Int. Astron. Union Vol. XXIV B 2000 Proc. 24th General Assembly, Manchester, 2000, Astron. Soc. Pacific. San Francisco.

Wells J. W. 1963 *Nature* **197**, 948-950.

Winkler G. M. R. 1968 *Memorandum submitted to the ad hoc group meeting at the International Bureau of Weights and Measures (BIPM)*, 30 May 1968.

Yoder C. F., Williams J. G., Dickey J. O., Schutz B. E., Eanes R. J., Tapley B. D. 1983 *Nature* **303**, 757-762.

Discussion

MIKHAIL MAROV: I am just curious: among other assumptions you just listed the communication problems. Is it still valid? Because we have, right now, so fast growing information technology and we need progress, so it seems that in reality wherever you are just at the moment, you could communicate easily through different kinds of channels.

DENNIS MCCARTHY: Most communication systems are based on interval and not epoch; those can always synchronise and they don't need to have a change in epoch from those that are right. There are some communication systems though that actually are based on time synchronisation in which the introduction of the second, if not done properly at the two sites causes confusion. We have instances of these things actually happening, where we'll have telephone cells drop out because somebody doesn't introduce the leap second the way they are supposed to, somebody forgets, or somebody does it and puts in the wrong sum, or changes their clock in the wrong direction – those are the problems. And then we have drop-outs which usually last only for a short time ... for 20 minutes while they adjust the clocks.

MYLES STANDISH: You said right at the very end that we're getting away from the rotation of the Earth. When I try to explain what the different times are, I wish I could get rid of the word "time," and then for UT1 you'd use the rotation of the Earth, for UTC you'd use your wristwatch, etc., and I think there's something in that for me ...

DENNIS MCCARTHY: Well, I didn't have a chance to say it, but I think we need to get used to the fact that UT1 – UTC is just a way of expressing an angle, the Earth's rotation angle. It's no longer a time, UT1 is not really a time, it's a way to express the Earth's rotation angle and it should not be thought of as a real time scale. If we can get people to understand that, that's fine. So that we can neglect the Earth's rotation angle in our calculations, I suspect many people who write software and point telescopes are cheating by saying that UT1 is essentially equal to UTC. But in doing so they lose the accuracy of UT1. UT1 – UTC is known to an accuracy better than one second of time – a tenth of a microsecond – it's really a disservice to keep that going.

DON KURTZ: How much does the zero-point of the Earth's rotation angle change because of continental drift causing a change in the reference frame?

DENNIS MCCARTHY: The two time systems have to be maintained. The definition of the terrestrial reference frame is designed so that there is no drift, so there's no risk. We compensate for it essentially by taking it out. At some level that affects the observations of UT1, but this is going to be a 5-, 10-, 15-year sort of thing, so it's pretty low.

NICOLE CAPITAINE: Are you sure that now the Earth is accelerating? Are there some signs that this will change?

DENNIS MCCARTHY: As we speak the Earth is decelerating, but in the past it's been accelerating. We would look for it to change, we have had very substantial periods of acceleration in the past hundred years. They've always turned right around after a while and we get the question, "Is it ever going to unwind?" The answer is, "Yes, it probably will." It has in the past, it unwinds. It'll catch up, that problem is real.

MYLES STANDISH: No negative leap seconds?

[laughter]

DENNIS MCCARTHY: No, we don't anticipate negative leap seconds. When we first thought this up we tried to tell people there would be leap seconds at the rate of two or three a year, and since that time we haven't had a leap second, so ...

[general laughter]

Diane, Dennis and Dierdre McCarthy

Conference tour of Stonyhurst College Observatory with Fintan O'Reilly

Transits of Venus: New Views of the Solar System and Galaxy
Proceedings IAU Colloquium No. 196, 2004
D.W. Kurtz, ed.

© 2004 International Astronomical Union
doi:10.1017/S1743921305001389

Thomas Henderson and α Centauri

Brian Warner

Department of Astronomy, University of Cape Town, Rondebosch 7700, South Africa
email: warner@physci.uct.ac.za

Abstract. The first observations containing evidence of a measurable stellar parallax were made by Thomas Henderson at the Cape of Good Hope in 1832/33. Although his response to Manuel Johnson's discovery in early 1833, that α Cen has a large proper motion, was to intensify observations for his remaining month at the Cape, Henderson apparently saw no urgency to reduce his observations. Instead he laboured through more routine matters, producing a catalogue of declinations of southern stars, improvements of refraction tables, and work on the solar and lunar parallaxes. It was Bessel's announcement of a determination of the parallax of 61 Cyg that finally stirred Henderson into action in late 1838.

1. Introduction

Thomas Henderson was born in Dundee in 1798 and showed early talent. At the age of fifteen he was apprenticed to a solicitor and in his spare time began to study astronomy. In 1819 he moved to Edinburgh and completed his legal studies, meeting the local astronomers and being allowed to use the observatory on Calton Hill. From the age of 26 on Henderson published a number of papers, mostly on computational methods, and he narrowly missed being appointed to the astronomy Professorship at Edinburgh and to the Superintendency of the Nautical Almanac Office. Although technically an amateur astronomer he was elected a Fellow of the Royal Society.

When the Directorship of the Cape Observatory came available through the death of Fearon Fallows in 1831, he was persuaded to take the post, largely as a way in which to enter the professional arena of astronomy.

While becalmed in Falmouth, on his way to the Cape, he wrote to his friend and fellow amateur astronomer, Thomas Maclear, that "...I received my official instructions. They prescribe a sufficient amount of work. In the Southern hemisphere I am asked to do as much as now occupies twelve or twenty Observatories in the northern. However I am not at all daunted. Patience and perseverance can do much. I am required to pay particular attention to the Transit of Mercury; but as nothing is said with regard to those of Venus, I conclude that it is held a settled point that I shall not be alive at the first one that happens, which I believe is in 1874" (Henderson 1832).

Henderson held the position of His Majesty's Astronomer at the Cape of Good Hope for a mere 13 months, arriving at the Observatory on 8 April 1832 and leaving for England on 28 May 1833, soon to become the first Astronomer Royal for Scotland. During his time at the Cape he assiduously observed the brighter southern stars with the mural circle, resulting in a catalogue of precise declinations for 172 stars, which was dated at Edinburgh on 30 September 1836 (Henderson 1838a). As he naturally included α Centauri among his selected objects he serendipitously made measurements that were the first to be affected by a detectable parallax. However, the mural circle had problems, which had caused Fearon Fallows, Henderson's predecessor at the Cape, great anguish (Warner 1979, 1995) and which Henderson himself investigated with considerable care soon after arriving at the Cape† (Henderson 1835). Sheepshanks & Airy (1833), using observations made by Fallows, and Henderson from his own observations, all concluded that a reliable measurement could be obtained if a mean was taken of the six microscopes attached to the mural circle. Henderson's successor, Thomas Maclear, together with John Herschel, also investigated the circle and eventually arranged for it to be exchanged with its twin, which was in use at Greenwich and arrived at the Cape in July 1839. (On investigation in London the steel collar on one of the pivots of the Cape mural circle was found to be loose: Warner 1979)

It is natural to expect that Henderson would be cautious about any claim to what would be the first true detection of stellar parallax, particularly when Brinkley had had his fingers burnt on the subject some fifteen years earlier (Brinkley 1815; Pond 1817). But the delay, which was of over four years, was a surprising length of time, considering the apparent importance of the matter and that Henderson had already told Maclear in June 1834 that "It is certain that α^1 and α^2 Centauri have an annual proper motion of $3''\!.6$ in space...They should be watched to see if parallax be visible in them" (Henderson 1834b).

Henderson's paper on the parallax of α Cen was dated in Edinburgh 24 December 1838; it was read at the RAS meeting on 3 Jan 1839 and appears in Volume 11 of Mem RAS, dated 1840 on the title page (when the volume was completed). In his paper Henderson states "It was only about the termination of my residence at the Cape, that I learned from Mr Johnson the fact of the great proper motion, which first made me suspect that there might be a sensible parallax. Had I been aware of the proper motion at an earlier period, a much greater number of observations, and as such as would have been better adapted for ascertaining the parallax, would have been made, so that a greater degree of probability would have attended the result".

2. Manuel Johnson

Manuel John Johnson had been appointed in 1823 to the South Atlantic island of St Helena as part of the St Helena Artillery, which was detailed to protect the tomb of Napoleon Bonaparte. On St Helena Johnson was taught land surveying and showed such proficiency that he was selected to erect an observatory to provide accurate time for the ships at Jamestown. To increase his knowledge of meridian astronomy he was permitted to visit the Royal Observatory at the Cape of Good Hope, where Fearon Fallows was in the process of erecting the Observatory. Johnson visited Fallows in early 1826 and again in late 1828. From Fallows he was able to copy the plans of the central section of the Cape observatory, which he adapted for the observatory on Ladder Hill in St Helena. On his second visit he was able to practice with the meridian instruments that had just been installed at the Cape.

By April 1833 Johnson had been at St Helena for 10 years and was entitled to furlough with pay. His application stated that he intended to "visit the principal European

---

† Consider what Maclear, on his way to the Cape in 1833, must have felt when he received Henderson's warning about the mural circle, which was one of the two principal instruments that Maclear was expected to employ in making the most accurate possible measurements of star positions: "Be not alarmed although you find sudden and violent changes in the readings of the top and bottom Microscopes of the Circle when directed to the same Star on two successive nights, for what one gains, its opposite will lose, and the mean of both and of all the six will remain the same" (Henderson 1833). After he had examined his extensive set of microscope comparisons, Henderson was able to assert that "...the accuracy of the Astronomical Observations is not impaired by the anomalies of the Instrument, and that it is little (if at all) inferior in accuracy to the best Instruments of similar construction hitherto made" (Henderson 1834a).

Table 1. Dates of Mural Circle observations of α Centauri

α^1	Direct	1832: 23/5, 11/6, 19/8, 23/9, 11/10, 9/11, 15/12
		1833: 12/1, 8/4, 10/4, 13/4, 21/4, 28/4, 30/4, 2/5, 9/5, 16/5
	Reflected	1832: 27/7, 5/9, 14/9, 21/9, 25/9, 3/10, 7/10, 31/10, 8/11
		1833: 28/1
α^2	Direct	1832: 22/5, 28/5, 6/7, 16/7, 12/9, 24/9, 30/10, 12/11, 13/12,
		1833: 9/1, 14/1, 9/4, 12/4, 20/4, 25/4, 29/4, 1/5, 4/5, 11/5, 20/5
	Reflected	1832: 6/8, 20/8, 11/9, 22/9, 14/10, 16/10, 29/10, 7/11
		1833: 8/1

Observatories, an advantage... I never have had an opportunity to enjoy" (Warner 1982). This was granted on 13 May and he left within a few days (the precise date of sailing is not known).

By the time that Johnson left St Helena he had accumulated sufficient observations to form a catalogue. Johnson intended to return and extend his work, but in August 1833, while he was away, the East India Company's garrison on St Helena was dispersed when an Act of Parliament transferred the island to control by the British Government. His reductions were evidently in an advanced state, for shortly after his arrival in England he presented the Royal Astronomical Society with a catalogue of declinations of 598 stars, the manuscript of which is still in the Archives of the RAS and is dated as received 14 June 1833. The final catalogue, expanded to include both right ascensions and declinations of 606 stars, was privately printed in 1835 (Johnson 1835). The RAS, using Henderson as a referee, judged the catalogue of such high quality that they awarded Johnson the Society's Gold Medal in January 1835. The draft introduction to the catalogue still exists, with comments in Henderson's hand (Oxford).

From the Introduction to the catalogue we learn that it was while making comparisons between his own catalogue and those of others, especially Lacaille (from observations made in 1751 and published by Baily in 1831 (Baily 1833)), that Johnson discovered the large proper motion of α Cen. Although there were other differences between the two catalogues, that for α Cen was too large to ascribe to anything other than proper motion. Henderson left the Cape on 28 May 1833, and had received Johnson's letter about the proper motion not long before departing, so Johnson must have started work on his reductions (at least the declinations) and sent his letter to Henderson from St Helena some time before he left his own observatory. The letter, which would be of some historic interest, has not been found. The immediate delivery on arrival in England to the RAS of Johnson's preliminary catalogue again suggests that he had completed at least the declination catalogue before he left St Helena, the work for which revealed the large proper motion of α Cen.

3. Henderson's measurements of α Centauri

Henderson's parallax paper (Henderson 1840) contained just 17 measurements of α^1 Cen and 20 of α^2 Cen made by 'direct' observation, and 10 of α^1 Cen and 9 of α^2 Cen made by 'reflected' observation. The latter used a trough of mercury in which to see the reflected image of the star – a relatively new process at the time that enabled accurate declinations to be determined, independent of knowledge of latitude, from averages of the direct and reflected altitudes. The observations were made on the dates shown in Table 1.

The declination measurements were made personally by Henderson with the mural circle of the Cape Observatory. Over the same period of time, April 1832 to May 1833, measurements were obtained with the Observatory's transit instrument, by the assistant,

Lieutenant William Meadows, which provided 24 observations of α^1 Cen and 25 of α^2 Cen. Henderson and Meadows observed α Cen simultaneously on only two nights, presumably because they worked separately through the list of stars to be observed.

The total number of independent observations was therefore 105, spaced unevenly over just one Earth orbital period, with (as seen in Table 1) a heavy concentration in April and May 1833 until 8 days before Henderson left the Cape – which suggests that he received Johnson's letter about a month before he embarked. The only two ships arriving from St Helena during that period were the Meta, which left St Helena on 26 February and arrived at the Cape on 21 March, and the Lord Hobart, which left on 20 March and arrived on 14 April. It presumably was the latter that brought Johnson's letter. Despite the increased frequency of transit observations made by Johnson and Meadows in April and May 1833, on only the first (8 April) and penultimate (16 May) observing nights did they make simultaneous observations. Although the earliest observations in April might appear to have anticipated the arrival of Johnson's letter, there had been no measurements made in February or March, and Henderson and Meadows were merely starting their observations again at a time similar to the first observations of the previous year.

The result from this set of observations, which were poorly distributed, under-sampled and generated by a slightly suspect instrument, was a parallax $\pi = 1''\!.16 \pm 0''\!.11$, obtained from six independent groups of measurements (viz. each star observed with the transit instrument and by direct and by reflected observation with the mural circle)†. The spread was from $0''\!.48$ to $1''\!.96$, but the fact that all the measurements were positive certainly suggested that a parallax effect had been detected. Main (1842), in reviewing the state of parallax measurements in 1840, reanalyzed Henderson's measurements and found that for the deviations from the mean (at least for α^1 Cen) "there is scarcely an exception to the proper change of sign, according to the change of sign of the coefficients of parallax", and therefore accepted the evidence for parallax "until some distinct reason, independent of parallax, shall have been assigned for the changes in the declinations. Such I do not consider impossible, having before my eyes the results which Dr Brinkley derived, in the cases of certain stars, with the Dublin circle)". (The central problem in parallax measurement with meridian circles is that one is looking for a modulation in position with a period of a year – and this may occur merely from the summer/winter variations in observing conditions.)

4. The delay

From a modern perspective the delay between Henderson's receipt of Johnson's letter about α Cen in 1833 and his working on the issue in 1839 seems perplexing. It is clear from the following comment in a letter to Maclear at the end of 1838 that it was Bessel's announcement (in a letter to John Herschel, dated at Königsberg 23 October, 1838: Bessel 1838) of his detection (with the use of a heliometer) of a parallax in 61 Cyg that finally moved Henderson into action: "The question of the parallax of the Stars is again making a stir, from Bessel having announced that he makes a parallax of $\frac{1}{3}^d$ of a second in 61 Cygni. I have sent off a Paper on the parallax of α Centauri to the Astronomical Society, and I enclose a copy" (Henderson 1838b).

In his α Cen paper, Henderson's explanation of the cause of the hiatus was that "I delayed communicating the result till it should be seen whether it was confirmed by the

---

† Henderson was among the first astronomers to use Gauss's method of least squares and to derive probable errors therefrom.

observations of Right Ascension made by Lieutenant Meadows". This rather sounds as if he had to wait for the RA measurements to be sent to him, but in fact he took copies of them with him when he left the Cape in 1833. What had actually happened was that he had set himself a program of work and publication of the Cape observations. This can be traced through successive letters to Maclear (RGO Archives), for example, 29 Sep 1834 – "I am devoting as much time as possible to the reduction of the Cape Observations, but the labour is immense, and my interruptions have been frequent. I am investigating several points which have been scarcely attended to in English Observations, for instance the magnitude of the numerical Coefficients which enter into the formulae of refraction. I expect to have the first great division of the work, the Declinations, ready about the new Year. The Right Ascensions will follow next"; *two years later*, on 20 September 1836 – " I am now engaged with the Memoir upon the Cape Declinations. When it is finished I shall prepare one shewing the observations of refractions of Stars near the Horizon, and their results. The calculations are ready. Next I intend to investigate the constant quantity of the Moon's parallax from my Cape observations compared with those of Greenwich and Cambridge.... Then I shall devote myself to the Right Ascensions of the Stars". In addition to these projects he worked on the solar and lunar parallaxes and many minor issues.

It is evident that, because of this plan of action, combined with the problems of getting the observatory in Edinburgh into operation, Henderson did not get to the extensive reductions required for the right ascensions, the results of which would be needed for use on α Cen, until the perceived more urgent matters were completed. He published a series of papers in the RAS Memoirs on these more pressing topics. Nowhere in his correspondence before the end of 1838 is there any awareness of urgency to seek parallax in α Cen.

Henderson, although having finally investigated his and Meadows's observations, and having committed himself to the public announcement in 1839, wrote to Sheepshanks that he had "not very great confidence in my parallax of α Centauri; but I thought it might be right to let it be known that my observations were better satisfied by the hypothesis of parallax†. I am anxious to learn in due time the result of additional observations. I wrote to Maclear two months ago with a copy of my Paper" (Henderson 1839). In the letter to Maclear, Henderson said that "I recommend to you to make as numerous a set of observations as you can for the space of a twelvemonth at least, to ascertain beyond doubt whether there is a sensible parallax... perhaps your most convenient mode of observation is the measurement of the double Altitudes of the two Stars, with the Mural Circle" (Henderson 1838b).

5. Maclear's measurements of α Centauri

Maclear was sufficiently impressed with Henderson's work to tell Sir John Herschel "Henderson sent me a copy of his paper on the parallax of α^1 & α^2 Centauri. The result appears palpable enough although the observations are not so numerous nor well placed as they might be. Therefore I have taken up the subject" (Maclear 1839). Maclear's observations were sent to Henderson for reduction and resulted in Henderson's second paper on the subject, read to the RAS on 8 April 1842 (Henderson 1842). This contained observations from 26 March 1839 to 12 August 1840, comprising 124 individual

---

† This statement, which is that of a true scientist rather than a mere measurer and cataloguer, suggests that Henderson had, as with so many other scientists of the time, been influenced by John Herschel's treatise published a few years earlier (Herschel 1830).

measurements of α^1 and α^2 Cen with the original mural circle and 483 with the replacement mural circle, sent out from Greenwich. These provided 134 measurements of double altitudes of α^1 Cen and 138 of α^2 Cen, made at same transits, distributed around the year but with higher concentration near the months of April and October when the parallax factors in declination are maximal for α Cen.

This was a very substantial amount of observational work, especially because the double altitude method required extra effort. As Maclear mentioned to Herschel: "Considerable dexterity is required to take each star of a double star by direct vision & reflexion at the same transit; the errors of observation are greater. A micrometer has to be read twice independent of the bisections at or near to the second wire – again, two bisections at the 4th..." (Maclear 1840). Henderson's deduced result from Maclear's generous effort was $\pi = 0.''9128 \pm 0.''0640$. The modern value is $0.''785$, which confirms that it certainly was within reach of the meridian instruments of the time.

6. Conclusion

Henderson evidently thought that his small number of observations of α Cen would not show the presence of parallax, and therefore left it to Maclear to carry out his 1834 recommendation to follow up on the clue of large proper motion. In 1838, with nothing forthcoming from Maclear, Henderson, inspired by Bessel's success, analysed his α Cen observations for the first time — which would not have taken much time because by then he had all the basic reductions of the Cape declinations and right ascensions under control. It probably came as a surprise to find such clear evidence for parallax, which gave him the confidence to publish immediately.

References

Baily, F. 1833 *Mem R.A.S.* 5, 93.
Bessel, F.W. 1838 *MNRAS* 4, 152.
Brinkley, J. 1815 *Trans. Roy. Irish Acad.* 12, 33.
Henderson, T. 1832 *Letter, 25 January 1832, TH to T. Maclear*, RGO Archives.
Henderson, T. 1833 *Letter, 24 October 1833, TH to T. Maclear*, RGO Archives.
Henderson, T. 1834a *Letter, 31 January 1834, TH to T. Maclear*, RGO Archives.
Henderson, T. 1834b *Letter, 14 June 1834, TH to T. Maclear*, RGO Archives.
Henderson, T. 1835 *Mem. R.A.S.* 8, 141.
Henderson, T. 1838a *Mem. R.A.S.* 10, 49.
Henderson, T. 1838b *Letter, 26 December 1838, TH to T. Maclear*, RGO Archives.
Henderson, T. 1839 *Letter, 23 February 1839, TH to Richard Sheepshanks*, RAS Archives.
Henderson, T. 1840 *Mem. R.A.S.* 11, 61.
Henderson, T. 1842 *Mem. R.A.S.* 12, 328.
Herschel, J.F.W.H. 1830 *A Preliminary Discourse on the Study of Natural Philosophy*, Longmans, London.
Johnson, M.J. 1835 *A Catalogue of 606 Principal Fixed Stars in the Southern Hemisphere*, Privately printed, London.
Maclear, T. 1839 *Letter, 3 April 1839, TM to J. Herschel*, Library of The Royal Society of London.
Maclear, T. 1840 *Letter, 19 February 1840, TM to J. Herschel*, Ibid.
Main, R. 1842 *Mem. R.A.S.* 12, 1.
Oxford *Museum of the History of Science*, Papers of the Radcliffe Observatory.
Pond, J. 1817 *Phil. Trans. Roy. Soc.* 107, 158.
Sheepshanks, R. & Airy, G.B. 1833 *Mem. R.A.S.* 5, 325.

Warner, B. 1979 *Astronomers at the Royal Observatory, Cape of Good Hope*, Balkema, Cape
    Town.
Warner, B. 1982 *Vistas in Astr.* 25, 383.
Warner, B. 1995 *Royal Observatory, Cape of Good Hope 1820-1831*, Kluwer, Dordrecht.

Discussion

WAYNE ORCHISTON: Is there any evidence to show why, in fact, Maclear wasn't co-author
on the 1832 paper, because there's not any documentation online about that?

BRIAN WARNER: No. The two were very, very close friends. Henderson died in 1846 and
Maclear was dreadfully hard hit by that. Maclear was doing it as a friendly gesture;
certainly Maclear was way over his head with his observational work and stuff at the
Cape Observatory. He had not himself the time to go through what was a very accurate
set of reductions, so I think he generously gave to Henderson. And if you look at those
times, there was not a great deal of co-authored papers at that time; it was usually a
senior person who was using other people's data, so I don't think there was anything in
it than that. Maclear was asked a favour; he did a favour. Henderson actually had a huge
amount of work doing the reductions and he says in the title it was based on Maclear's
observations. I don't think one should read too much into that.

CORYN BAILER-JONES: You alluded to the answer just now, but one perspective, of
course, is why not look at the parallax anyway? Just do the data reduction; but of course
it does take long, or it would then have taken long. Do you have any idea of how long
that kind of analysis would have taken for that data set?

BRIAN WARNER: Henderson's declination catalogue contains 196 stars, including α Cen,
and it looks to me as though there was nearly one whole year's hard labour; it took him
five years to do it. It was all done with logarithm tables, it was a huge amount of manual
labour, but for someone who was observing at night, trying to build an observatory in the
daytime, it was a huge amount of manual labour. I think that's why it was so slow. The
whole idea that there might be a detectable parallax in those first observations evidently
didn't occur to him at all. I'm sure he was quite astonished to find what was done; but
when you needed to do it, all the data reductions were there, all you had to go back
and do was add the parallax factor, He would have removed aberration, he would have
removed nutation, he would have removed everything except parallax and proper motion.
He probably didn't remove proper motion for β and α Cen. So that part probably took
him a week – it was easy, once the notion was in his mind, but the concept wasn't there in
1833 – to look for effective parallax – and he would've been nervous, as I said. Everybody
got it wrong up until that moment.

NICOLE CAPITAINE: Are there other stars for which the parallax would have been detectable at that time?

BRIAN WARNER: Bessell came up with 61 Cygni and at about the same time Struve
did – what was it? – γ Draconis. There were 3 measurements all announced at the same
time. Struve's observations had been made in 1838-1839; and as I said, Henderson's
were actually made in 1832-33. And they used different techniques of course. You see,
Henderson's was an absolute measurement of position, of declination; what Struve and
Bessell did were differential measurements, which is potentially much more accurate. He
was using a micrometer in the eyepiece to measure the difference in position between,
say, 61 Cygni and a nearby star in the same field. That's the right way to do it. You

may remember Bradley tried to do that in the 1720s to measure parallax and discovered aberration. So there were three, suddenly. Henderson then published a paper showing that there was no detectable parallax in six or eight other bright stars, so he only got one positive result.

Brian Warner at Carr House

Anne Lemaitre, Myles Standish and Menios Tsiganis on a brisk walk from Stonyhurst College to the Shireburn Arms

The Shireburn Arms was a favourite "watering hole" of J. R. R. Tolkien; it is locally believed that much of the *Lord of the Rings* was imagined in and around Hurst Green and Stonyhurst College, where Tolkien's sons taught.

Part 3
TRANSITS
THE SOLAR SYSTEM
AND
EXTRA-SOLAR PLANETS

Vladimir Elkin, Mikhail Marov and Lena Pitjeva

Ignas Snellen talking to Chris Allen and Clive Davenhall; Gert Zech

Transits of Venus: New Views of the Solar System and Galaxy
Proceedings IAU Colloquium No. 196, 2004
D. W. Kurtz, ed.

© 2004 International Astronomical Union
doi:10.1017/S1743921305001390

Mikhail Lomonosov and the discovery of the atmosphere of Venus during the 1761 transit

Mikhail Ya. Marov

Keldysh Institute of Applied Mathematics, Russian Academy of Sciences,
Miusskaya sq. 4, Moscow 125047, Russia   email: marov@keldysh.ru

Abstract. The atmosphere of Venus was discovered for the first time by the Russian scientist Mikhail V. Lomonosov at the St Petersburg Observatory in 1761. Lomonosov detected the refraction of solar rays while observing the transit of the planet across the disk of the Sun. From these observations he correctly inferred that only the presence of refraction in a sufficiently thick atmosphere could explain the appearance of a light ('fire') ring around the night disk of Venus during the initial phase of transit, on the side opposite from the direction of motion. Lomonosov described this phenomenon, which carries his name, as the appearance 'of a hair-thin luminescence', which encircled a portion of the planet's disk that had not yet contacted the solar disk. He also observed a bulge set up at the edge of the Sun during the egress phase of the Venus transit. 'This bears witness to nothing less than the refraction of solar rays in the Venusian atmosphere', he wrote. This paper is based on the original Lomonosov publications and describes historical approaches to the study involving procedure, drawings, and implications.

---

1. Introduction

On 8 June 2004 and on 6 June 2012 mankind has the chance to witness a historical celestial event when the silhouette of Venus crosses the face of the Sun. Obviously, a transit of Venus across the Sun's disc can occur only when the planet is at the inferior conjunction at one of the points in its orbit where the orbital plane crosses the plane of the ecliptic (along the line of nodes) in every June and December. However, because of the necessary orbital configurations and significant eccentricity of the Earth's orbit, such an astronomical oddity is comparatively rare, repeating in cycles whose intervals are 8 yr, 105.5 yr, 8 yr, and 121.5 yr. The last four occurred in 1761, 1769, 1874, and 1882. Thus, after 2012 our descendants will not have an opportunity to observe the next pair again from the Earth until December 2117.

The first observed transits have been of the great value because they allowed us to determine the solar parallax, and thereby the scale of the solar system, with an uncertainty of astronomical unit only a few million km. Although being incomparable with the today's AU accuracy (about 5 m), it was a great achievement in the 18th century. Additionally, during the transit of 1761 an atmosphere around Venus was discovered.

The first evidence of solar rays' refraction in the Venusian gas envelope was supported by the follow-up observations. It was found that contours of the visible disk are indeed considerably influenced by an atmosphere. Refraction shifts the position of the terminator by producing a penumbral zone, where one can observe a gradual decrease of illumination from the day hemisphere to the night hemisphere. One can also see the effect of the atmosphere in the elongation of the horns of the crescent when Venus is near inferior conjunction (Fig. 1). Because of refraction, the lighted arc of the limb appears to be greater than 180°, and each horn seems to be shifted towards the dark half of the limb (see, e.g., Sharonov 1953). In the extreme case, when the crescent is very thin, a complete

Figure 1. The crepuscular arc around Venus. The elongation of the horns is clearly visible, almost encompassing the disk.

ring of light is observed surrounding the planet. In addition to their intrinsic beauty, these phenomena have provided useful insights into some properties of the atmosphere, including the existence of high-altitude clouds.

The beginning of study of Venus' nature is attributed to Galileo Galilei, who first established the phases of the planet, analogous to the well-known phases of the Moon. Many valuable observations of planets were carried out medieval astronomers using rather primitive instruments, long before photographic plates were available. In particular, Johannes Kepler predicted that Venus would pass in front of the Sun on 6 December 1631, but it was not observed at that time from Europe. In England Jeremiah Horrocks (who observed jointly with his friend William Crabtree) had predicted the Venus transit on 4 December 1639 based on mathematical calculations that turned out even more accurate than those Kepler used, and this was the first recorded sighting of the event.

When addressing that historical time, it is worth noting that the beginning of successful astronomical observations and growing understanding of solar system configuration developed by Nicholas Copernicus naturally led some astronomers to surmise that neighboring planets might have atmospheres and that life might exist on them. In particular, such a possibility was seriously considered by Giordano Bruno who wrote: 'There are countless suns and countless earths all rotating around their suns in exactly the same way as the seven planets of our system. We see only the suns because they are the largest bodies and are luminous, but their planets remain invisible to us because they are smaller and non-luminous. The countless worlds in the universe are no worse and no less inhabited than our Earth' (Bruno 1584).

2. Historical highlights

That Venus does indeed have an atmosphere was first determined by the distinguished Russian scientist Academician Mikhail Vasil'evich Lomonosov at the St Petersburg Observatory. Lomonosov (Fig. 2) was born on 8 November 1711 in Cholmogori village near Arkhangelsk in the northern part of European Russia, and died on 4 April 1765 in St Petersburg. He graduated first from Slavonic-Greek-Latin Academy in Moscow and then from Marburg and Freiburg Universities in Germany. He was recognized in the different

Figure 2. Mikhail V. Lomonosov. Painting after C.-A. Wortman (1757).
Museum of M.V. Lomonosov, St Petersburg.

fields of natural sciences (first of all in physics and chemistry), as well as in philosophy, history and philology, and he was the founder of Moscow University, established in 1755.

Historically, the passage of Venus across the face of the Sun in 1761 drew the attention of numerous astronomers throughout the world. More than 40 different sites were selected and 112 astronomers observed the event. Beginning in 1760 many calculations of the visible path of Venus were undertaken. In Russia, the study was made by the physics professor France-Ulrich-Theodore Epinus, Director of the St Petersburg Observatory. He published in the October 1760 issue of the magazine 'Writings and Translations for Profit and Entertainment' his paper 'News on the Forthcoming Venus Transit Between the Sun and the Earth' accompanied by the three engraved drafts (Epinus 1760a).

Lomonosov, however, found the third of these drafts depicting the path of Venus across the solar disc insufficiently correct, and drew this fact to Epinus' attention. Because Epinus disagreed and argued for his being right in the 'Comments' that followed his first publication (Epinus 1760b), Lomonosov presented to the Academy Assembly on 18 December 1760 the special 'Note on the F. Epinus complaints that Lomonosov called in question his paper "News on the Forthcoming Venus Transit between the Sun and the Earth"' (Lomonosov 1760). The essence of the arguments of the critics was that, as Lomonosov stated, in his drawings Epinus:

1) did not take into account nearly 10° difference of the ecliptic position to the local horizon when observing the visible path of Venus across the Sun's disk in St Petersburg and also the respective corrections that should be introduced for Venus' entry/emergence in Siberia where expeditions were intended to be sent;

2) did not account for the declination of Venus' orbit relative to the ecliptic by $3°22'$ (a bit different from the contemporary value of $3°23'39''$) that influences the ephemerides evaluation and

3) did not refer to the original calculations of the Venus transit in different parts of the world, as it was shown in the special map earlier published in Paris.

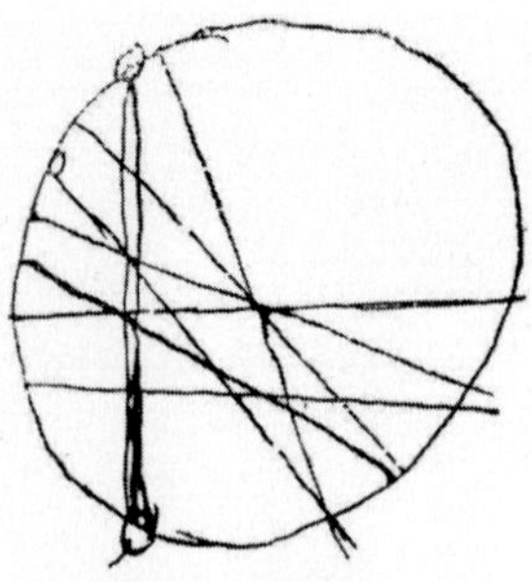

Figure 3. Lomonosov's drawing of different angles to the horizon under which an observer will see the actual Venus path for several sites in Siberia, accounting for local horizon and solar zenith distance and the proposed method to draw the path for every observer's position using a special device – the hodograph astrolabe.

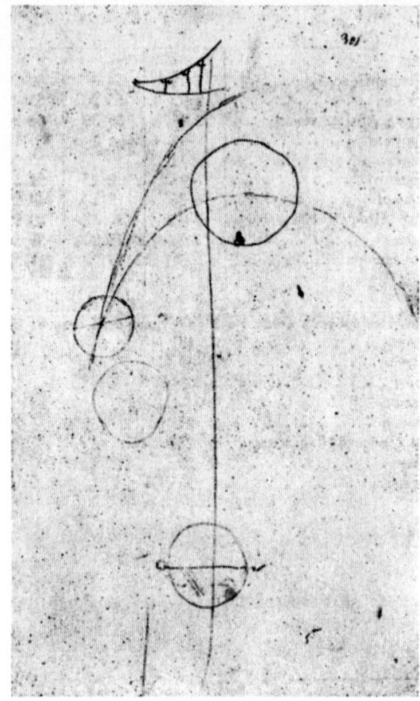

Figure 4. Lomonosov's drawing of an enhanced curvature of the Venus passage to an observer caused by changing of both ecliptic inclination and the path of Venus relative to the local horizon.

3. Observations and results

The discussion encouraged Mikhail Lomonosov to pay more attention to the event. During a few months of early 1761 he wrote a paper dealing with calculations of the Venus transit accompanied by a note with tables. The title of the paper was 'Venus Path Along the Solar Plane as It will be Seen to Observers in Different Parts of the World on May 26, 1761 According to the Calculations of Russian Academy of Science Adviser and Swedish Royal Academy Member' (Lomonosov 1761a). Eliminating the Epinus' errors and using the (Manfredi 1750) astronomical tables containing ephemerides of celestial bodies from 1751 to 1762, Lomonosov found the times when the Venus entry to the Sun and emergence at the St Petersburg site would occur. He thus determined that the total time of the event was to be $6^{h}33^{m}$ corresponding to angular distance on the Earth's surface $98°25$, the Venus path having a small curvature slightly pronounced to

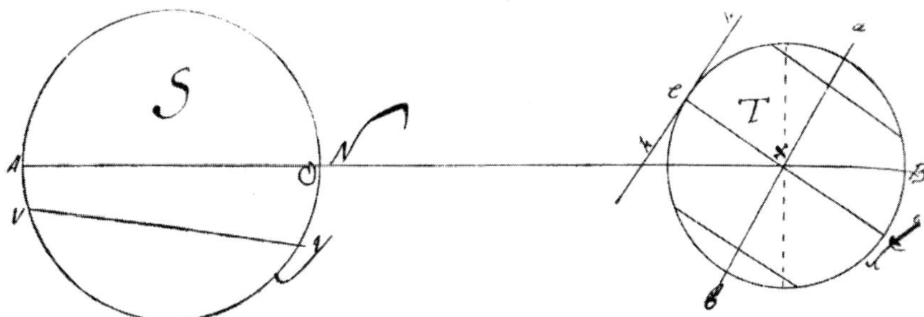

Figure 5. Lomonosov's drawings of Venus' transit across the Sun's disk. S is the Sun and T is the Earth, ANB – ecliptic running along the Sun's and Earth's centers, VY – Venus path as it is seen from the center of the Earth, ab – the Earth's axis, ef – equator, x – site on the Earth's surface at the equator pointing to the center and orthogonal to AB.

the observer. This is because both ecliptic inclination and the Venus path change relative to the horizon in the course of its passage depending on the local positions on the Earth's surface. The original Lomonosov drawings of the event are shown in Figs 3 and 4.

In his evaluation Lomonosov specially emphasized the importance of accurate position and time of the planet's entry to the solar disk in order that the observers would not miss the very beginning of the first contact. This is why he calculated the time of Venus-Sun first contact and emergence for several sites in Siberia accounting for the local horizon and solar zenith distance, in other words, different angles to the horizon at which an observer will see the actual Venus path.

The results of these calculations are shown in Fig. 5. He suggested a special device to draw the Venus path wherever the observer's position is (see Fig. 6), which he called the hodograph astrolabe, applicable also for observing solar eclipses (Lomonosov 1761a).

Based on this preparatory work, astronomical observations of Venus transit at the St Petersburg Observatory were carried out by the very skilled and experienced astronomers Andrey D. Krasilnikov and Nikolay G. Kurganov using a 6-feet-long focus telescope. The contact of the rear side of Venus' disk occurred at $4^{\mathrm{h}}26^{\mathrm{m}}39^{\mathrm{s}}$, the exit of its front side occurred at $10^{\mathrm{h}}19^{\mathrm{m}}04^{\mathrm{s}}$, and exit of its rear side (Venus left the Sun) at $10^{\mathrm{h}}37^{\mathrm{m}}00^{\mathrm{s}}$. From these measurements the diameter of the Sun was deduced to be as large as $0°31'36''$, and the diameter of Venus as large as $01'02''$. The total time of the planet's conjunction with the Sun was found to be $7^{\mathrm{h}}43^{\mathrm{m}}05^{\mathrm{s}}$, and inclination angle of its path to the eastward longitude circle was $81°29'00''$ (Lomonosov 1761a). The results of these observations gave rise to the solar parallax value $8.''49$, not too different from the contemporary value $8.''794148$.

In turn, Lomonosov himself focused on the observations of physical phenomena that accompanied the Venus transit. For this purpose he used a sort of spyglass – the 4.5-feet two-lenses tube covered with a smoky glass. The results of these remarkable observations he described in his paper 'The Appearance of Venus on Sun as It was Observed at the St Petersburg Emperor's Academy of Sciences on May 26, 1761' (Lomonosov 1761b; Lomonosov 1761c). We shall quote here a few paragraphs from this paper referring also the drawings in Fig. 7:

'Following the ephemerides I had, I waited for about forty minutes until the solar edge in the place of Venus ingress, which was clearly seen before (see 'B' in Fig. 7.1), became unexpectedly vague and obscured. I thought initially that my tired eye caused an obscuration, but when looking again in a few seconds I found a black indentation from the coming Venus, which replaced the former vague spot. I continued to look attentively how the trailing side of the planet approaches the

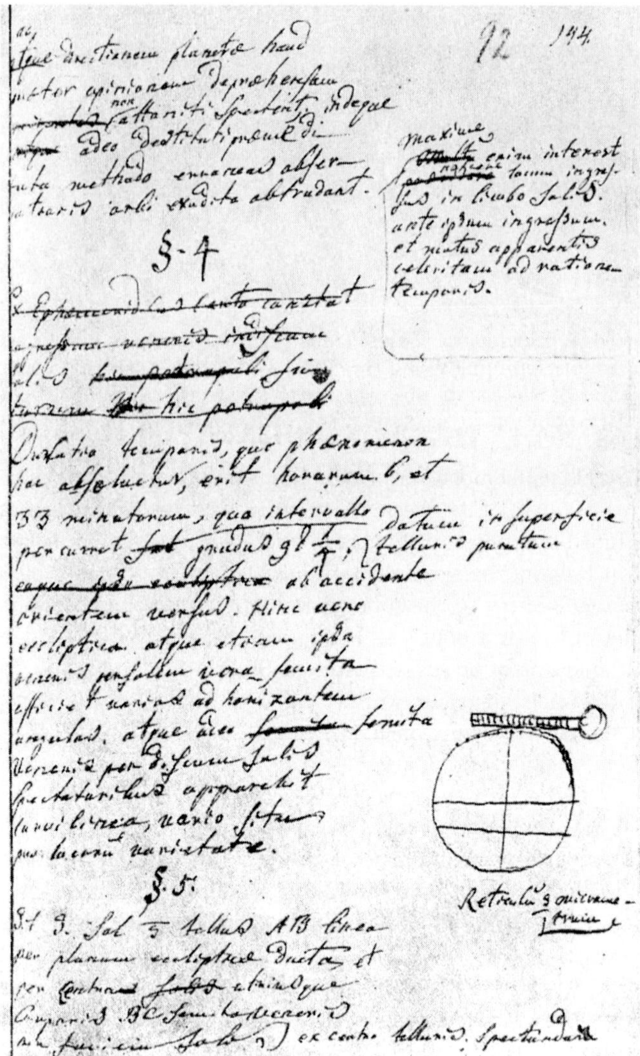

Figure 6. A sample page from the manuscript of the Lomonosov's paper 'Venus Path Along the Solar Plane...'. In the drawing a special device called hodograph astrolabe is shown.

Sun; suddenly, a hair-thin bright radiance (luminescence) between Venus' trailed side and solar edge appeared that lasted only less than a second.

Before the Venus ingress, when its front side approached the solar edge at about one tenth of the planet's diameter, a bulge set up (see 'A' in Fig. 7.1) which progressively became more pronounced as Venus came to leave the Sun (see Fig. 7.3 and 7.4, where LS is the solar edge and mm is a bulging Sun). Soon after that the bulge disappeared and instead, Venus appeared with no edge (see nn in Fig. 7.5). Similar to the ingress phase, the last touch of the planet's trailing side at the emergence was also accompanied by a small break and solar edge obscuration.

Because colors caused by the light rays refraction appeared stronger as larger was a Venus offset from the tube center X (see Fig. 7.2), I permanently maintained the tube in the position with Venus in the tube center. It provided a clear image of the planet with no colors at its edges.

Based on these observations I conclude that the planet Venus is surrounded by a distinguished air atmosphere similar (or even possibly larger) than that is poured

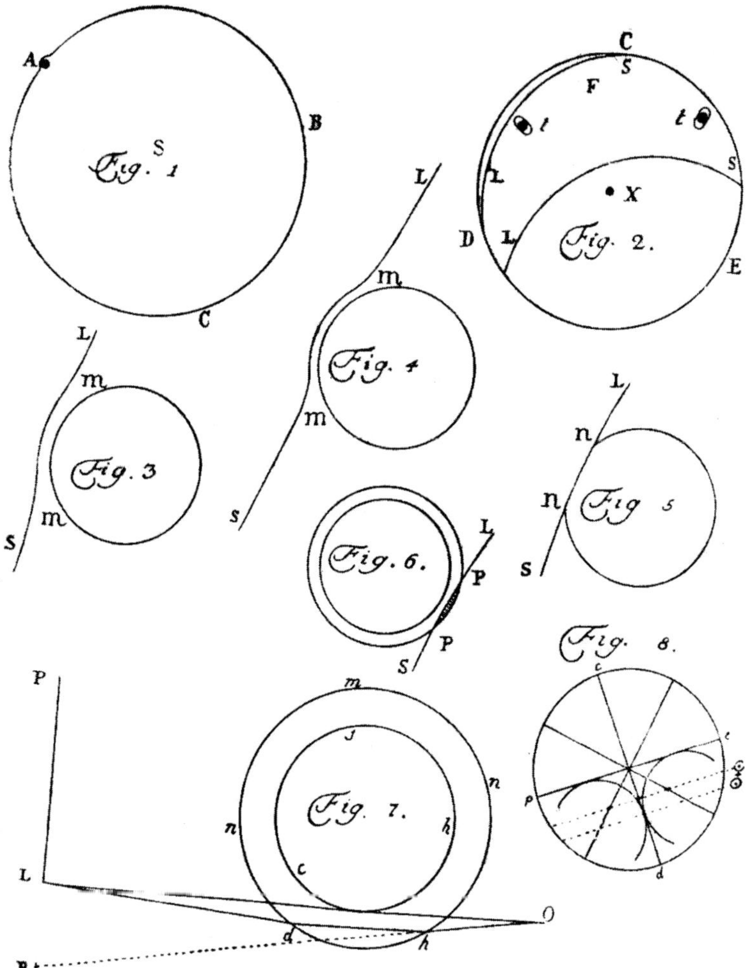

Figure 7. Lomonosov's drawings of the sequence of physical events accompanied the Venus transit across the Sun's disc at the entry and exit phases: 1 – losing clearness at the solar disc edge just before Venus entered the solar surface 'B' and bulge set up 'A' in the place of Venus leaving the Sun; 2 – colors caused by the light rays' refraction appeared offset from the tube center X; 3 and 4 – a bulge progressively becomes as more pronounced as Venus leaves the Sun, LS is the solar edge and mm is a bulging Sun; 5 – the planet appeared with no edge soon after the bulge disappeared (see nn); 7 – the bulge formed close to the solar disc edge at the time of Venus' emergence for which the refraction of solar rays in the Venus atmosphere should be responsible. Here L in the line PL is the end of the Solar diameter (solar disc edge), cjh is the Venus solid body, mnn is its atmosphere, LO is the tangent to the solid Venus body along which an observer O would see the solar disc edge at the point L in the absence of atmosphere, LdhO is the path along which the solar light propagate due to atmospheric refraction at dh, and OR is offset of solar edge L as it is visible to an observer.

over our Earth. This finding is supported by the following arguments. Firstly, losing clearness at the solar disc edge ('B' in Fig. 7.1) just before Venus entered the solar surface means that the edge was obscured by the Venus atmosphere. Secondly, for the bulge formed close to the solar disc edge at the time of Venus emergence, the refraction of solar rays in the Venus atmosphere should be responsible. Indeed, let us address the Fig. 7.7 where L in the line PL denotes the end of the Sun diameter (solar disc edge), cjh is the Venus solid body, and mnn is its atmosphere. In the absence of the atmosphere an observer O would see the solar disc edge at the point L along the line LO that is the tangent to the solid Venus. In the case of the atmosphere, however, the solar light will propagate along the curve LdhO due to refraction at the path dh. It is known from optics that an eye see along the incident line; therefore, the solar edge at L turns out shifted at the point R along the virtual line OR. An excess LR compared to the real solar radius explains the bulge formation before the front side of Venus at the time of its emergence from the Sun.'

In the second part of his paper Lomonosov discusses philosophical aspects of this important discovery in the context of how astronomy has progressed since ancient times and how it impacted on both the world outlook and practical needs such as navigation. He addresses many problems involving the possibility of numerous habitable worlds in the Universe (in parallel to Giordano Bruno's paradigm cited above) and correlation between nature and religion that he considers as two Main Books given to mankind by God. The discussion is even accompanied by two short poems written by himself in support of the heliocentric system of Nicholas Copernicus which was still obscured by the church.

4. Discussion

Lomonosov detected the refraction of solar rays while observing the Venus transit across the disk of the Sun during both entry and emergence phases in 1761. Only the presence of refraction in a sufficiently thick gas envelope could explain the appearance of light ('fire') ring around the night disk of Venus at the entry phase and a bulge formation close to the solar disc edge during the Venus emergence. It should be noted, however, that Lomonosov's description of the observed phenomenon during the initial phase of transit, on the side opposite from the direction of motion, is not fully accurate: it seemed to him as 'a hair-thin bright radiance (luminescence) between Venus' back side and solar edge'. In reality, it was a bright thin rim which encircled a portion of the planet's disk that had not yet contacted the solar disk and thus separated the Venus limb from the sky's background. Interestingly enough, some observers of the following Venus transits in 1874 and 1882 similarly reported that a 'fire ring' seemed projected on the planetary disc as well. Nonetheless, Lomonosov's interpretation of losing clearness at the solar disc edge, just before Venus entered the solar surface, in terms of its atmosphere occurrence was correct, and even more conclusive was his thorough explanation of the bulge appearance when the planet left the Sun.

In general, Lomonosov properly described the physical mechanism underlying his observations and came to the right conclusion that Venus possesses an atmosphere which could be comparable to, or even more dense than that of the Earth. It fully paralleled the basics of the refraction theory he earlier studied with implications for accuracy of navigation at sea. According to Lomonosov (1759), 'the rate of refraction corresponds to the transparent matter, i.e. air, thus the amount of matter that a ray propagates is the rate of refraction'. Only in the middle of 1960s did it become clear that his assumption that Venus' atmosphere can be 'even more dense than that of the Earth' was far-sighted, because Venus' atmosphere turned out nearly two orders of magnitude thicker than the terrestrial standard indeed (see, e.g., Marov & Grinspoon 1998).

The paper of Lomonosov, where his observations and discovery of Venus' atmosphere were described, was submitted for publication on 4 July 1761 and 200 copies were published in Russian on 17 July 1761 (Lomonosov 1761b). In parallel, Lomonosov prepared a German translation of the paper which appeared in August 1761 (Lomonosov 1761c) and hence, became more available to the world scientific community. However, when

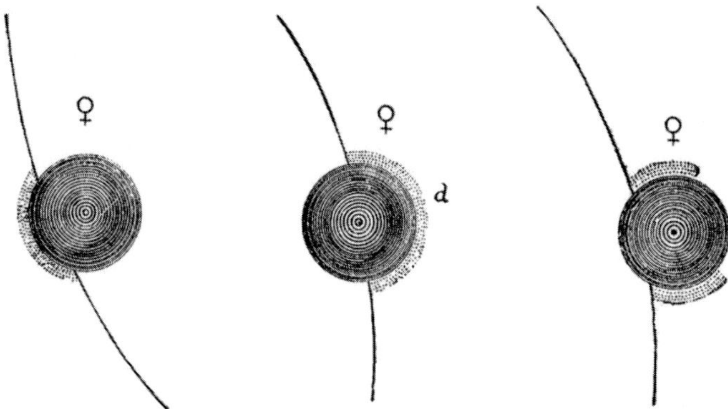

Figure 8. Bergman's drawings in support of the finding of Venus' atmosphere. The three sequence from the light illumination of Venus' trailed side before it fully crossed the solar disk, to the second and third ones showing the more pronounced bright light at the front side of Venus at the emergence, which eventually attenuated and ultimately only horns were left (Bergman 1762).

approximately 30 years later the German astronomer Johan Hieronymus Schroeter and the English astronomer Frederick William Herschel discovered the above-mentioned crepuscular phenomena on Venus and came to the correct conclusion that they result from the scattering of solar rays in the upper portion of the planet's atmosphere, there were attempts to ascribe to them the finding of Venus' atmosphere and thus to lapse into silence on Lomonosov's priority. The history of the discovery, including the discussion between Schroeter and Herschel on some hot points, and later comments of French astronomer Dominique Francois Arago in support of the Venus atmosphere existence with tribute to the priority of the Lomonosov's discovery, can be found elsewhere (Perevozshikov 1865).

It is worth mentioning that Lomonosov himself was aware that it was not he alone who studied physical phenomena which accompanied the Venus transit. He wrote, 'The great atmosphere that we found around Venus was also noticed by other observers in Europe'. Among them, the most advanced observations were made in Uppsala by Tobern Olof Bergman who reported the results at the meeting of Royal Society in London on 19 November 1761. He argued for the finding of Venus' atmosphere based on three drawings (see Fig. 8). One may emphasize, however, that this and other observations and their analysis were much less detailed compared to what was made by Lomonosov. Also, these results were published later and in no way cast doubt on Lomonosov's priority.

5. Summary

The discovery of Venus' atmosphere was a great astronomical achievement in the 18th century. The Russian scientist Mikhail Lomonosov was the first who found it from the study of the solar rays' refraction patterns while observing the Venus transit across the solar disk in 1761. He argued correctly that only refraction in a sufficiently thick gas envelope could explain the appearance of a light ('fire') ring around the night disk of Venus at the entry phase and a bulge formation close to the solar disc edge during the Venus emergence. The Venus atmosphere discovery, in addition to astrometric measurements, clearly manifested the great importance of transit phenomena for astronomical science, with the basics of this technique being now readdressed to the study of extrasolar planets.

References

Bergman, T.O. 1762. *Philosophical Transactions* **52**, 227–230.

Bruno, G. 1584. In: *De L'infinito Universo E Mondi.*

Epinus, F.U.T. 1760a. 'News on the Forthcoming Venus Transit between the Sun and the Earth.' In: *Writings and Translations for Profit and Entertainment.*

Epinus, F.U.T. 1760b. 'Comments.' *Submitted to Acad. of Sci. Office, St Petersburg.*

Lomonosov, M.V. 1759. '*Meditationes de via Navis in Mari Certius Determinanda Praelectae in Publico Conventu Academiae Scientiarum Imperalis Petropolitanae die VIII Mai, A.C. 1759 Auctore Michaele Lomonosow Consilario Academico.*' In: *Lomonosov, M.V. (1955). Memoirs in Physics, Astronomy and Instruments Building, USSR Acad. of Sci., Leningrad.*

Lomonosov, M.V. 1760. 'Note on the F. Epinus complaints that Lomonosov called in question his paper "News on the Forthcoming Venus Transit Between the Sun and the Earth"'. In: *Lomonosov, M.V. (1955). Memoirs in Physics, Astronomy and Instruments Building, USSR Acad. of Sci., Leningrad.*

Lomonosov, M.V. 1761a. 'Venus Path Along the Solar Plane as It will be Seen to Observers in Different Parts of the World on May 26, 1761 According to the Calculations of Russian Academy of Science Adviser and Swedich Royal Academy Member.' In: *Lomonosov, M.V. (1955). Memoirs in Physics, Astronomy and Instruments Building, USSR Acad. of Sci., Leningrad.*

Lomonosov, M.V. 1761b. 'The Appearance of Venus on Sun as It was Observed at the St Petersburg Emperor's Academy of Sciences on May 26, 1761. Acad. of Sci. Office, St Petersburg.' In: *Lomonosov, M.V. (1955). Memoirs in Physics, Astronomy and Instruments Building, USSR Acad. of Sci., Leningrad.*

Lomonosov, M.V. 1761c. '*Erscheinung der Venus vor der Sonne beobachtet bey der Kayserlichen Akademie der Wissenschafften in St. Petersbourg den 26 May 1761.*' Aus dem Russischen ubersezt.

Manfredi, E. 1750. *Introductio in Ephemerides cum Opportunis Tabulis ad Usum Bononiensis Scientiarum Instituti, Bononiae.*

Marov, M.Ya. & Grinspoon, D.H. 1998. *The Planet Venus.* Yale University Press.

Perevozshikov, D.M. 1865. 'Study of Lomonosov in Physics and Physical Geography.' *Raduga* **4**, 176–201.

Sharonov, V.V. 1958. 'The Nature of Planets.' *Fizmatlit*, Moscow.

Discussion

SUZANNE DÉBARBAT: Is the Lomonosov observatory on the top of the Academy where the Lomonosov museum is in St Petersburg? Or was it in another place? From where did he make his observations?

MIKHAIL MAROV: No. The original St Petersburg observatory was at the top of the Academy building indeed (Pulkovo observatory was set up much later, in 1839), while the Lomonosov museum is just a branch of the Academy Institute of Natural History located in one of two wings of the Academy building.

ED BUDDING: Did he measure the scale of the enhancement of the disc – the small bump which you mentioned – in relation to the size of the disc of Venus?

MIKHAIL MAROV: No – just what I describe. This is the bulge; it's the solar disc . . . when Venus was just leaving the sun. It is because of the refraction that to you it looks like an enhancement, like it's a bulge in the solar disc.

ED BUDDING: Yes, we are agreed about that. What I ask is: what is the number which you put on the size of this bulge?

MIKHAIL MAROV: Well, I can't answer your question quantitatively, but I can tell you that it was sufficient to be discernable when observing with the eye. I tried just to do the same yesterday! It was a rare opportunity, but because of bad luck with the weather conditions it turned out impossible essentially – I saw nothing.

WAYNE ORCHISTON: An interesting paper: Do we have any of the original records, the observations and archival material of his work, or is the only thing that survived the actual publication of his results?

MIKHAIL MAROV: Yes, we have a very good archive of Lomonosov. It is very well preserved, and this is just in some picture book. I had only limited access to these archives, so I mostly used just what was published later. The most comprehensive from the archives are eight volumes of work written by Lomonosov that were published by the Soviet Academy of Science in 1955; this is everything: this is science, this is poetry, this is history. These volumes were quite accurately made by the people who went through all the available manuscripts. The manuscripts themselves are accessible usually only to specialists who study history.

Mikhail Marov at Alston Hall

Transits of Venus: New Views of the Solar System and Galaxy
Proceedings IAU Colloquium No. 196, 2004
D.W. Kurtz, ed.

© 2004 International Astronomical Union
doi:10.1017/S1743921305001407

Probing extrasolar planet atmospheres through transits

Ignas Snellen

Institute for Astronomy, University of Edinburgh, Blackford Hill, Edinburgh EH9 3HJ, UK
and
Sterrewacht Leiden, Postbus 9513, 2300 RA, Leiden, The Netherlands
email: snellen@strw.leidenuniv.nl

Abstract. A revolution is taking place in the research of extra-solar planets with the discovery of the first exoplanets only a decade ago to the more than 100 systems known to date. Almost all of these extrasolar planets have been discovered using the radial velocity technique. Unfortunately, this limits the amount of information which can be obtain from these systems, with a $\sin i$ ambiguity in the planet's mass, and no further measurements of fundamental parameters as long as these planets can not be detected directly. This situation is very different in the rare case that the orbit of a planet has an inclination such that it occults its host star, as in the case of HD 209458b. Not only can the mass and radius of the planet be accurately determined, but it makes the system also suitable for many detailed follow-up studies, in particular atmospheric transmission spectroscopy. This has resulted in the detection of the atmosphere of HD 209458b in Sodium, and the discovery of an evaporating exosphere in Hydrogen, Carbon and Oxygen using the Hubble Space Telescope. In this paper I briefly review transiting exoplanets and methods to probe their atmospheres, with the emphasis on a new method of transmission spectroscopy making use of the Rossiter effect, which may be more suitable for the large *ground-based* telescopes.

1. The extra-solar planet revolution

The many talks during this conference about the 18th and 19th century voyages to observe the transits of Venus, show that it must have been a very exciting time to be an astronomer. In a few hundred years from now people may think in the same way about the early 21st century, at least from a scientific perspective, with a genuine revolution taking place in extra-solar planet research.

We still virtually know nothing about other solar systems: what are the characteristics of extra-solar planets? How many planets are typical? What are their masses, compositions, orbits, and does their occurence depend on properties of the host star or certain environmental effects? However, it is expected that we will be able to answer many of these questions within the next few decades. Much of the driving force behind the boom in this area of research is the prospect of finding earth-like planets within our lifetimes, including possible evidence for carbon-based life.

Formation processes of stars and planets are so complex that they are almost impossible to derive from first principles. The developments of more and more powerful telescopes, instruments, and observation techniques are therefore crucial. For obvious reasons, direct detection of extra-solar planets is highly challenging and, so far, beyond reach. However, indirect methods have already been very successful, in particular the radial velocity technique which measures the reflex motion of a star about the star-planet barycentre (for extensive reviews on detection techniques see Perryman 2000; Sackett 1999). The first exoplanets around solar-type stars were discovered using this method almost a decade ago (Mayor & Queloz 1995; Marcy & Butler 1996), and more than 100 systems are known to

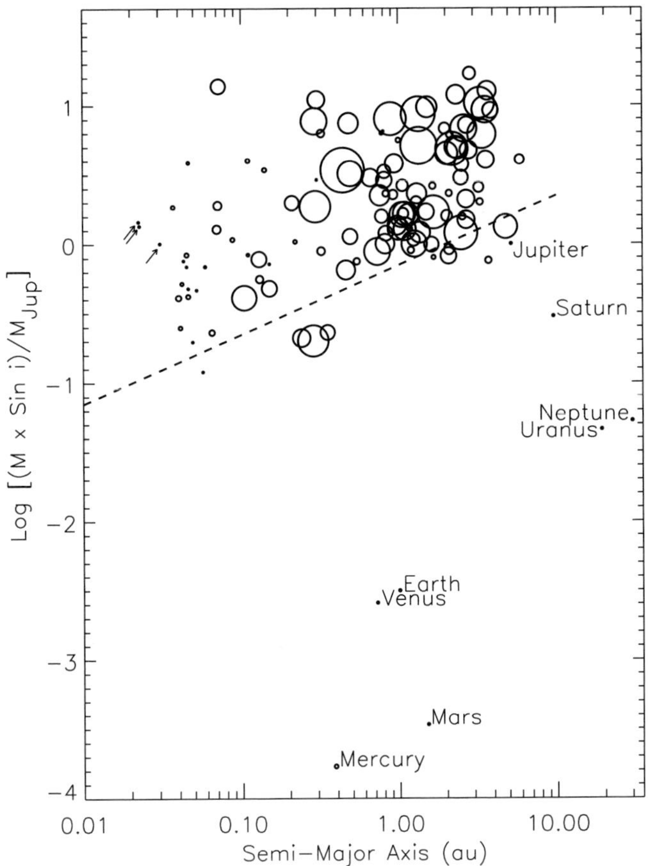

Figure 1. Current status of the extra-solar planet population discovered around main sequence stars, with orbital semi-major axis on the x-axis and mass on the y-axis (data from http://www.obspm.fr/encycl/encycl.html). The symbol sizes reflect the eccentricity of the orbits. Note that a large fraction of the exoplanet population discovered so far have highly eccentric orbits. The arrows indicate those planets discovered by their transits.

date (http://www.obspm.fr/encycl/encycl.html). An overview of the masses and orbits of the planets currently known is shown in Fig. 1. The limited data available to date have already resulted in several surprises. Firstly, the existence of the so-called Hot Jupiters, planets with orbital periods of only a few days, was not anticipated. Secondly, the fact that a high fraction of the orbits are strongly eccentric was also not expected. In addition it has been found that on average the stars hosting planets have a significantly higher metal content than the average solar-type star in the solar neighbourhood (Gonzales & Laws 2000). These early unexpected results show that the predictive power of physical theory in this area of research has indeed so far been rather limited.

Although the Doppler wobble technique has been a great success, it also has its limitations. Since the inclinations of the planetary orbits are unknown, the amplitude of a radial velocity wobble strictly only results in a lower limit to the planetary mass, and not much else can be learned about the basic parameters of the exoplanet. A great advance could be made if starlight reflected from the planets were to be detected. The phase-function of this signal depends on the orientation of the orbit and could therefore solve the $\sin i$ ambiguity. The detection of reflected light, even from Hot Jupiters, is, however, highly challenging, with expected contrasts of 10^{-4} to 10^{-5}, or less. Deep spectroscopic observations have resulted in typical upper limits in geometric albedo of $<20\%$, implying that these Hot Jupiters only possess a fraction of the reflectivity of Jupiter (e.g. Leigh et al. 2003). There is good hope that observations with the MOST satellite will yield the first positive results (see Matthews et al., these proceedings).

2. Transiting exoplanets

The Doppler wobble technique has been very successful in providing a large database of extra-solar planets. Unfortunately, detailed investigations of these systems individually can not yet be undertaken. This situation is very different in the rare case that the orientation of the orbit is such that the planet transits its host star. Not only can the mass and radius of the exoplanet be accurately determined, the transit itself is also suitable for many detailed follow-up studies. During transit, a Jupiter-size planet would block off in the order of $\sim 1\%$ of the starlight, a factor of ~ 100 more than Venus. Unfortunately, the chances for this to happen are slim. From a random position in space, the chance that a planet in our solar system would transit the Sun ranges from about 1% for Mercury, to 0.1% for Jupiter. Fortunately, for Hot Jupiters this chance is more like 10%.

To date, 14 Hot Jupiters with an orbital radius of $a < 0.05$ have been discovered using the Doppler technique, and indeed one, HD 209458b, has been found to transit its host star (Charbonneau et al. 2000; Henry et al. 2000; Mazeh et al. 2000). It blocks off about 1.6% of the star-light for a period of $\sim 3\,\mathrm{hr}$, every 3.52 d. The combination of radial velocity data and accurate photometric observations with the HST have resulted in a precise determination of the fundamental parameters of HD 209458b, showing it has a mass of $0.69\,\mathrm{M_{Jup}}$ and radius of $1.35\,\mathrm{R_{Jup}}$ (Brown et al. 2001), and proving beyond any doubt that this is indeed a planet.

2.1. Extra-solar planet transmission spectroscopy

One of the great successes in recent astronomy has been the detection of an atmosphere around HD 209458b. This has been achieved using transmission spectroscopy. At particular wavelengths, corresponding to transitions of common atoms or molecules in the planetary atmosphere, the photometric transit is stronger due to extra absorption of star-light in the outer layers of the planetary atmosphere. Therefore, by measuring the transit signal as function of wavelength the atmospheric constituents can be determined. Atmospheric transmission spectroscopy was first used for a particular type of binary stars, ζ Aurigae, where the chromospheric structure of a large, cool star is studied when it eclipses its small companion (e.g. Wilson & Abt 1954). This technique has also been used extensively for the planets in our own solar system, in particular using spacecraft, such as the Voyager I & II, that observe the setting or rising of the sun or a star behind the limb of a planet (Smith & Hunten 1990).

Detailed models indicated that the dominant features in the transmission spectra of Hot Jupiters are the Sodium and Potassium lines in the optical, and Helium and the CO, H_2O and CH_4 molecules in the near infrared (e.g. Seager & Sasselov 2000; Brown 2001; Hubbard et al. 2001). In addition, it was predicted that the close-in Hot Jupiters would have an extended exosphere in the form of a cometary tail, caused by the strong irradiation of the host star (Schneider et al. 1998).

It was Charbonneau et al. (2002) who first detected the atmosphere of HD 209458b in Sodium, by monitoring 6 transits with the Hubble Space Telescope. They found that in a region of 12Å around the two Sodium D lines the transit depth was $0.023\pm0.0057\%$ deeper

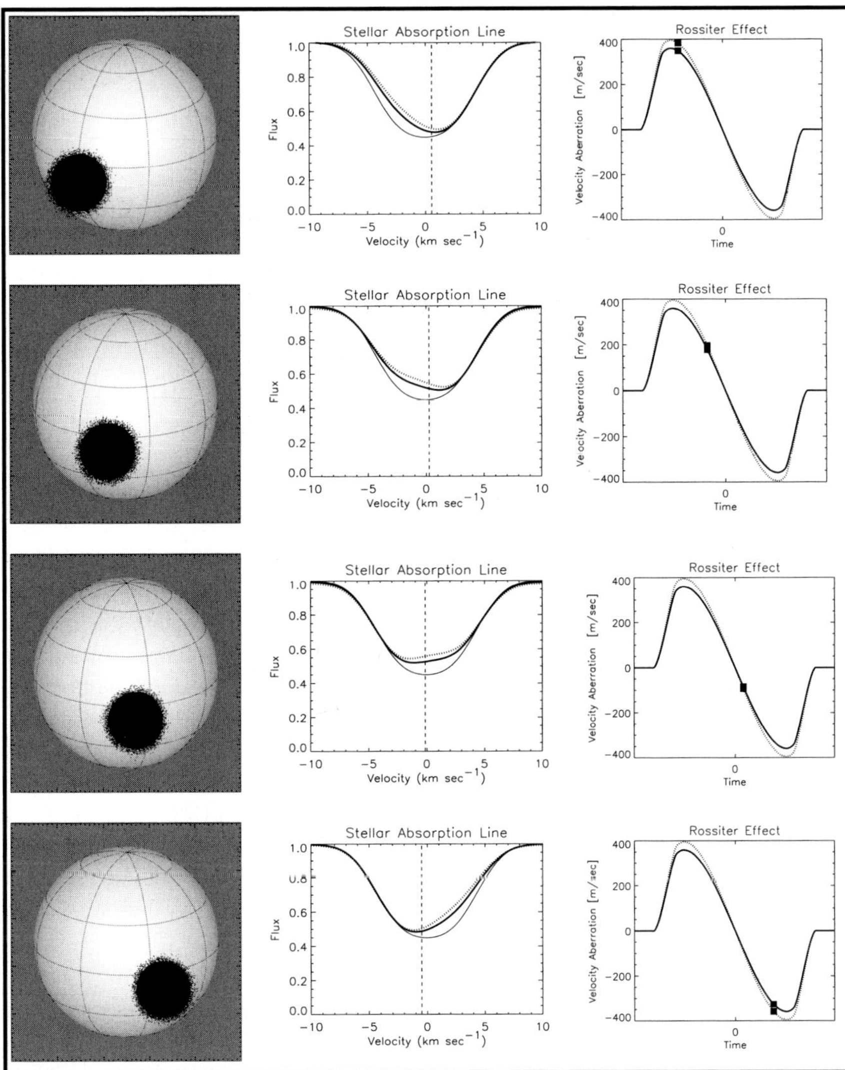

Figure 2. Illustration of the Rossiter effect, caused by a dark object transiting a rotating star. Since the object subsequently blocks off the approaching and receding parts of the stellar surface, a change in the profile of the rotationally broadened stellar absorption line occurs, which results in an effective red and blue shifting of the line. The perturbed and unperturbed absorption line profiles are shown in the middle panels by the thick and thin solid lines respectively. The resulting Rossiter effect is shown in the panels to the right. The effects of extra absorption caused by a possible atmosphere is indicated by the dotted lines. Note that for illustrative purposes the transiting object is significantly larger than HD 209458b.

than in the surrounding continuum. HST observations in the ultraviolet subsequently measured very strong absorption features, at a level of $5 - 10\%$, from Hydrogen, Oxygen, and Carbon (Vidal-Madjar et al. 2003, 2004). The depth and width of these features suggest that this absorption comes from an evaporating exosphere of HD 209458b.

Attempts to detected the atmosphere or exosphere using ground-based telescopes have so far been unsuccessful. Observations targeting the Sodium D feature in the optical have resulted in upper limits of typically $1 - 2\%$ (Brown et al. 2000; Bundy & Marcy 2000; Moutou et al. 2001). Similar results have been obtained for the main absorption features in the near-infrared (Brown et al. 2002; Moutou et al. 2003). More stringent upper limits have been obtained in the infrared by Harrington (2002) using the secondary eclipse, by comparing the light from the star plus planet with that of the star only when the planet is occulted.

The main reason why ground-based observations do not yet reach the required precision seems to be the fact that atmospheric transmission spectroscopy relies on the comparison of spectra on and off transit, with the weather and instruments being insufficiently stable over timescales much greater than an hour.

3. A new method of transmission spectroscopy

We have developed a new method for probing the atmospheres of transiting exoplanets, making use of the Rossiter effect (Snellen 2004), which may be more suitable for ground-based telescopes.

The Rositter effect

The Rossiter effect was first observed for the binary star β Lyrae, thereby showing that stars rotate (Rossiter 1924). If an object transits a rotating star, it will subsequently block off light from the approaching and receding parts of the stellar surface (Fig. 2). The resulting effect is an apparent red and blueshift of the rotationally broadened stellar absorption lines during the transit. Rossiter's observations showed this effect in β Lyrae to have an amplitude of $10 - 15\,\mathrm{km\,s^{-1}}$. Schneider (1999) suggested that the Rossiter effect could also be observable for transiting exoplanets, and this was observed during a transit of HD 209458b by Queloz et al. (2000). The exact radial velocity pattern can constrain the relative orientation of the orbit. As expected, in the HD 209458 system the the planetary orbit is in the same direction as the stellar rotation, with a measured upper limit to the angle between the orbital plane and equatorial plane of the star of $30°$.

The amplitude of the Rossiter effect, determined by Queloz to be $\sim$$35 - 40\,\mathrm{m\,s^{-1}}$, is mainly dependent on the rotation velocity of the host star and the relative size of the planetary disk. Due to this latter dependence, this effect can be used to probe the atmosphere of HD 209458b in the same way as with conventional transmission spectroscopy. At particular wavelengths the amplitude of the Rossiter effect will be higher due to the extra absorption in the planetary atmosphere, causing an increase in the effective size of the planetary disk. The main advantage of using the Rossiter effect over a straightforward photometric signal is that it only relies on the *relative* positions of the stellar lines in the same spectra, and is therefore much less hampered by calibration issues. Particularly from the ground this means that potentially a much higher accuracy could be achieved.

The resulting excess in the Rossiter effect depends on the width and depth of the atmospheric absorption feature. If the absorption in Sodium were to be homogeneous over the 12Å wavelength range used by Charbonneau et al. (2002), it would result in an excess in the Rossiter effect in the Sodium D lines of $\sim$$0.5\,\mathrm{m\,s^{-1}}$ compared to other stellar lines. If the absorption is peaked at smaller scales it could lead to an excess of up to $\sim$$2\,\mathrm{m\,s^{-1}}$ (Snellen 2004).

Results from UVES archival data

The new method was tested using an archival dataset from the UV-Visual Echelle Spectrograph (UVES) on the Very Large Telescope (VLT). The data comprised 4 on-transit datasets, centered on the Sodium D absorption lines at 5990 and 5996 Å, with 15×400-s

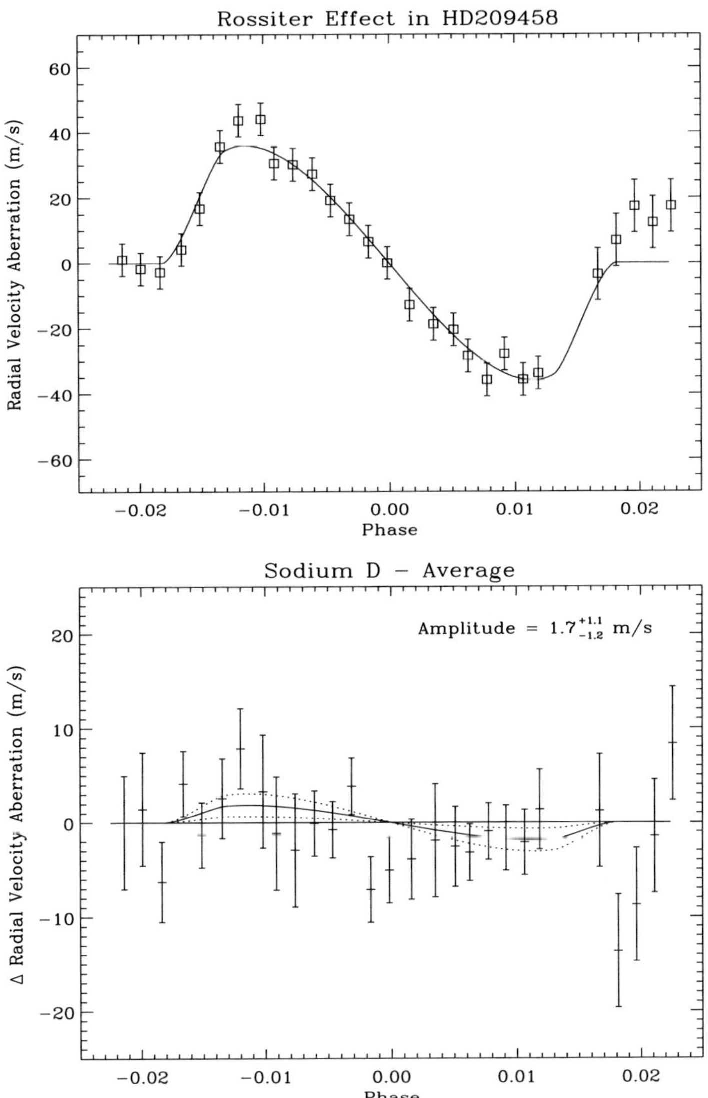

Figure 3. Upper panel: The Rossiter effect as observed for HD 209458 using an archival UVES/VLT dataset covering 4 transits. Lower panel: The relative excess in the Rossiter effect for the two Sodium D lines. The 1σ confidence limits are indicated by the dotted lines. The best-fit model shows an excess of $1.7^{+1.1}_{-1.2}\,\mathrm{m\,s^{-1}}$ (1.4σ).

exposures covering each transit. Unfortunately, a significant fraction of the transit time was spent for this method on unnecessary calibration stars, and in most cases only a part of the transit was observed. This has significantly reduced the accuracy.

The reduction and analysis of the data are discussed in detail in Snellen (2004). A typical positional accuracy of $\sim 1 \times 10^{-4}$ Å (0.006 pixel) was obtained for the strongest lines. This corresponds to a precision in relative radial velocity of $\sim 5\,\mathrm{m\,s^{-1}}$ for each exposure. The combined dataset of 4×15 exposures was binned and fitted with a model for the Rossiter effect, which was constructed by numerically simulating the transit of a dark object over a limb-darkened star. The best fitting model shows an excess in the Rossiter effect of $1.7^{+1.1}_{-1.2}\,\mathrm{m\,s^{-1}}$ (1.4σ – Fig. 3). Assuming a homogeneous absorption over the width of the stellar line, this corresponds to a photometric signal using conventional atmospheric transmission spectroscopy of $7 \pm 5 \times 10^{-4}$. Although this archival dataset is far from ideal, it shows that for ground-based observations the use of the Rossiter effect instead of a photometric signal results in an order of magnitude improvement in precision. This, while a further factor $2-3$ in accuracy can be gained for observations specifically designed for this method, covering complete transits without unnecessary gaps for calibration purposes.

4. Future prospects

Plans for further observations of HD 209458 are in place to explore the full potential of the Rossiter effect. Since observations of complete transits are necessary to reach the highest possible precision, only $6 - 7$ transits per year are optimally timed for observatories in the Northern hemisphere, and only half of that for those in the south. Obvious absorption features to target in the optical are the Sodium and Potassium lines, for which with the improved accuracy we would expect a firm detection. Comparison of a detection using the Rossiter method, which is most sensitive to atmospheric absorption on a sub-Ångstrom scale, with the conventional photometric method as achieved with the HST, could for the first time constrain the profile of the Sodium absorption feature. Also, the so-far illusive Helium feature at $10830\,\text{Å}$ (e.g. Moutou et al. 2003) will be targeted. Not only has it been predicted to be one of the planet's strongest atmospheric features at a level of $\sim 0.25\%$ (Seager & Sasselov 2000), Helium is also expected to be a major component in the planet's extended exosphere, with the potential to significantly boost the absorption signal to an excess in the Rossiter effect of $5 - 15\,\mathrm{m\,s^{-1}}$. The He I feature itself is relatively weak in solar type stars, and also a tracer of coronal activity with a patchy surface distribution, which may result in some rotational velocity jitter. However, fortunately just $2.5\,\text{Å}$ blueward of it is the deep stellar absorption line of Si I, within the width of the expected atmospheric absorption, and centered on the expected evaporating exosphere of HD 209458b.

Finding new transiting exoplanets

So far HD 209458b is the only realistic target for atmospheric transmission spectroscopy. The three exoplanets discovered recently in the OGLE-III survey (Udalski 2002a,b) are all transiting much fainter stars (OGLE-TR-56b, OGLE-TR-113b and OGLE-TR-132b; Konacki et al. 2003, 2004; Bouchy et al. 2004). However, in the light of probing exoplanet atmospheres using the Rossiter effect, finding new planets transiting bright stars would be very exciting. Transiting planets around rapidly rotating stars and/or late M stars would be of particular interest. Firstly, the faster a star rotates, the larger the Rossiter effect, and therefore the larger the possible effect of an atmosphere. Secondly, for late-type M stars with a low enough surface temperature the near-infrared molecular absorption features in the planet's atmosphere, such as H_2O, CO and CH_4, will start to cover many stellar absorption lines. Since the measurement of the Rossiter effect could then be combined over many stellar lines, again a significant increase in precision could be achieved. Note that a Jupiter-sized planet transiting a rapidly rotating M dwarf (say with $v = 10 - 20\,\mathrm{km\,s^{-1}}$), would produce a Rossiter effect with an amplitude of

$1 - 2\,\mathrm{km\,s^{-1}}$! Potentially, with the strength and profiles of the main atmospheric absorption lines can not only the chemical composition be determined, but also constraints be put on planetary rotation and temperature structure, cloud properties and winds in the upper atmosphere.

A very promising survey to find such planets is SuperWASP. Conducted by a consortium of UK institutions it aims at the robotic operation of up to 8 ultra-wide-field cameras, each with a field of view of $7^\circ\!.8 \times 7^\circ\!.8$, to obtain well sampled light curves of a large sample of stars in the magnitude range $\sim$7 to 13 at better than 1% precision (Street et al. 2003).

References

Brown T.M., Butler R.P., Charbonneau D., Noyes R.W., Sasselov D., Libbrecht K.G., Marcy G.W., Seager S., Vogt S.S., 2000, 197th AAS Meeting, 11.05; BAAS, Vol. 32, p.1417

Brown T.M., Charbonneau D., Gilliland R.L., Noyes R.W., Burrows A., 2001, ApJ 552, 699

Brown T.M., 2001, ApJ 553, 1006

Brown T.M., Libbrecht K.G., Charbonneau D., 2002, PASP 114, 826

Bouchy F., Pont F., Santos N., Melo C., Mayor M., Queloz D., Udry S., 2004, A&A, 421, L13

Bundy K.A., Marcy G.W., 2000, PASP 112, 1421

Charbonneau D., Brown T.M., Latham D.W., Mayor M., 2000, ApJ 529, 45

Charbonneau D., Brown T.M., Noyes R.W., Gilliland R.L., 2002, ApJ 568, 377

Gonzales G. Laws C, 2000, AJ 119, 390

Harrington J., Deming D., Matthews K., Richardson L.J., Rojo P., Steyert D., Wiedemann G., Zeehandelaar D., 2002, 201st AAS Meeting, 46.04; BAAS, Vol. 34

Henry G.W., Marcy G.W., Butler R.P., Vogt S.S., 2000, ApJ 529, L41

Hubbard W.B., Fortney J.J., Lunine J.I., Burrows A., Sudarsky D., Pinto P., 2001, ApJ 560, 413

Leigh C., Collier Cameron, A., Udry, S., Donati, J., Horne, K., James, D., Penny, A., 2003, MNRAS, 346, L16

Konacki M., Torres G., Sasselov D., Pietrzynski G., Udalski A., Jha S., Ruiz M.-T., Gieren W., Minniti D., 2004, ApjL 609, L37

Konacki M., Torres G., Jha S., Sasselov D., 2003, Nature, 421, 507

Marcy G.W., Butler R.P., 1996, ApJ 464, L147

Mayor M., Queloz D., 1995, Nature, 378, 355

Mazeh T., Naef D., Torres G., Latham D.W., Mayor M., Beuzit J.-L., Brown T., Buchlave L., Burnet M., Carney B.W. et al. 2000, ApJ 532, L55

Moutou C., Coustenis A., Schneider J., St Gilles R., Mayor M., Queloz D., Kaufer A., 2001, A&A 371, 260

Moutou C., Coustenis A., Schneider J., Queloz D., Mayor M. 2003, A&A 405, 341

Sackett, P. 1999, in "Planets Outside the Solar System: Theory and Observations." Eds J.-M. Mariotti and D. Alloin. Dordrecht ; Boston : Kluwer Academic Publishers, p.189

Perryman, M. 2000, Rep. Prog. Phys., 63, 1209

Queloz, D., Eggenberger, A., Mayhor, M., Perrier, C., Beuzit, J.L., Naef, D., Sivan, J.P., Udry, S., 2000, A&A, 359, L17

Rossiter, R.A., 1924, AJ, 60, 15

Seager S., Sasselov D.D., 2000, ApJ 537, 916

Schneider J., Rauer H., Lasota J.P., Bonazzola S., Chassefiere E., 1998, in "Brown dwarfs and extrasolar planets", eds R. Rebolo, E.L. Martin; M.R. Zapatero Osorio, ASP Conference Series 134, 241

Schneider J., 1999, in VLT opening symposium, Antofagasta, ed. Paresce F., Springer, p. 499

Smith G., Hunten D., 1990, Reviews of Geophysics, vol. 28, p. 117

Snellen I.A.G. 2004, MNRAS, MNRAS, 353, L1

Street R., D.L. Pollacco, A. Fitzsimmons, F.P. Keenan, Keith Horne, S. Kane, A. Collier Cameron et al., 2003, in "Scientific Frontiers in Research on Extrasolar Planets", ASP

Conference Series, Vol 294, Eds. Drake Deming and Sara Seager. (San Francisco: ASP) p. 405-408

Udalski A, Zebrun K., Szymanski M., Kubiak M., Soszynski I., Szewczyk O., Wyrzykowski L., Pietrzynski G., 2002a, Acta Astronomica, 52, 115

Udalski A, Szewczyk O., Zebrun K., Pietrzynski G., Szymanski M., Kubiak M., Soszynski I., Wyrzykowski L., 2002b, Acta Astronomica, 52, 317

Vidal-Madjar A, Lecavelier des Etangs A., Désert J.-M., Ballester G.E., Ferlet R., Hébrard G., Mayor M., 2003, Nature 422, 143

Vidal-Madjar A, Désert J.-M., Lecavelier des Etangs A., Hébrard G., Ballester G.E., Ehrenreich D., Ferlet R., McConnell J.C., Mayor M., Parkinson C.D., 2004, ApJ, 604, L69

Wilson O., Abt H., 1954, ApJS 1, 1

Discussion

MIKHAIL MAROV: In searching for Earth-like like planets, do you think – just based on our current knowledge – that in a configuration completely different from what we have for the solar system the closely-located planets to the star would allow the preservation of Earth-like planets? Would they disturb the whole situation near the mother star so tremendously that there would be no chance at all for survival of an Earth-like planet?

IGNAS SNELLEN: No, I really don't think that. Yet, if you have a hot Jupiter, then it would be very unlikely that you will find an Earth-like planet. But that is only a reasonably small fraction of the total extra-solar planet population.

JAYMIE MATTHEWS: Ignas, does your model include the tidally-locked rotation of the planet's atmosphere, as well, to order of magnitude. I'm assuming there's something like a km/s, or so, in the atmosphere in both directions, and the diameter of the planet relative to the star is large enough that there's actually a gradient in the projected radial velocity of the star. Will that blur out the signal, and is that part of your calculation?

IGNAS SNELLEN: Do you mean: would the planetary absorption be too narrow in wavelength?

JAYMIE MATTHEWS: No. What I'm saying is that the planet is rotating, so the absorbing – let's say for the sake of argument the sodium – atmosphere has the trailing edge moving towards you and the leading edge moving away with about a km/s, and the equatorial velocity of the star is maybe 10, or so, km/s, so it's not insignificant in terms of the effects and the absorption in the radial velocity parameter space. I am just curious if that's part of the calculation.

IGNAS SNELLEN: No. At this moment I haven't concentrated on these kinds of models, and how broad you would expect the actual absorption to be. The only thing I can say is that if you just take the face value of HST observations, and just would expect exactly the same absorption for this object, then you would expect an amplitude of about 30 cm/s, or so. What I measure here is 1.7 m/s, so it's much higher. However, this is what I would expect because the HST signal is averaged over 11 Å, while this Rossiter effect is basically within the absorption of the sodium line which is smaller than 1 Å, so I would really expect that secondary Rossiter effect would be a factor of 5 or so larger – say on the Å scale – than it is at the actual HST scale, but the exact details are not really clear.

FRITZ BENEDICT: A comment: While you rightly point out that all of these planets have been discovered by radial velocity techniques, there's at least been one measured – confirmed to be a planetary mass – and this was through astrometry with the Hubble

Space telescope (Gliese 876). The connection here is that we had to remove the parallax with exquisite precision in order to see the wobble due to the planet.

MARY BRÜCK: Did you observe yesterday's Transit of Venus? Is somebody doing it with this technique that you are using?

IGNAS SNELLEN: No, I don't think so, but the actual Rossiter effect is very well known. But to actually do this for Venus would be much more difficult because you have to measure the total spectrum of the sun and to do this ...

MARY BRÜCK: Well, I would have thought this would have been a jolly good test, wouldn't it after all?

IGNAS SNELLEN: No, the actual amplitude you expect for Venus – what is it? a 1000 times smaller? a mm/s? So I don't think it's going to be observed, but it may be interesting; maybe in eight years I will be doing that.

MARY BRÜCK: Yes, good.

FRITZ BENEDICT: There's actually a group in Hawaii who were looking at the moon yesterday to see the effect of the Transit of Venus on the integrated light of the sun from the moon, so it's not the Rossiter effect, but at least they were looking for a transit!

Stonyhurst College Observatory tour

Transits of Venus: New Views of the Solar System and Galaxy
Proceedings IAU Colloquium No. 196, 2004
D. W. Kurtz, ed.
© 2004 International Astronomical Union
doi:10.1017/S1743921305001419

Precise determination of the motion of planets and some astronomical constants from modern observations

E. V. Pitjeva

Institute of Applied Astronomy, Russian Academy of Science, 10, Kutuzova emb.,
St. Petersburg, 191187, Russia
email: evp@quasar.ipa.nw.ru

Abstract. The accomplishments of space flights and introduction of new astrometric methods (radar ranging, lunar laser-ranging, VLBI measurements) in the 1960s required considerably more precise planetary ephemerides than it was possible with classical analytical theories by Leverrier, Hill, Newcomb and Clemence. On the other hand, these modern data made possible the creation of such ephemerides. Two series of numerical ephemerides of planets most complete up to now, and of the same level of accuracy, are considered in this paper. There are the well-known numerical DE ephemerides of JPL as well as the EPM (**E**phemerides of **P**lanets and the **M**oon) ephemerides produced at the Institute of Applied Astronomy. The description of the dynamical models, the brief characteristics of DE118, DE200, DE403, DE405, DE410, EPM87, EPM98, EPM2000, EPM2004 ephemerides, and the comparison between DE410 and EPM2004 are given. The latest DE410 and EPM2004 ephemerides have resulted from a least squares adjustment to observational data totaling about 300 000 position observations (1911–2003) of different types. The accurate radar observations of planets and spacecraft have made it possible not only to improve the orbital elements of planets but to determine a broad set of astronomical constants from the value of the astronomical unit (AU) to parameters of PPN formalism. Recent estimates of different astronomical constant are presented, and progress is shown in the improvement of the AU value, the parameters β, γ, as well as possible variability of the gravitational constant G.

---

1. Dynamical models of planetary motion of DE and EPM ephemerides

Until the 1960s, classical analytical theories of planets (Leverrier, Hill, Newcomb, Clemence) were constantly being improved in order to meet practical needs. The accomplishment of space flights and introduction of new astrometric methods (radar ranging, lunar laser-ranging, VLBI measurements) required considerably more precise planetary ephemerides than those possible with classical theories. On the other hand, these modern data made possible the creation of such ephemerides.

Currently, the creation of these ephemerides is an easier process by using numerical integration of the equations of the motion of the planets and the Moon. At the same time, to ensure space flights the construction of numerical planetary ephemerides was undertaken by several groups in the USA and Russia. Two dynamical models of planet motion having the same level of accuracy, and being most complete up to now are considered in this paper. They are the well-known numerical DE ephemerides of JPL and the EPM (**E**phemerides of **P**lanets and the **M**oon) ephemerides produced at the Institute of Applied Astronomy.

As for analytical ephemerides of planets, the most precise analytical theories of planets and the Moon are the series of French ephemerides VSOP (Bretagnon & Francou 1988) and ELP (Chapront & Chapront-Touzé 1987), produced in BDL and IMCCE. Recently, significant progress has been achieved for the new analytical ephemeris VSOP2002b (Fienga & Simon 2004), where perturbations from the Moon, the 300 main belt asteroids, the solar oblateness and relativistic corrections have been accounted for. However, a comparison of this ephemeris with numerical ephemerides which began to be constructed in IMCCE shows differences between them up to 100 m over three decades. Furthermore, the values of the initial conditions of these ephemerides were obtained by fitting to DE200, DE403, DE405 rather than by fitting to the observation data.

Common to all DE and EPM ephemerides is a simultaneous numerical integration of the equations of motion of the nine major planets, the Sun, the Moon and the lunar physical libration performed in the Parameterized Post-Newtonian metric for the harmonic coordinates $\alpha = 0$ and General Relativity values $\beta = \gamma = 1$.

The various ephemerides differ slightly in

- the modeling of the lunar libration,
- the reference frames,
- the accepted value of the solar oblateness,
- the modeling of the perturbations of asteroids upon the planetary orbits,
- the sets of observations to which ephemerides are adjusted.

Some characteristics of DE118, DE200, DE403, DE405, DE410, EPM87, EPM98, EPM2000, EPM2004 ephemerides are given in Table 1.

Earlier ephemerides have been aligned onto the FK4 reference frame, then onto the dynamical equator and equinox and now ephemerides are oriented onto the **I**nternational **C**elestial **R**eference **F**rame (ICRF) by including in the adjustment the ICRF-based VLBI measurements of spacecraft near to planets.

The solar oblateness causes a secular trend in the planetary elements except for the semi-major axis and eccentricity (for example, see Brumberg 1972). Starting with DE405 (Standish 1998) a nonzero value of the solar oblateness $J_2 - 2 \cdot 10^{-7}$ obtained from some astrophysical estimates was accepted for integrating of DE and EPM ephemerides. Now the value of the solar oblateness is determined while processing the observations.

A serious problem in the construction of planetary ephemerides arises due to the necessity to take into account the perturbations caused by minor planets. In DE200 (Standish 1990) and our previous versions, the perturbations from only three or five biggest asteroids were accounted for. The experiment showed that the fitting of these ephemerides to the Viking lander data is poor. The perturbations from 300 asteroids were taken into account in the ephemerides starting with DE403 (Standish et al. 1995), and EPM98 (Pitjeva 2001) ephemerides. EPM2000 and EPM2004 have been produced by simultaneous numerical integrations of equations of motion of all planets and the 300 main belt asteroids, therefore the perturbations of these asteroids are accounted for all planets. However, masses of many of these asteroids are quite poorly known, and as shown by Standish & Fienga (2002), the accuracy of the planetary ephemerides deteriorates due to this factor. Masses of few most massive asteroids which more strongly affect Mars and the Earth can be estimated from observations of martian landers and spacecraft orbiting Mars. The five of 300 asteroids proved to be double and their masses are known now. The masses of Eros (433) and Mathilda (253) have been derived by perturbations of the spacecraft during the NEAR flyby. Unfortunately, the classical method of determining masses of asteroids for which close encounters occur is limited by uncertainty in masses of the large asteroids, perturbations by others, unmodeled asteroids, and the quality of observations. Perhaps masses of many asteroids will be obtained by high accuracy observations during

Table 1. Ephemerides DE and EPM

ephemeris	interval of integration	ref. frame	mathematical model	type	data number	interval
DE118 (1981) ⇓ DE200	1599→2169 ⇓	FK4 ⇓ dynamic. frame	Integrating of Sun,Moon,9 planets +perturbations from 3 asteroids (Keplerian ellipses)	optical radar spacecraft LLR total	44755 1307 1408 2954 50424	1911-1979 1964-1977 1971-1980 1970-1980 1911-1980
EPM87 (1987)	1700→2020	FK4	Integrating of Sun,Moon,9 planets +perturbations from 5 asteroids (Keplerian ellipses)	optical radar spacecraft LLR total	48709 5344 – 1855 55908	1717-1980 1961-1986 – 1972-1980 1717-1986
DE403 (1995) ⇓ DE404	−1410→3000 ⇓ −3000→3000	ICRF	Integrating of Sun,Moon,9 planets +perturbations from 300 asteroids (mean elements)	optical radar spacecraft LLR total	26209 1341 1935 9555 39057	1911-1995 1964-1993 1971-1994 1970-1995 1911-1995
EPM98 (1998)	1886→2006	DE403	Integrating of Sun,Moon,9 planets 5 aster.+ perturb. from 295 asteroids (mean elements)	optical radar spacecraft LLR total	– 55959 1927 10000 67886	– 1961-1995 1971-1982 1970-1995 1961-1995
DE405 (1997) ⇓ DE406	1600→2200 −3000→3000	ICRF	Integrating of Sun,Moon,9 planets +perturbations from 300 asteroids (integrated)	optical radar spacecraft LLR total	28261 955 1956 11218 42410	1911-1996 1964-1993 1971-1995 1969-1996 1911-1996
EPM2000 (2000)	1886→2011	DE405	Integrating of Sun,Moon, 9 planets, 300 asteroids	optical radar spacecraft LLR total	– 58076 24587 13500 96163	– 1961-1997 1971-1997 1970-1999 1961-1999
DE410 (2003)	1901→2019	ICRF	Integrating of Sun,Moon,9 planets +perturbations 300 ast.integrated	optical radar spacecraft LLR total	39159 978 154685 9555 204377	1911-2003 1964-1997 1971-2003 1970-1995 1911-2003
EPM2004 (2004)	1886→2011	ICRF	Integrating of Sun,Moon, 9 planets, 301 asteroids, asteroid ring	optical radar spacecraft LLR total	46064 58116 197271 15590 317041	1913-2003 1961-1997 1971-2003 1970-2003 1913-2003

the Gaia mission, but it will not be soon. So at present masses of the rest of the 301 asteroids have been estimated by the astrophysical method. The latest published diameters of asteroids based on infrared data of IRAS (**I**nfra **R**ed **A**stronomical **S**atellite) and MSX (**M**idcourse **S**pace **E**xperiment), as well as observations of occultations of stars by minor planets and radar observations have been used in this paper. The mean densities for C,S,M taxonomy classes have been estimated while processing the observations.

At the several meters level of accuracy the orbit of Mars is very sensitive to perturbations from many minor planets. These objects are mostly too small to be observed from the Earth, but their total mass is large enough to affect the orbits of the major planets. The main part of these celestial bodies moves in the asteroid belt and their instantaneous positions may be considered homogeneously distributed along the belt. Thus, it seems reasonable to model the perturbations from the remaining small asteroids (for which individual perturbations are not accounted for) by computing additional perturbations from a massive ring with the constant mass distribution in the ecliptic plane (Krasinsky et al. 2002). Two parameters that characterize the ring (its mass and radius) are included in the set of solution parameters for EPM2004.

The quality of ephemerides, i.e. their accuracy, mostly depends upon improvement of quality and increase of amount of observational data. The section below maintains the description of the database of modern ephemerides. However, we should stress here the uniqueness of the EPM87 ephemeris (Krasinsky et al. 1993), whose parameters were fit to a large variety of observational data for the time span of XVIII–XX centuries, including observations of 32 transits of Mercury (1723–1973) and 4 transits of Venus (1761–1881) across the solar disc, used in order to correct the adopted Brouver system of differences between ET and UT back into the past till 1715, to estimate variations of the solar radius, the Mercury perihelion and node advance.

Along with the planetary ephemerides the ephemerides of the orbital and rotational motions of the Moon were being produced and improved on by processing LLR observations at JPL and IAA RAS. The last versions of the lunar theory of the JPL are given in Williams & Dickey (2002) and of the IAA — in Krasinsky 2002 where a number of subtle selenodynamical effects is described.

The lunar-planetary integrator for EPM is embedded into the program package (Krasinsky & Vasilyev 1997) ERA-7 (**ERA: E**phemeris **R**esearch in **A**stronomy) developed to support scientific research in dynamical and ephemeris astronomy. Integrating and other computations have been done by using this package.

2. Processing radar and optical data

Since the creation of DE405 in 1997, the quality and quantity of observational data have vastly improved. New high accuracy observations have evolved: ranging and Doppler measurements from the martian lander Pathfinder, range and VLBI points from MGS and Odyssey, CCD astrometric data of the outer planets and their satellites. So, the latest DE410 and EPM2004 ephemerides have resulted from a least squares adjustment to the observational data totaling about 300 000 position observations (1911–2003) of different types. Data used for the production of the EPM2004 ephemeris were taken from databases of the JPL website (http:/ssd.jpl.nasa.gov/iau-comm4/) created and kept by Dr. Standish, the database of optical observations of Dr. Sveshnikov and extended to include Russian radar observations of planets (on the website of IAA http: //www.ipa.nw.ru/ PAGE/DEPFUND/LEA/ENG/englea.htm). All the observations used are described in Tables 2 and 3.

The reduction of the radar measurements maintains all the relevant corrections, including the modeling of Mercury, Venus and Mars topography. However, uncertainties due to planetary topography do remain in radar ranging despite the modeling of topography. So, observations of the martian Viking, Pathfinder landers, MGS and Odyssey data which are free from these uncertainties are of great importance.

Unfortunately, observations of MGS and Odyssey as distinct from two frequencies measurements of Viking were carried out at one frequency and the effect of the solar corona delay was considerable, particularly near superior solar conjunctions in 1998 and

Table 2. Radiometric observations used in the ephemeris solutions.

MERCURY

station, object	type	time interval	number of obs.	normal points	a priori accuracy
Millstone	τ	1964	5	—	7.5–75 km
Haystack	τ	1966–1971	217	—	3 km
Arecibo	τ	1964–1982	341	323	3–30 km
Goldstone	τ	1971–1997	259	138	1.5–3 km
Goldstone cl.p.	τ	1990–1997	40	—	0.15–2.5 km
Crimea	τ	1980–1995	75	23	1.2–4.8 km
Mariner-10	τ	1974–1975	—	2	0.1 km

VENUS

station, object	type	time interval	number of obs.	normal points	a priori accuracy
Millstone	τ	1961–1967	135	—	1.5–120 km
Haystack	τ	1966–1971	219	—	1.5 km
Arecibo	τ	1964–1970	319	—	3–15 km
Goldstone	τ	1964–1990	512	—	1.5–6 km
Crimea	τ	1962–1995	1139	170	0.15–22.5 km
Magellan	$\alpha\delta$	1990–1994	—	18	$0.''001$–$0.''004$

MARS

station, object	type	time interval	number of obs.	normal points	a priori accuracy
Haystack	τ	1967–1973	3801	133	0.075–12 km
Arecibo	τ	1965–1973	1680	43	0.075–45 km
Goldstone	τ	1969–1994	48989	149	0.075–0.6 km
Crimea	τ	1971–1995	381	78	0.15–4.8 km
Mariner-9	τ	1971–1972	643	—	15–270 m
Viking-1	τ	1976–1982	1161	—	7–12 m
Viking-1	$d\tau$	1976–1978	14980	—	0.16–3.2 m
Viking-2	τ	1976–1977	80	—	7–10 m
Pathfinder	τ	1997	90	—	10–22 m
Pathfinder	$d\tau$	1997	7576	—	0.012 m
Phobos	τ	1989	—	1	0.2 km
MGS	τ	1998–2003	110538	4930	2–7.5 m
Odyssey	τ	2002–2003	62093	1715	2–3 m
spacecraft	$\alpha\delta$	1984–2003	—	44	$0.''0003$–$0.''006$

JUPITER

station, object	type	time interval	number of obs.	normal points	a priori accuracy
spacecraft, VLA	α	1979–1995	—	4	$0.''003$–$0.''046$
spacecraft, VLA	δ	1979–1995	—	4	$0.''005$–$0.''2$
spacecraft	τ	1973–1995	—	6	1–6 km
spacecraft	$\alpha\delta$	1996–1997	—	24	$0.''007$–$0.''012$
Arecibo s 3,4	τ	1992	—	4	3–14 km

2002. The following model was used for the solar corona reduction:

$$N_e(r) = \frac{A}{r^6} + \frac{B + \dot{B}t}{r^2}$$

where $N_e(r)$ is the electron density.

The parameters B and $\dot{B}$ determined from observations were different for different solar conjunctions. Although residuals decreased considerably, remaining influence of the corona has been seen in them (see rms of MGS and Odyssey in Fig. 3). Moreover, the parameters of the corona are correlated with other estimated parameters of the Mars orbit and deteriorate their determination. This fact should be taking into account for highly precise astrometrical measurements of future space missions.

The residuals of all radiometric data are shown in Figs 1–3. The rms residuals of ranging for the Mercury are 1.4 km, for Venus and Mars are 0.7 km, for Viking are 8 m, for Pathfinder 4.4 m, for MGS and Odyssey 1.4 m. The estimations of biases due to uncertainties of the calibration for Viking ranges (about 20 m) and for Odyssey ranges (about 2 m) have been obtained from the observations the way it has been done for the DE410 ephemeris.

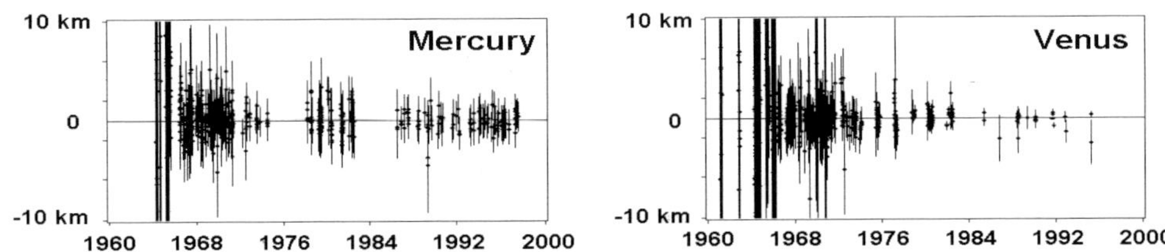

Figure 1. Ranging residuals for Mercury and Venus.

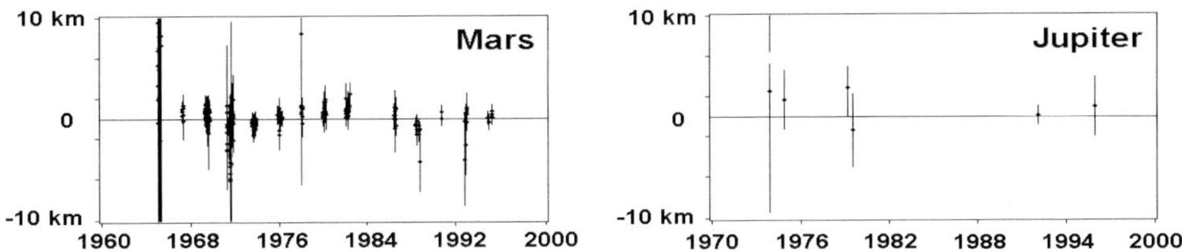

Figure 2. Ranging residuals for Mars and Jupiter.

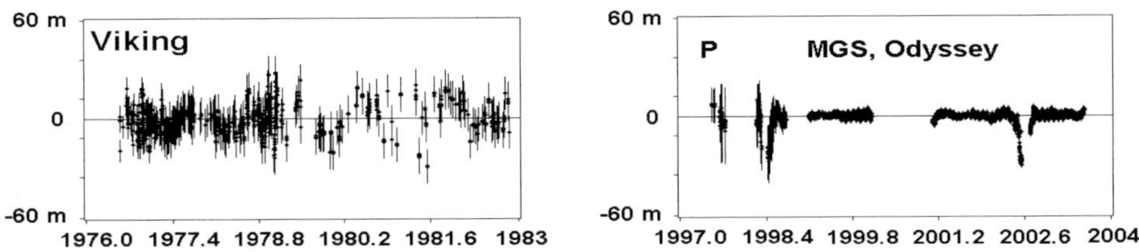

Figure 3. Ranging residuals of Viking, Pathfinder-P (1997), MGS (1998–2003), Odyssey (2002–2003).

The observations of satellites of Jupiter and Saturn are of great importance for optics, as they are more accurate than the observations of their parent planets and practically free from the phase effect. CCD data, obtained at Flagstaff observatory, whose observational program started in 1995 and is still being continued are the most accurate. Another group of high accuracy data is photographic observations of satellites of Jupiter, Saturn, as well as Uranus and Neptune planets obtained at Nikolaev observatory during 1962–1998. Combination of the data from Flagstaff and Nikolaev has been successfully used to improve the planet ephemerides. Residuals of all the observations of the outer planets are shown in Fig. 4. Unfortunately, observations of Pluto are mainly photographic and of quite poor accuracy, so their rms residuals are worse.

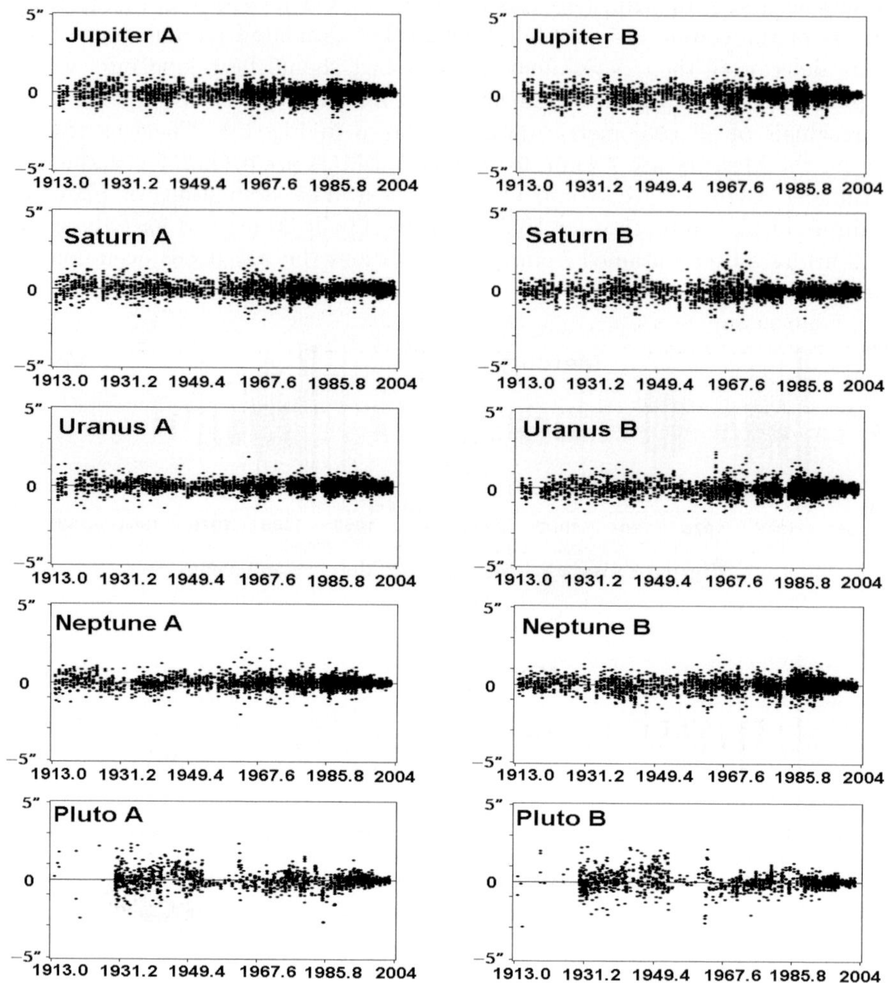

Figure 4. Residuals of the outer planets 1913–2003 in $\alpha \cos\delta$ (A) and in δ (B), the scale $\pm 5''$.

Ephemerides EPM were oriented onto the International Celestial Reference Frame (ICRF). The most precise optical data of the outer planets and their satellites, obtained at Flagstaff, Nikolaev, La Palma have already been referenced to the ICRF. The remaining optical observations, referenced to different catalogues, at first were transformed to the FK4 systems by Sveshnikov. Then they were referenced to the FK5 using known formulae (see as the example Standish et al. 1995), and were finally transformed to the ICRF using the values of the three angles of the rotation between the HIPPARCOS and FK5 catalogues, J2000 in mas (Mignard 2000):

$$\varepsilon_x = -19.9, \ \varepsilon_y = -9.1, \ \varepsilon_z = 22.9.$$

Orbits of the four inner planets (with the exception of angles of the orientation) are determined entirely by the ranging observations of planets and spacecraft. The orientation of this system is provided by the using the ICRF-base VLBI measurements of spacecraft. The orientation has been improved considerably due to an addition to previous Magellan and Phobos data of new VLBI points of MGS and Odyssey. The angles of the rotation

Table 3. Optical and radio observations used in the ephemeris solutions of the outer planets.

JUPITER

station, object	planet satellite	type	time interval	number of obs.	a priori accuracy
USNO	p	transit	1913–1994	4388	$0\rlap{.}''5$
Tokyo	p	ph-e transit	1963–1988	568	$0\rlap{.}''5$–$0\rlap{.}''8$
La Palma	s 3,4	ph-e transit	1986–1997	1316	$0\rlap{.}''25$
Nikolaev	s 1,2,3,4	photo	1962–1998	2628	$0\rlap{.}''2$
Flagstaff	s 1,2,3,4	CCD	1998–2003	2408	$0\rlap{.}''2$
Mountain	s 1,2,3,4	CCD	2002–2002	16	$0\rlap{.}''5$

SATURN

station, object	planet satellite	type	time interval	number of obs.	a priori accuracy
USNO	p	transit	1913–1982	3054	$0\rlap{.}''5$
Tokyo	p	ph-e transit	1963–1988	506	$0\rlap{.}''5$–$0\rlap{.}''8$
Bordeaux	s 6,8	ph-e transit	1987–1993	238	$0\rlap{.}''25$
La Palma	s 5,6,7,8	ph-e transit	1987–1997	1460	$0\rlap{.}''25$
Nikolaev	s 3,4,5,6,8	photo	1973–1997	1264	$0\rlap{.}''2$
Flagstaff	s 3,4,5,6,7,8	CCD	1998–2003	4014	$0\rlap{.}''2$
Mountain	s 3,4,5,6,7,8	CCD	2002–2003	628	$0\rlap{.}''15$
VLA	p	radio	1984	8	$0\rlap{.}''03$–$0\rlap{.}''06$

URANUS

station, object	planet satellite	type	time interval	number of obs.	a priori accuracy
USNO	p	transit	1913–1993	4244	$0\rlap{.}''5$
Tokyo	p	ph-e transit	1963–1988	366	$0\rlap{.}''5$–$0\rlap{.}''8$
Bordeaux	p	ph-e transit	1985–1992	330	$0\rlap{.}''25$
Bordeaux	p	CCD	1997	34	$0\rlap{.}''2$
La Palma	p, s 4	ph-e transit	1984–1997	2072	$0\rlap{.}''25$
Nikolaev	p	photo	1961–1998	440	$0\rlap{.}''2$
Flagstaff	p, s 3,4	CCD	1995–2003	2324	$0\rlap{.}''2$
Mountain	p, s 3,4	CCD	1998–2003	174	$0\rlap{.}''15$
VLA,ring occ.	p	radio	1977–1985	16	$0\rlap{.}''03$–$0\rlap{.}''2$

NEPTUNE

station, object	planet satellite	type	time interval	number of obs.	a priori accuracy
USNO	p	transit	1913–1993	3804	$0\rlap{.}''5$
Tokyo	p	ph-e transit	1963–1988	320	$0\rlap{.}''5$–$0\rlap{.}''8$
Bordeaux	p	ph-e transit	1985–1993	366	$0\rlap{.}''25$
Bordeaux	p	CCD	1997	28	$0\rlap{.}''2$
La Palma	p	ph-e transit	1984–1998	2212	$0\rlap{.}''25$
Nikolaev	p	photo	1961–1998	436	$0\rlap{.}''2$
Flagstaff	p, s 1	CCD	1995–2003	1888	$0\rlap{.}''2$
Mountain	p, s 1	CCD	1998–2003	120	$0\rlap{.}''15$
VLA,ring occ.	p	radio	1981–1997	22	$0\rlap{.}''03$–$0\rlap{.}''2$

between EPM2004 and the ICRF have been obtained (in mas):

$$\varepsilon_x = 1.9 \pm 0.1, \varepsilon_y = -0.5 \pm 0.2, \varepsilon_z = -1.5 \pm 0.1,$$

which are close to the rotation angles between DE405 and DE410.

Table 3. Optical and radio observations used in the ephemeris solutions of the outer planets.
(continued)
PLUTO

station, object	planet satellite	type	time interval	number of obs.	a priori accuracy
Different stat.	p	photo	1914–1967	1164	$0\rlap{.}''5$–$1''$
Different stat.	p	photo	1969–1988	674	$0\rlap{.}''5$–$1''$
Different stat.	p	photo	1989–1995	82	$0\rlap{.}''5$–$1''$
Pulkovo	p	photo	1930–1993	416	$0\rlap{.}''5$
Tokyo	p	photo	1994	24	$0\rlap{.}''3$
Bordeaux	p	ph-e transit	1996	12	$0\rlap{.}''3$
Bordeaux	p	CCD	1995–1997	64	$0\rlap{.}''2$
La Palma	p	ph-e transit	1986-1998	760	$0\rlap{.}''25$
Flagstaff	p	CCD	1995–2003	1152	$0\rlap{.}''2$
Mountain	p	CCD	2000–2003	68	$0\rlap{.}''15$

Table 4. The formal standard deviations of elements of the planets

planet	a [m]	$\sin i \cos\Omega$ [mas]	$\sin i \sin\Omega$ [mas]	$e\cos\pi$ [mas]	$e\sin\pi$ [mas]	λ [mas]
Mercury	0.105	1.654	1.525	0.123	0.099	0.375
Venus	0.329	0.567	0.567	0.041	0.043	0.187
Earth	0.146	—	—	0.001	0.001	—
Mars	0.657	0.003	0.004	0.001	0.001	0.003
Jupiter	639	2.410	2.207	1.280	1.170	1.109
Saturn	4222	3.237	4.085	3.858	2.975	3.474
Uranus	38484	4.072	6.143	4.896	3.361	8.818
Neptune	478532	4.214	8.600	14.066	18.687	35.163
Pluto	3463309	6.899	14.940	82.888	36.700	79.089

Table 5. The parameters of the Mars rotation.

$\dot V$ [°/day]	I_q [°]	$\dot I_q$ [″/year]	Ω_q [°]	$\dot\Omega_q$ [″/year]
350.891985294	25.1893930	-0.0002	35.437685	-7.5844
$\pm$ 0.000000012	$\pm$0.0000053	$\pm$0.0007	$\pm$ 0.000021	$\pm$ 0.0015

3. Results obtained

The formal standard deviations of the orbital elements of planets are shown in the Table 4, where a – the semi-major axis, i – the inclination of the orbit, Ω – the ascending node, e – the eccentricity, π – the longitude of perihelion, λ – the mean longitude. Note that the uncertainties, given in this paper, are formal standard deviations; realistic error bounds may be an order of magnitude larger.

Accurate radar observations of planets and spacecraft have made it possible not only to improve the orbital elements of planets but to determine a broad set of astronomical constants as well: AU, parameters of Mars' rotation including its precessional rate, the masses of Ceres, Pallas, Vesta, Iris, Bamberga, Juno; the estimation of the total mass of the main asteroid belt, parameters of the PPN formalism (β, γ), the variability of the gravitational constant G, the solar quadrupole moment. Values for them are in Tables 5, 6 and 7.

Table 6. Masses of Ceres, Pallas, Juno, Vesta, Iris, Bamberga in $(GM_i/GM_\odot)\cdot 10^{-10}$.

(1)Ceres	(2)Pallas	(3)Juno	(4)Vesta	(7)Iris	(324)Bamberga
4.753	1.027	0.151	1.344	0.063	0.055
±0.007	±0.003	±0.003	±0.001	±0.001	±0.001

Table 7. The solar quadrupole moment, the radius and the mass of the asteroid ring, the total mass of the main asteroid belt.

J_2 10^{-7}	R_{ring} AU	M_{ring} $10^{-10}M_\odot$	M_{belt} $10^{-10}M_\odot$
1.9±0.3	3.13±0.05	3.35±0.35	15.0±2.0

Table 8. Progress in the determination of the parameters of PPN formalism and $\dot{G}/G$.

year	β	γ	$\dot{G}/G(10^{-11}\mathrm{yr}^{-1})$
1985	0.76±0.12	0.87±0.06	4.1 ±0.8
1994	1.014±0.070	1.006±0.037	0.28 ±0.32
2004	1.0000±0.0001	0.9999±0.0001	0.001 ±0.005

The value of the precession constant for Mars is close to the recent value obtained from Viking, Pathfinder landers and MGS radio tracking data (Yoder et al. 2003):

$$\dot{\Omega}_q = [-7.''597 \pm 0.''025(10\sigma)]/\text{year}.$$

Lately some papers have appeared where impossibility of a separate determination of PPN parameters and the solar oblateness from ranging observations is stated. However, the PPN parameters (β, γ) and the solar oblateness cause secular and periodic effects among different planets and other orbital elements, therefore the estimations of the this parameters and possible variability of the gravitational constant have been simultaneously obtained.

Three main factors influence the progress in the improvement of parameters — reductions of the observational data, dynamical models of planet motion, and the observational data themselves. As an example, the progress in the determination of the AU (in km) in Russia from ranging (Table 9), as well as in the estimation of parameters of PPN formalism and the time variation of the gravitation constant (Table 8) is given.

The last AU value of Table 9 differs from the value of the DE410 (Standish 2003) 149 597 870.6974 ± 0.0003 by 1.4 m, which is the real error of the determination of the AU.

4. Comparison DE410 and EPM2004

The differences between various ephemerides are useful to know since they are indicative of the realistic accuracies of the ephemerides. Comparison between the latest versions of DE410 and EPM2004 over the time interval 1970–2010 has been made. These ephemerides are based on the similar data and the mathematical models, but distinguish by different way of taking into account asteroids, by their masses as well as corrections for the topography of planet surfaces and the solar corona.

Coordinates of Mercury and Venus have been obtained from fitting radar observations of these planets, having the uncertainties mainly about km, so the maximum differences of geliocentric distances up to 258 m for Mercury and up to 139 m for Venus (Fig. 5) can be considered acceptable.

Table 9. Progress in the determination of the AU (in km) in Russia from ranging.

149599300	±600	Kotelnikov et al. 1962
149597867.3	±0.3	Akim et al. 1986
149597870.62	±0.18	Krasinsky et al. 1993
149597870.6960±0.0001		Pitjeva, 2004

The maximum differences of geliocentric distances for the Earth and the Mars are considerably less: up to 12.8 m for the Earth and up to 35.7 m for Mars, which is not surprising as data of MGS and Odyssey used have the uncertainties at the two-meter level. The differences may be explained by different account of asteroids and solar corona.

The availability of a number of spacecraft Jupiter data (besides optical observations) allow its ephemerides to be known better than those of other outer planets. The distance differences for Jupiter are less than 10 km. For the other four outer planets the only observations are optical. Moreover, for Neptune and Pluto a single orbital period has not yet elapsed since the time when more or less accurate observations of these planets appeared. The distance differences amount for Saturn to 180 km, for Uranus to 410 km, for Neptune to 1200 km, for Pluto to 14000 km. Those are the current accuracies of the modern ephemerides.

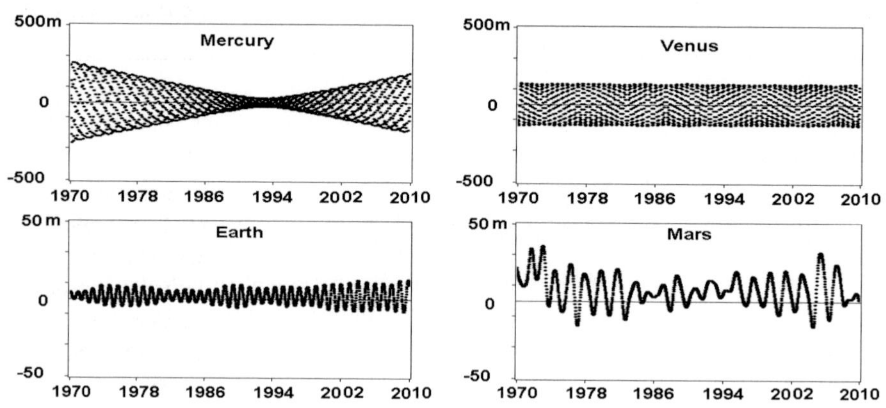

Figure 5. DE410-EPM2004: heliocentric distance differences for the inner planets, 1970–2010.

References

Akim, Eh. L., Brumberg, V. A., Kislik, M. D., Koljuka, Yu. F., Krasinsky, G. A., Pitjeva, E. V., Shiskov, V. A., Stepanianz, V. A., Sveshnikov, M. L. & Tikhonov, V. F. 1986 In *Proceedings of the IAU Symp. N 114, Relativity in Celestial Mechanics and Astrometry* (ed. J. Kovalevsky & V. A. Brumberg), Kluwer, Dordrecht, 63–68.

Bretagnon, P. & Francou, G. 1988 *A&A* **202**, 309–315.

Brumberg, V. A. 1972 *Relativistic Celestial Mechanics.* Izdatel'stvo "Nauka", Moskva, 382p. (in Russian)

Chapront, J. & Chapront-Touzé, M. 1987 *A&A* **190**, 342–452.

Fienga, A. & Simon, J.-L. 2004 *A&A*, in press.

Kotelnikov, V. A, & Dubrovin, V. M., Kislik, M. D., Korenberg, E. B, Minashkin, B. P., Morozov, V. A., Nikitsky, N. I., Petrov, G. M., Rzhiga, O. N. & Shachovskoy, A. M. 1962 *Dokl. AN SSSR* **145**, 1035–1038.

Krasinsky, G. A., Pitjeva, E.V., Sveshnikov, M. L. & Chunajeva, L. I. 1993 *Celest. Mech. & Dyn. Astr.* **55**, 1–23.

Krasinsky, G. A. & Vasilyev, M. V. 1997 In *Proceedings of the IAU Coll.165, Dynamics and Astrometry of Natural and Artificial Celestial Bodies* (ed. I.M.Wytrzyszczak, J.H.Lieske, & R.A.Feldman), Dordrecht, Kluwer, 239–244.

Krasinsky, G. A., Pitjeva, E. V., Vasilyev, M. V. & Yagudina, E. I. 2002 *Icarus* **158**, 98–105.

Krasinsky, G. A. 2002 *Communication of IAA RAS* **148**, 27p.

Mignard, F. 2000 In *Towards models and constants for sub-microarcsecond astrometry* (ed. Johnston K. J., McCarthy D. D., Luzum B. J. & Kaplan G. H.), U.S. Naval Observatory, Washington, DC, USA, 10–19.

Pitjeva, E. V. 2001 *Celest. Mech. & Dyn. Astr.* **80**, N 3/4, 249-271.

Standish, E. M. 1990 *A&A* **233**, 252–271.

Standish, E. M. 1998 *Interoffice Memorandum,* **312.F-98-048**, 18p.

Standish, E. M. 2003 *Interoffice Memorandum,* **312.N-03-009**, 16p.

Standish, E. M. & Fienga, A. 2002 *A&A* **384**, 322–328.

Standish, E. M., Newhall XX, Williams, J. G. & Folkner, W. M. 1995 *Interoffice Memorandum* **314.10-127**, 22p.

Yoder, C. F., Konoplev, A. S., Yuan, D. N., Standish, E. M. & Folkner, W.M. 2003 *Science* **300**, Issue 5617, 299–303.

Williams, J. G. & Dickey, J. O. 2002 In *13th International Workshop on Laser Ranging*, 17p. Washington, D.C.

Discussion

NICOLE CAPITAINE: Can you say for which kind of parameters the observations of transits of planets are useful?

ELENA PITJEVA: This observation may be used for the estimation of the variation of the solar radius. There are some examples where the variation of the solar radius has been estimated from transits.

JESUS DE ALBA: Do you use your model to calculate the ephemeredes of potentially dangerous near-Earth objects?

MYLES STANDISH: I will comment on that question: It would not be practical to use her model for that computation because the equations of motion are extremely complicated; it would terribly long to integrate. What you do is use her ephemeris after it is computed as perturbations on a different integration.

Transits of Venus: New Views of the Solar System and Galaxy
Proceedings IAU Colloquium No. 196, 2004
D.W. Kurtz, ed.

© 2004 International Astronomical Union
doi:10.1017/S1743921305001420

The black-drop effect explained

Jay M. Pasachoff[1], Glenn Schneider[2], and Leon Golub[3]

[1]Williams College-Hopkins Observatory, Williamstown, MA 01267, USA
email: jay.m.pasachoff@williams.edu
[2]Steward Observatory, University of Arizona, Tucson, Arizona 85721, USA
[3]Harvard-Smithsonian Center for Astrophysics, Cambridge, MA 02138, USA

Abstract. The black-drop effect bedeviled attempts to determine the Astronomical Unit from the time of the transit of Venus of 1761, until dynamical determinations of the AU obviated the need for transit measurements. By studying the 1999 transit of Mercury, using observations taken from space with NASA's Transition Region and Coronal Explorer (TRACE), we have fully explained Mercury's black-drop effect, with contributions from not only the telescope's point-spread function but also the solar limb darkening. Since Mercury has no atmosphere, we have thus verified the previous understanding, often overlooked, that the black-drop effect does not necessarily correspond to the detection of an atmosphere. We continued our studies with observations of the 2004 transit of Venus with the TRACE spacecraft in orbit and with ground-based imagery from Thessaloniki, Greece. We report on preliminary reduction of those data; see http://www.transitofvenus.info for updated results. Such studies are expected to contribute to the understanding of transits of exoplanets. Though the determination of the Astronomical Unit from studies of transit of Venus has been undertaken only rarely, it was for centuries expected to be the best method. The recent 8 June 2004 transit of Venus provided an exceptionally rare opportunity to study such a transit and to determine how modern studies can explain the limitations of the historical observations.

---

1. History

With *De Revolutionibus* of Nicolaus Copernicus, published in 1543, our modern solar-system came into consideration. From Copernicus's famous diagram, it was clear that only Mercury and Venus have orbits interior to that of Earth, with the result that only Mercury and Venus could transit across the sun's disk. The Copernican system reached England in 1576 with the appendix that Thomas Digges wrote to the *Prognostication Everlasting* book written by his father, Leonard Digges. This edition marked the first appearance in English of a Copernican diagram (Fig. 1).

The Mars data of Tycho Brahe, in the hands of Johannes Kepler, led to Kepler's laws of planetary orbits, which we use today to understand the orbits of not only the planets around the sun but also double stars and extra-solar planets (exoplanets). In 1609, in his *Astronomia Nova*, Kepler advanced (though not explicitly) his first two laws: (1) that the planets orbit the sun in ellipses, with the sun at one focus; and (2) that the line joining the sun and a planet sweeps out equal areas in equal times, which we now recognize as a consequence of conservation of angular momentum. Most important for the story of the determination of the Astronomical Unit is Kepler's third law, advanced in his *Harmonices Mundi* of 1619. This law holds that the squares of the periods of the planets are proportional to the cubes of the semimajor axes of their orbits. For this work, we will take the Astronomical Unit to be the average distance from the Earth to the sun and equivalent to the semimajor axis of the Earth's orbit; a technical discussion of the definition of the Astronomical Unit is given in the review by Standish (these proceedings).

Figure 1. The first English-language Copernican diagram, from a 1596 printing of the *Prognostication Everlasting* of Leonard Digges with the Copernican appendix by Thomas Digges.

Though Kepler's third law provided the proportions of orbital semimajor axes – with that of Venus being about 70%, and that of Mars about 150%, that of Earth – no actual distances were known. Among other consequences of this deficiency was that the physical diameters of any of the planets or of the sun was not known. For example, Digges's book showed Venus much smaller than Earth.

Kepler's *Rudolphine Tables* (1627) was not an ephemeris; that is, it did not make specific predictions, though it provided tables to allow such predictions to be made. The story is told elsewhere in these proceedings of how Jeremiah Horrocks himself predicted and observed the 1639 transit of Venus, and brought his friend William Crabtree also into the sole pair of observers of that event.

The suggestion by Edmond Halley (1716) that transits of Venus can be used to determine the distance to Venus from its parallax led to dozens of international expeditions for the 1761 transit. Halley's method required accurate measurement of the duration of the transit from a given location in order to determine the length of the chord (Fig. 2). He thought that such determination could be made to approximately one second of time. But Bergman (1761), in observing the transit, reported that a ligature joined the silhouette of Venus to the dark background exterior to the sun (Fig. 3). This dark "black drop" meant that observers were unable to determine the time of contact to better than 30 s or even 1 min. The black-drop effect thus led to uncertainty in Venus's parallax, and thus the sun's parallax, and thus the Astronomical Unit.

Of the dozens of expeditions that went all over the world to observe the transit from different latitudes, perhaps the best known was that of James Cook (Fig. 4; see Orchiston, these proceedings). The British Admiralty hired the young lieutenant, gave him a ship, and sent him to Tahiti for the observations. His later peregrinations around New Zealand and the eastern coast of Australia are thus spinoffs of astronomical research. On board was an astronomer, Charles Green (a descendant of whose was excited, just after the 2004 transit, to hear her ancestor's work still being discussed at a site on a Greek island). Cook's and Green's observations (Fig. 5) clearly show the black-drop effect (Cook & Green, 1771).

Cook (1768–71) reported in his *Journal*:

This day prov'd as favourable to our purpose as we could wish, not a Clowd was to be seen the whole day and the Air was perfectly clear, so that we had every

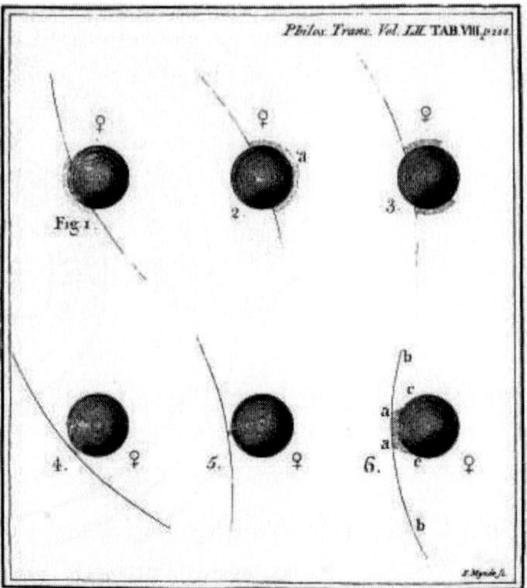

Figure 2. (left) The path of Venus across the face of the sun at the transit of 1761, from a book by Johann Doppelmayr (1742).

Figure 3. (right) The black-drop effect, as reported at the transit of 1761 by T. Bergman (1761) ©Royal Society

Figure 4. Captain James Cook, shown in a statue in the garden of his parents' house, which had been moved from England to a public park in Melbourne, Australia.

advantage we could desire in Observing the whole of the passage of the Planet Venus over the Suns disk: we very distinctly saw an Atmosphere or dusky shade round the body of the Planet which very much disturbed the times of the Contacts particularly the two internal ones. Dr Solander observed as well as Mr Green and my self, and we differ'd from one another in observeing the times of the Contacts much more than could be expected. Mr Greens Telescope and mine were of the

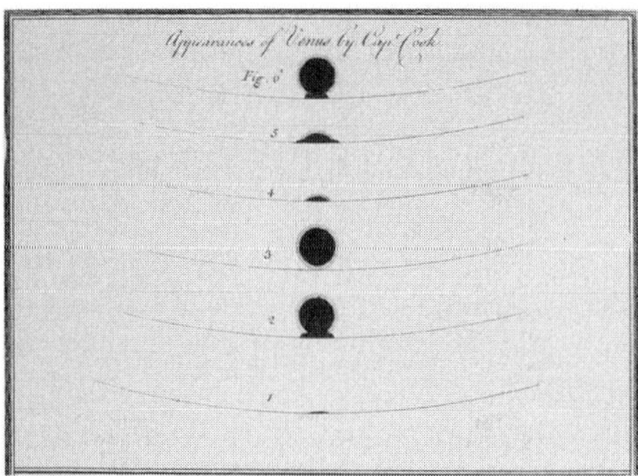
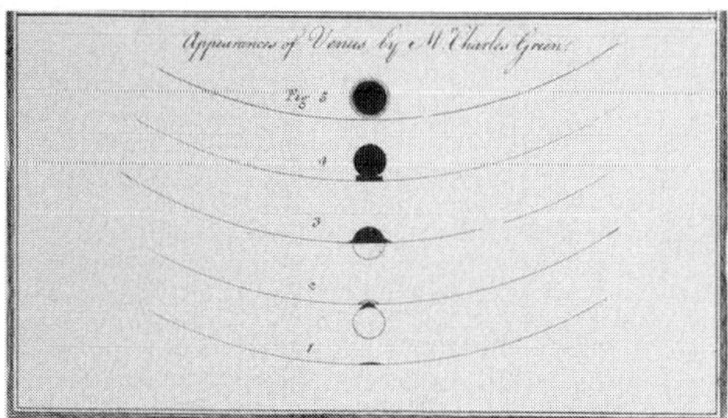

Figure 5. The 1769 black-drop observations of James Cook (left) and Charles Green (right) (Cook & Green 1771). ©Royal Society

same Mag[n]ifying power but that of the Dr was greater than ours. It was ne[a]rly calm the whole day and the Thermometer expose'd to the Sun about the middle of the Day rose to a degree of heat we have not before met with.

So Cook's report both correctly mentioned the existence of an atmosphere around Venus, which had been discovered at the 1761 transit, but then incorrectly attributed the inaccuracy in his ability to time the transit's contacts to that atmosphere.

The next pair of transits wasn't for another 105.5 years. The 1874 transit was clearly observed in Australia. The black-drop effect showed well and was widely reported and drawn (Fig. 6). Janssen attempted to photograph it with a "photographic revolver," a revolving plate of which three French examples (Flammarion, 1875) are preserved in France, and one example of a British derivative remains on display at the Sydney Observatory. Unfortunately, the widely-reproduced plate of his observations is not of the actual transit but rather of a previous test (Fig. 7), as shown by Launay & Hingley (2005). They showed that it was first published in 1891 and then reproduced in Janssen's *Oeuvres Scientifiques* of 1929.

The newfangled photography turned out not to be of use in resolving the problem of the black-drop effect. In general, the 19th-century techniques did not provide an advance over the 18th-century techniques (Meadows 1974). Some photographs survive from the 1882 transit expeditions (Fig. 8).

Many historic images of the black-drop effect appear at Chuck Bueter's Website at http://www.transitofvenus.org/blackdrop.htm. Pasachoff (2003b, 2004) reviewed recent books that discussed the historical transits of Venus.

2. The 1999 transit of Mercury

The rare transit of Venus expected in 2004 was anticipated (Pasachoff 2000, 2002, 2003a; Pasachoff & Filippenko 2004) as a remarkable event. A paper delivered before the History of Astronomy Division of the American Astronomical Society (Schaefer 2001) led us to realize the utility of existing observations of the 1999 Mercury transit with NASA's Transition Region and Coronal Explorer (TRACE) for analyzing the problem of the black-drop effect. Schaefer reported that most of the citations, even current ones, incorrectly attributed the black-drop effect to Venus's atmosphere. But TRACE was above the Earth's atmosphere and Mercury has essentially no atmosphere, so any Mercurian black drop could not arise from atmospheric factors.

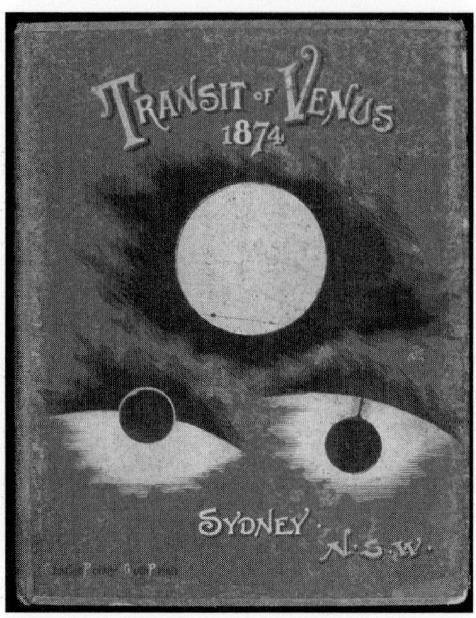

Figure 6. The cover of *Transit of Venus 1874* by Henry Chamberlain Russell, Government Astronomer for New South Wales, Australia. (Courtesy of Sydney Observatory, part of the Powerhouse Museum)

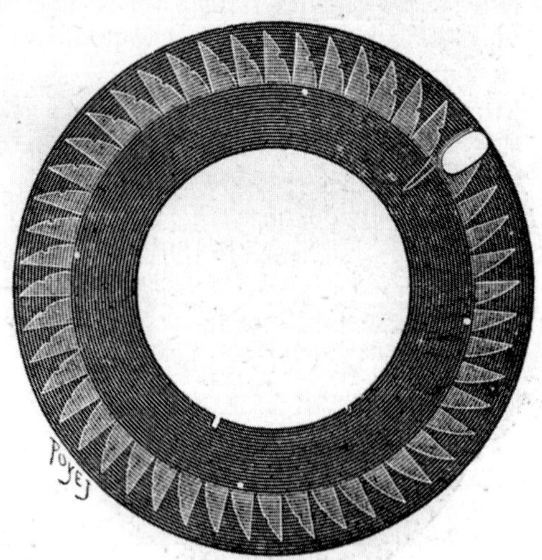

Figure 7. Engraving made in the 19th century from one of Janssen's tests. (Courtesy of Françoise Launay, Observatoire de Paris, Section de Meudon)

The 1999 Mercury transit was widely observed, including high-resolution observations from the Swedish Solar Telescope on La Palma, Canary Islands, Spain (Fig. 9); New Jersey Institute of Technology's Big Bear Solar Observatory; the University of Hawaii's Mees Solar Observatory; and the National Solar Observatory's Global Oscillation Network Group's telescopes around the world (GONG).

We used the white-light $(0.2 - 1.0\,\mu\text{m})$ imagery from TRACE, though various ultraviolet/extreme ultraviolet (UV/EUV) data also exist. TRACE provides steady seeing with a temporally stable point-spread function of $1''$ resolution, sampled with $0.''5$ pixels.

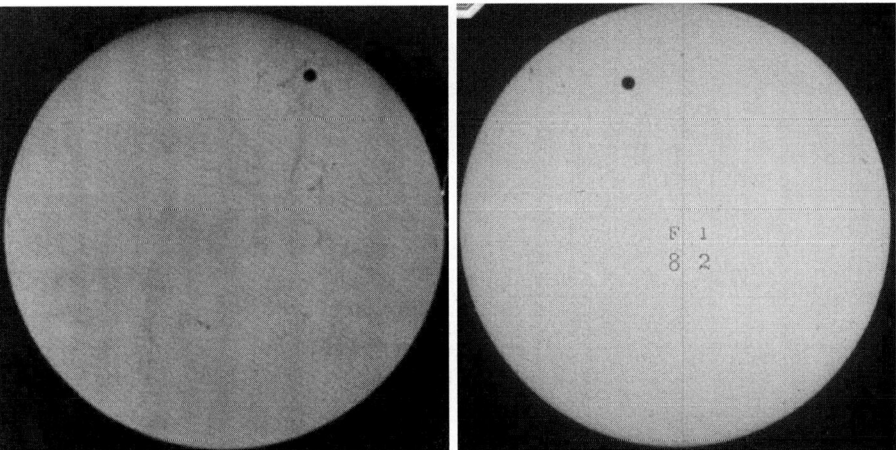

Figure 8. (left) The transit of Venus photographed at Vassar College by Prof. Maria Mitchell and her students. (Courtesy of Special Collections, Vassar College Library). (right) The transit of Venus from one of the U.S. Naval Observatory expeditions, with a superimposed grid to aid in data reduction. (U.S. Naval Observatory)

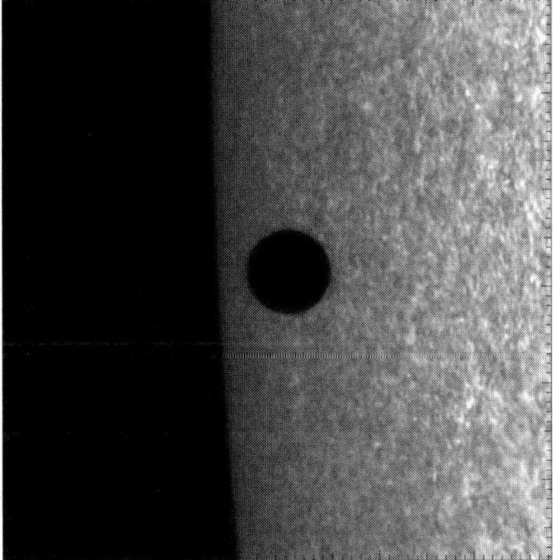

Figure 9. The 1999 transit of Mercury, photographed from the Swedish Solar Telescope. (Royal Swedish Academy of Sciences)

The results show a slight black-drop effect at ingress (Fig. 10); egress was not observed. Wittmann (1974) had shown that a black drop for a Mercury transit could be the result of blurring in the Earth's atmosphere, but we were observing from space.

We calibrated the CCD data in standard ways, and then applied additional corrections to remove pattern noise and other low-level effects (Fig. 11, left). The result clearly showed the black-drop effect (Fig. 11, right). Extensive modeling revealed that the point-spread function was not sufficient to explain the observed form of the black drop, and that the effect of the solar limb darkening had to be included (Fig. 12). Though the possibility of such a contribution had been alluded to by others (Sveshnikov & Sveshnikov 1996), in a conference proceedings of which we were unaware, we provided the first numerical model of observational data proving that the contribution indeed existed. Removing the two contributions left a circularly symmetric silhouette for Mercury, indicating that all causes of a measurable black-drop effect were accounted for (Fig. 13). We discussed our

Figure 10. Series showing the evolution of the black-drop effect at ingress during the 1999 transit of Mercury, observed from NASA's Transition Region and Explorer (TRACE) spacecraft. (Schneider, Pasachoff, and Golub/LMSAL and SAO/NASA)

results at meetings of the Division of Planetary Sciences of the American Astronomical Society (Schneider, Pasachoff & Golub 2001), the General Assembly of the International Astronomical Union (Schneider, Pasachoff, & Golub 2003), and the History of Astronomy Division of the American Astronomical Society (Pasachoff, Schneider, & Golub 2004). Our refereed paper on the subject appeared in *Icarus* (Schneider, Pasachoff, & Golub 2004b).

Data acquired by TRACE from the 2003 transit of Mercury were downlinked in compressed format that did not allow us to carry out the type of data reduction we had previously made.

3. The 2004 transit of Venus

We observed the 8 June 2004 transit of Venus with the TRACE satellite, following a detailed coordinated observing plan developed in collaboration with the TRACE team. The plan included the timing and cadence of the observations, the filter interleaving, the pointing of the spacecraft, and the data transmission method. To obtain ground-based observations for comparison, we observed from the Aristotelian University of Thessaloniki, Greece (http://www.astro.auth.gr), using especially their 20-cm f/15 refractor as well as a host of other telescopes and cameras. The whole duration of the transit, including both black drops, was to be visible from both the spacecraft and Greece. We also arranged for a colleague, Steven Souza, to observe third and fourth contacts with our Carroll spar 12.5-cm f/14 solar refractor at Williams College, Williamstown, Massachusetts (http://www.williams.edu/astronomy).

Two weeks prior to the transit, one of us had observed the solar chromosphere with the Swedish Solar Telescope of the Royal Swedish Academy of Sciences. This 1-m telescope is

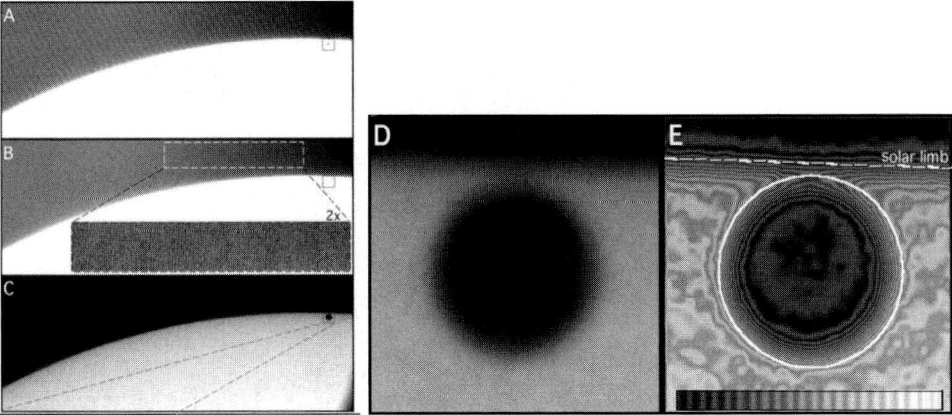

Figure 11. (left) Modeling of effects to improve the data quality of TRACE images from the 1999 transit of Mercury, observed from NASA's Transition Region and Explorer (TRACE) spacecraft. (right) The black-drop effect on a TRACE image from the 1999 transit of Mercury, observed from NASA's Transition Region and Explorer (TRACE) spacecraft. It is shown both as an image and as isophotes. (Schneider, Pasachoff, and Golub/LMSAL and SAO/NASA)

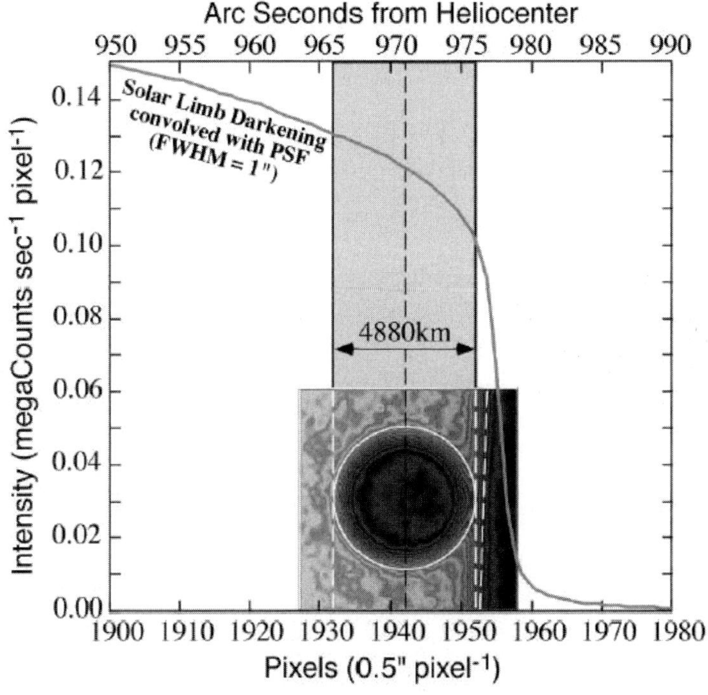

Figure 12. Modeling the contributions of the telescope's point-spread function and of solar limb darkening on TRACE images from the 1999 transit of Mercury, observed from NASA's Transition Region and Explorer (TRACE) spacecraft. (Schneider, Pasachoff, and Golub/LMSAL and SAO/NASA)

located on La Palma. We grew to understand the workings, the variability of the seeing quality – which at its best is unsurpassed in the world – and the need for a discrete, dark object in order to make the adaptive optics function consistently. Adaptive optics, made possible with their Wave Front Sensor, was achieved some of the time during the transit.

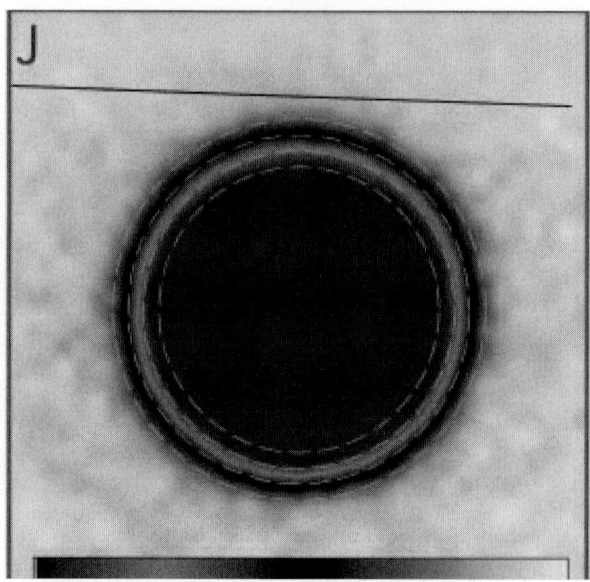

Figure 13. A TRACE image from the 1999 transit of Mercury, observed from NASA's Transition Region and Explorer (TRACE) spacecraft. The effect of the telescope's point-spread function and the solar limb darkening have been removed, revealing a symmetric Mercury silhouette. The position of the solar limb is marked. (Schneider, Pasachoff, and Golub/LMSAL and SAO/NASA)

One of their images from the transit of Venus appears in Fig. 14. Dan Kisselman of the Royal Swedish Academy of Sweden supervised the transit observations, and arranged for a new spectrograph to be installed and to operate.

Figure 14. Venus just after second contact in a ground-based image taken with the 1-m Swedish Solar Telescope on La Palma. The solar limb is at the top; the cicular arcs merely mark the edge of the field-of-view. (Royal Swedish Academy of Sciences)

We are very pleased with the series of transit of Venus observations obtained to our specifications with the TRACE satellite. Our data reduction significantly improves upon the image contrasts obtained with standard image processing (e.g., as shown on the TRACE website) by modeling and removing instrumentally ghosted and scattered light. We were able to follow the evolution of visibility of Venus's atmosphere at both ingress (Fig. 15, left) and egress with views up to a full Venus diameter from the solar limb. Presentation of the observations with a different stretch shows images more suitable for our study of the black-drop effect (Fig. 15, right). Our continuing data reduction appears on the Results page accessible from http://www.transitofvenus.info. These data and other images from our expedition are accessible through

http://www.williams.edu/astronomy/eclipse

One of our preliminary isophotal images of the Venus transit, with a limb darkening function removed, appears as Fig. 16.

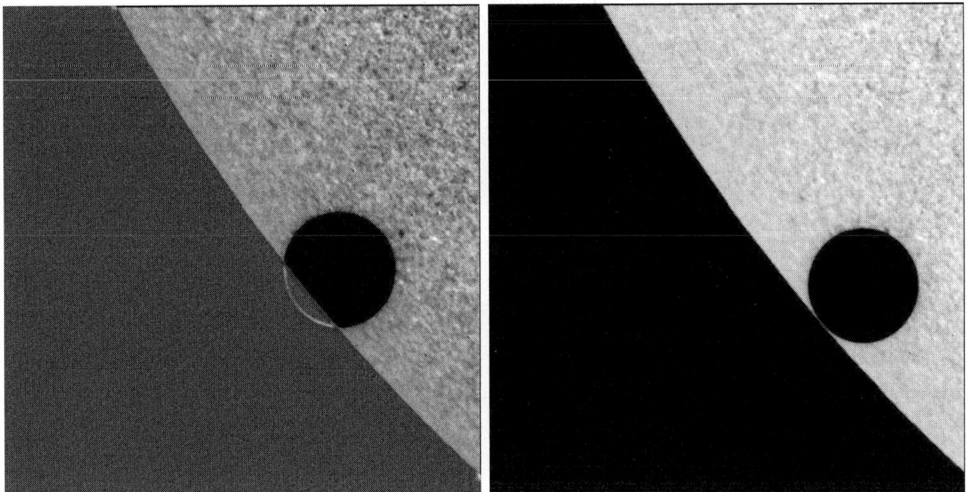

Figure 15. (left) Venus's atmosphere, seen between first and second contacts on 8 June 2004. (right) Venus in transit, seen just after second contact on 8 June 2004; from a series of images processed by us from TRACE data. (Schneider/Pasachoff/LMSAL/SAO/NASA)

Reports from observers all over the world indicate that the black-drop effect was less prominent than had been expected, though such comments had also appeared in reports of the 19th-century transits (Launay & Hingley 2005). Since prior knowledge showed that the black-drop effect was not intrinsic to Venus but was rather a combination of instrumental effects and effects to some degree in the atmospheres of Earth, Venus, and Sun, it is no surprise that the better telescopes we have now, since the time of Captain Cook, provide cleaner images and potentially less black drop. Some images posted on the Web, presumably with telescopes of lesser optical quality and/or worse seeing, indeed show the typical shape of the black drop. Some of our data from the 20-cm ground-based telescope in Greece clearly show the existence of a black-drop effect (Fig. 17), and we are making animations to follow its evolution. We can also follow the formation and evolution of a black-drop effect with scanned photographic observations obtained with a 1600-mm Nikon lens on slide film. We anticipate detailed data reduction of the CCD and photographic data.

Acknowledgements

The 2004 transit of Venus expedition was supported by a grant from the Committee for Research and Exploration of the National Geographic Society. We also acknowledge

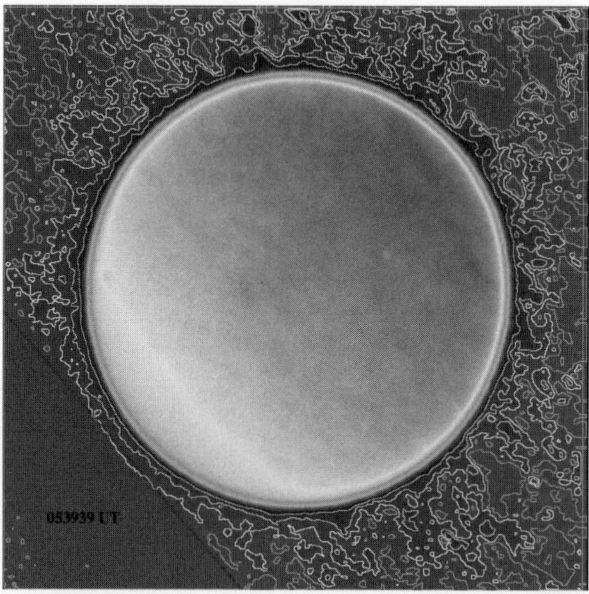

Figure 16. Isophotes on an image processed by us from TRACE data with a limb-darkening function removed. (Schneider/Pasachoff/LMSAL/SAO/NASA)

Figure 17. Images taken by the Williams College Expedition with the Aristotelian University of Thessaloniki's 20-cm refractor and an Apogee CCD, operated by Bryce Babcock, processed by Kayla Gaydosh with the assistance of Steven Souza.

the support for Joseph Gangestad's participation from Sigma Xi, the Scientific Research Society. The previous work on the transits of Mercury was supported in part by NSF grant ATM-000545; Pasachoff's work on eclipses has also been supported in part by the Committee for Research and Exploration of the National Geographic Society. His solar TRACE work is supported by NASA Guest Investigator grant NNG04GF99G. TRACE is supported by a NASA/GSFC contract to the Lockheed Martin Corp. Extensive use of analysis software developed by the NICMOS IDT under NASA grant NAG 5-3042 is acknowledged.

We thank especially Karel Schrijver at Lockheed Martin Solar and Astrophysics Laboratory and his colleagues there, and Edward DeLuca at the Harvard-Smithsonian Center for Astrophysics, for their collaboration in obtaining the transit of Venus observations with data recording and transmission to our specifications and requests.

References

Bergman, T., 1761, "An Account of the Observations Made in the Same Transit at Upsal in Sweden," *Phil. Trans.* **52**, 227-230.

Cook, James, 1768–71, "Endeavour: Captain Cook's Journal 1768-1771," http://endeavour. 8k.com/

Cook, J., & Green, C., 1771, "Observations made, by appointment of the Royal Society, at King George's Island in the South Sea; by Mr. Charles Green, formerly assistant at the Royal Observatory at Greenwich, and Lieut. James Cook, of His Majesty's Ship the Endeavour," *Phil Trans.* **61**, 397-421.

Flammarion, C., 1875. "Le passage de Vénus, Résultat des Expéditions Françaises," *La Nature,* May 8, 1875, 356-358.

Halley, E., 1716, "Edmond Halley's Famous Exhortation of 1716," in D. Sellers, *The Transit of Venus: The Quest to Find the True Distance of the Sun* (MagaVelda Press, 2001), pp. 214-217.

Launay, F., & Hingley, P. D., 2005, "Jules Janssen's 'Revolver Photographique' and its British Derivative, 'The Janssen Slide,' " *J. Hist. Astron.* **xxxvi**, 55-78.

Meadows, A. J., 1974, "The transit of Venus in 1874," *Nature* **250**, 749-752.

Pasachoff, J. M., 2000, *A Field Guide to the Stars and Planets* (Houghton Mifflin)

Pasachoff, J. M., 2002, *Astronomy: From the Earth to the Universe*, 6th ed. Brooks/Cole)

Pasachoff, J. M., 2003a, *The Complete Idiot's Guide to the Sun* (Alpha Books)

Pasachoff, J. M., 2003b, book reviews of Eli Maor, *Venus in Transit*; David Sellers, *The Transit of Venus: The Quest to find the True Distance of the Sun*; Michael Maunder and Patrick Moore, *Transit: When Planets Cross the Sun*; Donald Fernie: *Setting Sail for the Universe: Astronomers and Their Discoveries*; and Alex Soojung-Kim Pang, *Empire and the Sun: Victorian Solar Eclipse Expeditions*, in *The Key Reporter* (spring 2003), pp. 15, 17; http://www.pbk.org.

Pasachoff, J. M., 2004, book review of William Sheehan and John Westfall, *The Transits of Venus*, in *The Key Reporter* (summer 2004); http://www.pbk.org.

Pasachoff, J. M., & Filippenko, A., 2004, *The Cosmos: Astronomy in the New Millennium*, 2nd ed. (Brooks/Cole)

Pasachoff, J. M., Schneider, G., & Golub, L., 2004, "Explanation of the Black-Drop Effect at Transits of Mercury and the Forthcoming Transit of Venus," AAS Atlanta, January 2004, Special Session on the Transit of Venus.

Schaefer, B. E., 2001. "The transit of Venus and the notorious black drop effect," *J. History of Astronomy* **xxxii**, 325-336.

Schneider, G., Pasachoff, J. M., & Golub, L. 2001. "TRACE Observations of the 15 November 1999 Transit of Mercury," *Bull. Am. Astron. Soc.* **33**, 1037.

Schneider, G., Pasachoff, J. M., & Golub, L., 2003, "Space Studies of the Black Drop Effect at a Mercury Transit," IAU01082, presented at the Special Session on Mercury, Sydney, #1204, p. 156.

Schneider, G., Pasachoff, J. M., & Golub, L., 2004, "TRACE Observations of the 15 November 1999 Transit of Mercury and the Black Drop Effect for the 2004 Transit of Venus," *Icarus* **168** (April), 249-256.

Sveshnikov, M. L., & Sveshnikov, A. M., 1996, "The 'Black Drop' Phenomenon of Mercury Transits," in *Third International Workshop on Positional Astronomy and Celestial Mechanics*, 111-118 (Astronomical Observatory, University of Valencia).

Wittmann, A., 1974, "Numerical Simulation of the Mercury Transit Black Drop Phenomenon," *Astron. & Astrophys.* **31**, 239-243.

Transits of Venus: New Views of the Solar System and Galaxy
Proceedings IAU Colloquium No. 196, 2004
D.W. Kurtz, ed.
© 2004 International Astronomical Union
doi:10.1017/S1743921305001432

Orbit of a lunar artificial satellite: Analytical theory of perturbations

B. De Saedeleer and J. Henrard

Department of Mathematics, FUNDP, Namur, B-5000, Belgium
email: bernard.desaedeleer@fundp.ac.be, jacques.henrard@fundp.ac.be

Abstract. We are currently developing an analytical theory of an artificial satellite of the Moon. It is an interesting problem because the dynamics of a lunar orbiter is quite different from that of an artificial satellite of the Earth, by at least two aspects: the J_2 lunar gravity term is only $1/10$ of the C_{22} term and the third body effect of the Earth on the lunar satellite is much larger than the effect of the Moon on a terrestrial satellite. So we have to account at least for these larger perturbations. We use here the method of the Lie Transform as perturbation method. The Hamiltonian of the problem is first averaged over the fast angle, in canonical variables. The solution is developed in powers of the small factors linked to n_{C}, J_2, C_{22} and to the Earth's position. The Earth location is determined by the lunar theory ELP2000 (Chapront-Touzé & Chapront 1991) from which we take the leading terms. Series developments are made with our home-made Algebraic Manipulator, the MM (standing for "Moon's series Manipulator"). The results are obtained in a closed form, without any series developments in eccentricity or inclination. So the solution applies for a wide range of values, except for few isolated critical values. We Achieved, among others, second order results for the combined effect of J_2 and C_{22}. As a side result, we were able to check the second order generator $\mathcal{W}_2$ given by Kozai for the effect of the J_2 term on an artificial satellite.

1. Introduction

The case study of a satellite around the Moon is quite different from the one around the Earth on several aspects. First of all, the moon is a slowly rotating body and has no dense atmosphere. Secondly, as it is well known, the lunar gravity field is far from being central, nor does it exhibit any strong symmetry of revolution; see e.g. Konopliv (2001) for a recent model in spherical harmonics. The order of magnitude of the second order coefficients for the Earth (Kaula 1966) and the Moon (Bills & Ferrari 1980) is given in the Table 1. The Moon is much less flattened than the Earth, which makes the C_{22} coefficient to come closer to J_2 (at 1 order of magnitude instead of 3 in the case of the Earth); so it needs to be considered. Moreover, the effect of the Earth on the lunar satellite is much larger than the effect of the Moon on a terrestrial satellite; so the former effect is mixed to the effects of the shape of the lunar gravity field. In this paper, we will focus on the combined effect of the perturbations J_2 and C_{22}, and introduce the effect of the Earth considered as a third body.

Table 1. Some orders of magnitude for J_2 and C_{22}

	$C_{20} \equiv -J_2$	C_{22}
Earth $\oplus$	-10^{-3}	2.10^{-6}
Moon $\mathbb{C}$	-2.10^{-4}	2.10^{-5}

The following assumptions have been made: the orbit of the Moon is circular (we neglect $e_{\mathbb{C}} \approx 0.055$); the motion of the Moon is uniform (librations are neglected); the lunar equator lies in the ecliptic (we neglect the inclination of the lunar equator to the ecliptic of 1.5°, and the inclination of the lunar orbit to the ecliptic of about 5°); the perturbation of the Sun is negligible; and the longitude of the lunar longest meridian λ_{22} is equal to the longitude of the Earth $\lambda_{\oplus}$ (librations are neglected). This last assumption is one of the well known Cassini's laws (Cook 1988), stating that the Moon is in synchronous rotation: she rotates about her axis perpendicular to the plane of her orbit at an angular velocity $\gamma_{\mathbb{C}}$ that is equal to her mean angular velocity in her orbit $n_{\mathbb{C}}$; that is to say $\gamma_{\mathbb{C}} = n_{\mathbb{C}} = 2\pi/T_{\mathbb{C}}$ with the sidereal rotation period of the Moon being $T_{\mathbb{C}} = 27.321\,661\,5$ solar days.

2. Partial perturbative Hamiltonians

We work within the frame of the Hamiltonian formalism and use the classical Delaunay canonical variables $(q_i, p_i) = (l, g, h, L, G, H)$. The term $\frac{1}{2}v^2 - \frac{\mu}{r}$ is then simply written $\mathcal{H}_0^{(0)} = -\frac{\mu^2}{2L^2}$ (the unperturbed potential); next one has to develop the perturbations.

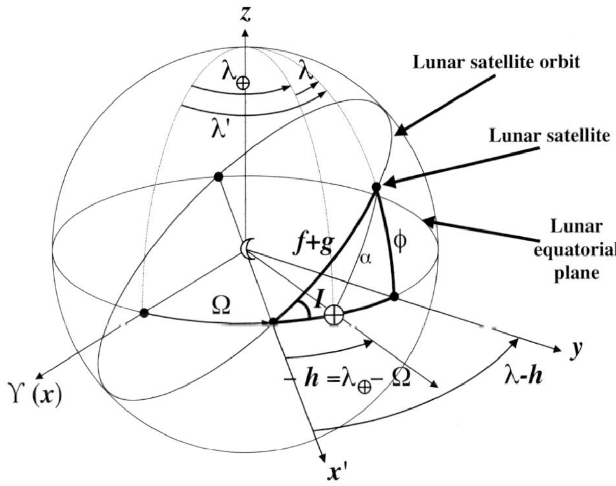

Figure 1. Simplified selenocentric sphere.

We define an inertial frame (x, y, z) as follows (see Fig. 1): the origin is taken at the center of the Moon; the x direction is the one of the first point of Aries Υ, the y is the direction normal to x and contained in the lunar equatorial plane containing x, and the z direction is the right-handed normal to (x, y). We define also the spherical coordinates (r, λ', ϕ). The zonal perturbation in J_2 is defined as usual by $\epsilon\,\mathcal{H}_2^{(0)} = \epsilon\,\frac{\mu}{r^3}P_{20}(\sin\phi)$ where we use $\epsilon = J_2 R^2$ and the Legendre Associated Functions P_{nm}. The argument $(\sin\phi)$ may partially be translated into Delaunay variables by way of spherical trigonometry (see Fig. 1, where the plane of the orbit is at an inclination I): $\sin\phi = \sin I \sin(f + g)$.

The sectorial perturbation in C_{22} requires us to define λ_{22} as the longitude of the lunar longest meridian (minimum inertia), which is $\lambda_{22} = \lambda_{\oplus}$. Since this angle rotates at the rate of the synchronous rotation $\dot{\lambda}_{\oplus} = n_{\mathbb{C}}$, we preferably introduce a rotating frame by defining $\lambda = \lambda' - \lambda_{\oplus}$ and also $h = \Omega - \lambda_{\oplus}$. A new term must then be added to the Hamiltonian in order to have $\dot{h} = \partial\mathcal{H}/\partial H = -n_{\mathbb{C}}$; we call this term $\mathcal{H}_1^{(0)} = -n_{\mathbb{C}}H$. Now the sectorial perturbation may be written as $\delta\,\mathcal{HB}_2^{(0)} = \delta\,\mu r^{-3}P_{22}(\sin\phi)\cos(2\lambda)$, where we define $\delta = -C_{22}R^2$. We then use again some spherical trigonometry to switch into the canonical variables and we introduce the useful shortcuts $(s, c) = (\sin I, \cos I)$.

And last, we express the perturbation of the Earth as usual for a third body by $\gamma\, \mathcal{HE}_2^{(0)} = \gamma\, a_\oplus^3 r_\oplus^{-3} r^2 P_{20}(\cos\psi)$, ψ being the angle between the Earth and the satellite, and with $\gamma = -\mu_\oplus a_\oplus^{-3}$. In summary, we will write in our case:

$$\mathcal{H}^{(0)} = \mathcal{H}_0^{(0)} + \mathcal{H}_1^{(0)} + \epsilon\, \mathcal{H}_2^{(0)} + \delta\, \mathcal{HB}_2^{(0)} + \gamma\, \mathcal{HE}_2^{(0)} \tag{2.1}$$

along with the definitions:

$$\mathcal{H}_1^{(0)} = -n_{\mathfrak{C}} H \tag{2.2}$$

$$\mathcal{H}_2^{(0)} = \frac{\mu}{4r^3}\left(1 - 3c^2 - 3s^2\cos(2f + 2g)\right) \tag{2.3}$$

$$\mathcal{HB}_2^{(0)} = \frac{3\mu}{4r^3}\Big\{2s^2\cos(2h) + (c+1)^2\cos(2f + 2g + 2h)$$

$$+ (c-1)^2\cos(2f + 2g - 2h)\Big\} \tag{2.4}$$

$$\mathcal{HE}_2^{(0)} = a_\oplus^3 r_\oplus^{-3} r^2 P_{20}(\cos\psi) \tag{2.5}$$

Note that we directly choose to class the perturbations by their order of magnitude: $n_{\mathfrak{C}}$ is first order, while the others are second order.

We come back to the choice of the variables now. There remains the variable r and f to be expressed as a function of (l, g, h) in order to be able to apply a canonical perturbation method. It turns out that the functions $r = r(l, g, h)$ and $f = f(l, g, h)$ cannot be expressed in a closed form; so we prefer to use the following set of auxiliary variables $(\xi, f, g, h, a, n, e, \eta, s, c)$, which is closed and allow high eccentricities:

$$
\begin{array}{llll}
\xi &=& \frac{a}{r} = \frac{1+e\cos f}{1-e^2} = \frac{1}{1-e\cos E} & \qquad f \\[2mm]
a &=& \frac{L^2}{\mu} & \qquad n = \frac{\mu^2}{L^3} \\[2mm]
e &=& \sqrt{1 - \left(\frac{G}{L}\right)^2} & \qquad \eta = \sqrt{1 - e^2} = \frac{G}{L} \\[2mm]
s &=& \sin I = \sqrt{1 - \left(\frac{H}{G}\right)^2} & \qquad c = \cos I = \frac{H}{G} \\[2mm]
g & & & \qquad h
\end{array}
\tag{2.6}
$$

The only drawback of this set (2.6) is that it is redundant and that we need to perform partial derivatives of them with respect to the canonical variables (l, g, h, L, G, H); but it is not too burdensome; the result is given in the Table 2. We have for example:

$$
\begin{aligned}
\frac{dA}{dl} &= \frac{\partial A}{\partial l} + \frac{\partial A}{\partial \xi}\,\frac{\partial \xi}{\partial l} + \frac{\partial A}{\partial f}\,\frac{\partial f}{\partial l} \\[2mm]
&= \frac{\partial A}{\partial l} + \frac{\partial A}{\partial \xi}\left(\frac{-\xi^2 e\sin f}{\eta}\right) + \frac{\partial A}{\partial f}\left(\xi^2\eta\right)
\end{aligned}
\tag{2.7}
$$

Note that the quantity $\frac{\partial f}{\partial l} = \xi^2\eta$ plays an important role, since it will allow us to switch the integration from l to f. In this new set of variables (2.6), the factor μr^{-3} appearing in (2.3) and (2.4) may be written $\xi^3 n^2$.

3. Development of the third body perturbation

We come back to the expression (2.5) of the perturbation of the Earth $\mathcal{HE}$. The $\cos\psi$ may be computed with:

Table 2. Table of partial derivatives

	$\partial/\partial L$	$\partial/\partial G$	$\partial/\partial H$	$\partial/\partial l$
ξ	$\dfrac{\xi^2\eta^2}{na^2e}\cos f$	$-\dfrac{\xi^2\eta}{na^2e}\cos f$	0	$-\dfrac{\xi^2 e}{\eta}\sin f$
a	$\dfrac{2}{an}$	0	0	0
n	$-\dfrac{3}{a^2}$	0	0	0
s	0	$\dfrac{c^2}{na^2\eta s}$	$-\dfrac{c}{na^2\eta s}$	0
c	0	$-\dfrac{c}{na^2\eta}$	$\dfrac{1}{na^2\eta}$	0
e	$\dfrac{\eta^2}{na^2e}$	$-\dfrac{\eta}{na^2e}$	0	0
η	$-\dfrac{\eta}{na^2}$	$\dfrac{1}{na^2}$	0	0
f	$\dfrac{1+\xi\eta^2}{na^2e}\sin f$	$-\dfrac{1+\xi\eta^2}{\eta na^2e}\sin f$	0	$\xi^2\eta$

$$\cos\psi = \frac{\vec{r}.\vec{r}_\oplus}{rr_\oplus} = A_\oplus(\cos h \cos(f+g) - c\sin h \sin(f+g))$$
$$+ B_\oplus(\sin h \cos(f+g) + c\cos h \sin(f+g)) + C_\oplus s\sin(f+g) \quad (3.1)$$

where $\vec{A}_\oplus = (A_\oplus, B_\oplus, C_\oplus)$ is the direction of the Earth from the Moon. We now use the lunar theory ELP2000 (Chapront-Touzé & Chapront 1991), which gives the position of the Moon with respect to the Earth, in the spherical coordinates (V, U, R), where V is the geocentric longitude (U and V are referred to the mean dynamical ecliptic and mean equinox of date).

The position of the Moon is described by a series of periodic functions mainly of the fundamental arguments D, l', l, F, but also of the arguments of the other planets Me, Ve, Te, Ma, Ju, Sa. All these arguments are taken as linear function of time. We recall that D is the secular part (nonperiodic part) of the difference between the mean longitude of the Moon and the geocentric mean longitude of the Sun, l' is the secular part of the geocentric mean anomaly of the Sun, l is the secular part of the mean anomaly of the Moon, and F is the secular part of the difference between the mean longitude of the Moon and the longitude of its ascending node on the mean ecliptic of date.

The authors of ELP2000 give the number of terms to take into account for each series in order to achieve several levels of precision; we follow the low precision recommendations, since we do not need a very long term accuracy: the duration of a lunar mission is typically much less than a century. So we will take only some leading terms and use the following simplified expressions for U and V:

$$V = L + \alpha \quad \text{and} \quad U = \beta \qquad \text{with} \qquad \alpha = S_V + 10^{-3}S'_V \quad \text{and} \quad \beta = S_U \quad (3.2)$$

and with

$$S_V = \sum_{n=1}^{N_V = 29} v_n \sin(i_{1,n}l + i_{2,n}D + i_{3,n}l' + i_{4,n}F)$$

$$S'_V = v'_1 \sin(18Ve - 16Te - l + 26.5426°) \tag{3.3}$$

$$S_U = \sum_{n=1}^{N_U = 14} v_n \sin(j_{1,n}l + j_{2,n}D + j_{3,n}l' + j_{4,n}F)$$

from which we can then deduce

$$
\begin{aligned}
A_\oplus &= -\cos U \cos V = -\cos\beta \,(\cos L \cos\alpha - \sin L \sin\alpha) \\
B_\oplus &= -\cos U \sin V = -\cos\beta \,(\sin L \cos\alpha + \cos L \sin\alpha) \\
C_\oplus &= -\sin U = -\sin\beta
\end{aligned}
\tag{3.4}
$$

Since the quantities α and β are small, they may be expanded into series developments like $\sin x \approx x - x^3/6$; the exact level of truncation must be determined. Now the series which gives $\cos\psi$ may be built, and we end up with about 32 044 terms (see Table 4). That series must further be put as argument into $P_{20}(x) = (3x^2 - 1)/2$, which generates even more terms.

4. Perturbation Method using several parameters

We use here the Lie Transform (Deprit 1969) as canonical perturbation method, with the parameter ϵ. The initial Hamiltonian (*input*) is written $\mathcal{H}^{(0)} = \sum_{i\geqslant 0} \frac{\epsilon^i}{i!}\mathcal{H}_i^{(0)}$; while the transformed Hamiltonian (*output*) is written $\mathcal{H}_0 = \sum_{i\geqslant 0} \frac{\epsilon^i}{i!}\mathcal{H}_0^{(i)}$. This transformation is symbolized by the Lie triangle, which may be adapted to the case (2.1):

$$
\begin{array}{llll}
\mathcal{H}_0^{(0)} & & & \\[4pt]
\mathcal{H}_1^{(0)} & \mathcal{H}_0^{(1)} & & \\[4pt]
\mathcal{H}_2^{(0)} & \mathcal{H}_1^{(1)} & \mathcal{H}_0^{(2)} & \\[4pt]
\mathcal{H}_3^{(0)} & \mathcal{H}_2^{(1)} & \mathcal{H}_1^{(2)} & \mathcal{H}_0^{(3)} \\[4pt]
\vdots & \vdots & \vdots & \vdots & \ddots
\end{array}
$$

$$
\begin{array}{llll}
\mathcal{H}_0^{(0)} = -\frac{\mu^2}{2L^2} & & & \\[4pt]
\mathcal{H}_1^{(0)} = -n_{\mathbb{C}}H & & \mathcal{H}_0^{(1)} & \\[4pt]
\epsilon\mathcal{H}_2^{(0)} + \delta\mathcal{H}B_2^{(0)} + \gamma\mathcal{H}\mathcal{E}_2^{(0)} & & \mathcal{H}_1^{(1)} & \mathcal{H}_0^{(2)} \\[4pt]
0 & & \mathcal{H}_2^{(1)} & \mathcal{H}_1^{(2)} & \mathcal{H}_0^{(3)} \\[4pt]
\vdots & & \vdots & \vdots & \vdots & \ddots
\end{array}
$$

Figure 2. The classical Lie triangle and our specific Lie triangle.

The triangle is filled by the way of the following recursive formula, where the $\mathcal{W}_k$ are the generating functions and $(A; B)$ is the Poisson parenthesis:

$$\mathcal{H}_i^{(j)} = \mathcal{H}_{i+1}^{(j-1)} + \sum_{k=0}^{i} C_i^k \left(\mathcal{H}_{i-k}^{(j-1)}; \mathcal{W}_{k+1} \right) \tag{4.1}$$

$$(A; B) = \sum_i \left(\frac{\partial A}{\partial q_i} \frac{\partial B}{\partial p_i} - \frac{\partial B}{\partial q_i} \frac{\partial A}{\partial p_i} \right) \tag{4.2}$$

We may write $\mathcal{H}_0^{(i)}$ as $\overline{\mathcal{H}}_0^{(i)}$ in order to remember that the fast angle l has been eliminated; we always put the periodic part in the generator $\mathcal{W}_i$.

For the first order: as $\mathcal{H}_1^{(0)}$ is already independent of l here, we have $\mathcal{H}_0^{(1)} = \mathcal{H}_1^{(0)} = -n_{\mathfrak{q}}H$ and $\mathcal{W}_1 = 0$. For the second order, we have:

$$\epsilon\,\mathcal{H}_0^{(2)} + \delta\,\mathcal{HB}_0^{(2)} + \gamma\,\mathcal{HE}_0^{(2)} = \epsilon\,\mathcal{H}_2^{(0)} + \delta\,\mathcal{HB}_2^{(0)} + \gamma\,\mathcal{HE}_2^{(0)} + \left(\mathcal{H}_0^{(0)};\mathcal{W}_2\right) \qquad (4.3)$$

and we choose:

$$\epsilon\,\overline{\mathcal{H}}_0^{(2)} + \delta\,\overline{\mathcal{HB}}_0^{(2)} + \gamma\,\overline{\mathcal{HE}}_0^{(2)} = \frac{1}{2\pi}\int_0^{2\pi}\left(\epsilon\,\mathcal{H}_2^{(0)} + \delta\,\mathcal{HB}_2^{(0)} + \gamma\,\mathcal{HE}_2^{(0)}\right)dl = \ldots \qquad (4.4)$$

while $\left(\mathcal{H}_0^{(0)};\mathcal{W}_2\right)$ reduces to $n\frac{\partial \mathcal{W}_2}{\partial l}$ which has then to be integrated with respect to l.

Higher orders may be achieved by the same way, provided we are able to compute the integrals. Some tricks are given in the literature, such as Aksnes (1971). The terms containing the factor $(f - l)$ receive a special treatment, as $(f - l)$ is indeed well known to play an important role in the problem of the artificial satellite (see Metris 1991). The third order will contain the combinations of perturbation parameters $(\epsilon n_{\mathfrak{q}}, \delta n_{\mathfrak{q}}, \gamma n_{\mathfrak{q}})$ and the fourth order will contain $(\epsilon^2, \delta^2, \gamma^2, \epsilon\delta, \epsilon\gamma, \delta\gamma, \epsilon n_{\mathfrak{q}}^2, \delta n_{\mathfrak{q}}^2, \gamma n_{\mathfrak{q}}^2)$. We can easily select an isolated effect by putting the other parameters to zero.

5. Results obtained by the symbolic manipulation software MM

We used a specific FORTRAN code called the MM, standing for "Moon's series Manipulator", which has been developed at our University. In this tool, each expression is given by a series of linear trigonometric functions, with polynomial coefficients. The concern to keep linear expression is motivated by the fact that we want to keep easy integrations. An example of such a series is given in the Table 3.

It is of course impossible to give all the results here, since the series may contain a lot of terms, but we give explicitly some of them here, and we comment the others. For example, we retrieve the known results for the first order effect of J_2 only. First, the expression of $\epsilon\overline{\mathcal{H}}_0^{(2)}$ is the same as the one given by $(-F_1^*)$ defined in (13) of Brouwer (1959), when setting $2k_2 = \epsilon$. Secondly, the expression of $\epsilon\mathcal{W}_2$ is also the same as the one given by $(-S_1)$ defined in (15) of Brouwer (1959). And thirdly, when writing the first order averaged equations of motion, we retrieve the two classical formulae giving the effect of J_2 on g and h (Szebehely 1989; Roy 1968; Jupp 1988). The associated peculiar value of the inclination which makes $\bar{g}$ vanish, known as the *critical inclination* $I_c = 63°26'$, is quite famous (Szebehely 1989).

Some partial results (the terms in $\epsilon^2, \epsilon\delta$ and δ^2) of the computation of the fourth-order averaged Hamiltonian are given explicitly in the Table 3; the series has been multiplied by a factor $\mathcal{F} = \frac{64a^2\eta^7}{3n^2}$ to make it more readable, and the variable n_δ represent the exponent in the perturbative parameter δ: for an expression of order $p+2$, we then have terms containing factors like $\epsilon^{p-n_\delta}\delta^{n_\delta}$. We retrieve also some known results for the effect of J_2^2 only: the expression of $\epsilon^2\overline{\mathcal{HX}}_0^{(4)}$ is the same as the one given by $(-2F_2^*)$ defined in (29) of Brouwer (1959). Moreover, the generator in ϵ^2 has been computed and validated: the exact equivalence with Kozai's S_2 (given by equation (3.2) of Kozai 1962) has been established elsewhere, using the relationships of Shniad (1970) for the correspondence between generators of von Zeipel (S_i) and the ones of Lie $(\mathcal{W}_i)$.

Table 3. The series $(\epsilon^2 \overline{\mathcal{HX}}_0^{(4)} + \epsilon\delta\, \overline{\mathcal{HY}}_0^{(4)} + \delta^2 \overline{\mathcal{HZ}}_0^{(4)}) \times \mathcal{F}$

	f	g	h	ξ	a	n	e	η	c	s	$(f-l)$	n_δ	coefficient
$\cos$	0	0	0	0	0	0	0	0	2	0	0	0	0.800×10^1
$\cos$	0	0	0	0	0	0	0	0	4	0	0	0	-0.400×10^2
$\cos$	0	0	0	0	0	0	0	1	0	0	0	0	-0.400×10^1
$\cos$	0	0	0	0	0	0	0	1	2	0	0	0	0.240×10^2
$\cos$	0	0	0	0	0	0	0	1	4	0	0	0	-0.360×10^2
$\cos$	0	0	0	0	0	0	2	0	0	0	0	0	0.500×10^1
$\cos$	0	0	0	0	0	0	2	0	2	0	0	0	-0.180×10^2
$\cos$	0	0	0	0	0	0	2	0	4	0	0	0	0.500×10^1
$\cos$	0	2	0	0	0	0	2	0	0	2	0	0	-0.200×10^1
$\cos$	0	2	0	0	0	0	2	0	2	2	0	0	0.300×10^2
$\cos$	0	0	2	0	0	0	0	0	0	2	0	1	-0.128×10^3
$\cos$	0	0	2	0	0	0	0	0	2	2	0	1	0.160×10^3
$\cos$	0	0	2	0	0	0	0	1	0	2	0	1	-0.480×10^2
$\cos$	0	0	2	0	0	0	0	1	2	2	0	1	0.144×10^3
$\cos$	0	0	2	0	0	0	2	0	0	2	0	1	-0.520×10^2
$\cos$	0	0	2	0	0	0	2	0	2	2	0	1	-0.200×10^2
$\cos$	0	2	2	0	0	0	2	0	0	0	0	1	-0.400×10^1
$\cos$	0	2	2	0	0	0	2	0	1	0	0	1	0.520×10^2
$\cos$	0	2	2	0	0	0	2	0	2	0	0	1	0.560×10^2
$\cos$	0	2	2	0	0	0	2	0	3	0	0	1	-0.600×10^2
$\cos$	0	2	2	0	0	0	2	0	4	0	0	1	-0.600×10^2
$\cos$	0	2	-2	0	0	0	2	0	0	0	0	1	-0.400×10^1
$\cos$	0	2	-2	0	0	0	2	0	1	0	0	1	-0.520×10^2
$\cos$	0	2	-2	0	0	0	2	0	2	0	0	1	0.560×10^2
$\cos$	0	2	-2	0	0	0	2	0	3	0	0	1	0.600×10^2
$\cos$	0	2	-2	0	0	0	2	0	4	0	0	1	-0.600×10^2
$\cos$	0	0	0	0	0	0	0	0	0	0	0	2	-0.176×10^3
$\cos$	0	0	0	0	0	0	0	0	2	0	0	2	0.384×10^3
$\cos$	0	0	0	0	0	0	0	0	4	0	0	2	-0.800×10^2
$\cos$	0	0	0	0	0	0	0	1	0	4	0	2	-0.720×10^2
$\cos$	0	0	0	0	0	0	2	0	0	0	0	2	-0.540×10^2
$\cos$	0	0	0	0	0	0	2	0	2	0	0	2	0.156×10^3
$\cos$	0	0	0	0	0	0	2	0	4	0	0	2	0.100×10^2
$\cos$	0	0	4	0	0	0	0	0	0	4	0	2	-0.800×10^2
$\cos$	0	0	4	0	0	0	0	1	0	4	0	2	-0.720×10^2
$\cos$	0	0	4	0	0	0	2	0	0	4	0	2	0.100×10^2
$\cos$	0	2	0	0	0	0	2	0	0	2	0	2	-0.360×10^2
$\cos$	0	2	0	0	0	0	2	0	2	2	0	2	0.600×10^2
$\cos$	0	2	4	0	0	0	2	0	0	2	0	2	0.300×10^2
$\cos$	0	2	4	0	0	0	2	0	1	2	0	2	0.600×10^2
$\cos$	0	2	4	0	0	0	2	0	2	2	0	2	0.300×10^2
$\cos$	0	2	-4	0	0	0	2	0	0	2	0	2	0.300×10^2
$\cos$	0	2	-4	0	0	0	2	0	1	2	0	2	-0.600×10^2
$\cos$	0	2	-4	0	0	0	2	0	2	2	0	2	0.300×10^2

All these checks and some symmetry considerations (especially by the way of the tracer η) give good confidence in the new results, which contain other effects than J_2; for example, the cross effect $\epsilon\delta$ may be rewritten in full as:

$$
\overline{\mathcal{HY}}_0^{(4)} \times \mathcal{F} = \left[
\begin{array}{rllll}
32 & & s^2(5c^2-4) & \cos(2h) \\
+48 & \eta & s^2(3c^2-1) & \cos(2h) \\
-4 & e^2 & s^2(5c^2+13) & \cos(2h) \\
-4 & e^2 & & (\cos(2g+2h)+\cos(2g-2h)) \\
+52 & e^2 & c & (\cos(2g+2h)-\cos(2g-2h)) \\
+56 & e^2 & c^2 & (\cos(2g+2h)+\cos(2g-2h)) \\
-60 & e^2 & c^3 & (\cos(2g+2h)-\cos(2g-2h)) \\
-60 & e^2 & c^4 & (\cos(2g+2h)+\cos(2g-2h))
\end{array}
\right]
\tag{5.1}
$$

A critical issue by making such computations is the rapid growth of the number of terms in the series (typically ten thousands of terms here). Needless to say, quite a big work of simplification has to be made, especially because of the redundancy $(\eta \leftrightarrow e)$ and $(c \leftrightarrow s)$ of the set (2.6). Among these simplifications are some series of the kind $\left(e^2\eta^{-7} - e^2\eta^{-9} + e^4\eta^{-9}\right)$ which in fact cancel to zero. Some substitutions like $(s^2 \to 1 - c^2)$ for example may cancel a lot of terms, while the reverse operation of combining some terms may also make the expressions more compact, all simplifications being made before. The third body perturbation series contains quite a lot of terms; the series expressing $\cos(\psi)$ only is already about $32\,044$ terms. We show in the Table 4 the extreme coefficients of the trigonometric variables and the extreme exponents of the polynomial variables which appear during the computation of $\cos(\psi)$.

Table 4. Extreme coefficients (or exponents) appearing in the computation of $\cos(\psi)$.

	f	g	h	L	D	l'	l	F	Ve	Te	c	s	*coefficient*
min	1	1	-1	-1	-12	-4	-9	-8	-54	-48	0	0	$0.20037083 \times 10^{-14}$
max	1	1	1	1	12	4	9	8	54	48	1	1	0.49740618×10^{0}

Numerical accuracy considerations are also required (we are working in double precision) and some terms must therefore be removed. We may also always reject during the computation the terms which are of lower magnitude than others, by a skilful choice.

6. Conclusions

We have achieved, among others, second order results for the combined effect of J_2 and C_{22}, and introduced the third body effect. The averaged Hamiltonians and the generators for the first and second order effects of J_2 have been validated by comparison with the results of Brouwer (1959) and Kozai (1962). More detailed results will be published in a forthcoming paper.

We could add even more perturbations (Sun, other lunar harmonics like C_{31}, librations, etc.) provided our symbolic manipulation software is able to handle them. We could also eliminate the next fastest angle (g or h) in order to obtain the secular motion. One future aim is to validate theoretically these results by numerical integrations of the several effects. An experimental validation could also be followed by a comparison with data of a real lunar satellite.

References

Aksnes, K. 1971, *Celest. Mech. & Dyn. Astr.* **4**, 119–121.
Bills, B., Ferrari, A. 1980, *J. Geophys. Res.* **85**, 1013–1025.
Brouwer, D. 1959, *Astron. J.* **64**, 378–397.
Chapront-Touzé, M., Chapront, J. 1991, *Lunar Tables and Programs 4000 BC to AD 8000.* Willmann-Bell.
Cook, A. 1988, *The Motion of the Moon.* IOP Publishing Ltd.
Deprit, A. 1969, *Celest. Mech. & Dyn. Astr.* **1**, 12–30.
Jupp, A. 1988, *Celest. Mech. & Dyn. Astr.* **43**, 127–138.
Kaula, W. 1966, *Theory of Satellite Geodesy.* Blaisdell Publishing Company.
Konopliv, A., Asmar, S., Carranza, E., Sjogren, W., Yuan, D. 2001, *Icarus* **150**, 1–18.
Kozai, Y. 1962, *Astron. J.* **67**, 446–461.
Metris, G. 1991, *Celest. Mech. & Dyn. Astr.* **52**, 79–84.
Roy, A. 1968, *Icarus* **9**, 82–132.
Shniad, H. 1970, *Celest. Mech. & Dyn. Astr.* **2**, 114–120.
Szebehely, V. 1989, *Adventures in Celestial Mechanics.* University of Texas Press.

Discussion

MIKHAIL MAROV: I am just wondering: You limited yourself only to the J2 harmonic. What about other even harmonics, say 4, say 6, and some others. The lunar field is quite well known at a very high order of expansion. Why did you neglect it?

BERNARD DE SAEDELEER: We can't always take it into account. In fact, I just submitted an article to *Celestial Mechanics* which computes the average Hamiltonian for the horizontal effect, even or odd. You can always introduce more and more terms, if you want. You just have to compute the corresponding Hamiltonian. It can be done.

MIKHAIL MAROV: But at the beginning you mentioned you also took into account the mascons. I guess at a level of J2 is hardly possible to account for the mascons.

BERNARD DE SAEDELEER: No. Of course, the J2 is just a one big feature of the potential. If you want really to have the resolution of the mascons ...

MIKHAIL MAROV: ...more subtle effects?

BERNARD DE SAEDELEER: Yes. If you want to have more precise effects, you go into more numerical analysis – it becomes more numerical. If you want to have a model 16×16, you have all those coefficients, and it becomes more and more complicated.

MIKHAIL MAROV: If you will limit yourself with such a theory, it will be hardly possible to predict accurately orbits – especially the low orbit lunar satellites.

BERNARD DE SAEDELEER: Yes, you can always refine it.

Transits of Venus: New Views of the Solar System and Galaxy
Proceedings IAU Colloquium No. 196, 2004
D.W. Kurtz, ed.

© 2004 International Astronomical Union
doi:10.1017/S1743921305001444

The Spin-orbit resonance of Mercury: a Hamiltonian approach

S. D'Hoedt and A. Lemaitre

Department of Mathematics, FUNDP, Namur, B-5000, Belgium
emails: sandrine.dhoedt@fundp.ac.be, anne.lemaitre@fundp.ac.be

Abstract. One of the main characteristics of Mercury is its 3:2 spin-orbit resonance, combined with a 1:1 resonance between the orbital node of its orbit and the angle describing the precession of the rotation axis, both measured on the ecliptic plane. We build an analytical model, using Hamiltonian formalism, that takes into account this phenomenon thanks to the introduction of three resonant variables and conjugated momenta. We calculate the equilibria corresponding to four different configurations, which means four completely different values of the (ecliptic) obliquity; in particular, we focus on the present (stable) situation of Mercury, and thanks to several canonical transformations, we obtain, near the equilibrium, three pairs of angle-action variables, and consequently, three basic frequencies. Let us note that the model is as simple as possible: the gravitational potential is limited to the second degree terms (the only ones for which a value can be presently given), and the orbit of Mercury is Keplerian. The numerical values obtained by our simplified model are validated by the coherence with existing complete numerical models.

1. Introduction

Motivated by the projects of space missions like BepiColombo† or MESSENGER‡, we have in mind to build an analytical resonant spin-orbit model for the *present* situation of Mercury. We use a very classical Hamiltonian formalism, starting with a kernel model, and adding step by step the different perturbations.

In this paper, we first present the kernel, in the form of a three-degrees-of-freedom Hamiltonian system, averaged over the short periods; it succeeds in describing the libration about the resonant spin-orbit 3:2 motion by three pairs of action-angle variables, obtained after a succession of simplifications and canonical changes of variables. The three basic frequencies of the system are in complete agreement with the very recent values obtained numerically by Rambaux & Bois (2003).

The basic hypotheses are the following: Mercury is a rigid non-spherical body, the gravity field is truncated after the second degree terms (the only ones for which we have significant numerical values), the orbital motion of Mercury is Keplerian. Let us remark that the obliquity is not put to zero and in the last part, the spin axis is not parallel to the third axis of inertia.

All the computations are performed with the software Mathematica with the parameters values taken in Anderson et al. (1987) and ESA-SCI (2000).

---

† Mission of the European Space Agency and ISAS, Japan's Institute of Space Astronautical Sciences
‡ MErcury Surface Space ENvironment, Geochemistry and Ranging, space mission of NASA

2. Coordinate Choices

We are going to work with four reference frames centered at Mercury's center of mass:

- (X_0, Y_0, Z_0) inertial frame with X_0 and Y_0 in the ecliptic plane (fixed at some epoch),
- (X_1, Y_1, Z_1) orbital frame with Z_1 perpendicular to the orbit plane,
- (X_2, Y_2, Z_2) with Z_2 pointing to the spin axis direction and X_2 directed along the ascending node of the equatorial plane on the ecliptic plane and
- (X_3, Y_3, Z_3) with Z_3 in the direction of the axis of greatest inertia and X_3 in the direction of the axis of smallest inertia.

The choice of the ecliptic frame as the inertial frame is motivated by our intention to introduce further planetary perturbations on the model.

These frames are linked together (see Fig. 1) by three sets of Euler's angles $(h, K, \_)$, (g, J, l) and $(\Omega_o, i_o, \omega_o)$ (where the subscript o stands for "orbital") with Ω_o the ascending node longitude, i_o the inclination and ω_o the pericenter argument.

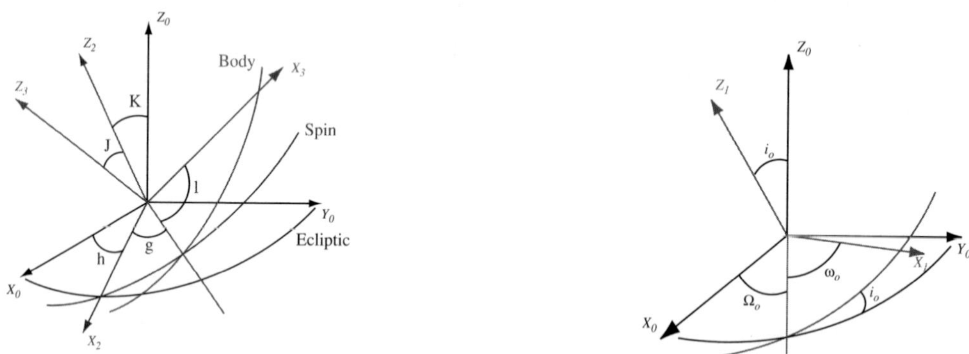

Figure 1. On the left side, the (X_0, Y_0, Z_0), (X_2, Y_2, Z_2) and (X_3, Y_3, Z_3) frames with $(h, K, \_)$ being Euler's angles related to (X_0, Y_0, Z_0) and (X_2, Y_2, Z_2) and (g, J, l) being Euler's angles related to (X_2, Y_2, Z_2) and (X_3, Y_3, Z_3). On the right side, the (X_1, Y_1, Z_1) frame with $(\Omega_o, i_o, \omega_o)$ being Euler's angles related to (X_0, Y_0, Z_0) and (X_1, Y_1, Z_1). All of them are centered at Mercury's center of mass.

In fact, to describe the motion of Mercury, we need two sets of canonical variables: the orbital ones, and the rotational ones.

A first classical orbital choice would be to take Delaunay's elements, related to the ecliptic plane (Mercury being a mass reduced to its center of mass):

$$L_o, G_o, H_o, l_o, g_o, h_o.$$

The capital letters designate the conjugated momenta associated to the angles l_o (mean anomaly), $g_o = \omega_o$ and $h_o = \Omega_o$, and are classically defined as

$$L_o = m\sqrt{\mu\, a}, \quad G_o = L_o\sqrt{1 - e^2}, \quad H_o = G_o \cos i_o,$$

with e the eccentricity and a the semi-major axis of Mercury's orbit. μ is $\mathsf{G}\,(m + M)$ in first approximation, G is the universal constant of gravitation, m is the mass of Mercury and M this of the Sun.

For the rotation, following Deprit (1967) and Kinoshita (1972), we adopt Andoyer's variables

$$L, G, H, l, g, h,$$

G being the norm of the angular momentum, $L = G\cos J$ the projection of the angular momentum on Z_3 and $H = G\cos K$ the projection of the angular momentum on Z_0.

The problem of these Andoyer's variables is that angular variables are not well-defined if K or/and J are zero but their sum is always well-defined. This is why we have chosen a new set of partially non singular variables $(\Lambda_1, \Lambda_2, \Lambda_3, \lambda_1, \lambda_2, \lambda_3)$ such that the angular variables are:

$$\lambda_1 = l + g + h, \qquad \lambda_2 = -l, \qquad \lambda_3 = -h$$

and their conjugated momenta:

$$\Lambda_1 = G, \quad \Lambda_2 = G - L = G\,(1 - \cos J), \quad \Lambda_3 = G - H = G\,(1 - \cos K).$$

Let us remark that the transformation from Andoyer's variables to the new one is canonical.

3. Model of rotation

Let us introduce our main hypothesis for this first approach: Mercury is assumed to move on a fixed elliptic orbit, with the present orbital parameters: $a_o = 57.9 \times 10^6$ km, $e_o = 0.206$, $i_o = 7°$. Without any planetary perturbation, the Hamiltonian can be written:

$$\mathcal{H} = -\frac{m^3\mu^2}{2L_o^2} + T(\Lambda_1, \Lambda_2, \Lambda_3, \lambda_1, \lambda_2, \lambda_3) + V_G(\Lambda_1, \Lambda_2, \Lambda_3, \lambda_1, \lambda_2, \lambda_3)$$

where T is the kinetic energy and V_G the gravity potential. One can show (Deprit 1967) that

$$T = \frac{(\Lambda_1 - \Lambda_2)^2}{2I_3} + \frac{1}{2}(\Lambda_1^2 - (\Lambda_1 - \Lambda_2)^2)(\frac{\sin^2\lambda_2}{I_1} + \frac{\cos^2\lambda_2}{I_2})$$

where I_1, I_2, I_3 are the principal inertia momenta with $I_1 < I_2 < I_3$.

As Peale (1974) and other authors did, and because only few data about Mercury are known, we will limit the development of V_G to the second order in spherical harmonics, the other ones will be considered as perturbations on the basic model:

$$V_G = -\frac{GMm}{r}\left(\frac{R_e}{r}\right)^2\left[C_2^0\,P_2(\sin\theta) + C_2^2\,P_2^2(\sin\theta)\cos 2\varphi\right]$$

where P_2 and P_2^2 refer to the second order Legendre's polynomial and associated polynomial, R_e is Mercury's equatorial radius, r the distance between the Sun and Mercury centers of mass, θ and φ the latitude and longitude of a Mercury's surface element, in the frame (X_3, Y_3, Z_3).

Let us now express V_G in terms of Cartesian coordinates:

$$\begin{cases} \bar{x}_3 = \cos\varphi\cos\theta \\ \bar{y}_3 = \sin\varphi\cos\theta \\ \bar{z}_3 = \sin\theta \end{cases}$$

where $(\bar{x}_3, \bar{y}_3, \bar{z}_3)$ is the unit vector in the direction of the perturbing body (the subscript "3" being used to recall that we are working in the frame (X_3, Y_3, Z_3)). Thus replacing the Legendre polynomials by their expressions, we find:

$$V_G = -\frac{GMm}{r^3}(R_e)^2 \left[\frac{C_2^0}{2}\left(2\bar{z}_3^2 - \bar{x}_3^2 - \bar{y}_3^2\right) + 3C_2^2\left(\bar{x}_3^2 - \bar{y}_3^2\right)\right] \tag{3.1}$$

However, the purpose being to write the potential in our canonical set of rotational $(\Lambda_1, \Lambda_2, \Lambda_3, \lambda_1, \lambda_2, \lambda_3)$ and orbital $(L_o, G_o, H_o, l_o, g_o, h_o)$ variables, let us express the vector $(\bar{x}_3, \bar{y}_3, \bar{z}_3)$ in the orbital frame:

$$\begin{pmatrix} \bar{x}_3 \\ \bar{y}_3 \\ \bar{z}_3 \end{pmatrix} = R_3(-\lambda_2)R_1(J)R_3(\lambda_1 + \lambda_2 + \lambda_3)R_1(K)R_3(-\lambda_3) \times$$

$$R_3(-h_o)R_1(-i_o)R_3(-g_o)\begin{pmatrix} \cos v_o \\ \sin v_o \\ 0 \end{pmatrix} \tag{3.2}$$

where v_o is the true anomaly, R_i are rotation matrices (their subscript is the number of the rotation axis).

4. Kernel model: 2 degrees of freedom

For this kernel model, we are going to take $J = 0$, which means that the spin axis Z_2 and the third principal axis of inertia Z_3 coincide. The coordinates change becomes:

$$\begin{pmatrix} \bar{x}_3 \\ \bar{y}_3 \\ \bar{z}_3 \end{pmatrix} = R_3(\lambda_1 + \lambda_3)R_1(K)R_3(-\lambda_3) \times$$

$$R_3(-h_o)R_1(-i_o)R_3(-g_o)\begin{pmatrix} \cos v_o \\ \sin v_o \\ 0 \end{pmatrix} \tag{4.1}$$

Thus replacing (4.1) in (3.1) and using the relation $L_o = m\sqrt{\mu a}$ and the well known developments in eccentricity, up to the order 3 of Brouwer & Clemence (1961) allow us to introduce missing orbital variables l_o and L_o in the potential expression V_G:

$$V_G = -\frac{GMm^7}{L_o^6}\mu^3(R_e)^2 \times$$

$$\left\{ \frac{1}{2}C_2^0\left[a_{000} + \sum_{i=1}^{5}(a_{00i}\cos(i\,l_o) + b_{00i}\sin(i\,l_o))\right.\right.$$

$$\left. + \sum_{j=1}^{2}\sum_{i=-5}^{5}(a_{0ji}\cos(j\lambda_3 + i\,l_o) + b_{0ji}\sin(j\lambda_3 + i\,l_o))\right]$$

$$\left. + 3C_2^2\left[\sum_{j=0}^{4}\sum_{i=-5}^{5}(a_{2ji}\cos(2\lambda_1 + j\lambda_3 + i\,l_o) + b_{2ji}\sin(2\lambda_1 + j\lambda_3 + i\,l_o))\right]\right\}$$

where a and b depend on K, h_o, i_o, g_o and e, with h_o, i_o, g_o and e considered as constants. So, in our basic model, V_G is only function of 3 momenta and 3 angles: Λ_1 and Λ_3 (through K), $L_o, \lambda_1, \lambda_3$ and l_o.

5. The main resonant angle

Let us recall again that Mercury is in a 3:2 spin-orbit resonance which can be expressed in our variables by:

$$\dot{\lambda}_1 = \frac{3}{2}\dot{l}_o.$$

The two sets of canonical variables (orbital and rotational) are then mixed in this commensurability.

Let us thus take

$$\sigma = \frac{2\lambda_1 - 3l_o}{2}$$

as resonant angle and Λ_1 as its conjugated momentum. In order to keep a canonical transformation, we must associate to l_o a new conjugated momentum $\Lambda_o = L_o + \frac{3}{2}\Lambda_1$.

If we replace by these new variables in (4.2) and after averaging on l_o, only a few terms stay ($<>$ stands for "averaged"):

$$< V_G > = -\frac{GMm^7}{(\Lambda_0 - \frac{3}{2}\Lambda_1)^6}\,\mu^3\,(R_e)^2$$

$$\times \left(\frac{1}{2}\,C_2^0\left(a_{000} + \sum_{k=1}^{2}(a_{0k0}\cos(k\lambda_3) + b_{0k0}\sin(k\lambda_3))\right)\right.$$

$$\left.+3\,C_2^2\left(\sum_{k=0}^{4}(a_{2k0}\cos(2\sigma + k\lambda_3) + b_{2k0}\sin(2\sigma + k\lambda_3))\right)\right)$$

6. Simplified Hamiltonian

Considering that the eccentricity $e \sim 0.206$, we are going to keep the terms in $(1 + \frac{3\,e^2}{2})$ and $(\frac{7\,e}{2} - \frac{123\,e^3}{16})$ but not in e^3 only; those terms will act as perturbations on our first model. And so, we obtain a two degrees of freedom Hamiltonian (see D'Hoedt & Lemaitre (2004) for details):

$$< \mathcal{H} > = \frac{\Lambda_1^2}{2\,I_3} - \frac{m^3\,\mu^2}{2\left(\Lambda_0 - \frac{3\,\Lambda_1}{2}\right)^2} - \frac{GMm^7\,\mu^3\,R_e^2}{\left(\Lambda_0 - \frac{3\,\Lambda_1}{2}\right)^6} \times$$

$$\left[\frac{1}{2}C_2^0\left(1 + \frac{3\,e^2}{2}\right)\left(\sum_{i=0}^{2}a_{0i}\cos(i\sigma_3)\right)\right.$$

$$\left.+3\,C_2^2\left(\frac{7\,e}{2} - \frac{123\,e^3}{16}\right)\left(\sum_{i=0}^{4}a_{2i}\cos(2\sigma_1 + i\sigma_3)\right)\right]$$

with $\sigma_1 = \sigma - h_o - g_o$, $\sigma_3 = \lambda_3 + h_o$ and where a_{0i} and a_{2i} depend on i_o and K. (The canonical transformation from old to new variables allows us to keep Λ_1 and Λ_3 as conjugated momenta to σ_1 and σ_3.)

After computation of the motion equations, we find that the equilibria for the couple (σ_1, σ_3) are: $(0, 0), (\pi/2, 0), (0, \pi), (\pi/2, \pi)$. Let us notice that the last two geometrical configurations are the same as the first two ones.

For each couple (σ_1, σ_3) we find the different values of the (ecliptic) obliquity K:

if $(\sigma_1, \sigma_3) = (0,0)$, it leads to four values of the ecliptic obliquity (two of them are dependent on the numerical values of the coefficients C_0^2 and C_2^2):

$$\sin(i_0 - K) = 0 \quad \text{or}$$

$$\cos(i_0 - K) = \frac{C_2^2 \left(\frac{7\,e}{2} - \frac{123\,e^3}{16} \right)}{C_0^2 \left(1 + \frac{3\,e^2}{2} \right) - C_2^2 \left(\frac{7\,e}{2} - \frac{123\,e^3}{16} \right)}$$

if $(\sigma_1, \sigma_3) = (\frac{\pi}{2}, 0)$, it also leads to four values of the ecliptic obliquity and we can see that the first two are the same ($K = i_o$ and $K = \pi + i_o$):

$$\sin(i_0 - K) = 0 \quad \text{or}$$

$$\cos(i_0 - K) = \frac{-C_2^2 \left(\frac{7\,e}{2} - \frac{123\,e^3}{16} \right)}{C_0^2 \left(1 + \frac{3\,e^2}{2} \right) + C_2^2 \left(\frac{7\,e}{2} - \frac{123\,e^3}{16} \right)}$$

Finally for each K, we can find numerically the value of Λ_1 (see Table 1) and *a fortiori* the value of Λ_3.

The numerical values shown in the table below are calculated with R_e, the equatorial radius of Mercury, M, the mass of Mercury, and the year chosen respectively as unities of length, mass and time.

(σ_1, σ_3)	$(i_o - K)$ (deg)	K (deg)	$\Lambda_1 \left(\frac{m\,Re^2}{year} \right)$
(0,0)	0	$K_1 = 7$	13.303
	180	$K_2 = 180 + 7$	13.303
	-95.332	$K_3 = 102.332$	13.303
	95.332	$K_4 = -88.332$	13.303
$(\frac{\pi}{2}, 0)$	0	$K_1 = 7$	13.303
	180	$K_2 = 180 + 7$	13.303
	-83.446	$K_5 = 90.446$	13.303
	83.446	$K_6 = -76.446$	13.303

Table 1. Equilibria in the case $\sigma_3 = 0$.

Let us notice that the values of Λ_1 seem equal, but they differ from each other when we consider more decimals.

We can check that the first equilibrium (the present situation of Mercury) is stable.

7. Frequency of the two angular variables

Of course, Mercury is not blocked at the exact 3:2 resonance, but performs a small libration about this equilibrium. To measure the frequency of this libration, we perform a translation to the equilibrium and after successive canonical transformations we can write our Hamiltonian in angle-action coordinates and find the frequencies of σ_1 and σ_3:

$$\nu(\sigma_1) = 0.396234 \, \frac{1}{year}, \qquad \nu(\sigma_3) = 0.00589928 \, \frac{1}{year}$$

In terms of periods, σ_1 has a period of 15.8573 yr and σ_3 a period of 1065.08 yr, which coincides with the results of Rambaux & Bois (2003).

8. Introduction of the third degree of freedom

Let us now take into account that $J \neq 0$, λ_2 is thus now well defined. The potential $< V_G >$ becomes:

$$< V_G > = -\frac{GMm^7}{(\Lambda_0 - \frac{3}{2}\Lambda_1)^6} \mu^3 (R_e)^2$$

$$\times \left(\frac{1}{2} C_2^0 \left(d_{000} + \sum_{k=1}^{2} d_{00k} \cos(k\sigma_3) + \sum_{k=0}^{4} d_{22k} \cos(2\sigma_1 + 2\lambda_2 + k\sigma_3) \right) \right.$$

$$+ 3 C_2^2 \left(\sum_{k=-2}^{2} d_{02k} \cos(2\lambda_2 + k\sigma_3) + \sum_{k=0}^{4} d_{20k} \cos(2\sigma_1 + k\sigma_3) \right.$$

$$\left. \left. + \sum_{k=0}^{4} d_{24k} \cos(2\sigma_1 + 4\lambda_2 + k\sigma_3) \right) \right)$$

with d depending on i_o, J, K and e.

If we express this third degree of freedom in terms of Cartesian coordinates (ξ, η) by this canonical transformation

$$\begin{cases} \sqrt{2\Lambda_2} \cos \lambda_2 = \xi \\ \sqrt{2\Lambda_2} \sin \lambda_2 = \eta \, , \end{cases}$$

we can show that $(\xi, \eta) = (0, 0)$ is an equilibrium and at this equilibrium we find the same equilibria as previously for the other 2 degrees of freedom.

Once more, Mercury isn't at the exact equilibrium, but performs a small libration about it. Thus, in the same way as previously for σ_1 and σ_3, we find the period of λ_2 which is 584 yr.

References

Anderson, J.D., Colombo, G., Esposito, P.B., Lau, E.L, Trager, G.B. 1987, 'The mass, gravity field and ephemeris of Mercury', *Icarus* 71, 337–349.

Brouwer, D. & Clemence, G.M. 1961, 'Methods of celestial mechanics' *Academic Press, New York*.

Deprit, A. 1967, 'Free Rotation of a Rigid body Studied in the Phase Plane', *American Journal of Physics* 35 (5), 424–428.

D'Hoedt, S. and Lemaitre, A. 2004, 'The spin-orbit resonant rotation of Mercury: a two degree of freedom Hamiltonian model', *Celestial Mechanics*, in press.

ESA-SCI 2000, 'BepiColombo, An Interdisciplinary Cornerstone Mission to the Planet Mercury' *System and Technology Study Report.*

Kinoshita, H. 1972, 'First-Order Perturbations of the Two Finite Body Problem', In *Publ. Astron. Soc. Japan* 24, 423–457.

Peale, S.J. 1974, 'Possible histories of the obliquity of Mercury' *The Astronomical Journal* 79, 722–744.

Rambaux, N. and Bois 2003, 'Theory of the Mercury's spin-orbit motion and analysis of its main librations', *Astronomy and Astrophysics* 413, 381–393

                S. D'Hoedt and A. Lemaitre

Discussion

JIM MESSAGE: Are there any observational results on these librations?

SANDRINE D'HOEDT: No, we are expecting two missions. Few things are known about Mercury – very few things.

MYLES STANDISH: Will a spacecraft get measurements of Mercury's libration?

SANDRINE D'HOEDT: Yes.

MYLES STANDISH: ...actual physical measurements, either visual or radar?

SANDRINE D'HOEDT: Yes.

WALTER BRISKEN: There's some evidence that the core of Mercury is rotating differently than the outer shell. Is there any way that this can be incorporated into your model?

SANDRINE D'HOEDT: Yes, our Brussels team works on the core of Mercury; we work on the rigid body and then we get the results.

Sandrine D'Hoedt and Anne Lemaitre

Transits of Venus: New Views of the Solar System and Galaxy
Proceedings IAU Colloquium No. 196, 2004
D.W. Kurtz, ed.

© 2004 International Astronomical Union
doi:10.1017/S1743921305001456

Observation and reduction of mutual events in the solar system

B. Noyelles[1], V. Lainey[2,1] and A. Vienne[1]

[1]IMCCE – Laboratoire d'Astronomie de Lille, UMR8028 du CNRS, 1 impasse de l'Observatoire, 59000 Lille, France   email: Benoit.Noyelles@imcce.fr

[2]Royal Observatory of Belgium, Ringlaan 3, B-1180 Brussels, Belgium

Abstract. Mutual event observations started in the early 1970s with the Galilean satellites. These observations were needed because of the Voyager spacecraft future arrival. Since 1979, IMCCE has organized observational campaigns for the Galilean satellites (called PHEMU), and since 1995 for the Saturnian satellites (also called PHESAT). Meanwhile, the reduction techniques have been greatly improved. Mutual event observations are one of the most accurate methods for obtaining positions of natural satellites, useful for detecting tidal effects. Hence mutual events of Jovian and Saturnian natural satellites are regularly observed around the world. This paper aims to describe mutual events and the advantages of this kind of observation besides the classical astrometric ones.

1. What is a mutual event?

There are two kinds of mutual events: occultations and eclipses. An occultation between two natural satellites occurs when the observer cannot distinguish the two satellites separately because a part of the further one is hidden by the nearest one (they are nearly aligned with the observer), whereas an eclipse occurs when the further satellite meets the umbra or the penumbra cone of the nearest one, so there is nearly an alignment between the Sun and the two satellites (see Fig. 1).

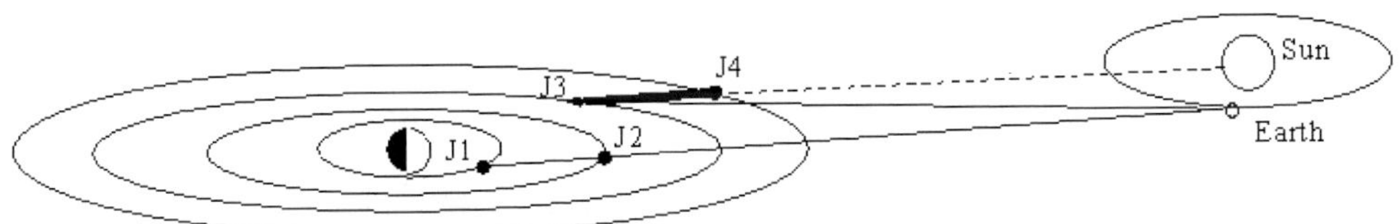

Figure 1. An eclipse and an occultation in the Jovian system.

The reduction of a mutual event observation provides a light curve; that means a time scale, and for each instant a photometric value, for which two quantities are important: the flux drop and the mid-light time. The flux drop is the maximum light-loss during the event, and the mid-light time is the instant corresponding to this light-loss. These two observed quantities may be linked to two geometrical ones: the distance minimum between the two satellites on the celestial sphere during the event, and the time corresponding to this distance (mid-time) which differs from the mid-light time by a few seconds due to light scattering by the surface of atmosphereless bodies, say all except Titan.

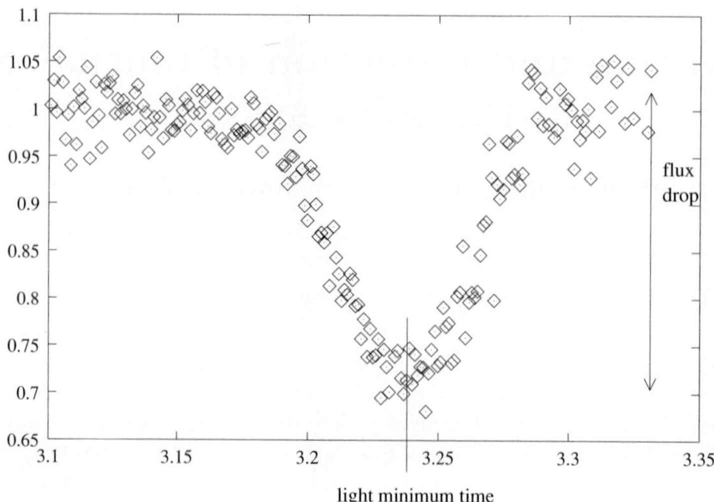

Figure 2. Observation of an occultation of S4 Dione by S3 Tethys at Pic-du-Midi Observatory on 21 September 1995. The flux is normalized and equal to 1 before and after the event. The abscissa is expressed in hours.

2. Why observe mutual events?

The mutual event observations have several advantages with respect to classical astrometric observations, for instance:

• **high accuracy** These observations are photometric instead of astrometric, so they are less dependant to the atmosphere turbulence.

• **easy to observe** Mutual events can be observed anywhere, even at low altitude sites, because some natural satellites, like the Galilean ones, are very easy to observe, their magnitudes being between 4 and 6. Since the data are easy to register, amateurs can take part in observation campaigns.

• **ephemerides improvement** One of the goals for getting such accurate observations deals with ephemerides improvement. Indeed, besides spacecraft needs, it is necessary to predict star occultations by natural satellites (see, e.g., Sicardy *et al.* 1999), or to detect a secular accelerations in the motion of these bodies, consequences of tidal effects.

3. Reduction method

The method consists of modelling a theoretical light curve depending on some parameters, and to adjust them, for instance by a non-linear least squares fitting. The first main point is to model the diffusion of solar light by the surface of the satellites. There are several classical laws:

• **Lambert law** This law represents an isotropic diffusion of the light; it is reliable for Titan, which has an atmosphere. This law does not introduce any shift between mid-time and mid-light time.

$$\frac{I}{F} = \frac{3}{2}p\cos i \tag{3.1}$$

• **Minnaert law** This is an empirical law introducing a limb-darkening parameter k, see Minnaert (1961) for further details.

$$\frac{I}{F} = B(\cos i)^{k(p)}(\cos e)^{k(p)-1} \tag{3.2}$$

• **Buratti-Veverka law** This law takes into account reflection by secondary sources on the body's surface. It comes from Lommel-Seeliger law which has been extended in Buratti & Veverka (1983).

$$\frac{I}{F} = A\frac{\cos i}{\cos i + \cos e}f(p) + (1 - A)\cos i \tag{3.3}$$

In this equation, f is a second-degree polynomial function. We used this law in Noyelles, Vienne & Descamps (2003) for S-1 Mimas to S-5 Rhea with photometric parameters taken from Buratti & Veverka (1984) and Devyatkin & Miroshnichenko (2001).

• **Hapke law** This law is not described in detail here, due to its complexity and to the fact that we did not use it. This is a theoretical law which has two expressions depending on the roughness of the body's surface, see Hapke (1984) and Hapke (1986) for further details.

In all these formulae, πF is the solar incident flux on a surface unit, I the incident flux, p the albedo, i and e respectively the incidence and emergence angles of the solar light on the body's surface (see Fig. 3), k the limb darkening, and A and B photometric parameters depending on the body.

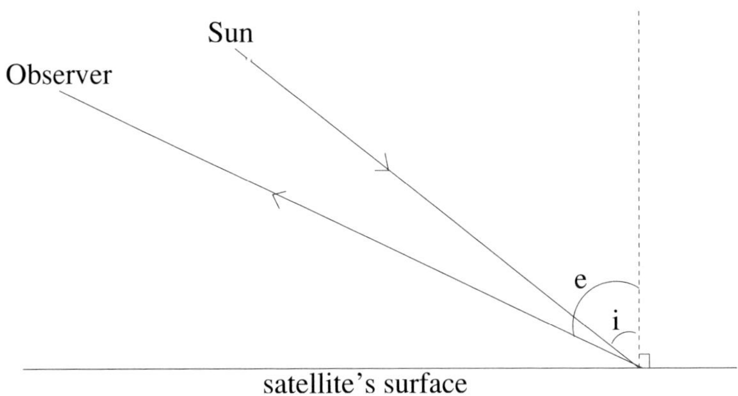

Figure 3. Description of the incidence angle i and the emergence angle e.

In the case of an eclipse, there is a little difficulty for estimating the incident solar flux for a surface located in the penumbra zone, because the solar luminosity is not uniform on the solar disk, we have to take the solar limb darkening into account. For this purpose, we used an empirical formula given by Hestroffer & Magnan (1998) giving a wavelength dependency of this darkening:

$$I(r) = \mu^{\alpha} \tag{3.4}$$

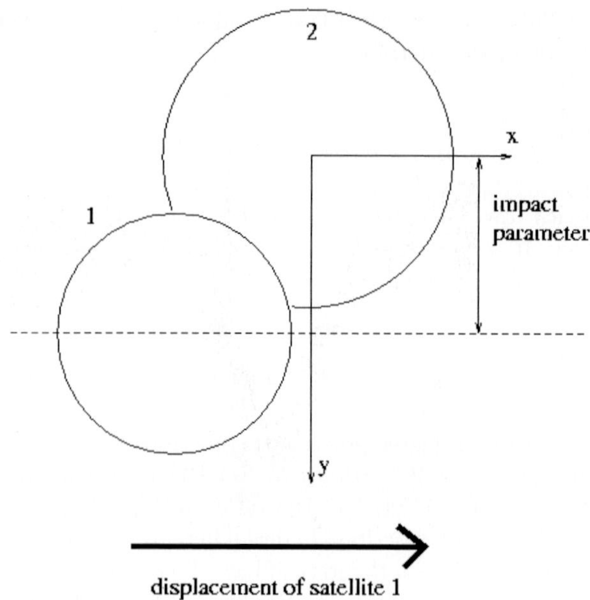

Figure 4. Geometrical representation of the impact parameter.

$$
\text{where:}
\begin{cases}
\alpha & \sim -0.023 + 0.292\lambda^{-1} \text{ if } \lambda \lesssim 2.4\mu m^{-1} \\
\alpha & \sim -0.507 + 0.441\lambda^{-1} \text{ if } \lambda \gtrsim 2.8\mu m^{-1} \\
\lambda & \text{wavelength in } \mu m \\
\mu & = \sqrt{1 - r^2} \\
r & \text{distance to the Sun's centre, } R_\odot = 1
\end{cases}
$$

Once the detected flux is modelled, we have to model the motion too. We considered that, during the event, the further satellite did not move and the nearest moved on a straight line like a train on a railway with a constant speed (see Fig. 4), this model gives good results when there is no elongation.

After modelling the theoretical event, the last step is to adjust it to the observed light curve (see Fig. 5). The fit gives us a mid-time and an impact parameter, linkable to differential astrometric coordinates $(\Delta\alpha \cos\delta, \Delta\delta)$.

4. Results

4.1. *Case of the Saturnian satellites*

The first astrometric results of Saturnian satellites were published by Aksnes *et al.* (1984) after observing 14 mutual events in 1980 around the world, and reducing without taking account of the anisotropy of light scattering by the surface of the satellites. Fifteen years later, the PHESAT95 campaign organized by the actual IMCCE collected 65 light curves (see Thuillot *et al.* 2001). In Noyelles, Vienne & Descamps (2003) we obtained 16 positions in which we are very confident and 32 in which we are confident, and re-reduced some light curves performed in 1980 in Japan by Soma & Nakamura (1982) already reduced by Aksnes *et al.* (1984). The residuals for the best observations are about $20 - 30$ mas.

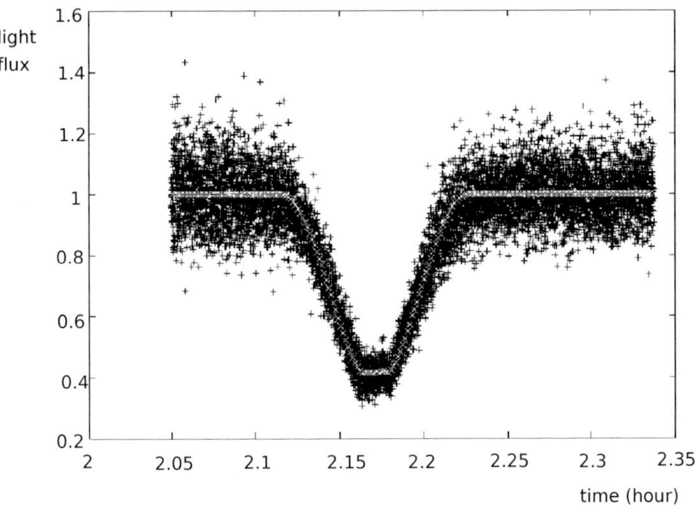

Figure 5. Fit of a theoretical light curve (bright curve) to an observed one (dark curve).

4.2. *Case of the Galilean satellites*

Now six observational campaigns have been done. Although the data are not reduced yet for the last one in 2002/2003, most observations have been introduced in the fit to the theories. For instance, Fig. 6 shows the residuals of PHEMU campaigns of 1985 and 1991 using L1 ephemerides (Lainey, Duriez & Vienne 2004). One can see that the residuals are clearly below $0.''1$ (which is the mean of the (O-C)s for the astrometric observations).

5. Improvements in the near future

5.1. *Uncertainty in the position angle $(\pm\pi)$*

The impact parameter determination gives us only the separation angle between the two satellites at mid-time. The ephemerides give us good accuracy for the direction of the relative motion of the two satellites. But when the impact parameter is small, it can be difficult to determine the position angle p between its actual value and $p \pm \pi$, more particularly for the Saturnian satellites, because the ephemerides are not as accurate as for the Galilean ones. In the future, the improvements in the dynamical theories will discriminate the good configuration.

5.2. *Knowledge of the satellites' photometry*

Thanks to the Galileo spacecraft mission, albedo maps of the four Galilean satellites are now available (see Vasundhara, Arlot, Lainey & Thuillot 2003 for details, and Fig. 7 for an example). It is clear that the albedo variation of the satellite's surface modifies the light curve in a non negligible way (considering the already high precision involved). The introduction of such an effect has been recently done in the PHEMU97 campaign, and is still in progress for the other ones. The expected improvements of such correction is estimated around $0.''02$.

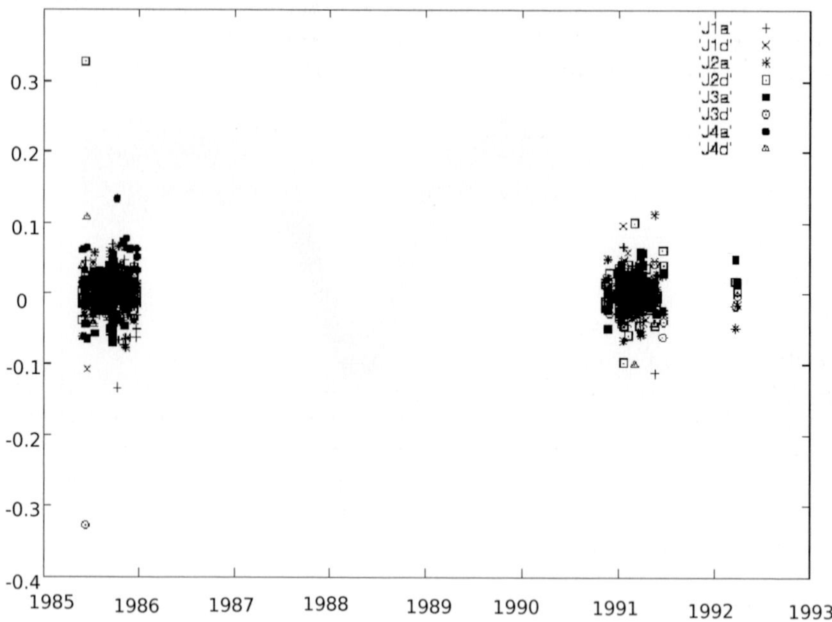

Figure 6. The (O-C)s of 1985 and 1991 PHEMU campaigns using L1 ephemerides. Y axis is expressed in arc seconds.

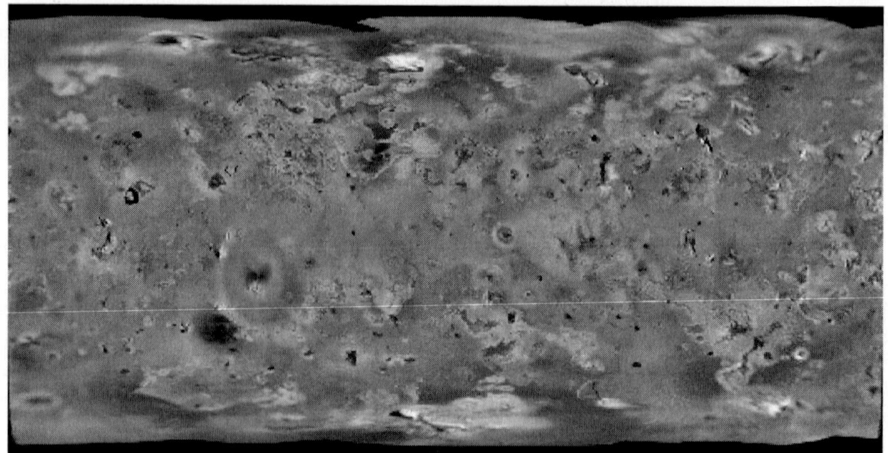

Figure 7. Io's albedo map (Geissler *et al.* 1999).

5.3. *Instrumental problems*

We said above that these observations are easy to perform; in fact it is easy to make good observations, but difficult to make very good ones, because some points are critical:

• The first one is the *time reference*. Lots of observers use their computer's clock, which has an accuracy not better than one second (inducing an error of around 10 km). Better accuracy would help to detect a shift in longitude, that means a secular acceleration coming from tidal effects.

- The second one is the *accuracy of the photometry*. The exposure has to be short enough to not saturate the detector, and it is better to have a comparison object in the field, which has constant luminosity, to correct the variations in atmospheric transparency.

- The last one is the *photometer calibration*. Indeed, the light drop is strongly correlated with the detection threshold. It seems very difficult to correct this effect as it completely depends on the detector.

6. Conclusion

The main interest in observing mutual events is that no other observation is as accurate. They are useful to detect small shifts in ephemerides – that means secular acceleration in the mean longitudes, induced by tidal effects. Such detections would provide better knowledge of the internal structure of the satellites, for instance the depth of oceans, where life could have appeared. The determination of these effects is still in progress.

The next campaigns are in 2009 (PHEMU) and 2010 (PHESAT). We encourage observers over the whole world to take part of it and we also hope that for the first time, mutual events involving Uranian satellites will be observed in a few years.

Acknowledgements

The authors are indebted to R. Vasundhara (IIA, India), J.-E. Arlot, P. Descamps and W. Thuillot (IMCCE, Paris) and N. Emelianov (SAI, Russia) for fruitful discussions.

References

Aksnes, K., Franklin, F., Millis, R. *et al.* 1984 *Astronomical Journal* **89**, 280–288.

Arlot, J.-E., Ruatti, C., Thuillot, W. *et al.* 1997 *Astronomy & Astrophysics Supplement Series* **125**, 399–405.

Buratti, B. & Veverka, J. 1983 *Icarus* **53**, 93–110.

Buratti, B. & Veverka, J. 1984 *Icarus* **58**, 254–264.

Devyatkin, A.V. & Miroshnichenko A.S. 2001 *Astronomy Letters* **27**, 193–197.

Geissler, P. E., McEwen, A. S., Keszthelyi, L., *et al.* 1999 *Icarus* **140**, 265–282

Hapke B.W. 1984 *Icarus* **59**, 41–59

Hapke B.W. 1986 *Icarus* **67**, 264–280

Hestroffer, D. & Magnan, C. 1998 *Astronomy & Astrophysics* **333**, 338–342.

Lainey, V., Duriez, L. & Vienne, A. 2004 *Astronomy & Astrophysics* **420**, 1171–1183

Minnaert, M. 1961 in *Planets and Satellites, ed. G.P.Kuiper & B.M.Middlehurst (The University of Chicago Press)* **213**.

Noyelles B., Vienne A. & Descamps P. 2003 *Astronomy & Astrophysics* **401**, 1159–1175

Sicardy, B., Ferri, F., Roques, F. *et al.* 1999 *Icarus* **142**, 357–390.

Soma, M. & Nakamura, T. 1982 *Tokyo Astronomical Bulletin Second Series* **267**, 3039.

Thuillot, W., Arlot, J.-E., Ruatti, C. *et al.* 2001 *Astronomy & Astrophysics* **371**, 343–349.

Vasundhara, R., Arlot, J.-E., Lainey, V. & Thuillot, W. 2003 *Astronomy & Astrophysics* **410**, 337–341.

Vienne, A. & Duriez, L. 1995 *Astronomy & Astrophysics* **297**, 588–605.

Discussion

MIKHAIL MAROV: Your talk is addressed to such effects in the solar system, but you mostly focused on the Galilean satellites. In recent years there has been an avalanche of discoveries of many new small-size satellites in the outer planets. Are there some applications of your model – your approach – in order to predict and observe such occultation effects for these new satellites?

BENOIT NOYELLES: Such satellites are faint, so observations are more difficult than, for instance, an occultation by a big satellite. For such faint ones the signal will be lost in the noise, so this observation would be difficult.

MIKHAIL MAROV: I disagree. These satellites were discovered with ground-based facilities. It's a quite powerful technique right now. So if your theory can predict such events, it is possible to observe them.

BENOIT NOYELLES: Yes, but not so easy as the Galilean satellites.

MIKHAIL MAROV: What kind of the model of scattering for the albedo features did you use? Lambert theory?

BENOIT NOYELLES: In fact, Lambert theory gives good results for Titan because it is has an atmosphere. But for the Saturnian satellites I use the Buratti-Veverka law which takes into account secondary reflections from the surfaces of the satellites. One can use the Minnaert law, but in this case it is better to fit the photometric parameters because it depends on the phase angle.

Sandrine d'Hoedt, Bernard De Saedeleer, Benoit Noyelles and Stephane Valk

Transits of Venus: New Views of the Solar System and Galaxy
Proceedings IAU Colloquium No. 196, 2004
D.W. Kurtz, ed.

© 2004 International Astronomical Union
doi:10.1017/S1743921305001468

Early dynamical evolution of the Solar System: constraints from asteroid and KBO dynamics

Kleomenis Tsiganis

Observatoire de la Côte d' Azur, CNRS UMR-6202 Cassiopée, Nice B.P. 4229, 06304 Nice
Cedex 4, France
email: tsiganis@obs-nice.fr

Abstract. The orbital distribution of asteroids and Kuiper-belt objects (KBOs) provides important information for the dynamical evolution of the solar system. Recent advances in modelling the dynamics of asteroids and KBOs have increased our understanding of the mechanisms responsible for the main features observed in the small-body belts. There are, however, several pieces left to complete the puzzle of the dynamical history of the solar system. In particular, we now understand that the solar system probably looked very different during the first $500 - 1000$ Myr than today, mainly because of two processes that took place during those times: (i) planetary formation and (ii) planetary migration. I will discuss observational and dynamical constraints that any model, attempting to reconstruct this early period, should obey. I will then present a new model for planetary migration, which successfully reproduces the observed orbital distribution of the trans-Neptunian objects. I will then discuss the implications of this model on the early evolution of the inner solar system, in particular the distribution of main-belt asteroids and the bombardment of the terrestrial planets by small bodies.

1. Introduction

The belief that our solar system has been, and for ever will be, in a steady dynamical state has been strong among dynamicists for several centuries. The prediction of periodic phenomena, such as the transits of Venus and Mercury across the solar disc, were considered as the great success of "old" celestial mechanics. More complex types of dynamical behaviour, although noted by Poincaré (1892), were considered merely as insignificant exceptions. Today, we know that the situation is most likely quite the opposite. Chaos is the rule rather than the exception in solar system dynamics. New methods and tools of celestial mechanics have been devised in order to cope with these phenomena, resulting in remarkable advances over the last 20 years. I mention only the discovery of the mildly chaotic motion of the planets (Sussman & Wisdom 1992; Laskar *et al.* 1993) and the explanation of the origin of the Kirkwood gaps in the distribution of asteroids (for a review see Moons 1997).

It is fair to say that we now understand quite well the dynamical evolution of the system over the last ~ 4 Gy. It is, though, intriguing that, despite our advanced analytical and numerical tools, a mathematical proof of the stability of the planetary system for such long time scales has not been obtained so far. On the other hand, our analytic and semi-analytic theories have been well tested by sophisticated numerical experiments, largely aiming to explain the "strange" motions of the thousands of asteroids and comets that reside in the system. Small bodies are mainly contained in two large reservoirs: the main asteroid belt and the Edgeworth-Kuiper belt. These belts provide a unique laboratory for dynamics, since the number of bodies observed over the last 100 years is very large and their orbital distribution is known with very good accuracy. The long term effects of gravitational perturbations, exerted by the planets on the small bodies, are imprinted in these distributions. The goal of "new" celestial mechanics is evident: to extract as much information as possible from these distributions, by "reading between the lines", in order to unveil clues about their dynamical evolution. As will be described in the following sections, the strongest long-term effects of the planets, especially on the asteroid belt, are now well understood. There are, however, many observed features, primarily in the distribution of trans-Neptunian objects, which cannot be understood in the framework of a steady-state planetary system. These dynamical features were most likely created during earlier epochs in solar system evolution.

There are two main phases of early solar system evolution during which the dynamics of the planets and the small-body belts were drastically different than over the last $\sim$4 Gy. They are related to two main processes: (i) planet formation and (ii) planet migration. The formation of the gas giants probably took place during the first $\sim$1 − 10 Myr, when the proto-solar gas nebula was still massive enough and the dynamics were dominated by the interactions of the planets with the gas disc (see Pollack *et al.* 1996; Lubow *et al.* 1999). Planetary migration could also have taken place during this phase (see Ward 1997; Lyn & Papaloizou 1986; Masset & Papaloizou 2003). However, whether this really happened (and to what extent) in our solar system can only be shown by extremely expensive numerical simulations, which are also quite dependent on a more or less accurate knowledge of the relevant physical parameters of the system at that time (e.g. mass of the disc, physical size, aspect ratio, initial positions and masses of the planets, etc.). This is the reason why this particular phase is the most difficult to tackle and remains still one of the "hottest" research areas of dynamics of planetary systems. Understanding the wide variety of physical and dynamical properties among the observed extra-solar systems, demands a better understanding of this first phase of evolution of planetary systems.

After the formation of the gas giants was over and most of the gas disc was dissipated, a new epoch of gas-free dynamics began. At that time the solar system contained (i) the four outer planets (Jupiter, Saturn, Uranus and Neptune), (ii) a large disc of solid *planetesimals* ($\sim$10-km-sized bodies, see Weidenschilling 2003), likely extending all the way from the inner edge of the solar system ($\sim$0.5 AU) to even beyond the currently observed Kuiper belt ($\sim$50 AU) and possibly (iii) a number of lunar-sized solid bodies, called planetary embryos, which are presumed to have been the building blocks of the terrestrial planets (Chambers 2001; also Petit *et al.* 2001). It is now widely accepted that Uranus and Neptune must have formed much closer to the Sun than is currently observed, in order for their core-accretion time scales to be have been shorter than the time scale of nebula dissipation ($\sim$10 Myr). Thommes *et al.* (1999) suggested that the initial orbital radii of these two planets must have been much smaller than $\sim$20 AU. Even if the exact locations of the giant planets at that epoch are poorly constrained, it is evident from the above results that they must have migrated a lot! A quite efficient mechanism of planetary migration, caused by the scattering of leftover planetesimals – contained in the interplanetary region and the external disc – by the planets, was first described by Fernandez & Ip (1984). Its efficiency was clearly demonstrated by the numerical simulations of Hahn & Malhotra (1999).

The question that we focus on now is the following: Is it possible for this early process of large-scale planetary migration to have left some traces behind and how can we identify them? In the following I am going to review recent results supporting the following answer to the above question: "Yes, at least in the orbital distributions of the small bodies". The relevant observational features and the constraints they imply, concerning the early dynamical evolution of the solar system, are presented in this paper.

2. The asteroid belt

The number of asteroids discovered in the inner solar system increases every day. The numbered objects, whose orbits are determined to an accuracy that allows us to speak of "well-known" objects, are now of the order of 100 000 (http://hamilton.unipi.it/cgi-bin/astdys/astibo). Spectroscopic and rotation properties are known for far fewer objects, but the results of on-going surveys are encouraging. The current status of our knowledge is well summarized in the *Asteroids III* book (edited by Bottke *et al.* 2002), so I will restrict myself here to presenting information that are most relevant to the present work.

The vast majority of asteroid orbits lie between Mars and Jupiter, in a region called the *main belt*. The semi-major axes of main-belt asteroids are $2.0 \leqslant a \leqslant 3.6$ AU, their eccentricities are typically smaller than $e \sim 0.3$ and the inclinations of their orbital planes with respect to the ecliptic are smaller than $i \sim 20°$. The current total mass of the main belt is estimated to be $\sim 5 \times 10^{-4}$ Earth masses (M_E), which is at least 1000 times smaller than what the minimum mass solar nebula model predicts (Petit *et al.* 2001). Thus, today's main belt is just the remnant of a largely depleted primordial belt. Petit *et al.* (2001) have shown that most of this depletion could have taken place during the first ~ 100 Myr of the solar system's life, during the formation of the terrestrial planets. It is important to remember that the main belt is the main source of the short-lived Near-Earth asteroids (NEAs), which leak out from the belt and spend ~ 10 Myr in the near-Earth space as a result of resonant perturbations, induced by the major planets, and close encounters with the terrestrial planets. This implies that the primordial, more-massive main belt was producing more NEAs, thus leading to a much larger number of impacts with the Earth than at present.

Fig. 1 shows the distribution of all numbered main-belt asteroids, as a function of their orbital elements, (a, e, i). This non-uniform distribution has been a puzzle for many years. The most prominent observable features are the Kirkwood gaps, which are associated with mean motion resonances (2:1, 7:3, 5:2, 3:1 and 4:1) between Jupiter and an asteroid. It has been shown that the dynamics in these resonances are chaotic, and primordial asteroids would have had plenty of time to slowly change their eccentricities and reach perihelion (aphelion) distances that would lead them to close encounters with the inner (outer) planets and subsequent ejection from the solar system. Other resonant phenomena, such as secular resonances between the precession frequencies of the asteroids and the planets, have also sculpted the distribution of asteroids. Note, however, that the locations of the resonances depend on the semi-major axes of the planets. Thus, to zeroth order, we could say that the current main belt distribution does not seem to contain information about the early phases of the solar system, in particular the era of planetary migration. As we will see in following sections, this is not true.

I should point out that recent results have revealed the importance of the long-term effects of two, "slow," still on-going processes: (i) chaotic diffusion, generated by a great number of thin resonances that criss-cross the belt and (ii) the Yarkovsky thermal force, which slowly changes the semi-major axis of small asteroids, thus leading them to cross several resonances. These two phenomena play an important role in transporting NEAs and meteorites to Earth, as well as in shaping the asteroid families (see the relevant chapters in *Asteroids III*).

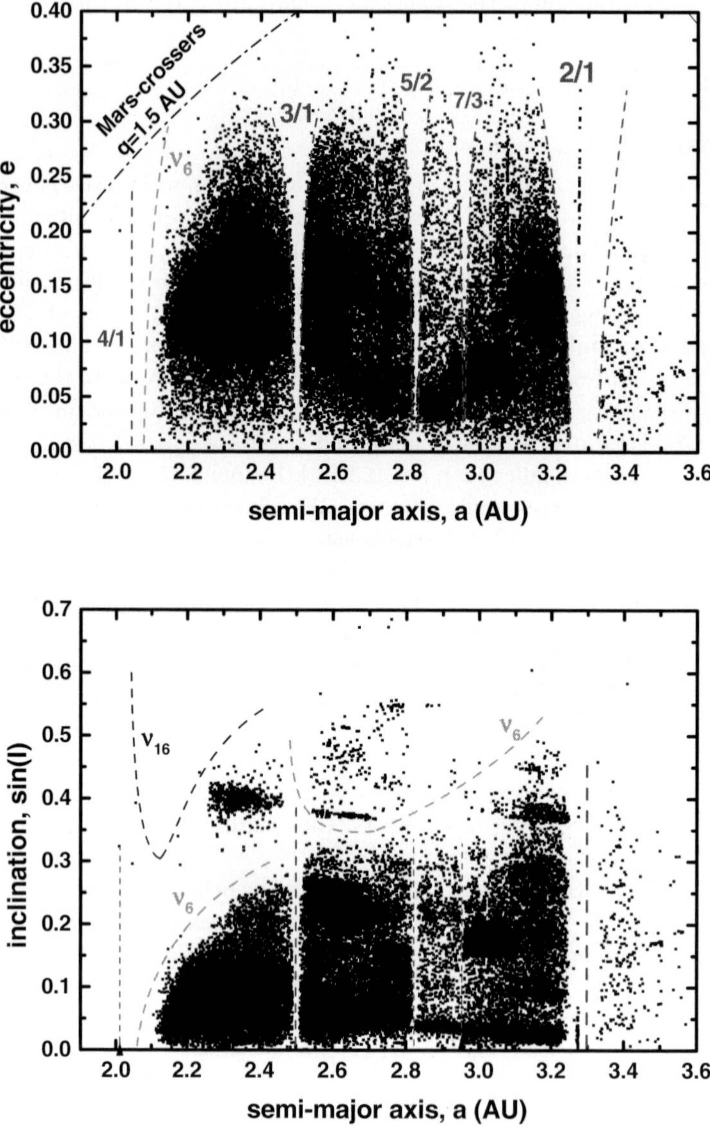

Figure 1. Distribution of main-belt asteroids in the (a, e) (top) and (a, i) (bottom) planes. The main mean motion resonances with Jupiter are indicated by "V"-shaped curves (top), which approximately define the width of each resonance zone. The location of first-order secular resonances (ν_i's), between the frequency of precession of the pericenter (grey) or the node (black) of an asteroid and the corresponding frequency of Saturn are also shown. Note how the locations of the resonances coincide with gaps in the distribution of asteroids.

3. The trans-Neptunian belt(s)

The first trans-Neptunian object was discovered by Jewitt & Luu (1993). Nowadays we know with a good accuracy the orbits of ~1000 objects. Their distribution is shown in Fig. 2. The trans-Neptunian objects are divided among several groups, depending on the dynamical characteristics of their orbits:

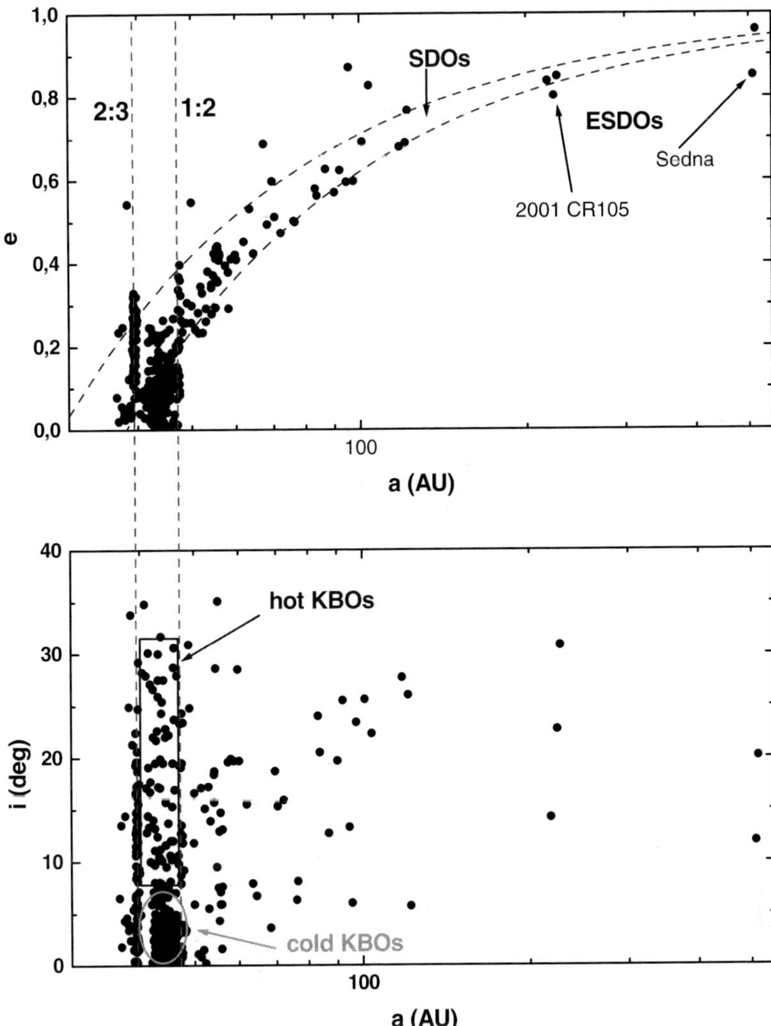

Figure 2. Distribution of the trans-Neptunian objects in the (a, e) (top) and (a, i) (bottom) planes. Lines of constant perihelion distance $q = 30\,\mathrm{AU}$ and $q = 39\,\mathrm{AU}$ are drawn in the top panel, to distinguish SDOs from ESDOs. The locations of the main mean motion resonances with Neptune are indicated by the vertical lines.

• The *"classical" Kuiper Belt objects* (KBOs) have orbits with $40 \leqslant a \leqslant 50\,\mathrm{AU}$ and $0 \leqslant e \leqslant 0.2$. These are stable orbits, since the KBOs are neither in a low-order resonance nor do they approach Neptune (Duncan *et al.* 1995). Frequently we distinguish between dynamically "cold" KBOs, which have $i \leqslant 5°$, and "hot" KBOs ($i > 5°$). This distinction seems to be also consistent with different spectroscopic properties (Brown 2001; Trujillo & Brown 2002). I should also point out that the apparent edge of the classical belt at $\sim 50\,\mathrm{AU}$ is most likely real. If a significant number of bodies in low-e orbits existed beyond 50 AU, the observational surveys would have detected at least a few of them by now. Several mechanisms have been proposed to explain the truncation of the primordial disc, including an early close stellar passage (Ida *et al.* 2000; Kobayashi & Ida 2001).

• The *Resonant objects* are trapped in a mean motion resonance with Neptune. The most well-known are the so-called *Plutinos* which, like Pluto itself, are captured in the 2:3 resonance. The resonant objects are estimated to be $\sim$7% of the total population. Malhotra (1995, 1996) was the first to show that the Plutinos could have been adiabatically captured into resonance, during Neptune's migration.

• The *scattered disc objects* (SDOs) have orbits with high semi-major axes and eccentricities and their perihelion distances are roughly between 30 and 39 AU. Thus, they can have close encounters with Neptune. In fact, this is considered to be the origin of the SDOs: they were scattered to high-e orbits by Neptune, as it was migrating outwards. I should note that SDOs were predicted theoretically by Duncan & Levison (1997), before being observed (Luu *et al.* 1997), as the most likely source of Jupiter-family comets.

• The *extended scattered disc objects* (ESDOs) also have high values of a and e, but their perihelion distances are larger than 40 AU. The most famous ESDOs are 2000 CR105 and *Sedna* (former 2003 VB12, Brown *et al.* 2004). Their large perihelion distances imply that they are not coupled to Neptune. That is exactly the problem of their origin: if they originated from the scattered disc, then an "extra" torque is required to decouple them from the planets. A number of possible solutions have been recently proposed by Morbidelli & Levison (2004).

It is interesting that, with the exception of the classical KBOs, the origin of all other groups is linked to the migration of Neptune. Thus, before discussing the formation of the KBOs, I will review recent results on planetary migration in our solar system.

4. Planetary migration

Let me briefly describe the process of planetary migration by planetesimal scattering. The principle of this process was first demonstrated by Fernandez & Ip (1984) and was later used by Malhotra (1995, 1996) to explain the origin of the Plutinos.

Suppose that a small particle with $a > a_{Nep}$ approaches Neptune, i.e. it has an eccentricity such that its orbit intersects the one of Neptune. Then, a close encounter with Neptune can decrease its semi-major axis. Conservation of angular momentum implies that Neptune should increase its own semi-major axis by a very small amount. A subsequent encounter may have the inverse effect, so that the net change of Neptune's semi-major axis would be zero. On the other hand, if the encounter is so effective that the particle is ejected to Oort cloud distances on a hyperbolic orbit, then the net change of Neptune's semi-major axis is negative.

However, when more than one planet exists in the system, the majority of the disc particles have a different evolution; this is true for our own system. The first encounter with Neptune, which decreases the particle's semi-major axis, can deliver the body to a Uranus-crossing orbit. Then, Neptune's small increase in a_{Nep} becomes permanent. Similarly, Uranus may hand over the particle to Saturn, and then Saturn to Jupiter. Jupiter, having a much larger mass than the other planets, ejects all particles to hyperbolic orbits. Thus, the net effect on the four-planets system is: Saturn, Uranus and Neptune move outwards, while Jupiter moves inwards. This was indeed observed in simulations, first by Hahn & Malhotra (1999). The speed of migration is roughly inversely proportional to the mass of the planet, but also depends on the orbital separation between planets, as the latter determines the efficiency of the "passing" mechanism. Thus, Neptune moves faster than any other planet, Uranus moves much slower than Neptune but still faster than

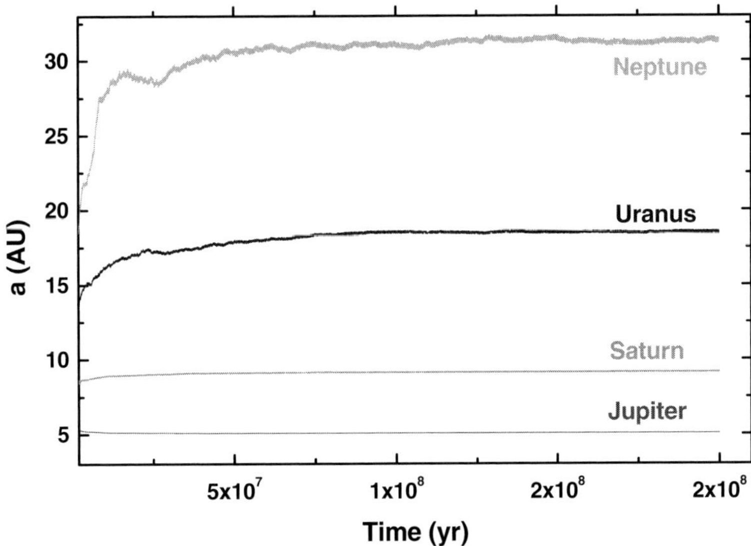

Figure 3. Evolution of the semi-major axes of Jupiter, Saturn, Uranus and Neptune as functions of time. Taken from an $N-$body simulation of planetary migration in a disc of 50 Earth masses (represented by 1000 equal-mass tracer bodies). The final values of a are very close to the observed ones for all four planets.

Saturn, while Jupiter has the smallest absolute variation in a. The result of a numerical simulation is shown in Fig. 3.

The migration of Neptune could generate the populations of SDOs and resonant objects. The most important component of the trans-Neptunian population – the KBOs – seems, at a first glance, not to be related to this process. Yet, its formation is still a mystery for solar system science. Fig. 4 shows a series of "snapshots", taken from a numerical simulation of planetary migration. The creation of a scattered disc, from an initially "cold" disc, is shown. Note also that, despite the fact that the disc was initially truncated at 30 AU, at the end of the simulation a few classical KBOs were also found in the 40 − 50 AU region. Obviously they were somehow transported outwards, during the migration of the planets.

5. Building the classical Kuiper belt

What is the origin of the classical KBOs? The obvious answer is: they were formed right where they are now. However, this solution opens the door to another problem (don't they always?). The total mass of the initial disc needed for the currently observed number of KBOs to have accreted *in situ* is too big (Stern 1996), compared to the observed mass (Jewitt *et al.* 1996; Chiang & Brown 1999). Two mechanisms have been proposed for getting rid of this extra mass: (i) dynamical elimination by Earth-sized planetary embryos (Morbidelli & Valsecchi 1997), or (ii) collisional grinding (Stern & Colwell 1997; Davis & Farinella 1998). However, they both lead to the same problem: if Neptune was migrating in this massive disc that was extending possibly beyond 50 AU, would it still stop at 30 AU?

According to the results of Gomes *et al.* (2004), the answer to the above question is no! For a disc as massive as necessary to form KBOs at 50 AU, Neptune would have migrated

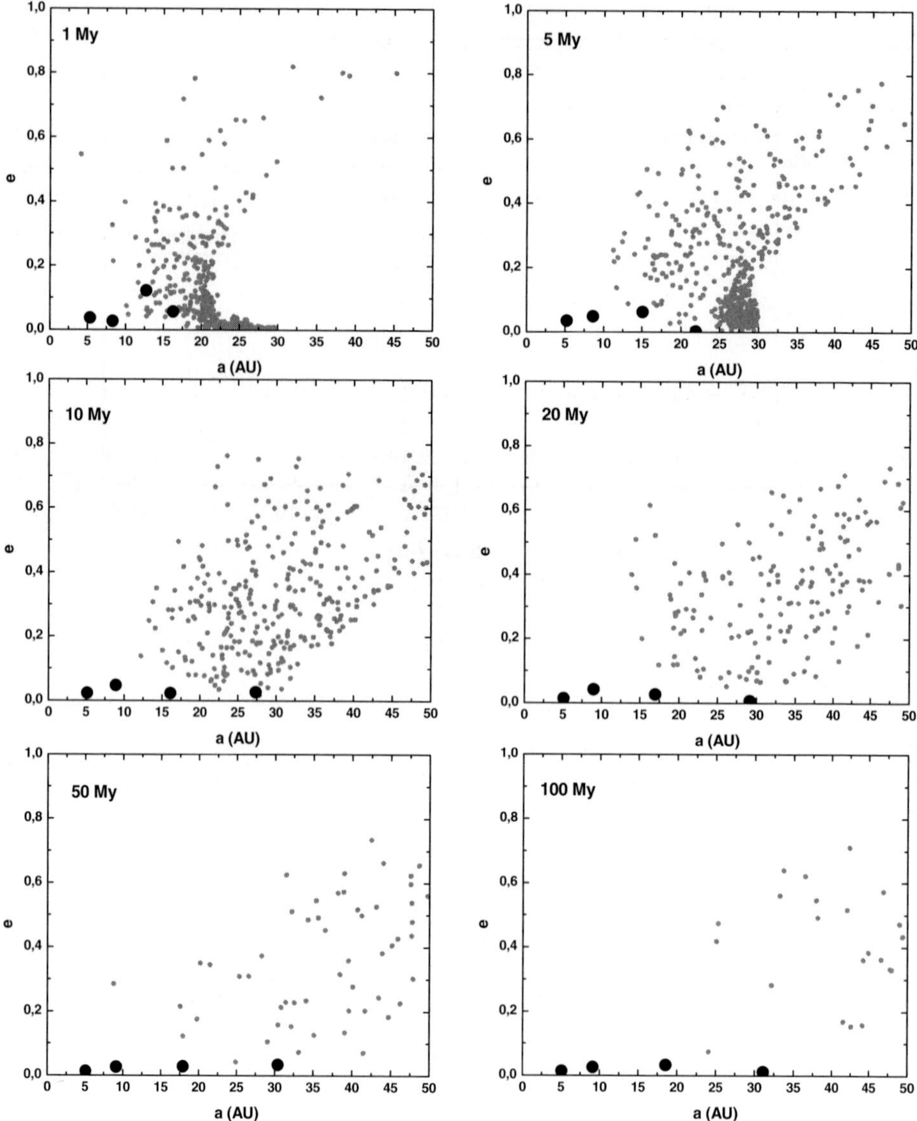

Figure 4. Planetary migration in a disc of planetesimals. Six snapshots of a simulation are shown: at $t = 1\,\mathrm{Myr}$, $5\,\mathrm{Myr}$, $10\,\mathrm{Myr}$, $20\,\mathrm{Myr}$, $50\,\mathrm{Myr}$ and $100\,\mathrm{Myr}$. The variation of the semi-major axes of the planets is apparent. In this simulation, the disc contained initially ~ 50 Earth masses. At the end of the simulation ($100\,\mathrm{Myr}$), less than ~ 4 Earth masses are still present, mostly in the scattered disc. Note that, despite the fact that the disc was originally truncated at $30\,\mathrm{AU}$, a few classical KBOs with $a > 40\,\mathrm{AU}$ seem to be present at the end of the simulation. They were most likely transported there by the mechanisms described in Gomes (2003) and Levison & Morbidelli (2003). Note however that the resolution of this experiment is not adequate to correctly represent resonance trapping (see Gomes *et al.* 2004). Thus, very few objects were found to be transported by resonances.

all the way to the edge of the disc. It is thus more likely that the planetesimals' disc was somehow truncated near $\sim 35\,\mathrm{AU}$, which means that there was never enough mass to form the KBOs *in situ*. An alternative idea for the origin of the KBOs, proposed by Gomes (2003) and Levison & Morbidelli (2003), is that they formed in the same region as Uranus and Neptune, i.e. much closer to the Sun than now, and were transported outwards during planetary migration. We note that the mechanism proposed by Gomes (2003) explains the origin only of the "hot" KBOs, while the mechanism of Levison & Morbidelli (2003) explains the origin of the "cold" KBOs.

Gomes (2003) performed numerical simulations, in which he observed that, as Uranus and Neptune migrated outwards, objects that were scattered away to high values of e and i were temporarily getting captured into resonance with Neptune. As Neptune was moving towards 30 AU, objects were constantly penetrating the "hot" classical KBOs region, being released from resonance at different values of e and i. The most unstable ones could not have survived for the age of the solar system. However a small, but sufficient, number of objects, with orbits similar to the ones of the "hot" KBOs, were produced.

Levison & Morbidelli (2003) started by asking themselves why the edge of the "cold" Kuiper belt almost coincides with the location of the 1:2 resonance ($\sim$48 AU) with Neptune. This observation led them to perform a number of numerical simulations, during which they observed that objects, initially following "cold" orbits (e, $\sin i \sim 0$) exterior to the one of Neptune but interior to 30 AU, were getting trapped and "pushed out" by the 1:2 resonance. Particles that were from time to time released from the resonance had eccentricities ranging from 0.8 down to zero. The low eccentricity objects developed stable orbits, very similar to those of the "cold" KBOs. Note that the 1:2 resonance does not excite the inclinations. Thus, mostly "cold" objects were created in the simulations of Levison & Morbidelli (2003).

The low efficiency of these two mechanisms, along with the results of Gomes *et al.* (2004), suggest a new scenario for planetary migration, in which Neptune was initially interior to $\sim$18 AU and the disc of planetesimals was truncated at $\sim$30 AU. The question now is: what is the effect of this planetary migration model on the orbital structure of the asteroid belt?

6. Depletion of the asteroid belt

As mentioned in section 2, the current mass of the asteroid belt is estimated to be at least 1000 times smaller than its initial one. The largest part of the implied depletion is supposed to have taken place during the formation of the terrestrial planets. Petit *et al.* (2001) have shown that proto-planetary embryos, residing at that time in the belt, would have excited the orbits of the asteroids, forcing more than 99% of these bodies to escape. The remaining asteroids would have achieved a distribution in the (a, e, i) space, similar to the current one.

Levison *et al.* (2001) have shown that, if the migration of the outer planets occurred early on when the asteroid belt was still dynamically "cold", the asteroid belt would have been almost completely depleted! The reason for this catastrophe is that, as the planets move, a strong 1:1 resonance between the perihelion precession frequency of an asteroid and the one of Saturn (called the ν_6 resonance) would have swept across the belt. The effect of this resonance would be to increase the eccentricity of any asteroid from 0 to nearly 1, thus placing the asteroid on a planet-crossing orbit. This phenomenon poses a serious problem for any migration scenario.

However, if the migration of the outer planets happened somehow late, the effect of the ν_6-sweeping may not have been the same. Assume that the belt was not dynamically "cold", but was already pre-excited by the mechanism described in Petit *et al.* (2001), during terrestrial planet formation and before the migration of the outer planets had started. Then, as the ν_6 resonance was sweeping the belt, different asteroids would

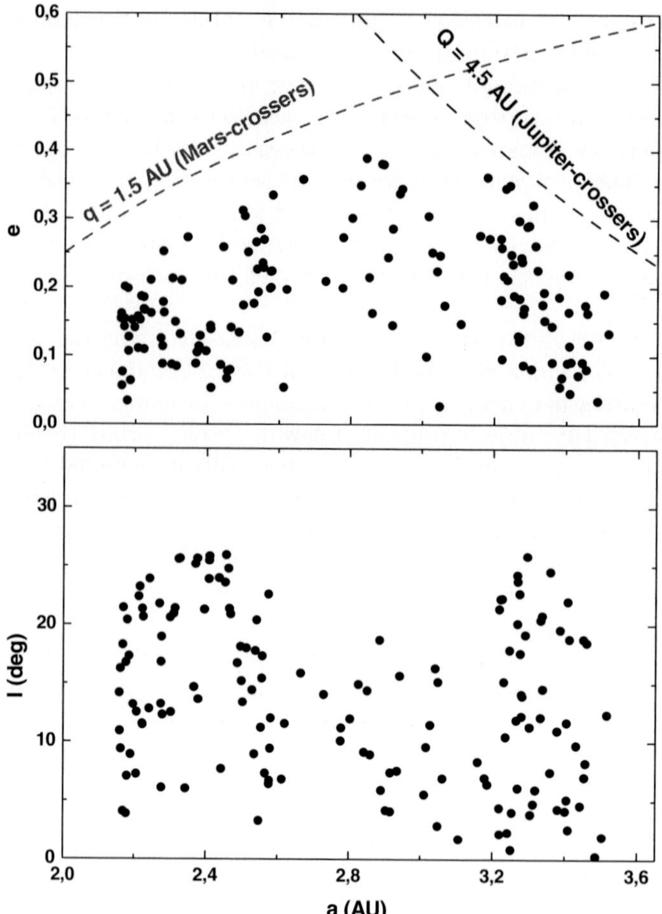

Figure 5. Depletion of the asteroid belt. The distribution of the fictitious asteroids, originally in pre-excited orbits, that survive planetary migration is shown. The shape of this remnant asteroid belt is close to the observed one. Main-belt asteroids reside on orbits that cross neither Mars's orbit, nor that of Jupiter, i.e. their perihelion distance is $q > 1.5\,\mathrm{AU}$, and their aphelion distance $Q < 4.5\,\mathrm{AU}$. The two limiting curves, $q = 1.5\,\mathrm{AU}$ and $Q = 4.5\,\mathrm{AU}$ are superimposed on the (a, e) plot.

encounter the resonance at different eccentricities and perihelion orientations. This "mixing" of initial conditions would lead to a "mixed" evolution, in which some asteroids would have had their eccentricities increased and some decreased. Provided that resonance sweeping occurred sufficiently fast, comparable numbers of high-e (unstable) and low-e (stable) asteroids would have been produced by this process. This orbital "mixing" could not have occurred if the belt was "cold" (all asteroids having $e \sim 0$); the resonance would have increased the eccentricities of all asteroids.

We simulated this process by integrating the orbits of Jupiter, Saturn and 1000 fictitious asteroids, initially on orbits with $2.0 \leqslant a \leqslant 4.0$, $0 \leqslant e \leqslant 0.3$ and $0° \leqslant i \leqslant 20°$.

An analytic prescription of the migration drag-force was built in the code (Malhotra 1996), in order to force the planets to migrate with the desired rate. For migration times of $\sim 10\,$Myr, we found that $\sim 10\%$ of the asteroid belt survived. More importantly, the orbital distribution of the surviving bodies was close to the one currently observed. This is shown in Fig. 5. Thus, we can conclude that, in order to be able to observe today an asteroid belt, with its current shape, the migration of the outer planets must have occurred late, after the formation of the terrestrial planets was almost complete, i.e. $\sim 100\,$Myr after the formation of the Sun.

7. Open problems

Despite the long list of exciting new results that have been recently published by several authors and I have presented in this paper, it would be naive for me to argue that we have unveiled the early dynamical evolution of the solar system. There are still a number of important issues to be answered that could help us to link all parts of the story and come up with a self-consistent model. In this last section I will only mention two that I believe are the most important ones and speculate on possible solutions, in agreement with the above mentioned results. These two problems are: (i) the observed orbital configuration of the outer planets, and (ii) the Late Heavy Bombardment of the Moon (LHB).

Although the "standard" migration scenario can explain the large-scale variation of the semi-major axes of the outer planets, it does not reproduce the eccentricities and mutual inclinations of their orbits, which go up to 9% and $2°$ respectively. The reason is that, during migration, the planetesimals exert on the planets a significant amount of dynamical friction, which damps the eccentricities and inclinations to zero. Thus, some excitation mechanism needs to act, during migration, in order to end up with the correct planetary orbital configuration in all three elements. This yet unidentified mechanism can be either (i) encounters between the planets, or (ii) a resonant interaction. A possible, but rather violent version of (i) has been studied by Thommes *et al.* (1999). However, it requires a very massive disc to stabilize the planetary system, which is not consistent with some of the results presented in the previous sections. The second mechanism (resonance) seems more appealing and has to be explored in detail.

Probably the most well known observational fact about the solar system is the existence of the large basins on the surface of the Moon. The analysis of rock samples taken by the Apollo mission suggests that the basins were formed during an intense phase of bombardment of the Moon, around 3.9 Gyr ago. This cataclysmic event is usually referred to as the Late Heavy Bombardment (LHB). However, other evidence does not support this cataclysm and seems to favour an alternative evolution scenario, in which the bombardment was very heavy even from the formation of the Moon, and suddenly ended around 3.9 Gyr ago. For a comprehensive review see Hartman *et al.* (2000; also Kring & Cohen 2002). It is clear that more evidence is needed to resolve this conflict. However, a few arguments can be made from the point of view of dynamics. It is frequently stated that the LHB is very difficult to reproduce by a dynamical model. This is not correct. Levison *et al.* (2001) have shown that a late formation and migration of Uranus and Neptune could provide the estimated mass of LHB projectiles, in the form of comet-like bodies originating from the outer planetesimal disc. Moreover, as shown in the previous section, an important amount of mass (a few times the current mass of the asteroid belt) would become NEAs during planetary migration. More results are needed in order to quantify the efficiency of mass delivery from both sources. More importantly, we still have to understand if and how planetary migration could be delayed, in order for the bombardment to occur ~600 Myr after the formation of the Moon.

As a conclusion, I think that the results reviewed in this paper are very encouraging. A comprehensive model for the late phases of planetary migration by planetesimal scattering and the sculpting of the small-body reservoirs is close to be completed. It seems that the community is close to understanding the early dynamical evolution of the solar system. That is, of course, if we consider as "time zero" the time at which no gas is left anymore in the solar system. The evolution of the planetary system during the gas-dominated phase, i.e. the first ~10 Myr after the formation of the Sun, still remains largely unknown. I believe that linking these two epochs is the big task for solar system dynamics in the next decade; it is a fascinating time to be in this line of work!

Acknowledgements

I would like to thank the organizers of this meeting for inviting me to present a talk at this historic IAU Colloquium. This work is supported by an EC Marie Curie Individual Fellowship (contract N^o HPMF-CT-2002-01972).

References

Bottke, W.F., Cellino, A., Paolicchi, P., Binzel, R.P. (eds.) 2002. Asteroids III. University of Arizona Press, Tucson.

Brown, M.E. 2001 *AJ* 121, 2804.

Brown, M.E., Trujillo, C.A., Rabinowitz, D.L. 2004 *International Astronomical Union Circular* 8304, 1.

Chambers, J.E. 2001 *Icarus* 152, 205.

Chiang, E.I., Brown, M.E. 1999 *AJ* 118, 1411.

Davis, D.R., Farinella, P. 1998 Lunar Planet Science Conf. 29, 1437.

Duncan, M.J., Levison, H.F. 1997 *Science* 276, 1670.

Duncan, M.J., Levison, H.F. Budd, S.M. 1995 *AJ* 110, 3073.

Fernandez, J.A., Ip, W.H. 1984 *Icarus* 58, 109.

Gomes, R. 2003 *Icarus* 161, 404.

Gomes, R.S., Morbidelli, A., Levison, H.F. 2004 *Icarus* 170, 492.

Hahn, J.M., Malhotra, R. 1999 *AJ* 117, 3041.

Hartmann, W.K., Ryder, G., Dones, L., Grinspoon, D. 2000 in *Origin of the Earth and Moon* (R. Canup and K. Righter, eds.), p. 493, Univ. of Arizona Press, Tucson.

Ida, S., Larwood, J. Burkert, A. 2000 *ApJ* 528, 351.

Jewitt, D., Luu, J. 1993 *Nature* 362, 730.

Jewitt D., Luu J., Chen, J. 1996 *AJ* 112, 1225.

Kobayashi H. Ida, S. 2001 *Icarus* 153, 416.

Kring, D.A., Cohen, B.A. 2002 *Journal of Geophysical Research (Planets)* 107, 4.

Laskar, J., Joutel, F., Robutel, P. 1993 *Nature* 361, 615.

Levison, H.F., Stern, S.A. 2001 *AJ* 121, 1730.

Levison, H.F., Dones, L., Chapman, C.R., Stern, S.A., Duncan, M.J., Zahnle, K. 2001 *Icarus* 151, 286.

Levison, H.F., Morbidelli, A. 2003 *Nature* 426, 419.

Lubow, S.H., Seibert, M., Artymowicz, P. 1999 *ApJ* 526, 1001.

Luu, J., Jewitt, D., Trujillo, C.A., Hergenrother, C.W., Chen, J., Offutt, W.B. 1997 *Nature* 387, 573.

Lyn, D.N.C. Papaloizou, J.C.B. 1986 *ApJ* 309, 846.

Malhotra, R. 1995 *AJ* 110, 420.

Malhotra, R. 1996 *AJ* 111, 504.

Masset, F.S. Papaloizou, J.C.B. 2003 *ApJ* 588, 494.

Morbidelli, A. Valsecchi, G.B. 1997 *Icarus* 128, 464.

Morbidelli, A., Levison, H.F. 2004 *AJ* 128, 2564.

Moons, M. 1997 *Celestial Mechanics and Dynamical Astronomy* 65, 175.

Petit J.M., A. Morbidelli, J. Chambers. 2001 *Icarus*, 153, 338.

Poincaré, H. 1892. *Les methodes nouvelles de la mecanique celeste. Gauthier-Villars et fils.* Paris.

Pollack, J.B., Hubickyi, O., Bodenheimer, P., Lissauer, J.J., Podolack, M., Greenzweig, Y. 1996 *Icarus* 164, 62.

Stern, S.A. 1996 *AJ* 112, 1203.

Stern, S.A. Colwell J.E. 1997 *AJ* 114, 841.

Sussman, G.J., Wisdom, J. 1992 *Science* 257, 56.

Thommes, E.W., Duncan, M.J., Levison, H.F. 1999 *Nature* 402, 635.

Trujillo, C.A., Brown, M.E. 2002 *ApJ* 566, L125.

Ward, W.R. 1997 *Icarus* 126, 261.

Weidenschilling, S. 2003 in: *Comets II*. Festou *et al.* (eds.). University of Arizona Press, Tucson.

Discussion

BERNARD DE SAEDELEER: Are the planets are still migrating?

KLEOMENIS TSIGANIS: Now? No, because there's no mass left to drive the migration. There are bodies encountering Neptune – this is the scattered disk. But the planets are now far apart, so the chain of passing particles from one planet to another is broken.

CHRISTINE ALLEN: I'm concerned by your saying that the disk gets truncated at 50 AU by a passing star. Work we have done in Mexico shows that passing stars will begin to truncate the distribution of semi-major axes at about 3000 AU for that age.

KLEOMENIS TSIGANIS: You mean by stars as they are now distributed in the galaxy?

CHRISTINE ALLEN: Or as they were millions of years ago.

KLEOMENIS TSIGANIS: Recent simulations show that if a star passed near enough within the Oort cloud, it could truncate the disk exactly at 50 [AU]. I haven't done the simulation myself, but it seems possible. Of course, you have to understand that people advocating these arguments assume that the Sun was most likely formed in a very small cluster, and that this truncation event happened very early.

MIKHAIL MAROV: With the recent discovery of the large bodies in the Kuiper Belt – bodies such as Quaoar and Sedna – do you think still that there has been mass depletion in the Belt? And that there is still something like less then the mass of the Earth for the total mass of the Belt? It is possible that a large-size body could be still found. I don't know whether you will designate them as cold or hot bodies.

KLEOMENIS TSIGANIS: They are scattered disk bodies. They have very eccentric orbits.

MIKHAIL MAROV: That's what I mean - bodies with very eccentric orbits. This is possibly the reason why we haven't found them yet.

KLEOMENIS TSIGANIS: I said that there would be no bodies outside 50 AU for the classical Kuiper Belt. With the surveys done, compiled and de-biased, it seems that if there were big bodies on low inclination eccentric orbits, they would already have been discovered. Now these bodies – Sedna, for example – are scattered disk bodies in highly eccentric orbits. The total mass of the scattered disk is estimated to be more or less the same as the Kuiper Belt, so a few hundredths to a few tenths of an Earth mass.

MIKHAIL MAROV: You did not mention that the idea that Uranus and Nepture formed very close to the Jupiter-Saturn position was first put forward by Safronov and his school at the beginning of the 80s. It was made because the very specific composition of Neptune and Uranus can hardly be explained in terms of the current position of formation of these bodies. A second comment: I disagree a little with the estimate which you drew in your presentation for the LHB projectiles. You quoted something like $7 - 10 \times 10^{22}$ g for Earth. In our publication we estimated a bit more; it's quite comparable to the mass of the Earth's ocean, so in the last phase of the heavy bombardment it can be possible to explain even the mass of the Earth's ocean.

KLEOMENIS TSIGANIS: Yes, the estimates that I used are from the recent paper by Levison *et al.* (2001). If I remember correctly, the estimate for the Earth and Mars are more or less the same, which would provide Mars with atmospheric water, but not the water of an ocean. The latest papers that I used as a base for this calculation provide something like 5% of the ocean.

MIKHAIL MAROV: It's important because we've found that it was even comparable in size to the mass delivered by the cometary-like bodies. So this is an explanation for the early ocean on all terrestrial bodies.

KLEOMENIS TSIGANIS: There is an alternative solution for the water on Earth which refers to the first phase of depletion of the asteroid belt – by the end of the formation of the Earth, a lunar-size body encounters the Earth and provides the oceans.

MIKHAIL MAROV: I am not in favour of the idea that it was direct bombardment from the Kuiper Belt to the inner planets. The major part comes first from Jupiter orbit-crossers.

KLEOMENIS TSIGANIS: Of course . . . Jupiter family comets coming from the disk.

FLOOR VAN LEEUWEN: During those first 100 million years the Sun itself goes through quite a violent phase of its evolution. If the observations of the Pleiades and Orion are anything to go by, then it has gone through a phase of very rapid spin-up, just before becoming an actual main sequence star, followed by a short period of release of angular momentum – very strong magnetic fields going through the whole system. This angular momentum is comparable to the whole angular momentum of the solar system. Where does that feature in your scenario?

KLEOMENIS TSIGANIS: Nowhere.

FLOOR VAN LEEUWEN: Nowhere? That's the problem! I have seen these scenarios being put out year after year. These phenomena have been known for years and it doesn't seem to penetrate this whole formation of the solar system story.

KLEOMENIS TSIGANIS: I skipped the first 10 million years, because even the gas alone with the quiet Sun is a big problem for doing a dynamical model that could be handled, even numerically. We are now trying to understand more deeply these effects of gas and radiation. But this is really an embryonic stage for the dynamicist, because we do not know these phenomena well enough to simulate them, for the moment. So slowly we are getting there.

Transits of Venus: New Views of the Solar System and Galaxy
Proceedings IAU Colloquium No. 196, 2004
D.W. Kurtz, ed.

© 2004 International Astronomical Union
doi:10.1017/S174392130500147X

Classical and modern orbit determination for asteroids

Giovanni F. Gronchi

Department of Mathematics, University of Pisa, Via Buonarroti 2, 56127 Pisa, Italy
email: gronchi@dm.unipi.it

Abstract. With the substantial improvements in observational techniques we have to deal with very big databases, consisting of a few positions of an object over a short time span; this is often not enough to compute a preliminary orbit with traditional tools. In this paper we first review a classical method by C.F. Gauss to compute a preliminary orbit for asteroids. This method, followed by a least squares fit to improve the orbit, still today gives successful results when we have at least three separate observations. Then we introduce the basics of a very recent orbit determination theory, that has been thought just to be used with modern sets of data. These data allow us in many cases to know the angular position and velocity of an asteroid at a given time, even though the radial distance and velocity $(r, \dot{r})$, needed to compute its full orbit, are unknown. The variables $(r, \dot{r})$ can be constrained to a compact set, that we call the *admissible region* (AR), whose definition requires that the body belongs to the Solar System, that it is not a satellite of the Earth, and that it is not a "shooting star" (i.e. very close and very small). We provide a mathematical description of the AR: its topological properties are surprisingly simple, in fact it turns out that the AR cannot have more than two connected components. A sampling of the AR can be performed by means of a Delaunay triangulation; a finite number of six-parameter sets of initial conditions are thus defined, with each node of the triangulation representing a possible orbit (a *virtual asteroid*).

1. Classical orbit determination: Gauss's method

An important problem in orbit determination is the computation of the orbital elements of a celestial body using a given set of angular observations (α_i, δ_i), that are, for example, the *right ascension* and the *declination* of the object on the *celestial sphere*.

This problem became particularly challenging when the Italian astronomer Giuseppe Piazzi discovered the first asteroid (Ceres) on 1 January 1801; he could follow this asteroid until 11 February 1801 writing for it 21 observations that covered only $3°$ in the sky (Foderà Serio *et al.* 2003).

At the beginning of the 19th century some methods for orbit determination had already been invented, but they did not allow the recovery of Ceres in the sky; the nature of the observations at disposal made the problem to be set as follows:

> *determinare orbitam corporis coelestis, absque omni suppositione*
> *hypothetica, ex observationibus tempus haud magnum complectentibus*
> *neque adeo delectum, pro applicatione methodorum specialium,*
> *patientibus* † (C. F. Gauss 1809)

The two astronomers Franz Von Zach and Heinrich W. Olbers independently recovered

---

† 'to determine the orbit of a celestial body, without making any hypothesis, from observations covering a space neither too large nor such as to allow the special methods to be applied'

Ceres (on 31 December 1801 and on 2 January 1802, respectively) following the indications of the German mathematician Carl Friedrich Gauss, who developed very important mathematical techniques just to deal with this problem (see Gauss 1809).

1.1. *Gauss's method for preliminary orbits*

The orbital elements, defining the Keplerian orbit of an object, are six, so that we can try to determine completely an orbit using only three angular observations $\{(\alpha_i, \delta_i), i = 1, 2, 3\}$.

Let us write three heliocentric positions $\vec{r}_k$ of a celestial body as

$$\vec{r}_k = \rho_k \vec{s}_k + \vec{R}_k \,; \qquad k = 1, \ldots, 3 \,, \tag{1.1}$$

where $\vec{R}_k = (R_{1,k}, R_{2,k}, R_{3,k})$ is the (known) heliocentric position of the observer at time t_k; $\vec{s}_k = (s_{1,k}, s_{2,k}, s_{3,k})$ is the (known) position of the body at time t_k on the unit sphere centered at the observer's position,

$$\vec{s}_k = (\cos \delta_k \cos \alpha_k, \cos \delta_k \sin \alpha_k, \sin \delta_k) \,,$$

and the three positive scalars ρ_k are unknown.

As a Keplerian orbit lies on a plane, the vectors $\vec{r}_k$ must be linearly dependent: there exist $c_1, c_2, c_3 \in \mathbb{R}$ such that

$$c_1 \vec{r}_1 + c_2 \vec{r}_2 + c_3 \vec{r}_3 = 0 \,. \tag{1.2}$$

Set $c_2 = -1$, so that

$$\vec{r}_2 = c_1 \vec{r}_1 + c_3 \vec{r}_3 \,,$$

and

$$\begin{cases} \vec{r}_1 \wedge \vec{r}_2 &= c_3 \vec{r}_1 \wedge \vec{r}_3 \\ \vec{r}_3 \wedge \vec{r}_2 &= c_1 \vec{r}_3 \wedge \vec{r}_1 \end{cases} .$$

If $\vec{r}_1 \wedge \vec{r}_3 \neq \vec{0}$ we can write the coefficients in terms of the f_k, g_k series (see Roy 1988) for $k = 1, 3$:

$$c_1 = \frac{g_3}{f_1 \, g_3 - f_3 \, g_1} \,; \qquad\qquad c_3 = \frac{-g_1}{f_1 \, g_3 - f_3 \, g_1} \,,$$

where f_k, g_k are such that

$$\vec{r}_1 = f_1 \vec{r}_2 + g_1 \dot{\vec{r}}_2 \,; \qquad\qquad \vec{r}_3 = f_3 \vec{r}_2 + g_3 \dot{\vec{r}}_2 \,.$$

Taking $t = t_2$ for the origin of the time, the values t_1, t_3 are comparably small; thus we can approximate the f_k, g_k series up to the third order in t_k:

$$f_k \approx 1 - \frac{\mu \, t_k^2}{2 r_2^3} \,; \qquad\qquad g_k \approx t_k - \frac{\mu \, t_k^3}{6 r_2^3} \,;$$

with μ the *total mass* of the two bodies. In case of an asteroid (a small body) orbiting around the Sun, μ can be taken as the mass of the Sun only. Note that the approximations of f_k, g_k depend only on r_2.

Substituting (1.1) in (1.2) we obtain

$$c_1 \rho_1 \vec{s}_1 + c_2 \rho_2 \vec{s}_2 + c_3 \rho_3 \vec{s}_3 + c_1 \vec{R}_1 + c_2 \vec{R}_2 + c_3 \vec{R}_3 = 0 \,, \tag{1.3}$$

that gives a system of 3 equations in the 4 unknowns $\rho_1, \rho_2, \rho_3, r_2$. Note that we also have the relation

$$r_2^2 = \rho_2^2 + 2 \rho_2 \vec{s}_2 \cdot \vec{R}_2 + R_2^2 \,, \tag{1.4}$$

that gives r_2 in terms of ρ_2.

Then we use the approximations

$$c_1 = c_1(r_2) \approx \frac{t_3}{t_3 - t_1}\left[1 + \frac{\mu}{6r_2^3}((t_3 - t_1)^2 - t_3^2)\right] := A_1 + \frac{B_1}{r_2^3} \ ;$$

$$c_3 = c_3(r_2) \approx -\frac{t_1}{t_3 - t_1}\left[1 + \frac{\mu}{6r_2^3}((t_3 - t_1)^2 - t_1^2)\right] := A_3 + \frac{B_3}{r_2^3} \ .$$

REMARK: the coefficients

$$A_1 = \frac{t_3}{t_3 - t_1} \ ; \quad B_1 = \frac{\mu\, t_3}{6(t_3 - t_1)}\left((t_3 - t_1)^2 - t_3^2\right) \ ;$$

$$A_3 = -\frac{t_1}{t_3 - t_1} \ ; \quad B_3 = -\frac{\mu\, t_1}{6(t_3 - t_1)}\left((t_3 - t_1)^2 - t_1^2\right)$$

are positive, and c_1, c_3 are functions of the variable r_2 only.

We can write equation (1.3) as follows

$$\begin{pmatrix} s_{1,1} & s_{1,2} & s_{1,3} \\ s_{2,1} & s_{2,2} & s_{2,3} \\ s_{3,1} & s_{3,2} & s_{3,3} \end{pmatrix} \begin{pmatrix} c_1\rho_1 \\ c_2\rho_2 \\ c_3\rho_3 \end{pmatrix} = -\begin{pmatrix} G_1 \\ G_2 \\ G_3 \end{pmatrix} \ ,$$

where $G_i = \sum_{k=1}^{3} R_{i,k}c_k$ and $R_{i,k}$ is the i-th component of $\vec{R}_k$. Computing $\mathcal{S}^{-1} := (\sigma_{i,k})_{i,k}$, the inverse of the matrix $\mathcal{S} := (s_{i,k})_{i,k}$, we obtain

$$c_k\rho_k = -\sum_{i=1}^{3} \sigma_{k,i}\, G_i \ ; \qquad k = 1,\ldots,3$$

and in particular

$$\rho_2 = \sum_{i=1}^{3} \sigma_{2,i}\, G_i \ . \tag{1.5}$$

By substituting the approximated values of c_1, c_3 in (1.5) we obtain

$$\rho_2 = \sum_{i=1}^{3} \sigma_{2,i} \sum_{k=1}^{3} R_{i,k}c_k = \widetilde{A}_2 + \frac{\widetilde{B}_2}{r_2^3} \ , \tag{1.6}$$

with

$$\widetilde{A}_2 = \sum_{i=1}^{3} \sigma_{2,i} \sum_{k=1}^{3} R_{i,k}A_k \ ; \qquad \widetilde{B}_2 = \sum_{i=1}^{3} \sigma_{2,i} \sum_{k=1}^{3} R_{i,k}B_k \ ,$$

where we have defined $A_2 = 1, B_2 = 0$.

Finally, substituting relation (1.6) into the equation (1.4), we obtain an 8th-degree equation for r_2:

$$p(r_2) = r_2^8 - \left[\widetilde{A}_2^2 + 2\widetilde{A}_2\, \langle \vec{s}_2, \vec{R}_2\rangle + \|\vec{R}_2\|^2\right] r_2^6 - 2\widetilde{B}_2\left[\widetilde{A}_2 + \langle \vec{s}_2, \vec{R}_2\rangle\right] r_2^3 - \widetilde{B}_2^2 = 0 \ , \tag{1.7}$$

whose positive solutions gives us preliminary orbits.

REMARKS ON THE SOLUTIONS OF (1.7):

(i) as $p(r_2)$ has only 4 terms, by Cartesio's sign rule we know that equation (1.7) cannot have more than 3 positive solutions;

(ii) the coefficient of the 6th-degree term is always non-positive because we have

$$\left[\widetilde{A}_2^2 + 2\widetilde{A}_2\, \langle \vec{s}_2, \vec{R}_2\rangle + \|\vec{R}_2\|^2\right] = \|\widetilde{A}_2\vec{s}_2 + \vec{R}_2\|^2 \geqslant 0 \ ;$$

(iii) the scalar product $\langle \vec{s}_2, \vec{R}_2 \rangle$ is positive if the observations are done in the region of the sky from quadrature to opposition;

(iv) the only coefficient of $p(r_2)$ that may change sign is the coefficient of the 3rd-degree term (the only odd term) so that, apart from vanishing of the coefficients, the sequence of signs of the coefficients of $p(r_2)$ can be

$$[+, \ -, \ +, \ -] \quad \text{or} \quad [+, \ -, \ -, \ -]$$

and, respectively, the sequence of signs of the coefficients of $q(r_2) := p(-r_2)$ can be

$$[+, \ -, \ -, \ -] \quad \text{or} \quad [+, \ -, \ +, \ -] \,.$$

From the previous remarks we notice that $p(r_2)$ has at most 4 roots, and at most 3 positive ones. In fact, the number of sign changes of either the coefficients of $p(r_2)$ or the coefficients of $q(r_2) := p(-r_2)$ is only one, so that Cartesio's rule says that either on the positive axis $r_2 > 0$ or on the negative axis $r_2 < 0$ there is exactly one root.

In several cases we obtain only one preliminary orbit from the method explained before; this orbit can be improved when additional observations are available and Gauss himself introduced the *least squares method* just to deal with this problem: the improvement procedure is usually called *differential corrections* and the classical algorithm for orbit determination, often used still today, consists of the two steps:

(a) search for a preliminary orbit;

(b) apply the differential corrections to improve it.

HINT: When the arc of the observations is short, or when it does not present significant geodetic curvature, then Gauss's method cannot be applied. In fact, the shortness of the arc corresponds to the smallness of the determinant of the matrix $\mathcal{S}$ that is inverted in this algorithm, and the uncertainty in the observations may allow the situation shown in Fig. 1 in the case of short arcs.

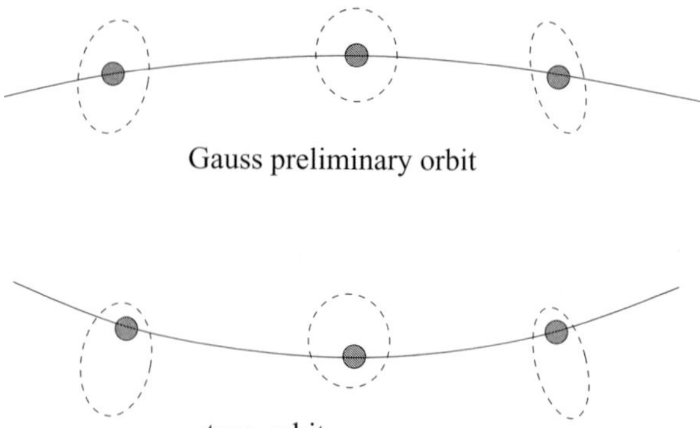

Figure 1. Qualitative sketch in which the orbit given by Gauss's method (preliminary orbit plus differential corrections) presents a curvature that is not reliable at all.

The observations of an asteroid/comet at Gauss's epoch (19th century) were not more than one per night, while today we have to deal with sets of observations all referred to the same night, in particular the hypotheses used by Gauss are usually *not* fulfilled for the modern sets of data.

2. Modern orbit determination: too short arcs

We introduce some definitions and some mathematical tools that can be used in the orbit determination problem when the hypotheses of Gauss's method are not fulfilled.

2.1. *Preliminary definitions*

A *sequence of observations* is a set of astrometric observations belonging to the same object:

$$t_i, \alpha_i, \delta_i, h_i\,; \qquad i = 1, \ldots, m\,; \qquad m \geqslant 2,$$

where α_i, δ_i are angles (RA, DEC) and t_i are times, with $t_i < t_{i+1}$. The quantities h_i are the (optional) values of the apparent magnitude.

A sequence of observations is called a *very short arc* if the observations are known to belong to the same object just because they can be fit together by some smooth curve, typically a low degree polynomial.

If either the preliminary orbit cannot be determined, or it can but the differential corrections procedure does not converge, we speak of a *too short arc*.

We recall the following fundamental concept, first introduced in (Milani *et al.* 2001):
Definition: we shall call *attributable* a vector

$$\xi = (\alpha, \delta, \dot{\alpha}, \dot{\delta}) \in [-\pi, \pi) \times (-\pi/2, \pi/2) \times \mathbb{R}^2$$

representing a sequence of observations at the average time $t^* = Mean(t_i)$.

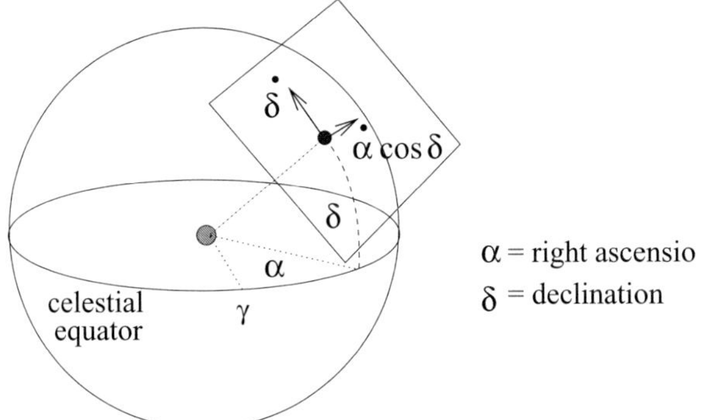

Figure 2. The *celestial sphere* and the components of an attributable.

The average apparent magnitude h^* is also part of the attributable if it is available, i.e. if at least one value h_i of the apparent magnitude is given.

Note that for a given attributable the (geocentric) radial position and velocity $r, \dot{r}$ of the body are *completely undetermined*.

2.2. *Attributables and orbit determination*

If the angular arc length is large enough it may be possible to compute a preliminary orbit by classical methods, e.g. by Gauss's method. Otherwise, an attributable can be used to perform the following operations:

(*a*) an *attribution*, that is a least squares fit identification of the object the attributable is referred to, with another object with a known orbit;

(*b*) a *linkage*, that is to join together two attributables, referring to two objects both without a full orbit, to form a set of observations allowing to compute an orbit;

(*c*) a *recovery* (resp. a *precovery*), that is to find *in the sky* (resp. *in the archives*) an object for which only an attributable is available; this operation may be difficult, but is possible for a short time after/before t^*.

In the following we shall try to answer these basic questions:

(i) *Can we give some dynamical and physical constraints on $r, \dot{r}$ in order to extract additional orbital information from each single attributable?*

(ii) *Can we represent this information in a suitable way for applications?*

3. The admissible region

Given an attributable we do not have, from the observations, any information on the values of the topocentric distance r and of the range rate $\dot{r}$; we shall make some assumptions on the nature of the object to constrain their values.

Definition: The *admissible region* (AR) is the subset of the $(r, \dot{r}) \in \mathbb{R}^+ \times \mathbb{R}$ defined using the following conditions:

(**A**) *The object is not a satellite of the Earth*, that is its geocentric energy

$$\mathcal{E}_{\oplus}(r, \dot{r}) \geq 0. \tag{3.1}$$

Note that this condition is meaningful only if

(**B**) *The object is inside the sphere of influence of the Earth:*

$$r \leq R_{SI}; \qquad\qquad R_{SI} \simeq 0.010044 \; AU,$$

that is we shall require that condition $[(\mathbf{A}) \cup (\mathbf{B})^{\mathbf{c}}]$ is fulfilled.

(**C**) *The object belongs to the Solar System* †, that is its heliocentric energy

$$\mathcal{E}_{\odot}(r, \dot{r}) \leq 0. \tag{3.2}$$

(**D**) *The object is not a shooting star* (i.e. very small and very close). We can assume that the absolute magnitude

$$H \leq H_{max}. \tag{3.3}$$

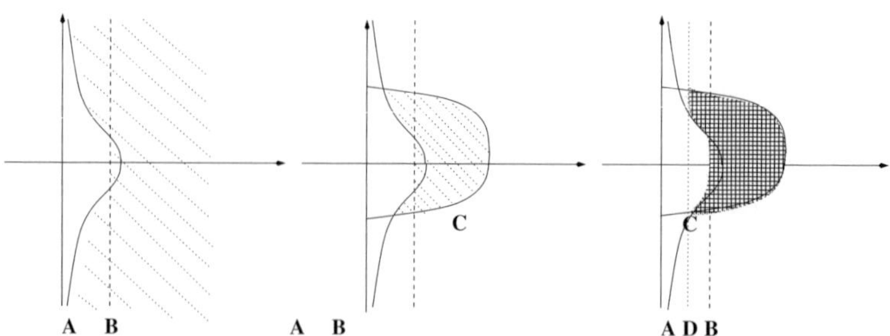

Figure 3. Qualitative sketch of an admissible region: it is defined by condition $[(\mathbf{A}) \cup (\mathbf{B})^{\mathbf{c}}] \cap (\mathbf{C}) \cap (\mathbf{D})$.

---

† We can easily modify this condition in order to obtain objects with semimajor axis less, for example, than a few hundreds of AU: the condition becomes $\mathcal{E}_{\odot} \leq -k^2/(2a_{max})$.

We shall examine in more details the constraints given by the previous conditions:

CONDITION $[(\mathbf{A}) \cup (\mathbf{B})^{\mathbf{c}}]$: NOT AN ARTIFICIAL SATELLITE OF THE EARTH
We can write condition (3.1) as

$$\dot{r}^2 \geqslant G(r) = \frac{2k^2 \mu_\oplus}{r} - \eta^2 r^2 \,, \tag{3.4}$$

where $\eta = \sqrt{\dot{\delta}^2 + \dot{\alpha}^2 \cos^2 \delta}$ is the *proper motion.*

The inequality (3.4) gives a constraint to $(r, \dot{r})$ only in the region where

$$r \leqslant R_{SI} = a_\oplus \left(\frac{\mu_\oplus}{3}\right)^{1/3} \,.$$

Note that $G(r)$ is positive for $0 < r < r_0 = (2k^2 \mu_\oplus / \eta^2)^{1/3}$ and that $r_0 \leqslant R_{SI}$ only for $\eta \geqslant \sqrt{6} n_\oplus$, very fast moving objects.

CONDITION $(\mathbf{C})$: NOT AN INTERSTELLAR OBJECT
Condition (3.2) is equivalent to

$$V(r) = P^2(r)\, S(r) \leqslant 4k^4 \,,$$

where $P(r)$ and $S(r)$ are 2nd-degree polynomials, so that we have a 6th-degree polynomial inequality.

The admissible region, defined by $[(\mathbf{A}) \cup (\mathbf{B})^{\mathbf{c}}] \cap (\mathbf{C}) \cap (\mathbf{D})$, may have in principle a complicated structure due to possible intersections of the curves defining its boundary. Luckily, and surprisingly enough, the structure of the AR is rather simple and it is possible to have a precise understanding of its topological features, necessary to perform a suitable sampling of it.

TOPOLOGY OF THE ADMISSIBLE REGION
The following results have been proven in (Milani *et al.* 2004a):

THEOREM 3.1. *The equation* $V(r) = 4k^4$ *has at most three positive roots. Thus the region* $\mathcal{E}_\odot \leqslant 0, r > 0$ *has either one or two connected components* †.

THEOREM 3.2. *For* $R_\oplus < r < R_{SI}$ *($R_\oplus = $ radius of the Earth)*

$$\mathcal{E}_\oplus \leqslant 0 \Rightarrow \mathcal{E}_\odot \leqslant 0 \,.$$

COROLLARY 1. *Intersections between the boundaries* $\mathcal{E}_\oplus = 0$ *and* $\mathcal{E}_\odot = 0$ *are excluded in the region with* $R_\oplus \leqslant r \leqslant R_{SI}$.

CONDITION $(\mathbf{D})$: NOT A SHOOTING STAR

Suppose the average apparent magnitude h^* is available. Then the absolute magnitude H can be computed:

$$H = h^* - 5\log_{10} r - x(r)\,.$$

† HINT: Multiple roots are also constrained.

In the approximation with $x(r) = 0$, condition (3.3) is

$$\log_{10} r \geqslant \frac{h^* - H_{max}}{5} = \log_{10} r_H \,, \tag{3.5}$$

hence the object needs to be very bright to be very close, otherwise it is just a *shooting star*.

Note that condition (3.5) does not change the topological features of the AR obtained by the two previous theorems, as it says that $(r, \dot{r})$ lies in the half–plane $r \geqslant r_H$.

4. Sampling the admissible region: virtual asteroids

Given an admissible region, we sample it with *virtual asteroids* (VAs):

1. The boundary is densely sampled. This requires us to follow all the portions of the boundary, defined by

$$\mathcal{E}_\oplus = 0 \,; \quad r = R_{SI} \,; \quad r = r_H \,; \quad \mathcal{E}_\odot = -k^2/(2a_{max}) \,.$$

2. A first triangulation is formed by joining the VAs along the boundary, selecting the most regular among the possible ones with these nodes (Delaunay's triangulation).
3. The triangulation is densified by adding points optimally selected to increase the regularity, until some target number of VAs is reached.

For each attributable, the *nodes* $N_i = (r_i, \dot{r}_i), i = 1, N$ are stored together with the *incidence relations*, that is the list of triples of nodes forming the triangles.

For each node N_i there is a VA, i.e. a complete 6-elements orbit.

5. Examples

We present the case of the main belt asteroid 2003 BH_{84}: in Fig. 4 we draw the triangulated admissible region defined by the attributable obtained from the only observations of the day of its discovery, on 25 January 2003.

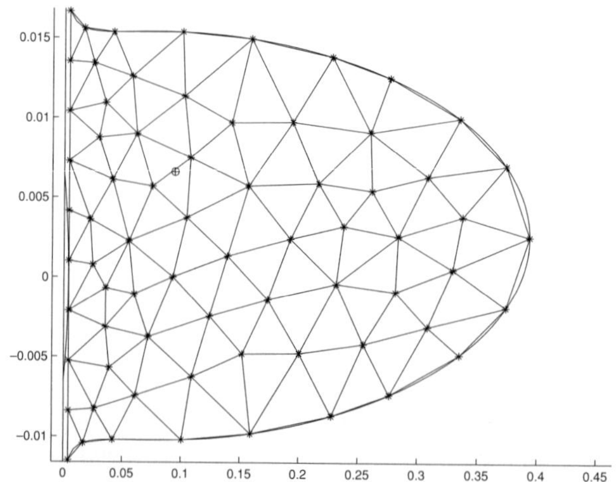

Figure 4. The AR for the asteroid 2003 BH_{84}: on the horizontal axis we plot $f(r) = 1 - e^{-r^2/(2s^2)}$, where $s = \max(r)$, in AU on the vertical axis we plot $\dot{r}$ in AU/day. The real position of the asteroid ($r \approx 1.9834, \dot{r} \approx 0.0066$) is marked with a plus.

TRIANGULATED EPHEMERIDES FOR 2003 BH_{84}

Once we have triangulated the AR, each of the nodes of the triangulation gives us a full orbit that can be propagated in the past or in the future. In Fig. 5 we draw the projections of the AR onto the planes $(\alpha, \delta), (\dot{\alpha}, \dot{\delta}), (\alpha, \dot{\alpha}), (\delta, \dot{\delta})$, after a propagation to 6 February 2003. We also plot with a circled asterisk the components of the attributable obtained from the only observations of 6 February, and we note that, while in the three planes $(\alpha, \delta), (\dot{\alpha}, \dot{\delta}), (\alpha, \dot{\alpha})$ these components lie inside the image of the triangulation, in the $(\delta, \dot{\delta})$ plane they fall outside it. Actually this feature can be explained only if we take into account the uncertainty of the orbits that we have propagated, as it is done in (Milani *et al.* 2004b).

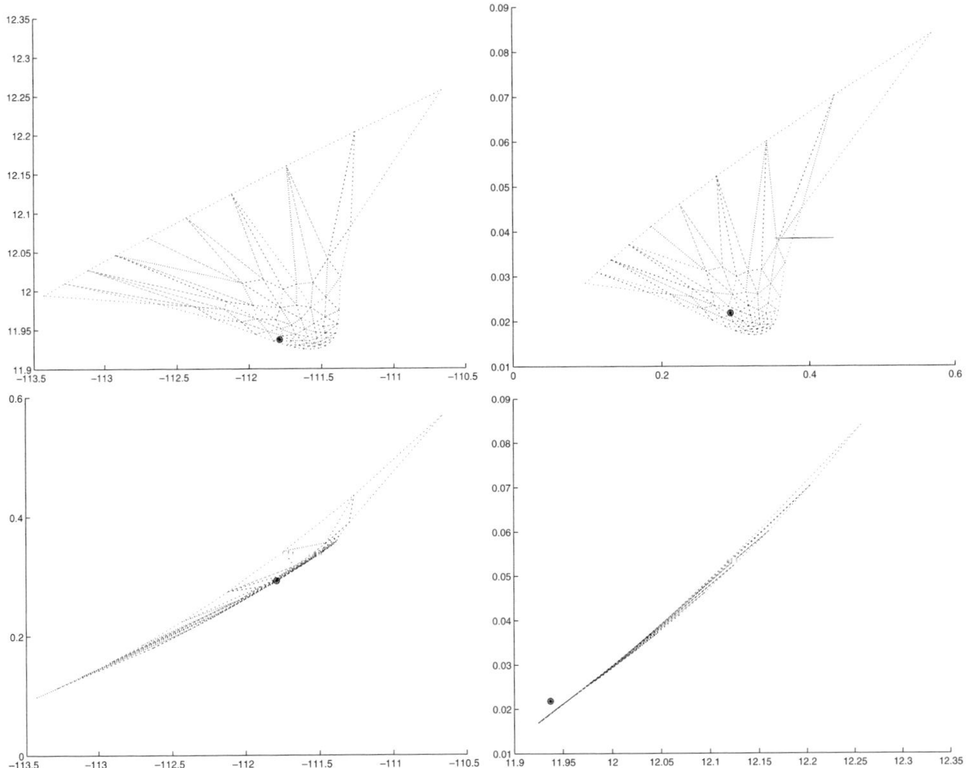

Figure 5. Triangulated ephemerides for 2003 $BH_8$4. Top left: $(-\alpha, \delta)$ plane; top right: $(-\dot{\alpha}, \dot{\delta})$ plane; bottom left: $(-\alpha, -\dot{\alpha})$ plane; bottom right: $(\delta, \dot{\delta})$ plane. Angles are in degrees and angular velocities are in degrees per day.

6. Conclusions and future work

Given a set of observations of an asteroid, for which it is not possible to write a full orbit, we can often define an attributable. The latter, under natural assumptions about the object, can provide useful orbital information through its related admissible region. In fact, this region can be suitably sampled and for each point of the sampling we have a full orbit that can be propagated together with its uncertainty up to a certain time. We think that these mathematical tools can be successfully used in performing orbit determination with too short arcs and we are working on the creation of algorithms to do that.

Acknowledgements

I have reported work done in collaboration with A. Milani, M. de' Michieli Vitturi and Z. Knežević; I'd like to thank here all of them. Thanks also to G. Tommei who helped me with suggestions and comments on this paper.

References

Foderà Serio, G., Manara, A. and Sicoli, P. 2003 *Giuseppe Piazzi and the Discovery of Ceres* in Asteroids III, W.F. Bottke Jr. et al. eds., Arizona University Press

Gauss, C. F. 1809, *Theory of the Motion of the Heavenly Bodies Moving about the Sun in Conic Sections* reprinted by Dover publications, 1963

Milani, A., Sansaturio, M. E. and Chesley, S. R. 2001, *The Asteroid Identification Problem IV: Attributions*, Icarus, **151**, 150–159

Milani, A., Gronchi, G. F., de' Michieli Vitturi, M. and Knežević, Z. 2004, *Orbit Determination with Very Short Arcs: I. Admissible Regions*, Cel. Mech. Dyn. Ast, **90**, 59–87

Milani, A., Gronchi, G. F., Knežević, Z., Sansaturio, M. E. and Arratia, O. 2004, *Orbit Determination with Very Short Arcs: II. Preliminary orbits*, in preparation

Roy, A. E. 1988, *Orbital Motion*, Adam Hilger, Bristol

Discussion

KLEOMENIS TSIGANIS: If you compute the admissible region – so the object can be either main belt, or centaur or whatever – could you try to calculate a more probable admissible region by using the current distribution of asteroids? We know that it is much more likely to get an asteroid between 2 AU and 4 AU, so if you could weight your admissible region by the distribution of some statistical model, then you get a more restrictive zone.

GIOVANNI GRONCHI: Yes, you are completely right. We have already thought about this possibility and would like to do something like that. Our first thought is a simple one; for example, I was speaking about the inclination which is quite indicative. But if we have an asteroid population model, it would be better, because we have in principle a probability distribution in this plane.

MIKHAIL MAROV: Last March the astronomical community dealing with the minor bodies was very much disturbed with an unexpected appearance of one of the quite small asteroids – it looked like it was going to impact the Earth. Does your procedure allow for an accurate prediction of NEO bodies, in a quite limited time frame?

GIOVANNI GRONCHI: Yes. We are working on adding other constraints to the admissible region that give a lower boundary for the impact time. We hope in the future to have some methods to be able to discard the possibility of an NEO impact at once.

SIMON MITTON: Following the last question, I want to make a public-understanding-of-science point, and in doing so I'm not criticising the work of you and your colleagues. But I think that where predictions are based on very short arcs, and they appear to show near misses with the Earth in 100, 200 or 300 years time – and we have examples of this – it's really not very helpful if this information is placed into the public domain in a very premature way which then a few days later has to be withdrawn on the basis that we have some more observations. It's quite important if a very short arc observation shows up a near-Earth object possibly impacting the Earth, the professional community should not behave in a way that gets the general public too excited and makes it front page news. Because it actually harms our profession if 10 days later all of this has to be withdrawn.

GIOVANNI GRONCHI: Yes, I agree completely. It's difficult to explain this to the general public because of the concept of probability. Most of these possible collisional impacts, luckily for us, disappear in a few days because of additional observations. So, yes, it should be confined to a restricted community.

David Clarke, Menios Tsiganis and Giovanni Gronchi
at Horrocks' supposed observing window in Carr House

Transits of Venus: New Views of the Solar System and Galaxy
Proceedings IAU Colloquium No. 196, 2004
D.W. Kurtz, ed.

© 2004 International Astronomical Union
doi:10.1017/S1743921305001481

Transits of Venus and Mercury: Patterns of occurrence, and near-resonance phenomena

P. J. Message

Department of Mathematical Sciences, University of Liverpool, Liverpool L69 3BX, UK
email: sx20@liverpool.ac.uk

Abstract. Transits of Venus occur in pairs 8 yr apart, the pairs separated by intervals of either 112 or 130 yr, because of the pattern of approximate orbital resonances. Transits of Mercury show a different pattern, partly because of the more eccentric orbit, and also because the structure of the approximate resonances is different. The near-resonance in Venus' axial rotation strongly suggests a tidal link with the Earth.

1. Introduction

Here are the dates of transits of Venus, over a period of about a millennium, all given in the Gregorian calendar:

1631 December 7
1639 December 4
1761 June 6
1769 June 3
1874 December 9
1882 December 6
2004 June 8
2012 June 5
2117 December 11
2125 December 8
2247 June 11
2255 June 9
2360 December 12
2368 December 10
2490 June 12
2498 June 9
2603 December 15

All these transits are in pairs 8 yr apart, the pairs separated by either 112 or 130 yr. A transit can only occur when an inferior conjunction of Venus with the Earth happens when Venus is very nearly in the plane of the Earth's orbit round the Sun, that is, when it is very near to one of the two nodes of its orbit. This list of dates shows clearly that one of these node crossings occurs in December (in fact at the ascending node), and the other in June. To understand more about these patterns, it is necessary to examine more closely the arithmetic of the orbits. (See the website prepared by H.M. Nautical Almanac Office (HMNAO 2004) for a longer list of transit dates, detailed predictions, and maps showing regions of visibility.)

Here is a sequence of dates of transits of Mercury, over about two centuries:

1802 November 8
1815 November 11
1822 November 4
1832 May 5
1835 November 7
1845 May 8
1848 November 9
1861 November 11
1868 November 4
1878 May 6
1881 November 7
1891 May 9
1894 November 10
1907 November 12
1914 November 6
1924 May 7
1927 November 8
1940 November 12
1953 November 13
1957 May 5
1960 November 6
1970 May 9
1973 November 9
1986 November 12
1999 November 14
2003 May 7

This shows a more intricate pattern, though it is plain that the corresponding node crossings are in November (in fact at the ascending node) and in May, and that those in November are more frequent.

2. Transit limit for Venus

It is necessary to be more precise about how near to a node the planet must be at conjunction for a transit actually to be seen from the Earth. We may define the "transit limit" (by analogy with "eclipse limit" for eclipses of the Sun or Moon) as the greatest distance of the planet in orbital longitude from the node, at conjunction, which is consistent with a transit being observable. Let us use the method given by Chauvenet (1863) for calculating the eclipse limits, as appropriately amended for transits. We need to note that Venus' orbit is inclined to the ecliptic at $3°.395$, that the mean angular motion of Venus in its orbit is about $1.6255\times$ that of the Earth, that the semi-diameter of the Sun, as seen from the Earth, is about $16'$, and that, since the major semi-axis of Venus' orbit is about $0.723\times$ that of the Earth, the apparent latitude of Venus as seen from the Earth, at conjunction, is about $2.614\times$ that seen from the Sun. (The orbits are sufficiently nearly circular for us to be able to neglect the variation in this to the precision we seek.) From these, the transit limit is found to be about $1°.743$

3. Approximate resonances, and the occurrence of transits of Venus

Let us now consider what governs the time needed to elapse, after one transit, before the circumstances occur leading to the next. For the next transit at the same node, we need the two planets to return approximately to the same direction from the Sun. This requires an approximate commensurability of orbital period, so let us seek these. The mean angular motion of Venus in its orbit referred to the equinox of date is (Seidelmann, Doggett & Deluccia 1974) $1°602\,130\,477\,119\ldots\,\mathrm{d}^{-1}$, and that of the Earth is $0°985\,609\,114\,533\ldots\,\mathrm{d}^{-1}$. Hence the ratio of the mean motion of Venus to that of the Earth is, also giving it in continued fraction form,

$$1.625\,523\,195\,144\ldots = 1 + \cfrac{1}{1+}\,\cfrac{1}{1+}\,\cfrac{1}{1+}\,\cfrac{1}{2+}\,\cfrac{1}{29+}\,\cfrac{1}{2+}\,\cfrac{1}{16+}\,\cfrac{1}{4+}\,\cfrac{1}{1+}\,\cfrac{1}{2+}\,\cdots \qquad (3.1)$$

Approximations ("convergents") to this ratio as rational fractions may be found by taking a finite number of terms of this continued fraction. Successive convergents are alternately greater than, and less than, the true value.

The third convergent is

$$1 + \cfrac{1}{1+}\,\cfrac{1}{1+}\,\cfrac{1}{1} = \frac{5}{3} = 1.\dot{6}. \qquad (3.2)$$

The fourth convergent is

$$1 + \cfrac{1}{1+}\,\cfrac{1}{1+}\,\cfrac{1}{1+}\,\cfrac{1}{2} = \frac{13}{8} = 1.625. \qquad (3.3)$$

This shows that 8 yr is close to thirteen Venus orbital periods and five synodic periods, so that, after five conjunctions, Venus and the Earth will have moved almost to the same radial line from the Sun. The conjunction line will in fact have moved through an integral number of revolutions, less $2°408866735\ldots$. This is less than twice the "transit limit", so that two successive such conjunctions can occur, each within the transit limit of the node. But, at the corresponding conjunction a further 8 yr later, the conjunction line will have moved out of the transit limit, so no transit will occur. Thus there will usually be a pair of transits, at the same node, 8 yr apart. (During 8 yr, the node of Venus' orbit moves backwards through about $0°002\,338\ldots$, which does not affect this conclusion. In fact, in 1388 there was a "near miss" at the end of a sequence of transits, the conjunction line having been moving gradually towards the edge of the transit region during the sequence, to be followed by the beginning of a new sequence 8 yr later.)

This near resonance may also be expressed by saying that linear combination $8L_V - 13L_E$ (where L_V is the mean longitude of Venus, and L_E is the mean longitude of the Earth), moves backwards through $0°004125328\ldots\,\mathrm{d}^{-1}$, and so completes a revolution in $238.9258\ldots$ yr. Thus, in the expressions for the perturbations of the orbit of Venus by the Earth, and the perturbations of the Earth by Venus, terms which involve sines or cosines of this combination of mean longitudes will have a period of about 238.9258 yr, and will be of enlarged amplitude because of the small denominator arising on integration with respect to time of the expressions for the time rate of change of the perturbations.

Airy (1828, 1832) predicted the occurrence of such terms from this cause. He showed that this term in the longitude of the Earth round the Sun (and so in the apparent motion of the Sun as seen from the Earth) had an amplitude of $2''\!.6$, and noted that the period of these terms is a larger multiple of the orbital periods of the planets concerned than for any other terms needing to be taken into account in the theory of perturbations in the solar system. (The "great inequalities" in the perturbations in the motions of Jupiter and Saturn, arising from the near 5:2 commensurability between the periods of these two planets, are of course of longer period – about 890 yr – but considerably less than 240× the periods of these planets.)

The fifth convergent is

$$1 + \cfrac{1}{1+} \cfrac{1}{1+} \cfrac{1}{1+} \cfrac{1}{2+} \cfrac{1}{29} = \frac{382}{235} = 1.6255319\ldots . \tag{3.4}$$

This shows that 235 yr is close to 382 Venus orbital periods, and 147 synodic periods, so that, after 147 conjunctions, Venus and the Earth will also be close to the same radial line through the Sun. The conjunction line will in that time have moved through an integral number of revolutions, plus $1°179\,318\,08\ldots$. This is less than the "transit limit", so, 235 yr after the second of a pair of transits 8 yr apart, the conjunction line will have moved forwards by this angle, so the conjunction will again be within the transit limit, and another transit can be expected. (During 235 yr, the node of Venus' orbit moves backwards through about $0°068\,671\,56\ldots$, which, again, does not affect the conclusion.) A pair of transits at the other node, of course also 8 yr apart, will usually occur very nearly midway between successive pairs of transits at the first node, since the two nodes are of course diametrically opposed, and the orbits are very nearly circular.

Let us now set out the dates of transits of Venus, putting those at the two nodes separately, and noting the interval in years between successive transits at the same node. First, those in December:

1631	December	7	8 yr
1639	December	4	235 yr
1874	December	9	8 yr
1882	December	6	235 yr
2117	December	11	8 yr
2125	December	8	235 yr
2360	December	12	8 yr
2368	December	10	235 yr
2603	December	15	

now those in June:

1761	June	6	8 yr
1769	June	3	235 yr
2004	June	8	8 yr
2012	June	5	235 yr
2247	June	11	8 yr
2255	June	9	235 yr
2490	June	12	8 yr
2498	June	9	

showing, in each case, the expected alternation between 8-yr and 235-yr intervals.

4. Transit limits for Mercury

We turn now to the case of Mercury, and begin by calculating the appropriate limits. We first note that the eccentricity of Mercury's orbit, 0.20563, is large enough so that we must take into account the difference between its distances from the Sun at different places in the orbit. The ecliptic longitude of the ascending node is $48°378\ldots$, so that, at a conjunction here, the ecliptic longitude of the Sun, seen from the Earth, will be about $228°378$, so such conjunctions will occur in November. Since the ecliptic longitude of the apse of Mercury is about $77°\!.518$, the true anomaly will then be $330°\!.860$, from which we find, using the fact that the major semi-axis of Mercury's orbit is about 0.3871 a.u., that the radial distance from the Sun will be about 0.3143 a.u., at which distance the apparent latitude of Mercury, at inferior conjunction, seen from the Earth, will be about $0.4583\times$ that seen from the Sun. Now Mercury's orbit is inclined to the ecliptic at about $7°\!.005$, and we find from all this that the transit limit; that is, the maximum distance from the node at conjunction consistent with a transit, is about $4°\!.832$.

The ecliptic longitude of the descending node is about $228°\!.378$, so that transits at this node will occur in May. The true anomaly is then $150°\!.860$, from which we find that the radial distance is about 0.4519 a.u., so that the apparent latitude of Mercury, seen from the Earth, at inferior conjunction, will be about $0.8245\times$ that seen from the Sun. Then we deduce that the transit limit at this distance is about $2°\!.686$.

The difference in size of the two transit limits leads us to expect more transits of Mercury in November than in May, as we do of course in fact observe.

5. Approximate resonances for Mercury

The mean angular orbital motion of Mercury is $4°\!.092\,338\,785\,462\ldots\,\mathrm{d}^{-1}$. The ratio of this to that of the Earth, also given as a continued fraction, is

$$4.152091053010\ldots = 4 + \cfrac{1}{6+}\cfrac{1}{1+}\cfrac{1}{1+}\cfrac{1}{2+}\cfrac{1}{1+}\cfrac{1}{4+}\cfrac{1}{1+}\cfrac{1}{92+}\cdots. \tag{5.1}$$

The second convergent of this is

$$4 + \cfrac{1}{6+}\cfrac{1}{1} = \frac{29}{7} = 4.\dot{1}4285\dot{7}, \tag{5.2}$$

which shows that 7 yr is close to 29 Mercury orbital periods, and 22 synodic periods, so that, after 22 conjunctions, Mercury and the Earth will have moved almost to the same radial line from the Sun. The conjunction line will have moved through an integral number of revolutions, less $7°\!.382227605\ldots$. This is more than twice the transit limit for May transits, so that there will not be transits of Mercury in May separated by 7 yr, but this angle is not far from the transit limit for November transits, so there will be some pairs of transits in November 7 yr apart.

The third convergent is

$$4 + \cfrac{1}{6+}\cfrac{1}{1+}\cfrac{1}{1} = \frac{54}{13} = 4.1538461\ldots, \tag{5.3}$$

showing that 13 yr is close to 54 Mercury orbital periods, and 41 synodic periods, so that, after 41 conjunctions, Mercury and the Earth will again have moved almost to the same radial line from the Sun. The conjunction line will have moved through an integral number of revolutions, plus $2°\!.605\,855\ldots$. This is less than twice either transit limit, so that there will normally be pairs of transits, at either node, 13 yr apart.

The fourth convergent is

$$4 + \cfrac{1}{6+}\cfrac{1}{1+}\cfrac{1}{1+}\cfrac{1}{2} = \frac{137}{33} = 4.\dot{1}\dot{5}, \tag{5.4}$$

showing that 33 yr is close to 137 Mercury orbital periods, and 104 synodic periods, so that, after 104 conjunctions, Mercury and the Earth will have again moved almost to the same radial line from the Sun. The conjunction line will have moved through an integral number of revolutions, less $2°170\,530\ldots$. So we will expect there to be pairs of transits, at either node, 33 yr apart.

The fifth convergent is

$$4 + \frac{1}{6+} \frac{1}{1+} \frac{1}{1+} \frac{1}{2+} \frac{1}{1} = \frac{191}{46} = 4.1521739\ldots, \tag{5.5}$$

showing that 46 yr is close to 191 Mercury orbital periods, and 145 synodic periods, so that, after 145 conjunctions, Mercury and the Earth will have again moved almost to the same radial line from the Sun. The conjunction line will have moved through an integral number of revolutions, plus $0°43532\ldots$. So we will usually expect the repetition of a transit, at either node, after 46 yr. (During 7 yr, the node of Mercury's orbit moves backwards through about $0°008791$, which does not affect any of these conclusions.)

Let us now set out the dates of transits of Mercury, putting those at the two nodes separately, and noting the interval in years between successive transits. First, those in November:

1802 November	8	13 yr
1815 November	11	7 yr
1822 November	4	13 yr
1835 November	7	13 yr
1848 November	9	13 yr
1861 November	11	7 yr
1868 November	4	13 yr
1881 November	7	13 yr
1894 November	10	13 yr
1907 November	12	7 yr
1914 November	6	13 yr
1927 November	8	13 yr
1940 November	12	13 yr
1953 November	13	7 yr
1960 November	6	13 yr
1973 November	9	13 yr
1986 November	12	13 yr
1999 November	14	

We notice that the pattern of occurrence of the 7-yr and 13-yr intervals leads to very frequent combinations giving both 33-yr and 46-yr intervals.

Then, the transits in May:

1832 May	5	13 yr
1845 May	8	33 yr
1878 May	6	13 yr
1891 May	9	33 yr
1924 May	7	33 yr
1957 May	5	13 yr
1970 May	9	33 yr
2003 May	7	

This time we notice that the pattern of occurrence of the 13-yr and 33-yr intervals leads to very frequent combinations giving 46-yr intervals.

6. Near-resonance in the rotation period of Venus

Successive determinations of the rotation period of Venus using radar soundings of formations on the surface, with increasing precision, have converged towards a period of 243.00 d, the rotation being retrograde, that is, in the opposite sense to most rotations and orbital revolutions in the Solar System (see, e.g., Gold & Soter 1969). Now, a period of 243.16 d would correspond to the resonant situation in which Venus presented the same face to the Earth at each successive conjunction. It is very unlikely that this is a pure coincidence, and the slight difference in the periods probably indicates the libration of the system about the exact "locked-in" situation. But such a situation would imply that Venus' departure from exact axial symmetry is great enough for a gravitational torque due to the Earth's attraction to have captured Venus' rotation in this way. Such a torque would of course be very small (though much greater at conjunction than at other times), so this situation is very surprising. There is an extensive literature considering possible ways in which this could have come about.

References

Airy, G.B. 1828 "Corrections in the elements of Delambre's Solar Tables required by the observations made at the Royal Observatory, Greenwich" *Phil. Trans. R. Soc. Lond.* **CXVIII**, 23–34.

Airy, G.B. 1832 "On an inequality of long period in the motions of Earth and Venus" *Phil. Trans. R. Soc. Lond.* **CXXII**, 67–124.

Chauvenet, Wm. 1863 "A Manual of Spherical and Practical Astronomy" vol. 1, p. 436. J.B.Lippincott, Philadelphia.

Gold, T. & Soter, S. 1969 "Atmospheric Tides and the Resonant Rotation of Venus" *Icarus* **11**, 356–366.

Seidelmann, P.K., Doggett, L.E. & Deluccia, M.R. 1974 "Mean Elements of the Principal Planets" *AJ* **79**, 57–60.

H.M. Nautical Almanac Office 2004 *http://www.nao.rl.ac.uk/nao/transit*

Discussion

DON KURTZ: How frequent are transits of Earth as seen from Mars?

JIM MESSAGE: I didn't finish that calculation, although I started it!

DON KURTZ: Is there any situation where the orbital periods are such that you would never see a transit of an interior planet from some exterior planet?

JIM MESSAGE: It is difficult to see how that could happen, but in principle, it might. [Added in proof:] This would happen if there were a resonance, and a critical angle associated with it were fixed, or oscillating about, an appropriate value. For example, if there were a near 3:1 orbital resonance, and the angle $3\times$ longitude of outer planet $-$ longitude of inner planet $- 2\times$ node longitude of inner planet $= 180°$, then conjunctions could only occur with the inner planet at $90°$ from the node, which would certainly prevent a transit from happening (unless the orbital inclination were very small, when there would be a transit every time!) If this configuration were in fact dynamically possible, heuristic considerations suggest that it would be a stable one.

NICK KOLLERSTROM: Between the two Venus transits I gather there are occultations of Venus by the Sun at four year intervals – like 2000, 2008, 2016 – so the sequence is something like 5 occultations, transit, occultations. Are these a bit like lunar eclipses turning up either side of the solar eclipse? Are they more frequent than the transits?

JIM MESSAGE: The transit limit will be different, because the angles seen from the Earth will be different as they relate to the orbital parameters. You'll have to do a fresh calculation as to when you would get such an event at the superior conjunction. I haven't done that. It would be very difficult to observe an occultation of Venus by the Sun, so that would be a challenge to the instrumentalist, wouldn't it?

STEVEN DICK: Just to comment on Don's question. Transits of Earth as seen from Mars: about every 100 years. There was one in 1984, and that's when Arthur C. Clarke wrote his famous short story called "Transit of Earth, " and I know that there is another one in 2084.

Jim Message taking notes at Stonyhurst Observatory

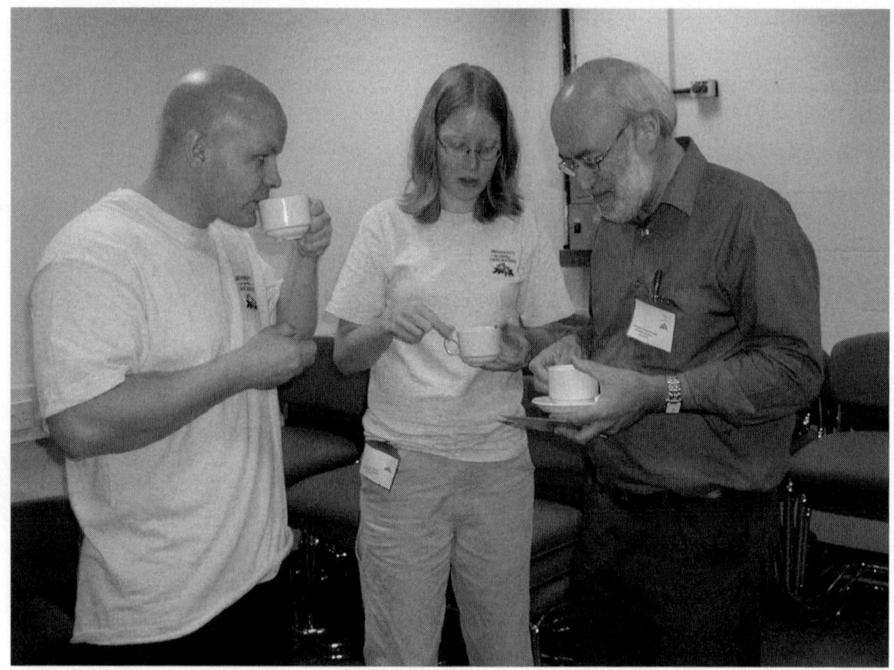

Mike Marsh, Hannah Worters and Gordon Bromage

Coffee-time discussion: Lena Pitjeva, Paul Marston, John Southworth, Mark Northeast, Kate Bird, Jacqueline and Simon Mitton

Part 4

THE JEREMIAH HORROCKS MEMORIAL PUBLIC LECTURE

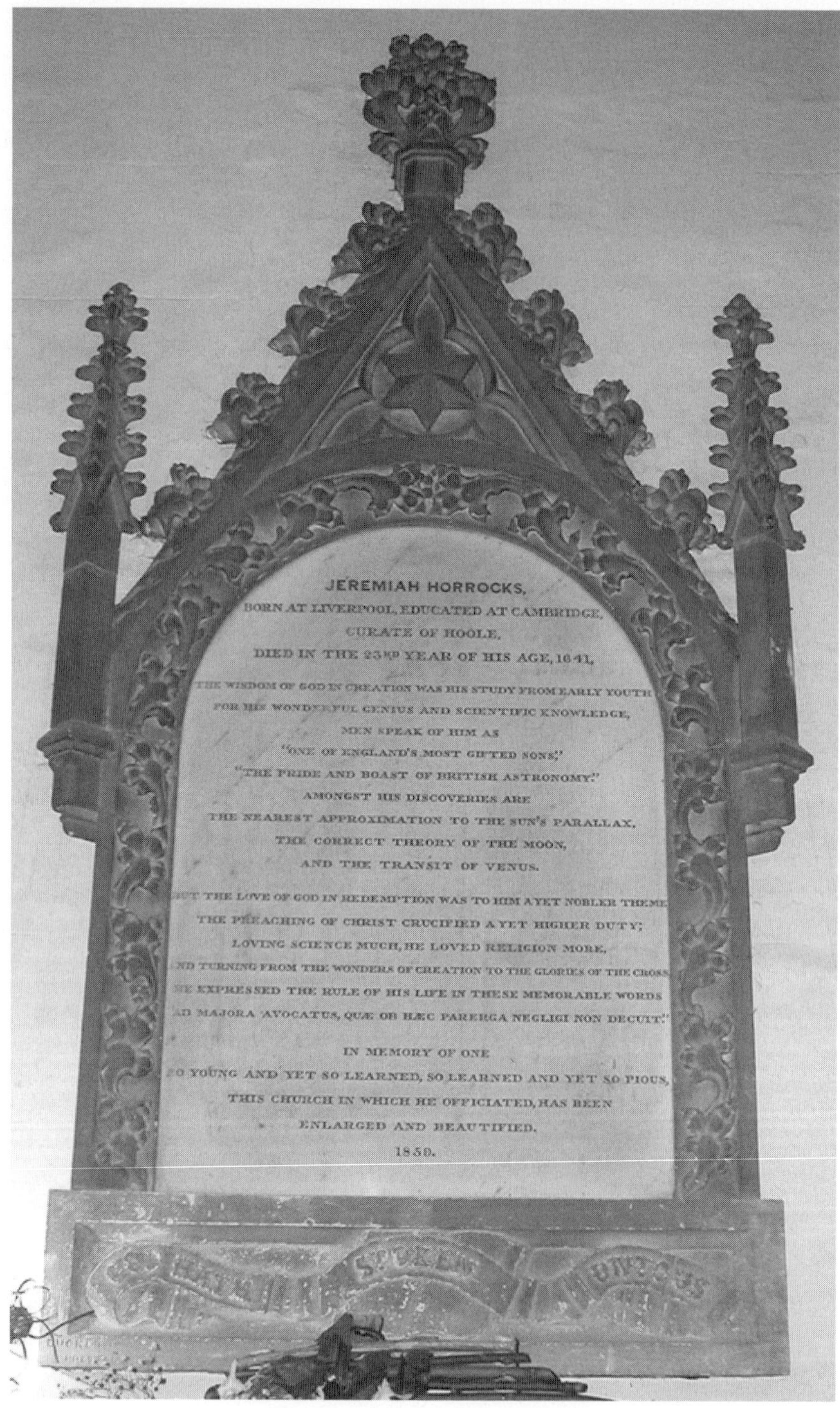

The Horrocks Memorial in Much Hoole church

Transits of Venus: New Views of the Solar System and Galaxy
Proceedings IAU Colloquium No. 196, 2004
D. W. Kurtz, ed.

© 2004 International Astronomical Union
doi:10.1017/S1743921305001493

Our Galaxy in three-dimensions:
the Jeremiah Horrocks Memorial Lecture

M. A. C. Perryman

Space Science Department, ESA/ESTEC, 2200AG Noordwijk, The Netherlands;
and Sterrewacht Leiden, The Netherlands

Abstract. This presentation has the following goals: (i) to explain, within a historical context, some of the difficulties in disentangling our Galaxy's three-dimensional structure; (ii) to summarise in broad terms what we presently know of this structure, concentrating on those features inferred from accurate distance measurements; (iii) to illustrate selected aspects in a visually stimulating manner; and (iv) to give a foretaste of the exciting results that lie ahead with future space astrometric experiments. The public lecture given in Preston, UK on 9 June 2004 employed 3-d visualisation techniques using polarised light to illustrate the talk.

1. Historical introduction

We have, of course, come a long way since mankind first began to reflect on the nature of the stars. Plato's contemporaries observed that the heavens rotated night after night with constant speed. Myriads of fixed stars shared this rotation, preserving their relative positions without change. Moving amongst them in a puzzling way were the seven wanderers: the Sun, the Moon, Mercury and the other planets.

In the early 1600s, Galileo turned his telescope to the Milky Way, and discovered that it could be resolved into innumerable faint stars. By the mid-eighteenth century Thomas Wright had described our Galaxy as a disk of stars in which the Sun is immersed, and Immanuel Kant postulated the existence of other 'island universes' distributed throughout space at enormous distances. William Herschel used star counts in different sky regions to deduce the relative dimensions of our Galaxy, based on the crucial but incorrect assumption that all stars had the same absolute brightness.

A little more than a hundred years ago, Kapteyn was able to use the medium of astronomical photography to initiate a new, quantitative discussion of our Galaxy's structure. Massive observational programmes, such as the great Durchmusterungen and the Carte du Ciel, were set up. The size of our Galaxy, and the distance scale within it, became issues of great debate. Interpretation was complicated by problems of interstellar extinction, and the different luminosity classes and kinematic populations.

We now know that billions of stars, as well as planets, interstellar gas and dust, radiation, and invisible material, are gravitationally bound to form our Galaxy – a magnificent disk spiral system, supported by rotation. Our Solar System lies far out in one of the spiral arms, 30 000 light-years from the centre. Great debates continue, with the nature of dark matter, and the details of galaxy formation, star formation, and spiral structure, amongst them.

Our current picture is that the Universe originated from a hot, dense, and smooth state, the Big Bang. Structure originated from fluctuations in energy density during an early inflationary period, growing by gravitational amplification as the Universe expanded. Cooling and condensation of gas in the cores of heavy halos, produced by hierarchical clustering of dark matter, somehow finally resulted in the dramatic diversity of galaxy types that we see today.

Remarkably, the origin, past evolution, present-day structure, and long-term future of our Galaxy can be examined by direct observation (such results are also of value as a template for cosmological studies on much larger spatial scales). Low-mass stars, below about $1\,\mathrm{M_\odot}$, live for longer than the present age of the Universe, and therefore preserve a chemical record of the elements and conditions from which they formed. Similarly, stellar motions reflect the large-scale components and kinematics of our Galaxy, trace out the otherwise unobservable distribution of dark matter, and contain a memory of events such as the epoch and rate at which our Galaxy has grown by devouring its smaller companions. Unravelling these clues requires distances to the stars, and the measurement of distance recurs throughout historical and current efforts to disentangle and interpret astronomical observations. Observational progress is still immensely constrained by the limited stellar samples at our disposal, by the range of populations for which good quality data are available, and by the precision with which distances and therefore the details of the stellar populations are known.

Observers, and even the most powerful ground and space telescopes, see celestial objects projected in two dimensions. Geometric distances can only be derived from the tiny oscillation in a star's apparent position arising from the Earth's motion around the Sun – the combination of this motion, and the motion of stars through space, then leads to a continuous evolution in a star's position. For two or more stars in the same field, their parallactic motions are in phase, and their relative angular separations yield only their relative distances, independently of how the stars are moving. However, superposing two widely separated fields on the sky allows absolute parallaxes and hence absolute distances to be measured. High accuracy, long-term monitoring of two-dimensional star positions therefore provides the pieces of a celestial jigsaw encapsulating a six-dimensional stereoscopic map of the stars and their motions through space.

What follows, directly from the observations or indirectly from modelling or theory, are absolute physical stellar characteristics, such as luminosities, radii, masses, and ages. Our goals are then to use these quantities to determine their physical composition and structure, to disentangle their motions and, eventually, to explain how our Galaxy formed, and how it will evolve in the future. We can then use these data to try to answer the questions posed so succinctly by Jan Oort: What is the distribution of matter? Why is it what it is?

Let me briefly place our present understanding of three-dimensional measurements in a broader historical context. 300–400 years ago, three main themes motivated the improvement of angular measurements: navigational problems associated with the determination of longitude on the Earth's surface, the comprehension and acceptance of Newtonianism, and understanding the Earth's motion through space. Even before 1600, there was general agreement that the crucial evidence needed to detect the Earth's motion was the measurement of trigonometric parallax. But for about 250 years, ambitious attempts to measure the first stellar distances were unsuccessful. In 1718 Edmund Halley, who had been comparing contemporary observations with those that the Greek Hipparchus and others had made, announced that three stars, Aldebaran, Sirius and Arcturus, were displaced from their expected positions by large fractions of a degree. Halley deduced that these stars had their own distinct velocity across the line of sight, or proper motion: stars were moving through space.

By 1725, angular measurements had improved to a few arcsec, making it possible for James Bradley to detect stellar aberration, as a by-product of his unsuccessful attempts to measure the distance to the bright star γ Draconis. This was an unexpected result – small positional displacements were detected, and correctly attributed to the vectorial addition of the velocity of light to that of the Earth's motion around the Sun. His observations supplied the first direct proof that the Earth was moving through space, and thus provided an unexpected confirmation both of Copernican theory, and Roemer's discovery of the finite velocity of light 50 years earlier. It also confirmed Newton's insight into the incomprehensible immensity of stellar distances, and showed that the measurement of parallax would pose a technical challenge of extraordinary delicacy.

During the eighteenth century the motions of many more stars were announced, and in 1783 William Herschel found that he could partly explain these effects by assuming that the Sun itself was moving through space. The early 1830s saw the formulation of criteria for estimating proximity, based on brightness and high proper motion, and eventually, the determination of the first parallaxes. Confirmation that stars lay at very great but nevertheless finite distances represented a turning point in the understanding of the Universe. John Herschel greeted the first distance measurements with the comment that *'the sounding line in the Universe of stars had at last touched bottom'*.

Measuring stellar distances remains exceedingly difficult, even within our Solar neighbourhood. William Herschel attempted to convey the unimaginable interstellar distance scales as follows: *'To drop a pea at the end of every mile of a voyage on a limitless ocean to the nearest fixed star, would require a fleet of 10 000 ships, each of 600 tons burthen.'*

One hundred and fifty years after the first stellar parallaxes were measured, and after a long period of painstaking ground-based efforts to map the Solar neighbourhood, space observations, beyond the effects of the Earth's atmosphere, are carrying this quantitative exploration forward. From a dense network of angular measurements, Hipparcos has provided tens of thousands of accurate distances and motions, with angular accuracies of around 1 milliarcsec. This is just the latest step in continuing efforts to measure distances across the enormous gulfs of space and time but, as we shall see, it places many old problems on a somewhat more secure observational footing.

2. The main components of our Galaxy

So let us look at some details of our Galaxy's structure as we understand them today. Various lines of argument show that on its largest scales, our Galaxy comprises two main structural elements, both roughly axially symmetric: a flattened disk, and a spheroidal component. On a clear dark night the projection of our own disk is perceived as a band of light across the sky, albeit heavily obscured by dust. The spheroidal component is generally pictured as comprising various sub-components, ranging from the nucleus of order 3 pc in size, through the bulge of order 3 kpc, out to the halo extending to 30 kpc or more. The relative importance of the disk and spheroidal components accounts for much of the observed variety in galaxy morphologies. Each contains quite different and characteristic stellar populations, and they have different compositions, kinematic and dynamic properties, evolutionary histories, and spatial sub-structures.

The halo consists of some billion old metal-poor stars, about 140 globular clusters, and a small number of satellite dwarf galaxies. Unless our present understanding of gravitation is incorrect, the flat rotation curves of gas in the outer parts of spiral galaxies imply that the entire systems are embedded in massive quasi-spherical outer halos of dark material. It appears to account for about 95% of the mass of the Galaxy and, for compatibility with standard nucleosynthesis, is believed to be non-baryonic. Otherwise it is of unknown composition and poorly known spatial distribution, and its mysterious character is one of the fundamental unanswered questions of our day.

From the southern hemisphere, the stellar bulge is partly visible to the naked eye in the direction of Sagittarius. The bulge is inferred to be triaxial, and appears to contain a bar-like structure. At its centre lies a black hole of about 3 million $M_\odot$. But studies are greatly hampered by its immense distance, by extreme reddening, and by the presence of the central parts of the disk and the halo, and it is unclear whether the bulge is a remnant of a disk instability, a successor or a precursor to the stellar halo, or a merger remnant.

2.1. *The Disk: large-scale structure*

The most conspicuous component of the Milky Way is, however, the flat disk, representing most of the stars seen by eye and in photographs. Here our knowledge is somewhat more certain, especially in the local Solar neighbourhood. About 200 pc thick, the disk contains some hundred billion stars orbiting the Galactic centre, extending to radial distances of about 30 kpc. This disk is thought to have been formed by the rapid collapse of a rotating, galaxy-sized gas cloud, about $8-12$ billion years ago. Star formation has been reasonably continuous since then, and as a result the disk contains stars with a range of metallicities, ages and kinematics.

The entire system is in a state of rapid differential rotation, with our Solar System completing a Galactic orbit every 240 million years. The prominent spiral arms are the primary locations of star formation, funneling mass from the gaseous component to the stars, and propagating such that the overall Galaxy appearance probably remains much the same for many rotation periods. Radio and mm observations of interstellar gas have delineated this spiral structure, and have mapped a significant warping of the Galactic disk outside the Solar orbit. While about half of all spiral galaxies display such warps, there is no detailed explanation of this common phenomenon. Hipparcos has provided hints of the spatial and kinematic distribution of the local Orion arm, but with accurate distances out to only a few hundred pc, we are far from having a detailed map. Our fuzzy perception of local spiral structure based on photometric distances should be replaced by a much sharper view once distance accuracies at tens of microarcsec become available.

The Sun and the nearby stars move with respect to our 'local standard of rest', which itself moves around the centre of the Galaxy with a velocity of about $220\,\mathrm{km\,s}^{-1}$, one revolution every 240 million years. The mean-free-time between collisions for the Sun in the Solar neighbourhood is of order 10^{13} years, very much longer than the present age of the Galaxy, so stars here behave as a collisionless gas. In the Solar neighbourhood, space velocities with respect to the local standard of rest tend to increase with time, due to accumulated perturbations from other stars, giant molecular clouds, and spiral arms. Free from kinematic biases, which in the past have plagued studies of local stellar kinematics, Hipparcos has revealed that the local stellar velocity distribution is highly structured and complex, with both the stellar warp outside the Solar circle, and the bar in the inner Galaxy, having left their imprints.

There are a number of curious aspects of our Sun's space motion within the disk. First, its velocity, $13.4\pm0.4\,\mathrm{km\,s}^{-1}$, is unusually small. Second, it lies at a vertical height of only 1012 pc from the Galaxy mid-plane, even though it must have oscillated up and down across the disk many times throughout its history. And as deduced from the Cepheid proper motions, our Sun lies very close (within 100 pc) to the Galaxy's co-rotation resonance, where the rotation velocities of the disk and of the spiral pattern coincide. These properties suggest that our Sun may have a privileged Galactic orbit, possibly with interesting habitability and anthropic implications.

What is called the thin disk accounts for about 90% of the visible light in the Milky Way, but only about 5% by mass. Since the 1980s we have come to recognise a faint, thick disk of old stars. Although the thick disk includes some of the oldest stars in the Galaxy, their provenance is uncertain: some may have formed *in situ*, some may have been scattered from the thin disk, while the majority may have originated from a galaxy which merged with our own around 9 billion years ago, leading to an abrupt increase in velocity dispersion. This is seen once the Hipparcos motions are combined with precise age estimates from isochrone modelling. The possible diversity of histories of the thin and thick disks, and their complex distribution of chemical abundances and kinematics, is presently very difficult to disentangle.

2.2. *The Disk: small-scale structure*

In practice, little is known about the stellar disk beyond $1-2\,\mathrm{kpc}$ from the Sun. This is due to significant interstellar extinction towards the central regions, and by our inability to determine accurate distances and space motions for objects so far away. Much more is known about the local small-scale structure, including star forming regions, open clusters, expanding associations, and stellar streams.

Open clusters contain a few tens to a few thousand stars, born in the same molecular cloud and gravitationally bound. They are tracers of the young and intermediate-age disk components of the Galaxy, with ages ranging from a few million to a few hundred million years. Members are coeval, have the same chemical composition as the parent cloud, share the same bulk space motion to within the internal velocity dispersion of the cluster (a few tenths of a $\mathrm{km\,s^{-1}}$), with members distributed across the whole mass spectrum, from brown dwarfs to A, B or O stars, depending on age.

In the case of the Hyades, at about $46\,\mathrm{pc}$, the cluster's proximity, high proper motion, and high space velocity with respect to its own local standard of rest, allow members to be identified with relative ease. Members form a coherent structure in velocity space, inheriting their bulk motion from the parent molecular cloud. Various methods can be used to identify these groups using their measured positions and velocities. The development of these coeval systems, both dynamically and with respect to temperature and luminosity, provide ideal laboratories for testing models of stellar evolution. They 'evaporate' and eventually disintegrate over tens or hundreds of millions of years due to mass loss by stellar evolution, due to gravitational encounters between stars in the cluster, and due to tidal forces between the cluster and the Galactic disk or giant molecular clouds.

The Hertzsprung-Russell Diagram for the Hyades is now beautifully delineated on the basis of the accurate individual trigonometric parallaxes, which resolve the depth of the cluster along the line of sight. Characteristics such as the helium abundance and cluster age can be inferred. Combining trigonometric distances with the proper motions, the HR diagram can be further tightened, with interesting consequences for the understanding of stellar evolution.

More than a thousand open clusters are known, but only half have reasonable distance estimates, with most closer than $2\,\mathrm{kpc}$. At about $110\,\mathrm{pc}$, even the Pleiades is too far away for individual distances to be measured with any degree of significance, but classifying stars according to bound members and evaporating objects starts to become possible.

O and B stars are not randomly distributed among the Galactic stellar populations, but generally as unbound associations of young stars. Members can again be detected kinematically because of their small internal velocity dispersion, although nearby associations have a large angular extent on the sky, which has traditionally limited membership determination to bright stars and early spectral types. Recent studies have led to significantly improved membership determination, including extension to later spectral types, and improved mean distances out to $600-700\,\mathrm{pc}$. Studying the formation, structure, and evolution of these young stellar groups and star-forming regions is important since as much as 90% of the stars in the Galaxy may have formed in OB associations. Like bound clusters, these slowly dissolve into the field population as a result of Galactic tidal forces.

Dense young clusters are also the birthplace of high-velocity 'runaway' stars. These may originate either via dynamical encounters, or in the unbinding of a binary after a supernova explosion of one component. The runaway stars AE Aur and μ Col, at distances of about 500 pc, are now some 100° apart on the sky, receding at about 100 km s^{-1}. Careful extrapolation backward in time provides strong evidence for a common origin. They were probably ejected in a dynamic encounter with the massive binary ι Orionis, in the heart of the Orion OB1 association, 2.2 million years ago. Nearly 200 runaway stars are known, often at very significant distances from young stellar groups, but current measurement accuracies do not generally make it possible to retrace their orbits meaningfully and understand their origins.

2.3. *The stellar halo*

The baryon halo (including blue horizontal branch stars, RR Lyrae variables, metal-weak red giants, and globular clusters) rotates very slowly compared to the disk – some have suggested that it may even be counter-rotating – and extends outwards to a radius of perhaps 100 kpc or more. Although comprising only some 2% of the light, and an even smaller fraction of the total mass, the stellar halo plays a key role in unravelling the sequence of events involved in galaxy formation.

Within the halo, globular clusters are dense swarms of about 100 000 stars, and are possibly relics of intergalactic ingredients added to the Milky Way after its creation. Far older than any fossil on Earth, older than any other structures in our Galaxy, some are so old that they have challenged age estimates of the Universe itself. Ages of the oldest, metal-poor globular clusters, in the inner and outer halo, the Large Magellanic Cloud, and the nearby Fornax and Sagittarius dwarf galaxies show a remarkable uniformity. To within about 1 Gyr the onset of globular cluster formation appears to have been well synchronised over a radius of more than 100 kpc.

Globular clusters are too far away for distances to be measured trigonometrically today. Instead, distances and hence luminosities of nearby field sub-dwarfs are used to estimate the distances of globular clusters through main-sequence fitting. For this, one needs a template main-sequence, from nearby metal-poor sub-dwarfs, of which Hipparcos has provided distances to about 100. Nevertheless, numerous complications make the chain of deduction from sub-dwarf distances to globular cluster ages far from straightforward.

3. Nearby stars

Our understanding of nearby stars has focussed on the 'Catalogue of Nearby Stars' maintained by Heidelberg astronomers since Wilhelm Gliese established his first such catalogue census there in 1957. Extending out to a distance of 25 pc, and cataloging some 2000 stars, the original compilations had to draw on painstaking ground-based parallax determinations extending over about one century, assembled from observations at many different observatories, and supplemented by spectroscopic or photometric distance estimates. The goal in such a census is to identify, from the two-dimensional projection of the hundreds of millions of stars on the celestial sphere, that small subset lying within 25 pc. While this corresponds to a vast region of space extending out to nearly 10^{15} km, it is minute on astronomical scales, extending to less than 0.3% of the distance to the Galactic centre, and including barely 10^{-8} of its total stellar content.

Seeing nearby space with enhanced stereoscopic vision we can simply pick out our nearest neighbours. Around 200 'new' nearby stars within 25 pc have become apparent, the nearest of these, HIP 103039, lying at a distance of only 5.5 pc. Several hundred stars previously suspected of being on our doorstep are now known to lie much further away. There are fewer main sequence stars, and a reduction in the number of giants within 20 pc by almost a factor of two. One white dwarf is expelled from the 5 pc sphere, and 37 previously unknown binary companions have entered the 25 pc census. The local stellar mass density has 'decreased' to a value of about $0.039 \, \mathrm{M_\odot \, pc^{-3}}$.

4. Distance scale

Cepheid variables play a crucial role as distance indicators. They are of high intrinsic luminosity, hence observable out to very large distances, but relatively rare, with few within reach of accurate trigonometric distance determinations. A plot of velocity in Galactic longitude versus Galactic longitude for the 220 Hipparcos Cepheids shows a pronounced sinusoidal structure beautifully illustrating Galactic rotation, with fainter and redder objects being more distant along the lines of sight through the Galaxy disk.

We can see how the same Cepheids participate in the overall rotation of our Galaxy by using distances from the period-luminosity-colour relation, velocities with respect to our own local standard of rest, and then extrapolating their instantaneous space velocities to circular motion over a period of, say, 100 million years. Effectively derived from this fairly complex pattern, the Oort parameters describing Galactic rotation imply a Galactic centre distance of about 8.5 kpc, with an uncertainty of about 5%. Transforming the measured velocities to a Galactic centre origin, a much more structured extrapolated motion is then evident, with the phenomenon of differential Galactic rotation clearly seen.

Well-behaved though these results appear, extending the trigonometric distance scale beyond our Galaxy, with any degree of precision, currently presents many puzzles. An early Hipparcos Cepheid-based distance for the Large Magellanic Cloud, of about 55 kpc, represented an upward revision of the extragalactic distance scale by nearly 10%. Others have since used different Hipparcos tracers, for example, the RR Lyrae stars, Mira vaiables, and red clump giants; and compared them with independent methods such as the SN 1987A ring and geometrically-derived orbital parallaxes. No clear consensus has yet emerged, and an improved understanding of the physics of these various objects is certainly called for. At this distance the parallax is about $20 \, \mu\mathrm{arcsec}$, so that the Large Magellanic Cloud will not escape the direct embrace of SIM and GAIA in the future.

5. Distribution of matter

All stars in the Galaxy move under the gravitational attraction of all other matter, and their motion therefore probes the gravitational signature of material irrespective of its physical manifestation. Visible stars and gas appear to make up only a small part of the total mass of galaxies, with most of it thought to be in some kind of dark form. The primary evidence is that disk galaxies rotate much faster in their outer parts than any reasonable model for the distribution of luminous matter allows. Our Galaxy's dark halo component may extend to at least 50 kpc, as traced by the Magellanic Clouds, and possibly to beyond Leo I at more than 200 kpc.

Locally, the distribution of mass is characterized by the mass per unit volume and, reflecting its scale height, its total surface density in a column perpendicular to the Galactic plane. These quantities are important for understanding chemical evolution, star formation, disk stability, and dark matter properties. Dynamical determination of the density of matter in the Solar neighbourhood, the 'K–z problem', aims to characterise the force law perpendicular to the Galactic plane. For a stellar population in equilibrium, its density and velocity distribution are connected via the gravitational potential, from which the dynamical density can be derived.

Hipparcos results have clarified the problem somewhat, based on a sample of 3000 A and F stars extending out to 125 pc, and including A stars such as Sirius, β Arietis and Castor. These stars are sufficiently luminous that they can be mapped at relatively large distances, whilst sufficiently numerous that they are well represented in the Solar neighbourhood. The sample provides a volume-limited and absolute magnitude-limited homogeneous tracer of stellar density and velocity distributions in the Solar neighbourhood, although these young stars may not be in gravitational equilibrium with the larger-scale Galactic potential. For a star making only a small excursion from the Galactic mid-plane, and moving with simple harmonic motion, the time taken to execute one oscillation is about 60 million years. The distance travelled from the mid-plane, for an initial velocity of $10\,\mathrm{km\,s^{-1}}$, is about 100 pc. Detailed analyses indicate that the local dynamical density of matter near the Galactic mid-plane, the Oort limit, is around $0.1\,\mathrm{M_\odot\,pc^{-3}}$. The vertical motion of the nearby A and F stars is controlled by a projected mass density of about $80\,\mathrm{g\,m^{-2}}$, equivalent to the thickness of normal writing paper.

Comparison with the sum of all observed local matter confirms that the bulk of the Galactic dark matter is distributed in the form of the halo, and not in the form of the disk. Uncertainties remain due to difficulties in detecting low-luminosity stars even very near the Sun, from the uncertain binary fraction among low mass stars, and from uncertainties in the stellar mass–luminosity relation. To summarise: the distribution of both visible and dark matter are still known with only modest accuracy, even in the Solar neighbourhood. Significant improvement requires accurate distances and velocities for a large sample of tracer stars to faint magnitudes, and to much larger distances.

6. Formation of the Galaxy

The Galactic disk is thought to have been formed by the rapid collapse of a rotating, galaxy-sized gas cloud, billions of years ago. The formation of the halo seems to be happening more slowly, and in a more piecemeal fashion.

Forty years ago, the high-velocity star Groombridge 1830 was discovered to belong to a moving group now passing through the Galactic disk. At a distance of only 10 pc, this is the third highest proper motion star, moving with a space velocity of $300\,\mathrm{km\,s^{-1}}$. At these high velocities, stars can travel many kpc from the Galactic plane before their motion is reversed by the overall gravitational field. All stars in the halo must pass through the disk periodically. Being in orbit around a disk star, we can observe halo visitors such as Groombridge 1830 during their rare visits to the Solar neighbourhood. In contrast, astronomers in orbit around a halo star would presumably get a magnificent view of the disk, but could only observe disk stars closely during their own rare passage through the Galactic disk.

Observations of objects like Groombridge 1830 were used to argue that the halo was created during the rapid collapse ($\sim 10^8$ yr) of a relatively uniform isolated protogalactic cloud shortly after it decoupled from the universal expansion. During the last two decades, more kinematic and abundance evidence has become available, and stellar archaeologists are guided more by models which argue that the halo has been built up over billions of years from infalling debris. Evidence includes coherent moving groups in the halo, such as that found towards the North Galactic Pole, and the elongated stellar stream of the Sagittarius dwarf spheroidal, which is moving through the plane on the far side of the Galaxy and is presently being disrupted by the Galactic tidal field.

Part of the evidence for this comes from satellite galaxies encircling our own, and apparently confined to two great streams across the sky. The Magellanic Clouds and the associated Magellanic Stream, as well as Ursa Minor, Draco, and Carina, may be the debris from a former, greater Magellanic Galaxy, which is expected to merge with our own in the distant future due to dynamical friction of the extended halo, somewhat analogous to the existence of comet and meteor streams in a single orbit around the Sun. The Fornax-Leo-Sculptor stream, along with some of the young halo globular clusters, may also be related through a common Galactic accretion event.

In our present picture of how the Galaxy materialised out of the hot, dense, early Universe, the outer parts of galaxies are thought to be accreting low-mass objects ($10^7 - 10^8$ M$_\odot$) even at the present time. Since the orbital time scales of stars in the outer parts of the Galaxy are several billion years, it is here in the baryon halo that we would expect to find fossil evidence of the surviving remnants of accretion, and hence the most compelling evidence to distinguish among competing scenarios for our Galaxy's formation.

Rather direct evidence for such a merging event has recently become available. From the proper motions of 275 metal-poor stars within 2.5 kpc of the Sun, distributed all over the sky with no obvious spatial or velocity structure, 13 metal-poor stars strongly clumped in angular momentum space have been identified Their velocities, of around $300 \, \mathrm{km \, s^{-1}}$, are similar, since they all have roughly the same kinetic energies, and roughly the same potential energy, all being in the Solar neighbourhood. These 13 stars are identified with a single coherent structure, which was captured during or soon after our Galaxy's formation. The satellite had a mass of about 4×10^8 M$_\odot$, and was in a highly inclined orbit, with a maximum distance of only 16 kpc, and a pericentric distance of 7 kpc. As it orbited the Galactic centre, it left a trail of debris over a period of 3 billion years. About 10% of the metal-poor stars in the halo appear to come from this single collision. This type of complex phase-space structure should be revealed in detail by the next generation of astrometric space missions.

7. Miscellaneous

I include some examples of other topics for which three-dimensional imaging of our Solar neighbourhood has provided interesting results.

7.1. *The age of the Universe*

A minimum age of the Universe, useful for constraining cosmological models, can be estimated by determining the age of the oldest objects in our Galaxy. Currently the best age estimates for the oldest stars is based on the absolute magnitude of the main-sequence turn-off in globular clusters, again requiring that the distance to the globular clusters be known. Results yield ages of the oldest clusters of around 11.5 ± 1.3 Gyr.

Independent ages of old objects can be obtained from nucleochronology, and from white dwarf cooling sequences, the latter method again needing precise distances. Higher accuracy in the future will provide accurate age estimates of the Galactic disk from stellar evolution modelling, and independently via white dwarf cooling curves, and their comparison should yield valuable constraints on any secular evolution of the gravitational constant, G.

7.2. *General Relativity*

The measurement of accurate distances and space motions is intimately tied up with the underlying fabric of space-time, and classical tests of light bending make use of the Solar eclipse configuration, in which star light is deflected by the gravitational field of our Sun. Hipparcos yielded a value of the light-bending term, γ, accurate to about 1 part in 10^3. To visualise the geometrical effect, first note that the presence of a gravitating mass deflects a stellar image radially outwards at any given instant. At a subsequent instant, when the mass has moved, displacements are again radially outwards – these radial displacements map into an intriguing time-dependent apparent motion as the gravitating mass moves. Similar stellar shifts would accompany the motion of halo black holes in the Solar neighbourhood, if dark matter exists in the form of primordial black holes formed before nucleosynthesis began.

7.3. *Extra-solar planets*

The idea that planets, and possibly intelligent life, exist beyond our Solar System, has stimulated scientific and popular imagination for centuries. Five years ago, the first extra-Solar planetary systems were discovered through the radial velocity variations of their photocentres. Any stellar companion – whether itself stellar, sub-stellar or planetary – will produce a periodic photocentric wobble in the system's linear barycentric motion through space. Superimposed on the star's linear proper motion, and parallactic ellipse, are additional perturbations due to any accompanying planets. Planet detections in their tens of thousands will be made by the next generation of global microarcsec astrometric satellites, yielding mass estimates independent of orbital inclination. Statistical access to planetary systems on this scale will advance our knowledge of planet formation, and should provide much insight into the formation of our own Solar System.

7.4. *Close passages and the Oort Cloud*

For stars passing near to the Sun, close passages through the Oort Cloud can deflect large numbers of comets into the inner Solar System, initiating Earth-crossing cometary showers and possible Earth impacts. Although the distribution of long-period cometary aphelia is largely isotropic, some non-random clusters of orbits do exist, and it has been suggested that groupings of long-period comet orbits record the tracks of recent stellar passages. Dynamical models suggesting typical decay times of around $2 - 3\,\mathrm{Myr}$. These studies can probe the link between comet showers and past impact events and mass extinctions on Earth. Gliese 710 is the most significant known perturber in the future. Now at 19 pc from the Sun and moving directly at us at $14\,\mathrm{km\,s^{-1}}$, it will pass through the Oort Cloud, at about 50 000 au from the Sun, in about 1 Myr. Algol, now receding from us, was our closest visitor in the recent past, coasting by at a distance of 2.5 pc about 7 Myr ago. Other past or future close passages will presumably be revealed once our knowledge of the kinematics of the local stellar population is more complete.

8. The future

Notwithstanding the huge effort that has been required to improve the stereoscopic mapping of our Solar neighbourhood, I suspect that future generations will look back at the start of the 21st century, and still consider our present knowledge of stellar distances to be remarkably limited. New measurements are now planned that will go very much further.

As we proceed to probe this fractal phase-space jigsaw of stellar motions a host of higher-order phenomena will be uncovered. At the microarcsec level, direct distance measurements will be possible right across our Galaxy and out to the Large Magellanic Cloud. In the process, dynamical consequences of dark matter will become more recognisable, and we will encounter new effects such as perspective acceleration and secular parallax evolution, additional metric terms, planetary perturbations, and astrometric micro-lensing. At this level, even the Sun's absolute acceleration with respect to the Galactic centre, of $2\,\text{Å}\,\text{s}^{-2}$, will have an observable effect on the apparent positions of distant quasars. Further in the future, effects of interstellar scintillation, ripples in space-time due to gravitational waves, and geometric measurements at redshifts of order unity, will enter at the nanoarcsec level.

The bulk of this seething motion is largely below our current observational capabilities, but it is there, waiting to be investigated. Later this decade, the European Space Agency's Gaia mission will carry these distance measurements forward. Gaia will not only provide distances to 10% for 15 mag stars at the distance of the Galactic centre, but will measure all stars in regions with up to about 3 million stars per square degree (i.e. everything down to 20 mag in Baade's Window), resulting in a catalogue of more than one billion objects complete to 20 mag, of which of order 100 million would have distance accuracies better than 5%.

Stereoscopic mapping at the microsecond level will certainly reveal many more important clues to our Galaxy's origin, structure, and future.

9. Illustrative material

This lecture, in the Harrington Lecture Theatre of the University of Central Lancashire, Preston, UK used a series of 3-d projections based on parallaxes, proper motions and derived parameters from the Hipparcos Catalogue. This employed a dual digital projection system, with orthogonally-polarised light beams and an 4m × 3m reflecting screen, with members of the audience equipped with appropriate polarised glasses. The lecture was an update of a similar talk that I gave as an Invited Discourse at the IAU General Assembly in Manchester in 2000.

Three-dimensional animations showed, for example, moving star fields over several thousand years, including Arcturus, the Hyades, Pleiades; reconstructed orbits of disk stars within the Galactic potential; motions of stars in a 20 pc cube centred on the Sun, simulations of the Galaxy's thin and thick disk populations; A and F stars moving perpendicular to the Galactic plane under the influence of the disk potential; the space motions of Groombridge 1830 and the approaching star Gliese 710; the fields of the planetary systems 51 Peg, 70 Vir, 47 UMa and υ Andromedae; the constellations of Orion's Belt and Ursa Major; and the Hubble Deep Field in three dimensions based on spectroscopic redshifts.

Acknowledgements

I acknowledge the many members of the Hipparcos collaboration whose work has made this contribution possible. My particular thanks go to my colleagues Karen O'Flaherty and Jos de Bruijne for much assistance with the visual effects, and to Don Kurtz for considerable assistance in the preparations needed for the 3d projections.

Discussion

BERNARD DE SAEDELEER: You showed at the beginning red stars going up and down. I just wonder how you can extrapolate this motion, because, as you said, we just measure an initial condition which is in the present, and then you try to extrapolate to the past, or to the future. Then how you can say that the star will come back up and down?

MICHAEL PERRYMAN: This is an area where you have to really dig into this in detail. It turns out that if you measure one particular star – just one star – you would have no idea, of course. But if you measure a large population of stars collectively – their distribution from the plane, and their instantaneous velocities – it turns out that you can reconstruct from that the probable model of the gravitational distribution of matter controlling their motion. So you can imagine that in a snapshot of this collective motion you have a lot of stars – several hundreds, several thousands – you can reconstruct from that the probable distribution of stars using Poisson's equation.

ANONYMOUS: Did you say that there are no globular clusters in our galaxy?

MICHAEL PERRYMAN: No. I said that they are far away, and there are not very many of them – something like a couple of hundred that we can see. This means that they are actually rather scarce, so there isn't one very close to us. Therefore we can't measure the structure very carefully, just because they are too far away.

ANONYMOUS: With respect to the motion stars: Do they obey Kepler's law in terms of sweeping out equal areas in equal times?

MICHAEL PERRYMAN: Stars don't quite move in the way that we would predict from our knowledge of Kepler's laws. It is because they don't move in agreement with certain basic laws of motion that we think there is the existence of dark matter. Now, many of you will perhaps know about dark matter; it is one of the big mysteries in astronomy at the moment and I have not referred to it at all. But it does appear that a large amount of matter must be there – because we see it's influence on the motion of stars – but we can't actually see it in terms of stars or any other matter that we can measure. It's the fact that stellar motions, spiral arms – actually galaxy rotation in general – deviate from the predictions of basic Newtonian gravity that we think, or other people think (I have no comment on this) . . .

[general laughter]

. . . that dark matter must exist.

ANONYMOUS: Thank you.

MICHAEL PERRYMAN: The elegant thing about measuring stellar motions is that whether you can see the gravitating matter, or not, it does have an influence on the stellar motions. This is one of the reasons why measuring stellar motions is very important, because it tells you about the distribution of matter irrespective of whether it's visible or whether it's invisible, whether it's dark or whether it's conventional, but the jury is still out, I think, on the presence of dark matter.

ANONYMOUS: You spoke earlier about being able to observe from the Earth arc seconds, and then Hipparcos was milli-arc seconds, and now a Gaia is going to be micro-arc seconds: Is there any theoretical minimum in visible light, or any other light, beyond which you cannot go in terms of angular measurement?

MICHAEL PERRYMAN: We can think of at least two reasons why you cannot go below a certain level. One of them is something called scintillation. It's important in the radio – interstellar, interplanetary scintillation because it perturbs the radio waves like a fluctuating screen. In the optical interstellar scintillation is going to come in at well below a nano-arc second, so it is a factor of another 1000 or 10 000 beyond Gaia before we have to start worrying about it. At around the nano-arc second level we believe that stochastic gravitational wave background energy becomes important. General relativity predicts that there are gravitational waves caused by massive bodies revolving around each other, and if we go well out into the galaxy we see black holes colliding, and binary stars going around each other – probably sending out gravitational wave energy. There's a mission under study by NASA and ESA called LISA which is intended to go into space and actually measure this gravitational wave energy. From what we know at the moment – at the level around a nano-arc second, or $\frac{1}{10}$ of a nano-arc second – probably this kind of noise coming from the gravitational wave background comes in. This means is that Gaia will be no where near the fundamental limit that you refer to. We can carry on measuring distances using trigonometric methods just by virtue of the earth going around the sun, right the way out to the nearest galaxies, right the way out to cosmological distances. We won't be doing that, I can confidently predict, in my lifetime, but perhaps in a 100 years. Who knows? They are the only two phenomena that I am aware of that would actually limit us. Perhaps structure on the surface of stars – star spots, convection, granulation, and so on – may put some limits in. We are not sure at the moment, but we are fairly confident we can go down to the micro-arc second level.

ANONYMOUS: You showed simulations of the Hyades and the Pleiades clusters. I'm thinking regarding our sun: Is there any evidence from Hipparcos whether we are in a system like some that you showed?

MICHAEL PERRYMAN: There is no evidence from Hipparcos, but that doesn't mean to say that it's ruled out. You would need to go to these more accurate missions to get that kind of information. It's an area of debate that comes up time and again: Is there another star, a fainter star going around the sun at very large distances? We don't have enough information to say that at the moment.

ANONYMOUS: I presume that one of the main reasons behind the search for planets in other solar systems is the search for life elsewhere. You sat very much on the fence at the end of your talk: You said we could be unique, or there could be life out there. Not wishing to put you in the position of seeing a headline in the paper tomorrow – "Leading scientist from ESA predicts life outside of the earth" – which side of the fence do you actually think you would ...

[general laughter]

MICHAEL PERRYMAN: I think life on earth is unique.

[stunned silence ... nervous laughter ... general clapping]

MICHAEL PERRYMAN: ... and perhaps if a few more people thought that life on Earth was unique we would be a little less pre-occupied with some of the crazy things that we do on this Earth.

GORDON BROMAGE: Ladies and gentlemen, I think you will agree that we have experienced a spectacular presentation and a wonderful lecture this evening. This public lecture tonight was co-sponsored by the University of Central Lancashire, The European Space Agency and the International Astronomical Union, and we have been privileged to hear it. Please show your appreciation in the usual way.

[sustained, enthusiastic clapping]

Michael Perryman and Marilyn Head at the Banquet

Part 5

NEW VIEWS OF THE GALAXY: PARALLAXES, DISTANCES AND IMPLICATIONS FOR ASTROPHYSICS

Dave and Alice Monet, Steve Dick; Jesús de Alba at the telescope

left: Fritz Benedict; right: Rick Collins and Steve Higgins

Transits of Venus: New Views of the Solar System and Galaxy
Proceedings IAU Colloquium No. 196, 2004
D.W. Kurtz, ed.
© 2004 International Astronomical Union
doi:10.1017/S174392130500150X

The (f)utility of ground-based parallaxes

David G. Monet

U. S. Naval Observatory, Flagstaff, AZ 86001, USA

Abstract. During the 25 years that the author has been involved in astrometry, the quality of ground-based parallaxes has increased by about a factor of 10 (from 3 mas to 0.3 mas), but the quantity has increased by only a few hundred. When asked, the average astronomer will cite the H^2 space missions (Hubble and Hipparcos) as the great advances in astrometry even thought the ground-based work has played a critical role in the understanding classes of stars such as the L- and T-dwarfs. The next decade promises stunning advances in both the quality and quantity of parallax measurements, and both ground- and space-based projects will play significant roles. The Gaia space mission and the various ground-based telescopes with large etendue (DMT, LSST, Pan-STARRS, etc.) will improve the quality or quantity (or both) by factors of a thousand or more. The situation will be discussed, and the author will express his hope that he might live long enough to see the fruits of these labors.

Discussion

CORYN BAILER-JONES: Dave, just to comment on one of your cynical observations, number 2. You rightly mentioned that the equivalent precision which one is trying to get to with Gaia is 50 micropixels, but, of course, the intention is not to get that in a single observation.

DAVE MONET: No, that's the end of mission. But that's what you are talking about: characterising your detector over roughly a five-year period such that the systematic errors are the only thing that you have left at that kind of a number.

CORYN BAILER-JONES: Correct, but it's not a single CCD.

DAVE MONET: Yes, those were not single exposure numbers. That was the desired accuracy after the end of the mission.

CORYN BAILER-JONES: Right, and that's presumably also the case for your USNO observations.

DAVE MONET: Yes, that's right. The playing field was as level as I could make that.

FRITZ BENEDICT: Is all the astrometry done on a single chip in PANSTARS? There's no chip-to-chip astrometry done?

DAVE MONET: My hope is that we can do wafer-to-wafer – that is to say each 4k × 4k piece of silicon should have the same thermal stretching blah, blah, blah – and we don't have to solve each 512^2 cells separately; that's the hope. But undoing this OT CCD guiding as a non-trivial sport because the charges have moved in different directions, so your bookkeeping has to be very good, and that's not something all programmers get right.

FLOOR VEN LEEUWEN: There's one area for ground-based astrometry that is completely out of sight for space-based astrometry, and that is globular cluster work. For globular clusters the density towards the core is far too high to deal with by satellites like Gaia. Are we developing ground-based facilities that can cope with globular clusters?

DAVE MONET: I guess I'm a bit curious, because I would have thought that in space where you just have λ/d for the size of your PSF ... from the ground you are lucky to get to a tenth of an arc-second. Isn't that about the Adaptive Optics limit for the Keck and other telescopes? Surely you can do better than that in space?

FLOOR VEN LEEUWEN: The density of stars per square degree is too high for Gaia.

DAVE MONET: Does confusion limit that?

FLOOR VEN LEEUWEN: Yes.

DAVE MONET: [inaudible comment from Fritz Benedict] ... I'm sorry. I won't give you that, Fritz, I'm sorry. There's no way SIM can handle that kind of density. It would be really cute if somebody could come up with the same kind of concept that we saw for MOST – flying a little suitcase that did the globular clusters correctly. That would be a tremendous ecological niche: build it and fly it for 10 million bucks, point it at 5 or 10 important globular clusters, and walk away with the science. That would be one *really* great opportunity.

RICKY SMART: $R = 24.3$ in the plane? You're just going to be confused out?

DAVE MONET: Well, that's where Moore's law comes in, right? I have no idea, no one has really gone to $R = 24$ on the ground. The key is how well can you do relative astrometry on difference imaging. One of the keys of the Pan-STARRS pipeline is to build a deep image of the sky, and then subtract that from your instantaneous frames. That highlights the supernovae, highlights the new objects there. That also helps with crowding a tremendous amount. The OGLE folks are doing that and finding all sorts of cute stuff in otherwise very crowded regions. So I think there's a lot of applied image processing before we give up completely on dense regions – like, if you solve the problem, you can do the globular clusters! Wouldn't it be nice to attract some young blood who still enjoyed computer programming, and got it right occasionally?

Transits of Venus: New Views of the Solar System and Galaxy
Proceedings IAU Colloquium No. 196, 2004
D.W. Kurtz, ed.
© 2004 International Astronomical Union
doi:10.1017/S1743921305001511

High-precision stellar parallaxes from *Hubble Space Telescope* fine guidance sensors†

G. Fritz Benedict and Barbara E. McArthur

McDonald Observatory, University of Texas, Austin, TX 78712, USA
email: fritz@astro.as.utexas.edu, mca@astro.as.utexas.edu

Abstract. We describe our experiences with on-orbit calibration of, and scientific observations with, the Fine Guidance Sensors (FGS), white-light interferometers aboard *Hubble Space Telescope*. Our original goal, 1 milliarcsecond precision parallaxes, has been exceeded on average by a factor of three, despite a mechanically noisy on-orbit environment, the necessary self-calibration of the FGS, and significant temporal changes in our instruments. To obtain accurate absolute parallaxes from these small fields of view ($3' \times 15'$) observations requires a significant amount of ancillary reference star information. These data also permit an independent estimate of interstellar absorption, critical in determining target absolute magnitudes, M_V, often the key result of a parallax program. With these techniques we and our collaborators have obtained absolute parallaxes for 21 astrophysically interesting objects. We briefly discuss a recent determination of the parallax of the Pleiades. *HST* routinely produces parallaxes with half the error of the best Hipparcos results, a precision that continues down to target $V = 15$. The FGS will remain a competitive astrometric tool for the generation of high-precision parallaxes until the advent of longer-baseline space-based interferometers (SIM), or the failure of some key *HST* component.

1. Introduction

We describe our experiences with on-orbit calibration and scientific observations with the Fine Guidance Sensors (FGS), white-light interferometers aboard *Hubble Space Telescope*. Our original goal, 1 milliarcsecond (mas) precision parallaxes, has been exceeded on average by a factor of three, but not without significant challenges. These included a mechanically noisy on-orbit environment, the self-calibration of an FGS (McArthur *et al.* 2002), and significant temporal changes in our instruments. Solutions included a denser set of drift check stars for each science observation, fine-tuning exposure times, overlapping field observations and analyses for calibration, and a continuing series of trend-monitoring observations.

To obtain accurate absolute parallaxes from these small field of view ($3' \times 15'$, shown in Fig. 1) differential astrometric observations requires a significant amount of ancillary reference star information. We employ combinations of visible ($BVRI$), near-infrared (JHK), and Washington-DDO (T_2, M, DDO51) photometry, along with MK spectral types and luminosity classes from classification-dispersion spectra. These data also permit an independent estimate of interstellar absorption, critical in determining target absolute magnitudes, M_V, often a key result of the parallax effort (e.g., Benedict *et al.* 2002a).

---

† Based on observations made with the NASA/ESA Hubble Space Telescope, obtained at the Space Telescope Science Institute, which is operated by the Association of Universities for Research in Astronomy, Inc., under NASA contract NAS5 26555

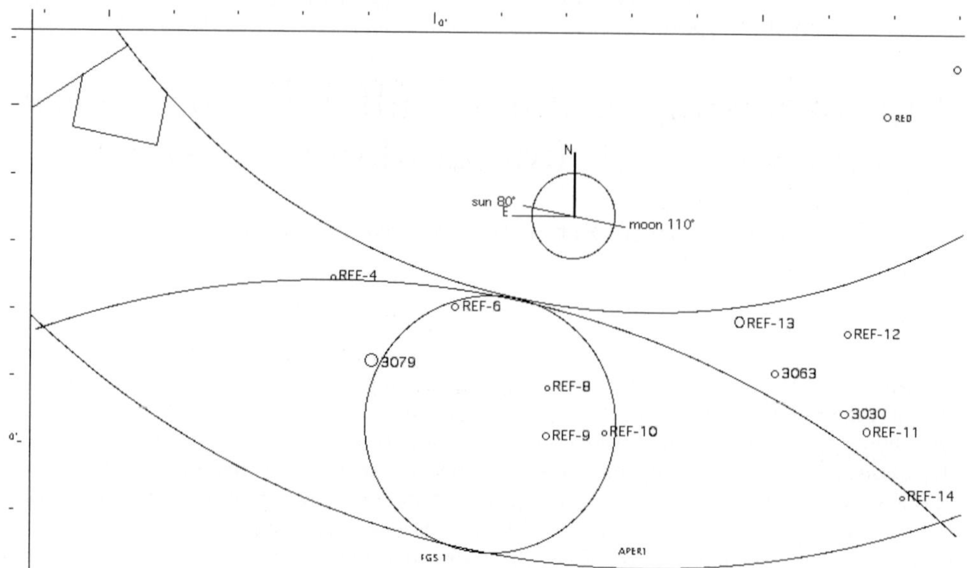

Figure 1. Field of regard of FGS 1 on the sky at the location of the Pleiades parallax targets 3030, 3063, and 3079 discussed in Section 5.1. Also shown are the reference stars, relative to which parallax and proper motion are obtained. The tick marks at top are spaced by $1'$.

2. The instrument and calibrations

Those with a deeper interest in the FGS instrument and calibration issues can find considerably more detail in Nelan & Makidon (2001) and McArthur *et al.* (2002), respectively.

2.1. *Anatomy of an FGS*

Each FGS is an interferometer. Interference takes place in a prism that has been sliced in half, had a quarter-wave retarding coating applied, and then reassembled. Most of the FGS consists of supporting optics used to feed the Koester's Prisms (top right, Fig. 2).

In particular the star selectors walk a $5''$ instantaneous field of view throughout the interferometer field of regard shown in Fig. 1. The output of each face of the Koester's Prism is measured by a PMT. These signals are combined

$$S = \frac{A - B}{A + B} \tag{2.1}$$

to form a signal, S, that is zero for waves exactly vertically incident on the Koester's Prism front face. Tilting the wavefront back and forth (equivalent to pointing the telescope slightly off, then on target, then slightly off to the other side) will generate a fringe pattern (Fig. 3). This technique of fringe scanning is often useful for resolved targets such as binary stars (Franz *et al.* 1998). For parallax work we obtain fringe tracking measurements. A series of measurements of the fringe zero-crossing position are obtained for each target and reference star. These are subjected to a median filter to produce a relative position within the FGS field of regard.

A perfect instrument would generate a perfectly symmetric fringe pattern. The significant spherical aberration of the as-built *HST* primary mirror, in the presence of internal FGS misalignments, produces a signature in the fringe which mimics coma. Coma causes decreased modulation and multiple peaks and valleys in a fringe. A replacement FGS installed in 1997 contains an articulated fold flat (FF3 in Fig. 2, center) that removes

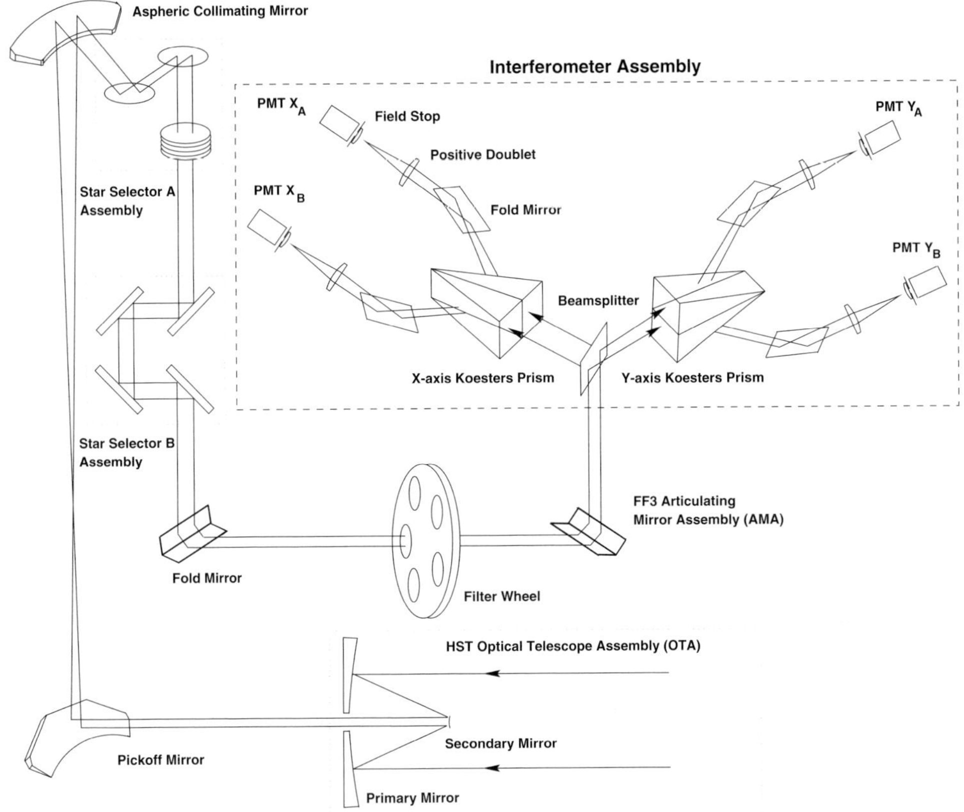

Figure 2. The optical layout of a Fine Guidance Sensor. We obtain fringe information for two orthogonal axes simultaneously.

most of the internal misalignments. This FGS (FGS 1r) produces nearly perfect fringes (Fig. 3), thus yielding far better fringe tracking and fringe scanning results.

2.2. *The optical field angle distortion calibration*

Optical distortions in the HST Ritchey-Chretien telescope and FGS combination have positional amplitudes in the focal plane exceeding $1''$. There was no existing star field with cataloged 1 mas precision astrometry, our desired performance goal. Our solution was to use FGS to calibrate itself with multiple observations of a distant star field (M35). A distant field was required so that during the two-day duration of data acquisition, star positions would not change. We obtained these data in early 1993 for FGS 3 and in 2000 for FGS 1r and reduced them with overlapping plate techniques to solve for distortion coefficients and star positions simultaneously. As a result of this activity distortions are reduced to better than 2 mas over much of the FGS field of regard. This model is called the Optical Field Angle Distortion (OFAD) calibration. Details can be found in McArthur *et al.* (2002). To date both FGS 3 and FGS 1r on *HST* have been calibrated

Once we have established a calibration we must maintain it: an FGS changes over months and years. For example, the FGS 3 graphite-epoxy optical bench was predicted to outgas for a period of time after the launch of *HST*, a process predicted to change the relative positions of optical components on the optical bench. As a consequence we revisit the M35 calibration field periodically to monitor these (scale-like) changes

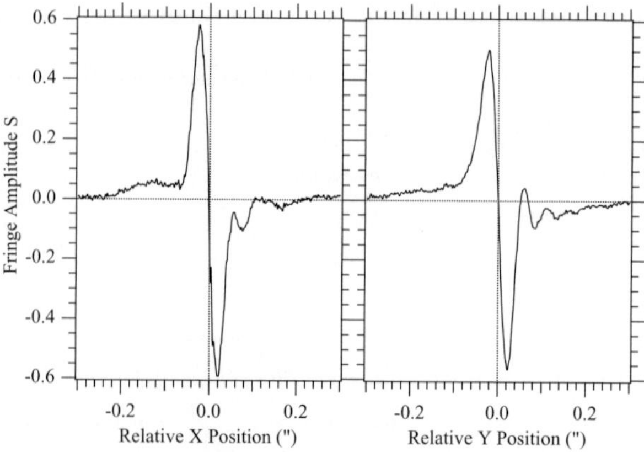

Figure 3. Fringe along the X and Y axes of FGS 1r. A series of fringe tracking measurements of the X and Y zero-crossing positions (the positions at which S = 0) are subjected to a median filter to produce a final X, Y position used for parallax work.

and other slowly varying non-linearities. This is the never-ending LTSTAB (Long-Term STABility) series. LTSTABs are required as long as it is desirable to do 1 mas precision astrometry with an FGS. The result of this series is to model and remove the slowly varying component of the OFAD, so that uncorrected distortions remain below 2 mas for center of an FGS. The character of these changes are generally monotonic with abrupt jumps in conjuction with HST servicing missions.

2.3. *Lateral color*

Because each FGS contains refractive elements (star selector A in Fig. 2), the position measured for a star can depend on its intrinsic color. This lateral color shift would be unimportant, as long as target and reference stars had similar color. However, this is certainly not the case for many of our science target stars (Table 1), hence our need for this calibration. For further details see Benedict *et al.* (1999).

2.4. *Cross filter*

The filter wheel in each FGS contains a neutral density filter with a 1% transmission (Nelan & Makidon 2001). This filter, designated FND5, provides 5 magnitudes of attenuation. This reduction of signal is required to obtain astrometry for stars that are brighter than $V = 8.5$, for which the count rate for the FGS PMTs would exceed the electronics capacity (Bradley *et al.* 1991). No filter has perfectly plane-parallel faces, an effect called filter wedge. Filter wedge introduces a slight shift in position when comparing an observation with the standard astrometry filter, F583W, with the FND5 filter. We required this latter filter to perform astrometry on RR Lyr, $V \sim 7.2$ and δ Cep, $V \sim 4$. To obtain milliarcsecond astrometry requires knowledge of the filter wedge effect to that precision or better.

Conceptually the calibration is simple. Observe in POS (fringe tracking) mode the same star with and without the FND5 filter and compare the positions. The shift so determined (Fig. 4) is then applied when comparing faint reference stars with bright science targets. The standard astrometry filter is F583W. As a consequence we actually measure differential filter wedge, because F583W is also a filter with non-parallel faces. Note that each filter is also a refractive element. Thus a star position will depend on the color of the star. This is an element of the lateral color effect discussed above.

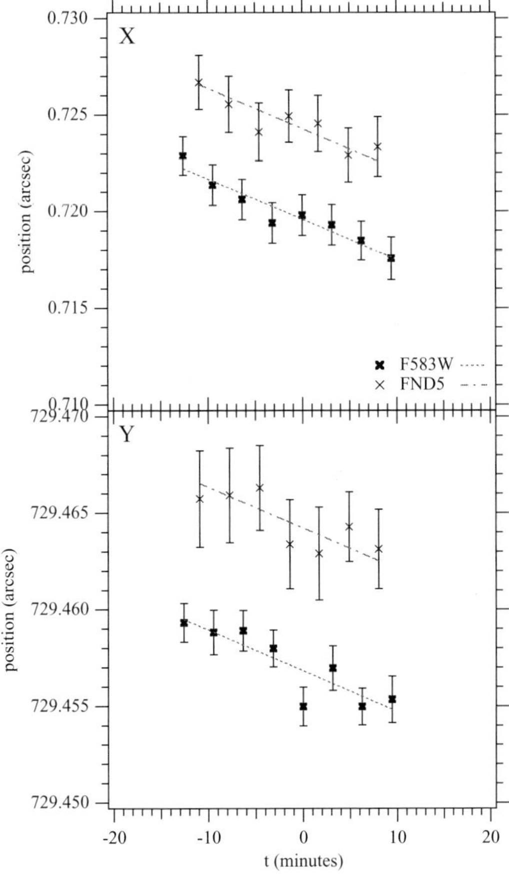

Figure 4. Cross filter calibration observations in 1998. Target is Upgren 69 in NGC 188. The plots show a shift in position between F583W and FND5 and typical intra-orbit drift in FGS 3.

3. Observations required for a parallax determination

3.1. *Observing with an FGS*

The issues summarized in this section are discussed at greater length in Benedict *et al.* (1998), particularly the intra-orbit observing strategies. A typical observation sequence has a duration of about 40 min and consists of a serial collection of from 10 to 30 time-series of positions sampled at 40 Hz. Each time series lasts from 30 to 300 s, depending on the target star brightness. We have identified sources of systematic and random position noise and discuss them from highest to lowest frequency.

We characterized the power spectrum of *HST* mechanical noise and determined that observing for 60 s or longer adequately sampled the frequency domain. A median filter was determined to abstract the best position (i.e., the median is a robust estimator for this system). The dispersion around the median provides an estimation of the observational error.

Over the course of an orbit guide stars autonomously drift as each FGS shifts slightly in its bay. Fig. 4 shows observations of a cross-filter calibration star over the span of 25 min. This behavior imposes additional overhead, reducing the time available within an orbit to do science. An observation set must contain multiple visits to two or more astrometric reference stars. Presuming no motion intrinsic to these stars over 40 min, one determines drift and corrects the reference frame and target star for this drift. As a result we reduce the error budget contribution from drift to less than 1 mas.

Lastly, we arrange to observe our science targets and associated reference stars near times of maximum parallax factor. Some results have been secured with as few as six orbits over 1.5 yr. Typically we obtain around ten orbits over that same time span. The extra orbits result in better parallax precision. By bracketing times of maximum parallax factor with a pair of observations spaced by a week or so we also insure against rare *HST* equipment glitches. We also generally observe a few times at intermediate parallax factor to distinguish parallax from proper motion.

3.2. *Ancillary observations*

3.2.1. *Spectrophotometric absolute parallaxes of the astrometric reference stars*

Because the parallax determined by an FGS will be measured with respect to reference frame stars which have their own parallaxes, we must either apply a statistically derived correction from relative to absolute parallax (van Altena, Lee & Hoffleit 1995, hereafter YPC95) or, preferably, estimate the absolute parallaxes of the reference frame stars (e.g. Harrison *et al.* 1999). With colors, spectral type, and luminosity class for a star one can estimate the absolute magnitude, M_V, and V-band absorption, A_V. The absolute parallax is then,

$$\pi_{abs} = 10^{-(V - M_V + 5 - A_V)/5} \tag{3.1}$$

The luminosity class is generally more difficult to determine than the spectral type (temperature class). However, the derived absolute magnitudes are critically dependent on the luminosity class. To confirm the luminosity classes we generally employ the technique used by Majewski *et al.* (2000) to discriminate between giants and dwarfs for stars later than $\sim$ G5, an approach whose theoretical underpinnings are discussed by Paltoglou & Bell (1994). The boundary between giants and dwarfs is 'fuzzy' and complicated by the photometric transition from dwarfs to giants through subgiants. This soft boundary is readily apparent in figure 14 of Majewski *et al.* (2000). However, objects above the boundry are statistically more likely to be giants than objects just below.

• Photometry – Our band-passes for reference star photometry include: $BVRI$, JHK (from 2MASS†), and Washington/DDO filters M, 51, and T_2 (obtained at McDonald Observatory with the 0.8-m Prime Focus Camera). We transform the 2MASS JHK to the Bessell (1988) system, using the transformations provided in Carpenter (2001).

• Spectroscopy – The spectra from which we estimated spectral type and luminosity class come from several sources. Classifications are obtained by a combination of template matching and line ratios, often by two independent co-investigators.

• Interstellar Extinction – To determine interstellar extinction we first plot our reference stars on several color-color diagrams. A comparison of the relationships between spectral type and intrinsic color against measured colors provides an estimate of reddening. Fig. 5 contains $V - R$ vs $V - K$ and $V - I$ vs $V - K$ color-color diagrams and reddening vectors for our RR Lyr campaign (Benedict *et al.* 2002a). Also plotted are mappings between spectral type and luminosity class V and III from Bessell & Brett (1988) and Cox (2000, hereafter AQ00), again with reddening vectors and the loci of luminosity classes V and III stars. Fig. 5, along with the estimated spectral types, provides measures of the reddening for each reference star.

Assuming an R = 3.1 galactic reddening law (Savage & Mathis 1977), we derive A_V values by comparing the measured colors with intrinsic $V - R$, $V - I$, $J - K$, and

---

† The Two Micron All Sky Survey is a joint project of the University of Massachusetts and the Infrared Processing and Analysis Center/California Institute of Technology

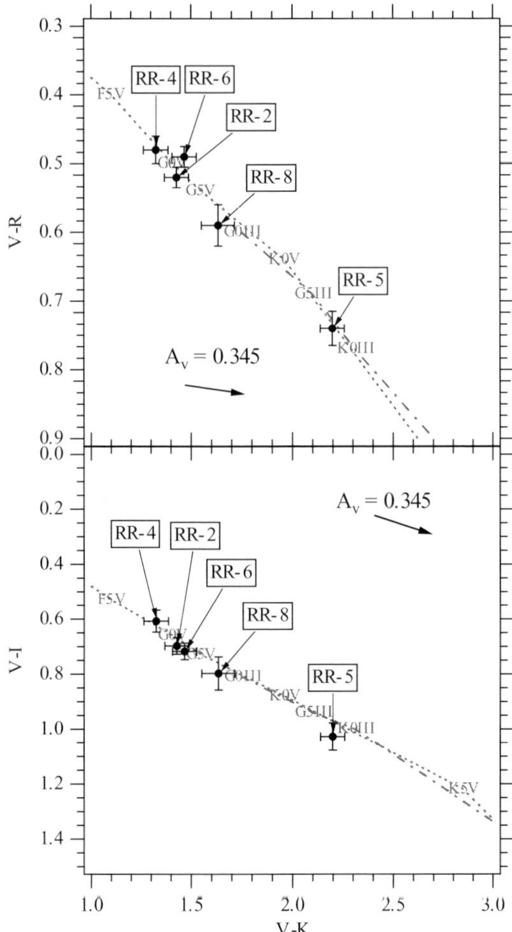

Figure 5. $V - R$ vs $V - K$ and $V - I$ vs $V - K$ color-color diagrams and reddening vectors, photometry used for our RR Lyr campaign (Benedict *et al.* 2002a).

$V - K$ colors from Bessell & Brett (1988) and AQ00. Specifically, we estimate A_V from four different ratios, each derived from the Savage & Mathis (1977) reddening law: $A_V/E(V - R) = 4.83$; $A_V/E(V - K) = 1.05$; $A_V/E(J - K) = 5.80$; and $A_V/E(V - I) = 2.26$. For some fields colors and spectral types are inconsistent with a field-wide average $<A_V>$. This was the case for the δ Cep field, where we ultimately adopted the reddening of the most proximate reference star as the target reddening (Benedict *et al.* 2002b).

3.2.2. *The inclusion of prior knowledge*

When possible, prior knowledge bearing on the determination of target parallax is included in our modeling process. This information passes to the model as observations with errors which weight their influence on the final outcome. This approach allows us to incorporate *any* measurements relevant to our investigation. Here are two examples of this quasi-Bayesian approach.

• In determining the parallax of δ Cep we had prior knowledge that reference star DC-2 was thought to be physically associated with δ Cep. This association was established through common proper motion. Also de Zeeuw *et al.* (1999) include both δ Cep and DC-2 in the Cep OB6 association. We constrained the difference in parallax between δ Cep and DC-2, using our prior knowledge of their association. From de Zeeuw *et al.* (1999) we estimated that the 1σ dispersion in Galactic longitude for the OB association thought to contain both δ Cep and DC-2 is $3°$. One can therefore infer that the 1σ dispersion in distance in this group is $3°$/radian $\sim5\%$. Hence, the 1σ dispersion in the parallax difference between two group members (e.g. DC-2 and δ Cep) is

$$\Delta\pi = 5\% \times \sqrt{2} \times 3.7\,\text{mas} = 0.26\,\text{mas} \tag{3.2}$$

where we have here adopted the mean parallax of Cep OB6, $\langle\pi\rangle = 3.7\,\text{mas}$, from de Zeeuw *et al.* (1999). The assumed zero parallax difference between δ Cep and DC-2 becomes an observation with an associated error $(\Delta\pi)$ fed to our model, an observation used to estimate the parallax difference between the two stars, while solving for the parallax of δ Cep.

• The reference star spectrophotometric absolute parallaxes are input as observations with errors, not as hardwired quantities known to infinite precision. The lateral color and cross-filter calibrations, as well as $B - V$ color indices, are entered into the model as observations with associated errors. We now also introduce proper motion data from UCAC2 (Zacharias *et al.* 2003) with typical input errors $5\,\text{mas}$ in each coordinate.

4. The astrometric model

With the positions measured by an FGS we determine the scale, rotation, and off-set "plate constants" relative to an arbitrarily adopted constraint epoch (the so-called "master plate") for each observation set (the data acquired at each epoch). Depending on reference frame characteristics (number and distribution of reference stars), we employ models with four to eight parameters (e.g., McArthur *et al.* 2001; Benedict *et al.* 2003) for those observations. In some of our earliest work we determined the plate parameters from reference star data only, then applied them as constants to obtain the parallax and proper motion of the science target. Usually we determine the plate parameters and the parallax and proper motion of the science target and reference stars simultaneously. Typically the reference stars have color indices that differ from the science target, and we apply the corrections for lateral color discussed in Benedict *et al.* (1999).

For all our astrometric analyses, we employ GaussFit (Jefferys *et al.* 1987) to minimize χ^2. The solved equations of condition are typically:

$$x' = x + lcx(B - V) \tag{4.1}$$

$$y' = y + lcy(B - V) \tag{4.2}$$

and

$$\xi = Ax' + By' + C + R_x(x'^2 + y'^2) - \mu_x\Delta t - P_\alpha\pi_x \tag{4.3}$$

$$\eta = -Bx' + Ay' + F + R_y(x'^2 + y'^2) - \mu_y\Delta t - P_\delta\pi_y \tag{4.4}$$

or

$$\xi = Ax' + By' + C - \mu_x\Delta t - P_\alpha\pi_x \tag{4.5}$$

$$\eta = Dx' + Ey' + F - \mu_y\Delta t - P_\delta\pi_y \tag{4.6}$$

where x and y are the measured coordinates from *HST*; lcx and lcy are the lateral color corrections; and $B - V$ are the $B - V$ colors of each star. A , B, D, and E are scale and rotation plate constants, C and F are offsets; R_x and R_y are radial terms; μ_x and μ_y are proper motions; Δt is the epoch difference from the mean epoch; P_α and P_δ are parallax

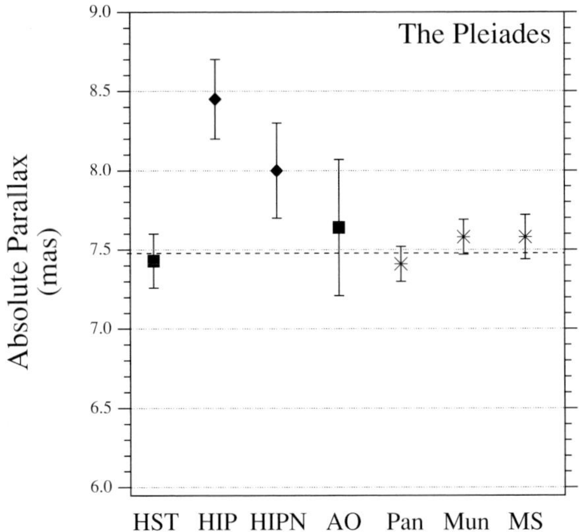

Figure 6. Absolute parallax determinations for the Pleiades. We compare astrometric parallax results (filled squares) from *HST* and Allegheny Observatory (AO, Gatewood *et al.* 2000) with Hipparcos (filled diamonds), both an older result (HIP, van Leeuwen 1999) and a very recent re-determination (HIPN, van Leeuwen 2004). Pan *et al.* (2004) have derived a dynamical parallax from long baseline interferometry of the binary star, Atlas. Munari *et al.* (2004) have determined a dynamical parallax using an eclipsing SB2 binary. MS denotes a parallax derived from main-sequence fitting (Pinsonneault *et al.* 1998). The horizontal dashed line is the weighted average of the *HST*, Pan, Munari, and AO measures, $\langle \pi_{abs} \rangle = 7.49 \pm 0.07$ mas.

factors; and π_x and π_y are the parallaxes in x and y. See Benedict *et al.* (2002b) for a model that includes cross-filter corrections. We obtain the parallax factors from a JPL Earth orbit predictor (Standish 1990), upgraded to version DE405. Before the existence of a reliable source for proper motions, we imposed the constraint that the reference star proper motions are random in direction by forcing their sum in x and y to be zero, $\sum \mu_x = \sum \mu_y = 0$. Orientation to the sky is obtained from ground-based astrometry (e.g., USNO-A2.0 catalog, Monet 1998) with uncertainties in the field orientation $\pm 0°.05$.

The solution process is allowed to adjust any input parameter by an amount depending on its variance to find the 'best' solution. These input parameters include previously measured colors, proper motions, reference star parallaxes estimated from spectrophotometry, the orientation of the reference plate to RA and Dec, and the lateral color and cross filter corrections.

5. Results

The choice of *HST* astrometry science targets is rightly determined by what can be done from the ground. The unique capabilities of *HST* must remain reserved for projects demanding them. It is by now clear that a significant investment of time and effort is required to obtain a parallax from FGS data. However, this investment guarantees parallax precision at or below 0.5 mas for objects as faint as $V = 15$.

With these techniques we have determined more precise absolute magnitudes for the distance-scale standards, RR Lyr and δ Cep. We have obtained absolute magnitudes and radii for two hot white dwarf stars, Feige 24 and the central star of the planetary nebula NGC 6853. We have generated precise distances with which to constrain theories of

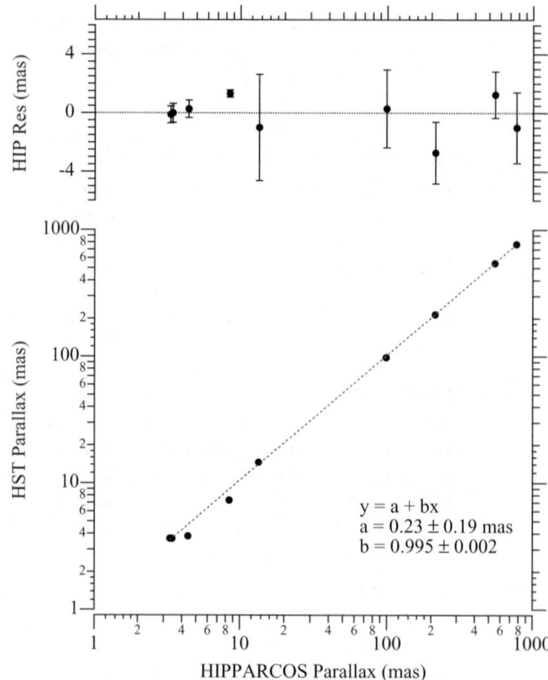

Figure 7. Bottom: *HST* absolute parallax determinations compared with Hipparcos for targets in common listed in Table 1. Top: The Hipparcos residuals to the dotted error-weighted impartial regression line that excludes the Pleiades. The error bars on the residuals are Hipparcos Catalog 1-σ errors.

the star-star interactions evidenced by cataclysmic variables (TV Col, RW Aur, WZ Sge, YZ Cnc, U Gem, SS Aur, SS Cyg, RU Peg, and EX Hya). Other investigations have treated the parallax as a nuisance parameter that must be removed with exquisite precision to determine the perturbation due to a planetary mass companion (Gl 876b), or to generate precise binary star orbital elements and component masses (Gl 791.2, Wolf 1062), from orbits whose dimensions on the sky are smaller than the typical seeing at an excellent ground-based observing site. Table 1 lists all objects for which we have derived parallaxes, including our soon to be published result for the Pleiades, which we now discuss in more detail.

5.1. *A parallax for the Pleiades*

We have recently completed a study of three members of the Pleiades. A full account of this study will appear shortly (Soderblom *et al.* 2004). Our original intent was to determine orbital parameters for several binaries thought to belong to the Pleiades. After realizing that FGS 1r could not adequately resolve these binaries, this dynamical parallax experiment changed to a standard parallax program for the three Pleiads in the field. Because of this the timing of the observation illustrated in Fig. 1 was not optimal for maximum parallax factor, having a Sun-target field separation of 80°, shown on the compass rose.

Six sets of astrometric data were acquired with *HST*, spanning 3.51 yr, for a total of 135 measurements of the three Pleiads (stars 3030, 3063, and 3179 in Fig. 1) and nine reference stars (see Fig. 1). Spectral types and luminosity classes were estimated from classification dispersion spectra. This information, with JHK photometry from 2MASS and *V* photometry from the FGS yielded reference frame absolute parallaxes through

Table 1. *HST* and Hipparcos Absolute Parallaxes. The last six parallaxes were obtained with FGS 1r. All others were obtained with FGS 3.

Object	*HST* mas	Hip mas	*HST* Reference
Prox Cen	769.7 ± 0.3	772.3 ± 2.4	Benedict *et al.* 1999
Barnard's Star	545.5 ± 0.3	549.3 ± 1.58	Benedict *et al.* 1999
U Gem	9.96 ± 0.37		Harrison *et al.* 1999
SS Aur	5.99 ± 0.33		Harrison *et al.* 1999
SS Cyg	6.06 ± 0.44		Harrison *et al.* 1999
RW Tri	2.93 ± 0.33		McArthur *et al.* 1999
Feige 24	14.6 ± 0.4	13.44 ± 3.62	Benedict *et al.* 2000a
Gl 791.2	113.1 ± 0.3		Benedict *et al.* 2000b
Wolf 1062	98.0 ± 0.4	98.56 ± 2.66	Benedict *et al.* 2001
TV Col	2.70 ± 0.11		McArthur *et al.* 2001
RR Lyr	3.60 ± 0.20	4.38 ± 0.59	Benedict *et al.* 2002a
δ Cep	3.66 ± 0.15	3.32 ± 0.58	Benedict *et al.* 2002b
HD 213307	3.65 ± 0.15	3.43 ± 0.64	Benedict *et al.* 2002b
Gl 876	214.6 ± 0.2	212.7 ± 2.1	Benedict *et al.* 2002c
NGC 6853	2.10 ± 0.48		Benedict *et al.* 2003
Ex Hya	15.50 ± 0.29		Beuermann *et al.* 2003
V1223 Sgr	1.96 ± 0.18		Beuermann *et al.* 2004
RU Peg	3.55 ± 0.26		Harrison *et al.* 2004
WZ Sge	22.97 ± 0.15		Harrison *et al.* 2004
YZ Cnc	3.34 ± 0.45		Harrison *et al.* 2004
Pleiades	7.43 ± 0.17	8.45 ± 0.25	Soderblom *et al.* 2004

equation 3.1. Reference star proper motions were obtained from UCAC2. A final piece of prior knowledge (Section 3.2.2) came from the assumption that our targets were in fact members of the Pleiades. This assumption generates an 'observation' asserting that the difference in parallax between our Pleiades member target stars is zero. The error associated with this 'observation' comes from the 1-σ angular extent of the Pleiades (1°, from Adams *et al.* 2001) and an assumption of spherical symmetry. From equation 3.2, we assert that the 1σ dispersion in distance in this group is 1°/radian =1.7%. Hence, the 1σ dispersion in the parallax difference between Pleiades members is $\Delta\pi = 0.20$ mas, where we have here temporarily adopted a parallax of the Pleiades, $\langle\pi\rangle = 7.7$ mas. The parallax dispersion among targets 3030, 3179, and 3063 becomes an observation with associated error fed to our model, an observation used to estimate the parallax dispersion among the three stars, while solving for their parallaxes. Neither loosening the cluster 1-σ dispersion to 2°($\Delta\pi = 0.38$ mas) nor adopting the HIPPARCOS parallax in equation 3.2 had any effect on the final average parallax.

Note that the priors did not include any previous direct observations of the parallax of the Pleiades. Our model used equations 4.1, 4.2, 4.5 and 4.6. We obtained an average parallax for the three Pleiades members $\pi_{abs} = 7.43 \pm 0.17$ mas. This result continues the trend of resolving the dispute between the main sequence fitting and the Hipparcos distance moduli in favor of main sequence fitting (Munari *et al.* 2004, Pan *et al.* 2004, Gatewood *et al.* 2000). Fig. 6 summarizes absolute parallax determinations for the Pleiades. Note that the very recent re-determination from Hipparcos raw data (van Leeuwen 2004, these proceedings) moves the parallax closer to our result and the average of the other determinations.

5.2. *HST parallax accuracy*

To assess our accuracy, or external error, we must compare our parallaxes with results from independent measurements. Table 1 includes an Hipparcos parallax, when available. We plot all parallaxes obtained with an FGS against those obtained by Hipparcos (Fig. 7). The dashed line is a weighted regression that takes into account errors in both input data sets and excludes the Pleiades. For this fit we obtain a reduced $\chi^2 = 0.265$. Including the Pleiades, we obtain a significantly poorer fit with reduced $\chi^2 = 0.551$, again, indicating a problem with the Hipparcos Pleiades parallax.

The regression seen in Fig. 7 indicates a 2.5σ scale difference between the Hipparcos and *HST* results. Measured proper motions provide an argument against the reality of this scale difference. Because it is desirable to reduce the impact of proper motion errors on an *HST* – Hipparcos comparison, we consider only two of the objects in Table 1, Proxima Cen and Barnard's Star. They have proper motion vector lengths exceeding $3800\,\mathrm{mas\,yr^{-1}}$. Comparing *HST* with Hipparcos, the average difference between these proper motion vectors is -0.01%, indicating a negligible scale difference.

Presently continuing parallax investigations include a determination of the Cepheid period-luminosity relation through precise parallaxes of eleven Cepheids. We have nearly completed several more extrasolar planet mass determinations by removing the signature of parallax from observations of v And and ϵ Eri. Finally we will assist in tightening up the lower main sequence mass-luminosity relation through astrometry of low mass binaries. In this case the parallax bears on the component masses and must be removed from orbital motion.

Acknowledgements

We thank our many co-investigators for their cheerful collaboration over the years: Klaus Beuermann, Art Bradley, Ray Duncombe, Otto Franz, Thierry Forveille, Geoff Marcy, Larry Fredrick, Paul Groot, Tom Harrison, Paul Hemenway, Todd Henry, Bill Jefferys, Ed Nelan, Ricky Patterson, Pete Shelus, Dave Soderblom, Bill Spiesman, Darrell Story, Bill van Altena, Qiangguo Wang, Larry Wasserman, and Art Whipple. We would like to acknowledge the support of Guaranteed Time Observer grants GO-06036, 06764, 06875, 06877, 06879, and 06880; and Guest Observer grants GO-08102, 08335, 08618, 09089, 09167, 09168, 09879; all administered through the Space Telescope Science Institute, which is operated by AURA, Inc., under NASA contract NAS5-26555. We are beholden to the support scientists at STScI who have assisted us over the years, especially Denise Taylor. Lastly we thank the many people in Danbury CT who have surfed the waves of change (once Perkin-Elmer, then Hughes Aerospace, then Raytheon, now Goodrich) and continue to support the FGS, especially Linda Abramowicz-Reed.

References

Adams, J. D., Stauffer, J. R., Monet, D. G., Skrutskie, M. F., & Beichman, C. A. 2001, *Astron. J.*, **121**, 2053

Benedict, G. F., McArthur, B., Nelan, E. P., Jefferys, W. H., Franz, O. G., Wasserman, L. H., Story, D. B., Shelus, P. J., Whipple, A. L., Bradley, A. J., Duncombe, R. L., Wang, Q., Hemenway, P. D., van Altena, W. F., & Fredrick, L. W. 1998, Proc. S.P.I.E., **3350**, 229

Benedict, G. F., McArthur, B., Chappell, D. W., Nelan, E., Jefferys, W. H., van Altena, W., Lee, J., Cornell, D., Shelus, P. J., Hemenway, P. D., Franz, O. G., Wasserman, L. H., Duncombe, R. L., Story, D., Whipple, A., & Fredrick, L. W. 1999, *Astron. J.*, **118**, 1086

Benedict, G. F., McArthur, B. E., Franz, O. G., Wasserman, L. H., Nelan, E., Lee, J., Fredrick, L. W., Jefferys, W. H., van Altena, W., Robinson, E. L., Spiesman, W. J., Shelus, P. J.,

Hemenway, P. D., Duncombe, R. L., Story, D., Whipple, A. L., & Bradley, A. 2000a, *Astron. J.*, **119**, 2382

Benedict, G. F., McArthur, B. E., Franz, O. G., Wasserman, L. H., & Henry, T. J. 2000b, *Astron. J.*, **120**, 1106

Benedict, G. F., McArthur, B. E., Fredrick, L. W., Harrison, T. E., Lee, J., Slesnick, C. L., Rhee, J., Patterson, R. J., Nelan, E., Jefferys, W. H., van Altena, W., Shelus, P. J., Franz, O. G., Wasserman, L. H., Hemenway, P. D., Duncombe, R. L., Story, D., Whipple, A. L., & Bradley, A. J. 2002a, *Astron. J.*, **123**, 473

Benedict, G. F., McArthur, B. E., Fredrick, L. W., Harrison, T. E., Slesnick, C. L., Rhee, J., Patterson, R. J., Skrutskie, M. F., Franz, O. G., Wasserman, L. H., Jefferys, W. H., Nelan, E., van Altena, W., Shelus, P. J., Hemenway, P. D., Duncombe, R. L., Story, D., Whipple, A. L., & Bradley, A. J. 2002b, *Astron. J.*, **124**, 1695

Benedict, G. F., McArthur, B. E., Forveille, T., Delfosse, X., Nelan, E., Butler, R. P., Spiesman, W., Marcy, G., Goldman, B., Perrier, C., Jefferys, W. H., & Mayor, M. 2002c, *Astrophys. J. Lett*, **581**, L115

Benedict, G. F., McArthur, B. E., Fredrick, L. W., Harrison, T. E., Skrutskie, M. F., Slesnick, C. L., Rhee, J., Patterson, R. J., Nelan, E., Jefferys, W. H., van Altena, W., Montemayor, T., Shelus, P. J., Franz, O. G., Wasserman, L. H., Hemenway, P. D., Duncombe, R. L., Story, D., Whipple, A. L., & Bradley, A. J. 2003, *Astron. J.*, **126**, 2549

Bessell, M. S. & Brett, J. M. 1988, *Pub Astron. Soc. Pacific*, **100**, 1134

Beuermann, K., Harrison, T. E., McArthur, B. E., Benedict, G. F., & Gänsicke, B. T. 2003, *Astron. & Astrophys.*, **412**, 821

Beuermann, K., Harrison, T. E., McArthur, B. E., Benedict, G. F., & Gänsicke, B. T. 2004, *Astron. & Astrophys.*, **419**, 291

Bradley, A., Abramowicz-Reed, L., Story, D., Benedict, G. & Jefferys, W. 1991, *Pub Astron. Soc. Pacific*, **103**, 317

Carpenter, J. M. 2001, *Astron. J.*, **121**, 2851

Cox, A. N. 2000, Allen's astrophysical quantities, 4th ed. Publisher: New York: AIP Press, Springer, 2000. Edited by Arthur N. Cox. (AQ00)

de Zeeuw, P. T., Hoogerwerf, R., de Bruijne, J. H. J., Brown, A. G. A., & Blaauw, A. 1999, *Astron. J.*, **117**, 354

Franz, O. G., Henry, T. J., Wasserman, L. H., Benedict, G. F., Ianna, P. A., Kirkpatrick, J. D., McCarthy, D. W., Bradley, A. J., Duncombe, R. L., Fredrick, L. W., Hemenway, P. D., Jefferys, W. H., McArthur, B. E., Nelan, E. P., Shelus, P. J., Story, D. B., van Altena, W. F., & Whipple, A. L. 1998, *Astron. J.*, **116**, 1432

Gatewood, G., de Jonge, J. K., & Han, I. 2000, *Astrophys. J.*, **533**, 938

Hanson, R. B. 1979, *Mon. Not. Royal Astron. Soc.*, **186**, 87

Harrison, T. E., McNamara, B. J., Szkody, P., McArthur, B. E., Benedict, G. F., Klemola, A. R., & Gilliland, R. L. 1999, *Astrophys. J. Lett*, **515**, L93

Harrison, T. E., Johnson, J. J., McArthur, B. E., Benedict, G. F., Szkody, P., Howell, S. B., & Gelino, D. M. 2004, *Astron. J.*, **127**, 460

Jefferys, W., Fitzpatrick, J., & McArthur, B. 1987, Celest. Mech., **41**, 39.

Majewski, S. R., Ostheimer, J. C., Kunkel, W. E., & Patterson, R. J. 2000, *Astron. J.*, **120**, 2550

McArthur, B., Benedict, G. F., Jefferys, W. H., & Nelan, E. 2002, The 2002 HST Calibration Workshop : Hubble after the Installation of the ACS and the NICMOS Cooling System, Proceedings of a Workshop held at the Space Telescope Science Institute, Baltimore, Maryland, October 17 and 18, 2002. Edited by Santiago Arribas, Anton Koekemoer, and Brad Whitmore. Baltimore, MD: Space Telescope Science Institute, 2002, p.373

McArthur, B. E. *et al.* 2001, *Astrophys. J.*, **560**, 907

McArthur, B. E., Benedict, G. F., Lee, J., Lu, C.-L., van Altena, W. F., Deliyannis, C. P., Girard, T., Fredrick, L. W., Nelan, E., Duncombe, R. L., Hemenway, P. D., Jefferys, W. H., Shelus, P. J., Franz, O. G., & Wasserman, L. H. 1999, *Astrophys. J. Lett*, **520**, L59

Monet, D. G. 1998, American Astronomical Society Meeting, 193, 112.003

Munari, U., Dallaporta, S., Siviero, A., Soubiran, C., Fiorucci, M., & Girard, P. 2004, *Astron. & Astrophys.*, **418**, L31

Nelan, E. & Makidon, R. 2001, Fine Guidance Sensor Instrument Handbook (version 10; Baltimore: STScI)

Paltoglou, G. & Bell, R. A. 1994, *Mon. Not. Royal Astron. Soc.*, **268**, 793

Pan, X., Shao, M., & Kulkarni, S. R. 2004, Nature, **427**, 326

Pinsonneault, M. H., Stauffer, J., Soderblom, D. R., King, J. R., & Hanson, R. B. 1998, *Astrophys. J.*, **504**, 17

Savage, B. D. & Mathis, J. S. 1979, *Ann Rev. Astron. & Astrophys.*, **17**, 73

Soderblom, D., Benedict, G. McArthur. B. Nelan, E. *et al.* 2004, in preparation

Standish, E. M., Jr. 1990, *Astron. & Astrophys.*, **233**, 252

van Altena, W. F., Lee, J. T., & Hoffleit, E. D. 1995, Yale Parallax Catalog (4th ed. ; New Haven, CT: Yale Univ. Obs.) (YPC95)

van Leeuwen, F. 1999, *Astron. & Astrophys.*, **341**, L71

van Leeuwen, F. , 2004, Transit of Venus: New Views of the Solar System and Galaxy, Proceedings IAU Colloquium No. 196, D.W. Kurtz & G.E. Bromage, eds. in press

Zacharias, N., Urban, S. E., Zacharias, M. I., Wycoff, G. L., Hall, D. M., Germain, M. E., Holdenried, E. R., & Winter, L. 2003, VizieR Online Data Catalog, 1289

Discussion

SUZANNE DÉBARBAT: Where did the idea for astrometry with the Hubble Space Telescope originate?

FRITZ BENEDICT: Why was there any place for astrometry on Hubble space telescope? Because they needed to point the damn thing! They needed to point it and they needed to hold it on target. The fine-guidance sensors' primary job was to acquire guide stars on either side of the targets so that they could take the pretty pictures. And if the fine-guidance sensors can hold things to within 2, 3, 4, 5, 6 milli-arcseconds, that's well smaller than the pixel size of the camera and you get those beautiful pictures. To echo Dave [Monet]: astrometry was "free". It wasn't a $100 million; it wasn't even $10 million. I will grant you it was probably about $5 million over the 25 years ... [slight pause with a sense of wonder at this length of time] ... that I've been working on this project.

Fritz "Hook-em Horns" Benedict

Transits of Venus: New Views of the Solar System and Galaxy
Proceedings IAU Colloquium No. 196, 2004
D.W. Kurtz, ed.

© 2004 International Astronomical Union
doi:10.1017/S1743921305001523

The Pleiades question, the definition of the zero-age main sequence, and implications

Floor van Leeuwen

Institute of Astronomy, University of Cambridge, Madingley Road, Cambridge, CB3 0HA, UK
email: fvl@ast.cam.ac.uk

Abstract. Determinations and measurements of the parallax of the Pleiades, obtained with ground-based studies and with Hipparcos data, are reviewed. A number of uncertainties in both sets of data are found. Although a further correction for abscissa correlations brings the Hipparcos determination closer to the ground-based value, the difference still seems too large. The new Hipparcos parallax determinations seem to reinforce a possible age related effect. A new reduction of the Hipparcos data is in progress. It reduces significantly the contribution of the attitude noise, providing higher accuracies and lower correlation levels for the brighter stars.

1. Introduction

The Pleiades cluster is the oldest recorded 'mini' constellation on the sky. In many ancient civilizations it played a crucial role in defining the seasons, and in today's astrophysics it has led the way to a number of crucial discoveries: the colour-magnitude relation by Hertzsprung (1915), for example, was based on his study of the Pleiades and Hyades stars; the discovery in 1980 of very fast rotation of G and K dwarfs shortly before becoming proper main sequence stars was first made in the Pleiades cluster (van Leeuwen & Alphenaar 1982; Stauffer *et al.* 1984; van Leeuwen *et al.* 1987). And now there is the Hipparcos parallax determination for the cluster, which seems to be at odds with the expected value by quite a margin (van Leeuwen 1999a; Robichon *et al.* 1999).

It should be no surprise that it is so often the Pleiades cluster that leads the way, or even causes problems. It is the same reason it had been so much noticed by ancient civilisations: the cluster stands out among the stars in the Northern hemisphere. It just so happened to fit nicely on a standard photographic plate size for an astrographic refractor, which made it a very suitable early candidate for proper motion studies (see, for example, Hertzsprung 1947; Vasilevskis *et al.* 1979). On the other hand, it is not too dense to generally cause problems in observations (unlike the Trapezium Cluster in Orion or most globular clusters), though it is situated in an area of dust clouds, showing as reflection nebulae and very significant variations in reddening for individual stars (van Leeuwen *et al.* 1986).

Since the first publication of the Hipparcos parallax determination for the cluster (van Leeuwen & Hansen-Ruiz 1997) many claims and counter claims have been produced on its validity. The value presented in that paper has since been revised downwards, on the basis of a better handling of the abscissa correlations in the data (van Leeuwen 1999a; Robichon *et al.* 1999). The arguments of both sides have generally not met: from the Hipparcos side it was claimed that aspects of the data reduction and catalogue construction, together with statistical tests performed on the reduced data, give very little leeway to account for the presence of widespread systematic errors in the published parallaxes (van Leeuwen 1999b, 2000) as was claimed to be the case by, for example, Narayanan & Gould (1999). On the other hand, models of stellar structure and stellar evolution seem to indicate that the differences between the positions of the main sequences for the Pleiades and other open clusters as based on the Hipparcos parallaxes are too large (see, for example, Pinsonneault *et al.* 2004, and references therein), although some authors claim from similar data that this may not be the case (Robichon *et al.* 2000; Castellani *et al.* 2002; Percival *et al.* 2003).

So how may we tackle this problem? In as far as the Hipparcos data are concerned the only way of tackling it is by going back to the original data and looking at all the individual steps and assumptions made along the way of the data reductions. That is a very big job, which has already taken several years (van Leeuwen & Fantino 2003). For all determinations that implement models, it is worthwhile to re-consider the accuracies and intrinsic (cosmic) noise levels associated with the calibrations on which they are calibrated. If data from the Pleiades cluster are used to calibrate model parameters, those model parameters should not be used to calibrate the distance of the cluster. Assumptions used in deriving reddening should be looked into: when reddening has been derived through shifting the data to an isochrone in the colour magnitude diagram, it is no longer an independent parameter. Reddening should only be derived through independent methods, such as colour-colour diagrams (although even there one has to be very careful) and polarization measurements. The uncertainty in the absorption needs to be fully taken into account. For example, the Pleiades are affected by differential reddening, and the most seriously affected star (HII 1084, situated near the Merope nebula) shows quite clearly deviating reddening directions in the two-colour diagrams of the Walraven photometry (van Leeuwen *et al.* 1986), indicating local, non-average dust properties.

It would be of interest not just to look at the situation for individual clusters, but also to consider the overall behaviour of the 10 clusters for which reliable parallaxes are now available: does the Pleiades problem stand on its own, or is it part of a more general feature? Discrepancies such as those observed for the Pleiades distance should be used as a motivation to reconsider critically ALL aspects from both sides of the medallion: the correct understanding of why a result is wrong is in the end far more relevant than the claim that someone else's result is wrong because it differs from your result, and you know you don't make any errors. Reading through the various claims and counter-claims it becomes clear that an expected result is not always a reliable result, and maybe an unexpected results is not necessarily a wrong result.

This paper consists of two main parts. The first considers aspects of ground-based distance determinations for the Pleiades cluster, looking at photometric methods, convergent point determinations, ground-based parallax determinations and double star analysis. Section 2 considers the Hipparcos data, with impressions from a completely new reduction of the data and looking at methods proposed to correct cluster parallaxes from the published data. As the new reduction is not finished yet, and within the Hipparcos tradition we do not make available preliminary results, it is too early at this stage to present any other conclusion than a general suggestion: be critical, not only to unexpected results, but also to studies that seem to confirm expected results so very well.

Hipparcos-based parallax values for open clusters as used in this paper are based on a revised analysis of the published data, which includes a better accounting for the differences in abscissa correlations for stars of different magnitudes. This correction, which appears to bring some Hipparcos results in closer agreement with ground-based results (giving a provisional value for the Pleiades parallax of 8 mas), is, however, not well determined.

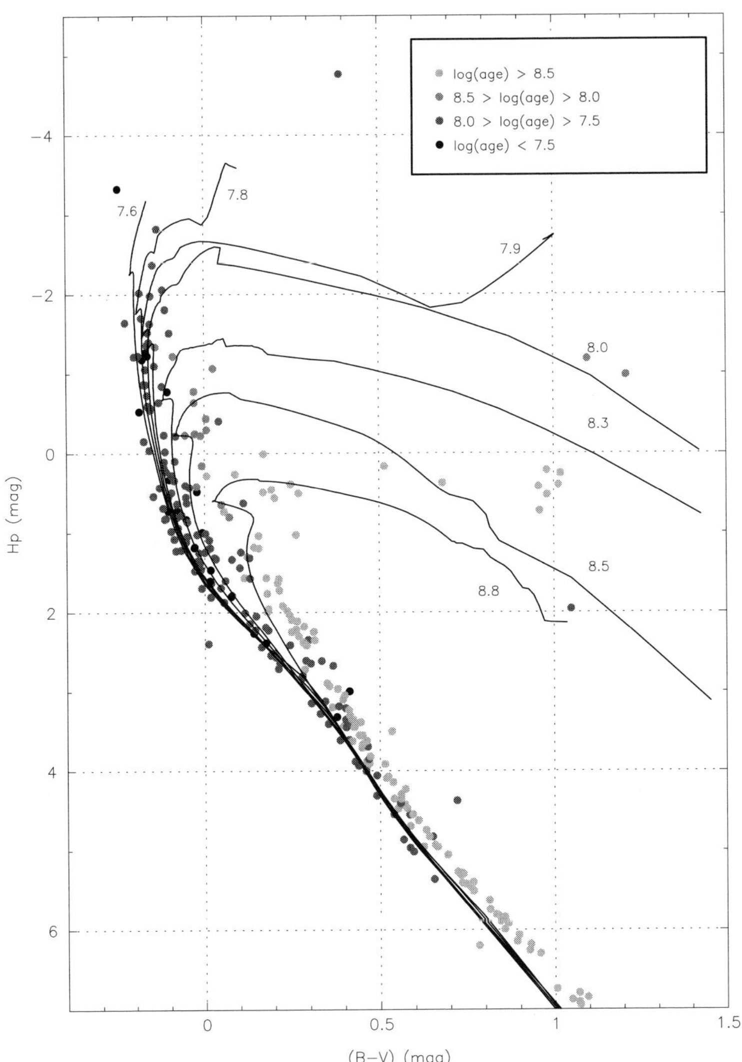

Figure 1. The combined HR diagram for 10 open clusters with Hipparcos parallaxes in a grey-scale according to cluster age. Reddening corrections have been applied. The differences between the young and the old clusters appear to start earlier than indicated by the isochrones (courtesy of Demarque & Kozhurina-Platais).

2. Ground-based studies

2.1. *Photometric determinations*

A photometric distance determination assumes a position for the main sequence (given an assumed or measured metallicity) and derives for that position a distance modulus; after correcting for reddening a distance is obtained. Often the reddening correction is included in the fitting procedure. As has been recognized by numerous authors before, it is not possible to use this method to detect an actual difference between the position of the Pleiades main sequence and mean main sequence to which it is calibrated: the assumption of the method is that there is no difference. For the same reason, all these determinations end up with nearly the same value for the distance modulus of the cluster, equivalent to a parallax of $7.59 + 0.14$ mas (see, for example, Pinsonneault *et al.* 1998).

For establishing a difference with the Hipparcos value these determinations are of rather limited value: you don't disprove a discrepancy by assuming that it doesn't exist.

Turning the argument around, it becomes the key issue: the relative positions of open cluster main sequences can provide essential information on calibration of models of stellar structure, evolution, and stellar atmospheres. Only a determination completely independent of these models can do so. Fig. 1 shows one way of looking at this problem. The photometric data, corrected for reddening, are shown for 10 open clusters with parallax determinations: although there may be a small discrepancy between the young (IC-2602, IC-2391, Pleiades) and the old clusters (Hyades, Coma Ber), there also appears to be good agreement within these age groups. The excess in blue for the faintest Pleiades stars is most likely due to chromospheric activity of these still relatively young objects (see also van Leeuwen *et al.* 1986; Percival *et al.* 2003; Stauffer *et al.* 2003).

2.2. *Convergent point determination*

The Pleiades cluster is poorly suited for a convergent point determination (CPD) such as used in the past for the Hyades cluster. The CPD compares the apparent contraction or expansion of the cluster, as observed from the proper motions of its members, with its radial velocity to derive a cluster distance. It relies on absolute proper motions (to enable measuring the expansion), a wide spread of cluster stars on the sky (to make the expansion large enough to obtain a significant result), and a large radial velocity (to produce a substantial expansion), and the Pleiades has none of these qualities. The radial velocity is only about $5.7\,\mathrm{km\,s^{-1}}$, which gives an expansion-related proper motion at $2°$ from the cluster centre of only $0.23\,\mathrm{mas\,yr^{-1}}$. Beyond $2°$ the density of cluster members is very low, and within $2°$ the effect is even smaller. These effects are also substantially smaller than the intrinsic velocity dispersion in the cluster (van Leeuwen 1984), which makes them very hard to determine accurately enough to provide a reliable distance estimate.

There are three papers referring to such a CPD for the Pleiades. One, by Madsen *et al.* (2002), shows that the expected error of a CPD for the Pleiades is considerably larger than the determination based on the parallaxes. Narayanan & Gould (1999) determined a CPD parallax of $7.5 \pm 0.6\,\mathrm{mas}$, which is at 1σ from the published Hipparcos value (less than 0.8σ from the new determination), even though the authors claim this to be a significant difference. Finally there is the study by Zhao Junliang and Li He, who claimed to have derived a parallax of $7.4 \pm 0.4\,\mathrm{mas}$ for the cluster. These authors, however, used for their determination the differential proper motions as published by Hertzsprung (1947), which do not contain the necessary information for a CPD. Also, they used only a $2°$-diameter field and radial velocities for 63 stars. The radial velocity effects in such a small field are at a level of a $1\,\mathrm{m\,s^{-1}}$ and less, well below anything detectable. A year after the publication of their study, one of the authors confirmed (private communication) that substantial errors had been made in this study (the CPD studies were referred to by Makarov (2002) as 'conclusive evidence' that there are problems with the Hipparcos parallax determinations in some parts of the sky). For the time being it seems to be clear that no sufficiently accurate CPD parallax for the Pleiades cluster can be expected to either confirm or contradict the Hipparcos result.

2.3. *Ground-based parallaxes*

The most recent ground-based determination of the Pleiades parallax is by Gatewood *et al.* (2000), who give a value of $7.64 \pm 0.43\,\mathrm{mas}$ for the mean parallax of 18 cluster members. Ground-based parallaxes are always differential, which means that they rely on assumptions made about the distances of the background stars that are used as

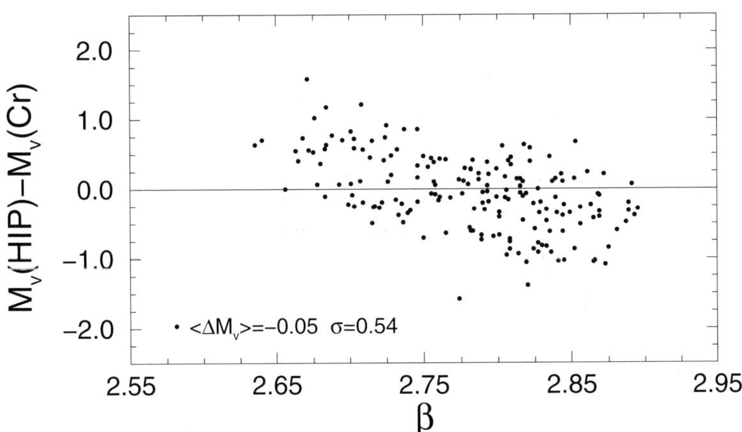

Figure 2. Differences between parallax- and photometry-based absolute magnitudes. The spread is due to several contributions: less than 0.2 mag from uncertainties in parallaxes, errors in colour indices, inaccuracies in photometric calibrations, rotation, cosmic dispersion (from Jordi *et al.* 2002.)

a reference frame. In Gatewood *et al.*'s study an iterative method is used in which the starting values of these distances (in the form of the assumed spectral types and observed apparent magnitudes) are adjusted until an acceptable solution is obtained. The difference with the current Hipparcos parallax determination for the Pleiades is less than 1σ and therefore not significant.

2.4. *Double stars*

Although double star determinations are often presented as the alternative to parallax measurements, in that they provide a direct measure of the distance of a system, this only applies if the determination is based purely on measurements of both orbital motions and radial velocities. For those Pleiades stars that have a known orbit, the radial velocities are difficult to measure with sufficient accuracy due to the rapid rotation and sparsity of spectral lines of the stars involved, and for those with a radial velocity curve no astrometric orbit data are available. Calculations based on only either orbital motion or radial velocity measurements still rely on stellar models. Two recent determinations (Pan *et al.* 2004; Munari *et al.* 2004) give distance estimates for two binaries in the Pleiades: HD 23850 (7.41 ± 0.11 mas) and HD 23642 (7.58 ± 0.11 mas). Both values are within 1 to $1.5\,\sigma$ from the determinations for these stars by Hipparcos. With their dependence on stellar-model parameters it seem premature to draw far-going conclusions from the difference with the parallax derived from the Hipparcos data. In particular, it is incorrect to quote these parallax values with the same formal errors as measurements of the cluster distance. When used to assess the cluster distance, the dispersion of parallaxes of cluster members along the line of sight has to be added to the formal error. As a comparison one may take the distribution of Pleiades stars on the sky, for which the projected centre is quite poorly determined (to a level of about 1 pc).

2.5. *Cosmic spread in stellar parameters*

In deriving distances from double star data as described above, assumptions are made about the reproduceability of stellar parameters as derived from observational quantities such as colour indices or spectral types. With the publication of the Hipparcos data it became possible to re-examine many of the basic calibrations of photometric systems and spectroscopic determinations, in particular where they concerned absolute magnitudes and their dependence on chemical composition (see, for example, Gomez *et al.* 1997, 1998; Jordi *et al.* 2002; Grenon 2002). A considerable cosmic dispersion is observed for photometric distance-modulus calibrations (see Fig. 2). It therefore seems appropriate to exercise some care when applying such calibrations to derive stellar parameters: whenever we are able to observe with sufficient detail the calibration relations appear to apply only globally, showing substantial deviations for individual objects. Application of old, pre-Hipparcos calibrations also generally introduces systematic errors in the estimated parameters (see, for example, Jordi *et al.* 2002).

2.6. *The lack of spread in parallax determinations for the Pleiades*

The opposite is observed for ground-based determinations of the Pleiades parallax or distance modulus: values produced in various studies generally agree very much better with the 'established value' than one would expect on the basis of the stated formal errors (see some of the figures quoted above), even when derived from completely unsuitable data.

3. The Hipparcos data

3.1. *The data*

The Hipparcos data have been published in great detail, allowing studies at a level that only very few would pursue (ESA 1997; van Leeuwen 1997). It has made possible the studies by, for example, Knapp, Pourbaix & Jorissen (2001), Knapp *et al.* (2003), Pourbaix & Jorissen (2000) and Pourbaix & Boffin (2003) on very red variables, where epoch information on the $V - I$ indices have led to more accurate astrometric parameters for these stars. Cluster parameter solutions whereby a single solution for the cluster parallax and proper motion is obtained from the abscissa measurements of the member stars (van Leeuwen & Evans 1998) is an other example of the use of these data. Allowing this has always been the intention of the publication of the Hipparcos data: within the time limits set for the publication of the data not all aspects could be fully investigated, but sufficient data should be provided to make *a posteriori* corrections possible. Still, to respond to some of the suggestions made as to what may be wrong with the cluster parallaxes, or even the general reliability of parallax data in the Hipparcos catalogue, it was required to refer back to the original, approximately 112 GByte of raw data. Only one, nearly complete, copy is still in existence: the data files used as input for the reductions by the Northern Data Analysis Consortium (NDAC) (Lindegren *et al.* 1992). A sufficiently detailed set of intermediate files still exists for the reductions of the FAST (Kovalevsky *et al.* 1992) consortium. Detailed analysis implies rewriting large amounts of software for a more modern hardware environment, and allowing more interactive control over the reduction processes.

In many ways reductions have also become much simpler. As a comparison, the original data arrived on nearly 1200 9-track tapes, filling a small office. In the NDAC consortium these data were transferred to optical disks, from which the data reductions were performed, mainly overnight and during weekends. One pass through all the data took several months. In the new application, the data are kept on 24 DVDs, and reductions are done on a laptop computer, taking two to three weeks to process all mission data. A complete set of new software is implemented to analyse the data for single stars, to provide input for double star analysis, and to provide a multitude of statistical tests and other data quality controls. Results reported below on data quality are mostly obtained with this software as applied on the raw data (see also van Leeuwen & Fantino 2003).

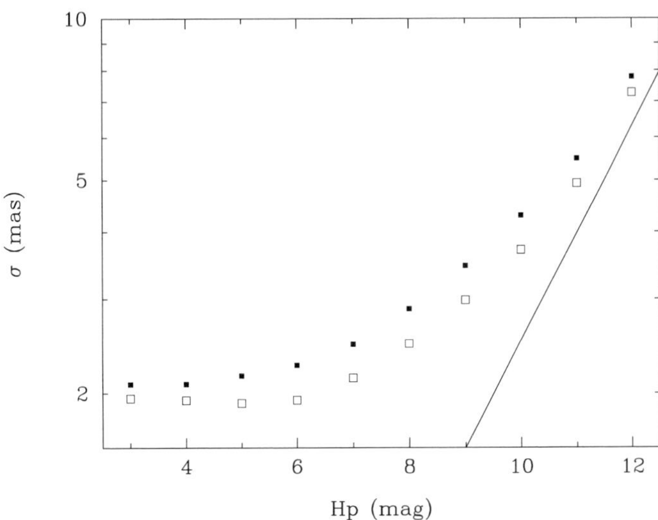

Figure 3. Standard deviations (σ in mas) for the abscissa residuals as a function of the Hp magnitude. Open symbols refer to FAST data, filled symbols to NDAC data. The diagonal line represents the expected relation for Poisson noise from the photon counts, given one and the same integration time for all intensities. The flattening of the relation towards brighter magnitudes shows the 2 mas noise contribution from primarily the along-scan attitude modelling contributions (from van Leeuwen 1997).

The data are subdivided into data sets, covering data collected over one orbit of the satellite, just under 11 hours. The mission results are based on the data collected for nearly 2300 data sets.

3.2. *One-dimensional astrometry*

The Hipparcos measurements provide one-dimensional positions: transit times of stellar images as measured on a modulating grid while the fields of view of the telescope describe a great circle on the sky. These one-dimensional measurements are referred to as abscissae or transit times. In fitting the along-scan attitude (rotation phase) of the satellite, the abscissa residuals for both fields of view are fitted as a function of time. The remaining abscissa residuals for individual stars are collected and averaged per field transit, and then fitted as a function of the five astrometric parameters (position, proper motion and parallax), relative to an *a priori* solution. Because of the relatively large errors on the *a priori* astrometric parameters of the program stars, these two parameter sets initially had to be estimated in one solution, referred to as the Great-Circle reduction (van der Marel & Petersen 1992). In the new solution this is no longer required, allowing improvements in the stability and accuracy of the along-scan attitude solution. By scanning each image in at least two different directions every 6 months, and accumulating data over 3.5 years, corrections to all five astrometric parameters can be solved.

The noise on the abscissa residuals contains two main contributions: attitude-fitting and Poisson noise. The first component dominates for bright stars, the second for faint stars. The attitude noise is more or less constant, though it will usually be somewhat higher in areas of the sky with few bright stars. In the published data, the effects of attitude noise start to get noticed for stars at 9th magnitude and brighter (see Fig. 3). In the new reduction the attitude noise level has been much reduced, and becomes noticeable only at 4th magnitude and brighter (van Leeuwen & Fantino 2003, and see below).

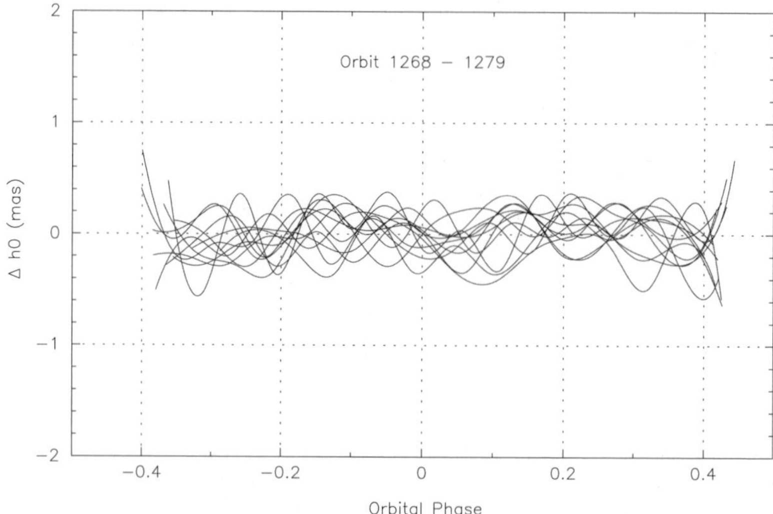

Figure 4. Noise levels on the reconstruction of the basic angle as observed from fitted residuals in the preceding and following fields of view. The data are for 12 successive orbits from May 1991. The rotation phases of successive circles at a given orbital phase change very little from one orbit to the next (from van Leeuwen & Fantino 2003).

3.3. *Parallax errors and basic-angle stability*

The basic angle of the Hipparcos telescope defines the angular distance on the sky between the projections from the two fields of view. The stability of the basic angle was a paramount condition for the success of the mission. In particular, a modulation of the basic angle with the rotation of the satellite could introduce a zero-point error in the parallaxes, and, it was thought, any systematic error in parallaxes locally on the sky had to be associated with systematic basic angle variations. Variations in the basic angle can be observed from systematic differences between the abscissa residuals in the two fields of view. No systematic modulation of the basic angle has been observed for 99% of the mission (see Fig. 4), but what was discovered is the presence of systematic drifts in the basic angle for 1% of the data sets (van Leeuwen & Penston 2003). Most of these drifts had not been detected for the preparation of the published data. All can be identified with known events of the mission. The most common of these was not identified as an anomaly, but could still be traced back in the mission reports. It was due to a communication problem between the on-board computer and the ground station, which could only be solved by turning off the entire payload. Unfortunately, this included the temperature control of the payload, and when it was switched on again, the change in temperature was noted from the value of the basic angle. The adjustment to a stable, controlled value usually took only a few hours. In the new analysis these drifts are solved for and corrected in the reduced data, except for secondary effects on the scale and rotation in the field of view.

Most importantly, no systematic modulations of the basic angle are observed down to a level of at least 0.1 mas. This confirmed earlier tests on the parallax zero-point, which was found to be less than 0.1 mas by Lindegren (1995); Arenou *et al.* (1995). This corresponds to a modulation level below 0.035 mas. Thus, there is no way in which lack of stability of the basic angle could have caused a difference of 0.5 to 1.0 mas in the parallax of the Pleiades cluster. In addition, as was shown by van Leeuwen (1999b), the accumulation of systematic errors over the 8° to 10°-diameter field of the Pleiades cluster is quite inefficient.

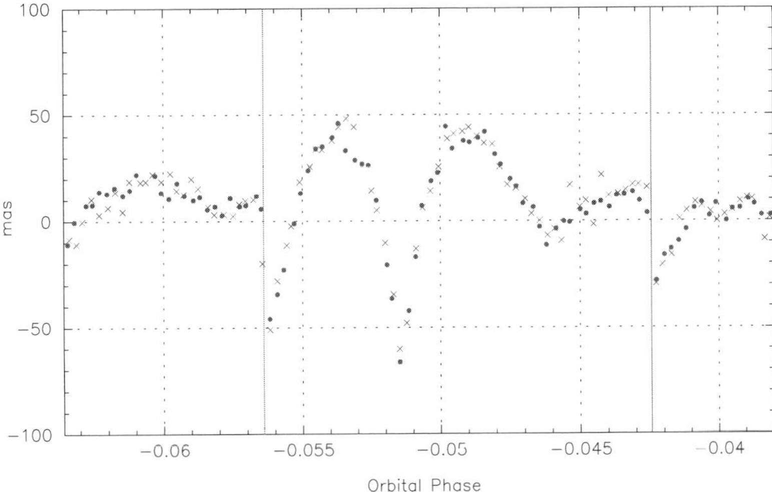

Figure 5. An example of an external hit of the satellite as recorded in the abscissa residuals before the final along-scan attitude fitting. The hit took place around phase -0.0515, and resulted in a change in the rotation velocity of the satellite by $4.7\,\mathrm{mas\,s^{-1}}$. Hits like these took place on average about once every two weeks. In the old solution such data were usually not recognized, in which case it would produce strongly correlated attitude errors for the data involved, or the data would be thrown away. In the new solution the discontinuity in the rate is fully accommodated and only a few seconds worth of data are lost (due to the uncertainty of exactly positioning the time of the hit).

3.4. *Data disturbances*

Other disturbances of the data have been noted too, but the amount of data affected is again small. The two main effects are external hits of the satellite (at a rate of roughly one every two weeks for an impact detection level of 7 to 8 nJ or higher, see Fig. 5), and scan phase discontinuities at times of thermal adjustments of the spacecraft, primarily around eclipses (see Volume 2, Chapter 11, in ESA 1997). Both of these effects were known to exist and had in some cases already been removed from the data by cutting out affected time intervals. There are now an estimated 1500 phase discontinuities recorded over the mission, affecting up to 1.5% of the data (either by it being thrown away or, at some stage, iterated out). In the new reduction the vast majority of these events are identified and accommodated in the reduction such that only about 0.01% of the data are lost. These effects, when not properly accounted for or removed, will have increased locally the noise level on the data, which may not always have been recognized.

3.5. *Statistical properties of the Pleiades parallax determination*

One of the problems we have had with accepting the idea that the Hipparcos determination for the Pleiades parallax is wrong comes from the statistical properties of the parallax solution. This solution is not based on just averaging the observed parallaxes of the probable members, but is an actual solution of the cluster parallax and proper motion based on the individual abscissa residuals for cluster members (van Leeuwen & Evans 1998). This solution takes into account correlations between the FAST and NDAC data and between abscissae measured in the same data set. The correlation levels have now been adjusted to account for differences for stars of different brightness, which brings the parallax to approximately 8.0 mas, back at the value that has been used for many decades. However, this value is sensitive to the exact implementation of the correlation levels for the bright stars, which are impossible to derive with great accuracy. In the new solution these correlation levels will be much reduced, as in general the attitude noise is lower by about a factor three. The current Pleiades parallax is based on 2246 abscissae for 56 stars, originating from 133 different data sets, spread over the entire mission. The stars occupy an area of $10°$ diameter on the sky. The unit-weight standard deviation of this solution is 1.08, which doesn't seem to indicate the presence of serious shortcomings in the solution, though there are indications that not all is well with the solution. These indications come from the proper motions rather than the parallaxes. The spread of the proper motions is too high, and those for the centre of the field fit poorly with ground-based studies of much higher precision (Vasilevskis *et al.* 1979).

3.6. *Absolute and relative parallaxes*

Not everything is always the way it looks. Makarov (2002, 2003) presented a method for correcting cluster parallaxes using stars observed simultaneously in the other FOV. In doing so he made several assumptions, some of which are impossible to verify from the published data, but can be verified from the raw data. Makarov applied two arguments for his study which contradict each other: he assumed that the apparent discrepancies in the Pleiades parallax are the result of the group being more or less detached from the catalogue, which may well be, at least partly, the correct explanation. But this situation would, for the same reason, not allow the use of stars from the other FOV to apply abscissa residual corrections, as the data for these stars is assumed to be poorly connected. On the other hand, the assumptions made on the way the measurements of these coinciding stars can be used to correct the Pleiades abscissae require an optimal connectivity of this field with the rest of the catalogue, which would make a correction unnecessary. The unsuitability of these data for a correction as applied by Makarov had already been pointed out in van Leeuwen (1999b).

The method as applied by Makarov results, amongst others, in transferring photon noise from transits in the other field of view on the abscissa measurements of the cluster stars, and will generally make those measurements worse (producing an 'expected' result is not a validation for the method). It was observed from the raw data that in the majority of cases the measurements of the stars in the other field of view are dominated (in amount of observing time assigned) by those not coinciding with the cluster stars. This confirms that the data for some cluster centres may not be optimally attached to the catalogue, possibly allowing for significant, local parallax zero-point offsets. How this affected the results will be different for the FAST and NDAC reductions, and as the published catalogue presents an averaged result, the final effect may not be very strong. In order to check the possible impact of this effect, the connectivity for each single star in the catalogue was determined, producing a map of potential soft spots for all those areas where the weight may have been generally higher than for most of the coinciding data in the other field of view. A comparison of this map with a map of the differences between the FAST and NDAC solutions shows a few small areas among the Southern OB associations with differences at the 2σ level and above. The Pleiades does show strongly in the 'soft spots' map, but not in the differences between FAST and NDAC. Neither solution will bring the Pleiades parallax down to a level of 7.3 to 7.7 mas. Also, the halo of the cluster is unaffected, and shows a parallax of around 8.0 mas. Thus, although it is likely that the centre of the Pleiades cluster was not solidly attached to the catalogue, the impression at this stage is that the error is probably no more than 0.2 to 0.4 mas.

The potential influence of the weight disparities between the two fields of view has been taken into account for the new reduction, where corrections are applied to reduce large disparities. The dynamical modelling of the attitude in the new solution, as well as

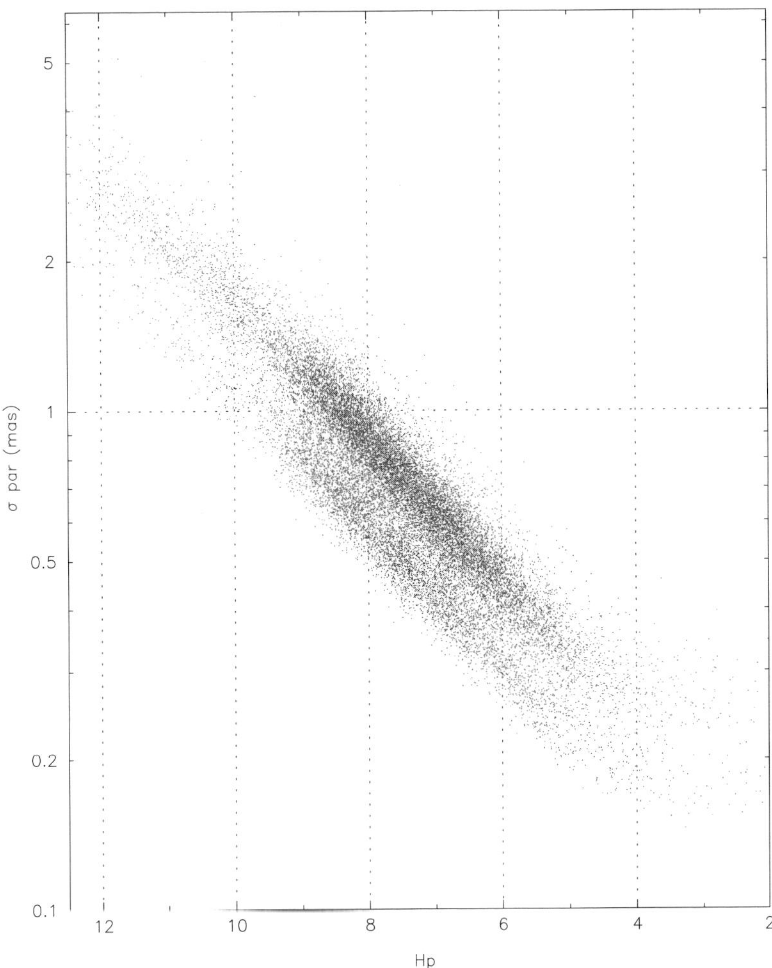

Figure 6. Formal errors on parallax determinations in the new solution. A significant reduction in the noise on the along-scan attitude reconstruction ensures that accuracies for a much larger fraction of the stars are dominated by Poisson statistics. Accuracies for the brightest stars have improved as a result by about a factor three. The bimodal character of the formal errors is due to the scanning strategy, and reflects different ecliptic latitudes.

the decoupling between the attitude and abscissa corrections also adds to reducing these effects.

3.7. *The new reduction*

The new reduction of the Hipparcos data is well on its way. It is based on a dynamic modelling of the satellite attitude and takes care of various other problems in the data. Preliminary results show an improvement in the attitude noise by a factor three or more, leading to a reduction in the errors of the astrometric parameters for bright stars by a factor 3 (see Fig. 6). This is currently only a measure of precision. A reduction in the number and size of negative parallaxes among the brighter stars indicates that at least some of this is indeed an improvement in accuracy. Determining by how much exactly the accuracies have improved is still one of the major tasks ahead. The publication of these data and the methods used to derive them is planned in the form of a book, describing the use of the data as well as aspects of how the data has been derived. As with the original publication of the Hipparcos data, there will be no release of preliminary results, and the first results of the new reduction are expected towards the end of 2004. This is likely to include new parallax determinations for the open clusters, much less affected by the uncertainties caused by abscissa-error correlations and weight disparities.

4. Conclusions

Problems with the Hipparcos data for some open cluster centres can most likely be attributed to high correlations in abscissa residuals for bright stars and possibly to weight disparities between two fields of view in the along-scan attitude reconstructions. A provisional adjustment of the abscissa error correlation levels for bright stars produces a reduction of the parallax to a value of around 8 mas, but the exact level of these corrections is difficult to determine. In the new solution the correlations are expected to play a much smaller role.

References

Arenou, F., Lindegren, L., Frœschlé, M., Gómez, A. E., Turon, C. Perryman, M. A. C., & Wielen, R. 1995, *A&A*, **304**, 52–60

Castellani, V., Degl'Innocenti, S., Prada Moroni, P. G., & Tordiglione, V. 2002, *MNRAS*, **334**, 193

ESA 1997, *SP*-**1200**

Gatewood, G., de Jonge, J. K., & Han, I. 2000, *ApJ*, **533**, 938–943

Gomez, A. E., Luri, X., Mennessier, M. O., Torra, J., & Figueras, F. 1997, in *Hipparcos - Venice '97*, ESA **SP-402**, 207–209

Gomez, A. E., Luri, X., Grenier, S., Figueras, F., North, P., Royer, F., Torra, J., & Mennessier, M. O. 1998, *A&A*, **336**, 953–959

Grenon, M. 2002, in *Highlights in Astronomy*, **12**, 680–683

Hertzsprung, E. 1915, *Astronomische Nachrichten* **200**, 137

Hertzsprung, E. 1947, *Annal.Sterrewacht Leiden*, **19**, 1–85

Jordi, C., Luri, X., Masana, E., Torra, J., Figueras, F., Domingo, A., Gómez, A. E., & Mennessier, M. O. 2002, *Highlights in Astronomy*, **12**, 684–687

Knapp, G. R., Pourbaix, D., Platais, I., & Jorissen, A. 2003, *A&A*, **403**, 993–1002

Knapp, G., Pourbaix, D., & Jorissen, A. 2001, *A&A*, 371, 222–223

Kovalevsky, J., Falin, J. L., Pieplu, J. L., Bernacca, P. L., Donati, F. *et al.* 1992, *A&A*, **258**, 7–17

van Leeuwen, F. 1984, in *Galactic and Solar System Optical Astrometry*, (ed. L. V. Morisson and G. Gilmore), Cambridge University Press, 223

van Leeuwen, F. 1997, *Space Sci. Rev.*, **81**, 201–412

van Leeuwen, F. 1999a, *A&A*, **341**, L71–74

van Leeuwen, F. 1999b, in *Harmonizing Cosmic Distance Scales in a Post-HIPPARCOS Era*, (ed. D.Egret & A. Heck), ASP Conf. Ser. 167, 52–71

van Leeuwen, F. 2000, in *Stellar Clusters and Associations: Convection, Rotation, and Dynamos*, (ed. R. Pallavicini, G. Micela, S. Sciortino), ASP Conf. Ser. 198, 85–94

van Leeuwen, F. & Alphenaar, P. 1982, *The Messenger* **28**, 15–18

van Leeuwen, F., Alphenaar, P., & Brand, J. 1986, *A&AS*, 65, 309–347

van Leeuwen, F., Alphenaar, P., & Meys, J. J. M. 1987, *A&AS*, **67**, 483–489

van Leeuwen, F. & Evans, D. W. 1998, *A&AS*, **130**, 157–172

van Leeuwen, F. & Fantino, E. 2003, *Space Sci. Rev.*, **108**, 537

van Leeuwen, F. & Hansen-Ruiz, C. S. 1997, in *Hipparcos Venice '97*, ed. M. A. C. Perryman and P. L. Bernacca, ESA, **SP-402**, 689

van Leeuwen, F. & Penston, M. J. 2003, *Space Sci. Rev.*, **108**, 471–497

Lindegren, L. 1995, *A&A*, **304**, 61–68

Lindegren, L., Høg, E., van Leeuwen, F., Murray, C. A., *et al.* 1992, *A&A*, **258**, 18–30

Madsen, S., Dravins, D. & Lindegren, L. 2002, *A&A*, **381**, 446–463

Makarov, V. V. 2002, *AJ*, **124**, 3299–3304

Makarov, V. V. 2003, *AJ*, **126**, 2408–2410

van der Marel, H. & Petersen, C. S. 1992 *A&A*, **258**, 60–69

Munari, U., Dallaporta, S., Siviero, A., Soubiran, C., Fiorucci, M., & Girard, P. 2004, *A&A*, **418**, L31–34

Narayanan, V. K. & Gould, A. 1999, *ApJ*, **523**, 328–339

Pan, X., Shao, M., & Kulkarni, S. R. 2004, *Nature*, **427**, 326

Percival, S. M., Salaris, M , & Kilkenny, D. 2003, *A&A*, **400**, 541–552

Pinsonneault, M. H., Stauffer, J., Soderblom, D. R., King, J. R., & Hanson, R. B. 1998, *ApJ*, **504**, 170

Pinsonneault, M. H., Terndrup, D. M., Hanson, R. B., & Stauffer, J. R. 2004, *ApJ*, **600**, 946–959

Pourbaix, D. & Boffin, H. M. J. 2003, *A&A*, **398**, 1163–1177

Pourbaix, D. & Jorissen, A. 2000, *A&AS*, **145**, 161

Robichon, N., Arenou, F., Mermilliod, J.-C., & Turon, C. 1999, *A&A*, **345**, 471–484

Robichon, N., Lebreton, Y., Turon, C., & Mermilliod, J.-C. 2000, in *Stellar Clusters and Associations: Convection, Rotation, and Dynamos*, ASP Conf. Ser. 198, 141–144

Stauffer, J. R., Hartmann, L., Soderblom, D. R., & Burnham, N. 1984, *ApJ*, **280**, 202–212

Stauffer, J. R., Jones, B. F., Backman, D., Hartmann, L. W., Barrado y Navascués, D., Pinsonneault, M. H., Terndrup, D. M., & Muench, A. A. 2003, *AJ*, **126**, 833–847

Vasilevskis, S., van Leeuwen, F., Nicholson, W., & Murray, C. A. 1979, *A&AS*, **37**, 333–343

Discussion

FRITZ BENEDICT: First of all I would like to apologise for not quoting the smaller number in my paper. There's still a 7σ difference, even with the smaller number. The HR Diagram which you showed has a considerable spread in the main sequence. Is some fraction of that due to binaries?

FLOOR VAN LEEUWEN: There's far more than the binaries; they are just $\frac{3}{4}$ of a magnitude. There are nearly two magnitudes here, most of which is due to evolution away from the main sequence. It appears for many different reasons that the evolutionary effects are much more severe, in terms of age, on the typical main sequence star, than seems to be represented in the isochrones. The question is, above all, to understand where the difference comes from. The question is *not* settled. It will be settled when we understand *quantitatively* where this difference comes from. We get two independent results from the same data, admittedly, but none of them is leading to a parallax of 7.3, 7.4 milli-arcseconds for the Pleiades.

RICKY SMART: Some of the problems that have come up recently with this Pleiades discussion has put the whole of the Hipparcos results into question. There was a paper that actually said this. What's your view on this?

FLOOR VAN LEEUWEN: I do not agree with the results of this paper. I consider that its conclusions are incorrect.

Poster Viewing

left: Floor van Leeuwen talking to Myles Standish; right: Ed Budding

Transits of Venus: New Views of the Solar System and Galaxy
Proceedings IAU Colloquium No. 196, 2004
D.W. Kurtz, ed.

© 2004 International Astronomical Union
doi:10.1017/S1743921305001535

The distance to the Pleiades from the eclipsing binary HD 23642

J. Southworth, P. F. L. Maxted and B. Smalley

Department of Physics and Chemistry, Keele University, Staffordshire, ST5 5BG, UK
email: jkt@astro.keele.ac.uk, pflm@astro.keele.ac.uk, bs@astro.keele.ac.uk

Abstract. The distance to the Pleiades open cluster is of fundamental importance to many aspects of stellar astrophysics but is currently controversial. The 'long' distance scale of 132 ± 3 pc is supported by main sequence fitting analyses, ground-based parallax observations and analysis of the astrometric binary Atlas. The 'short' distance scale of 120 ± 3 pc comes from parallaxes observed by the Hipparcos satellite. Munari *et al.* (2004) studied the detached eclipsing binary HD 23642 and found a distance of 132 ± 2 pc. We reanalyse the data of Munari *et al.* to explore the different methods of estimating the distance of an eclipsing binary system. We use the surface brightness relations of Kervella *et al.* (2004) to find a distance of 139 ± 4 pc, which is consistent with the 'long' Pleiades distance and in disagreement with the Hipparcos parallax distances to the Pleiades and to HD 23642 itself. By comparing the observed masses and radii of the binary with theoretical predictions we derive a metal abundance approximately equal to or slightly greater than solar. Further photometric observations of the binary are needed to improve the analysis.

1. Introduction

The Pleiades is a young open cluster situated in the solar neighbourhood. Due to its proximity the cluster has been exhaustively studied and its fundamental parameters, and distance, are considered to be well known. The 'long' distance scale of the Pleiades comes originally from Johnson (1957), who derived a distance of 129 pc from UBV photometry. Crawford & Perry (1976) found a distance of 128 pc from photoelectric $uvby\beta$ photometry of 83 Pleiades stars. Vandenberg & Bridges (1984) and Meynet, Mermilliod & Maeder (1993) found a distance of 132 pc by fitting model isochrones to the morphology of the Pleiades colour-magnitude diagram. An alternative method was adopted by O'Dell, Hendry & Collier Cameron (1994), who used the Barnes-Evans relation to find the angular diameters of Pleiades members. This technique is subject to different systematic errors, but these authors also found a distance of 132 pc.

However, observations by the Hipparcos satellite (Perryman *et al.* 1997; van Leeuwen, these proceedings) give a 'short' distance of 120 ± 3 pc, based on the entirely geometrical method of trigometric parallax. This is in significant disagreement not only with previous estimates of the Pleiades distance but also with the distance of 131 ± 7 pc given by Gatewood *et al.* (2000) from ground-based parallax measurements. van Leeuwen (1999) has confirmed the distance based on the Hipparcos observations, and found that five other young open clusters have main sequences as dim as the Pleiades if the Hipparcos distance estimates are accurate.

Pinsonneault *et al.* (1998) and Soderblom *et al.* (1998) investigated this discrepancy in distance determinations by performing a main sequence fitting analysis of the Pleiades and by attempting to discover nearby subluminous stars with solar metal abundance. They concluded that the Hipparcos measurements were most likely wrong and also noted that the most discrepant measurements belong to stars in the centre of the Pleiades, where the stellar density is the highest.

Grenon (1999) has suggested that the Pleiades distance problem could be resolved if the cluster was metal-poor, and has measured an iron abundance of [Fe/H] = −0.11 dex from Geneva photometry. This is in conflict with the solar iron abundance found by Boesgaard & Friel (1990) from high-resolution spectra of twelve F dwarfs in the Pleiades. Stello & Nissen (2001) used $uvby\beta$ photometry to determine the distance to the Pleiades by shifting its main sequence to coincide with the colours and brightnesses of nearby stars with a similar metal-sensitive Strömgren index, m_1. The resulting distance, 132 ± 2 pc, is in agreement with the 'long' distance scale and is not subject to metallicity effects. However, Castellani *et al.* (2002) were able to find a good agreement between the Pleiades main sequence, using Hipparcos parallax distances, and theoretical predictions from the FRANEC stellar evolution code, by using a subsolar metal abundance of $Z = 0.012$.

Makarov (2002) pointed out that the method of calculating the parallaxes of stars from Hipparcos observations may cause small systematic errors in regions of high stellar density. He reanalysed the Hipparcos Intermediate Astrometric Data for the Pleiades, including in the analysis some numerical terms which were neglected in previous analyses (Perryman *et al.* 1997; Robichon *et al.* 1999). The resulting distance of 129 ± 3 pc agrees well with the 'long' Pleiades distance scale and casts doubt on the original Hipparcos results.

Recent interferometric observations (Pan, Shao & Kulkarni 2004) of the astrometric binary Atlas (HD 23850) with the Palomar Testbed Interferometer give a distance in the range of 133 to 137 pc, supporting the 'long' Pleiades distance. Their results are also in good agreement with lunar occultation observations of Atlas. Munari *et al.* (2004, hereafter M04) measured the distance to the detached eclipsing binary HD 23642, a member of the Pleiades and found a distance of 131.9 ± 2.1 pc, in agreement with the 'long' distance scale.

1.1. *The eclipsing binary HD 23642 in the Pleiades*

Detached eclipsing binaries (dEBs) with double-lined spectra are one of the best sources of fundamental astrophysical data (Andersen 1991) as their absolute masses and radii can be measured to accuracies better than 1%. Such data can be used to provide a strict test of different stellar evolutionary models, and the distances to dEBs can be determined empirically to an accuracy of about 3%. dEBs in open clusters are particularly useful because the age and chemical composition of the cluster may be combined with accurate values of the masses and radii of the dEB to provide an even more exacting test of theoretical models. Alternatively, the astrophysical parameters of dEBs in open clusters can be compared with theoretical evolutionary models to determine the chemical composition of the cluster. For example, Southworth, Maxted & Smalley (2004a) studied the early-type dEBs V615 Per and V618 Per, in the open cluster h Persei, and determined a metal abundance of $Z \approx 0.01$. Southworth, Maxted & Smalley (2004b) analysed the high-mass dEB V453 Cyg, a member of the open cluster NGC 6871, in order to test theoretical predictions of the masses, radii, effective temperatures and central condensation of early-B stars.

HD 23642 (Table 1) was discovered to be a double-lined spectroscopic binary by Pearce (1957) and Abt (1958). Abt & Levato (1978) provided a spectral classification of A0 Vp (Si) + Am, where the metallic-line character of the secondary star relies on the presence of strong Fe I lines. Torres (2003) discovered shallow secondary eclipses in the Hipparcos photometric data of HD 23642 and also presented an accurate spectroscopic orbit.

Table 1. Identifications and astrophysical data for the HD 23642 system.

		Reference
Henry Draper number	HD 23642	1
Hipparcos number	HIP 17704	2
Bonner Durchmusterung	BD +23° 430	3
Hz number	Hz 540	4
α_{2000}	03 47 29.5	2
δ_{2000}	+24 17 18	2
Hipparcos distance (pc)	111 ± 12	2
Spectral type	A0 Vp (Si) + Am	5
B	6.898 ± 0.015	8
V	6.819 ± 0.015	8
J	6.673 ± 0.030	6
H	6.625 ± 0.030	6
K	6.631 ± 0.030	6
Orbital period (days)	2.46113400(34)	7
Reference time (HJD)	52903.5981(13)	7

References: (1) Cannon & Pickering (1918); (2) Perryman *et al.* (1997); (3) Argelander (1903); (4) Hertzsprung (1947); (5) Abt & Levato (1978); (6) 2MASS photometry transformed to the SAAO JHK system (see Moro & Munari, 2000); (7) M04; (8) Tycho data from Perryman *et al.* (1997), transformed to the Johnson system using the results of Bessell (2000).

M04 derived precise absolute masses and radii of both components from five high-resolution échelle spectra, and complete BV photoelectric light curves. From consideration of the effective temperatures (T_{eff}) and bolometric magnitudes of the components, and using the Wilson-Devinney light curve fitting code (Wilson & Devinney 1971; Wilson 1993) they found the distance to HD 23642 to be 131.9 ± 2.1 pc, where the quoted error is the formal error of the fit. We have reanalysed their data (which they have graciously provided on the internet) to investigate the advantages and disadvantages of different methods of estimating the distance of eclipsing systems.

2. Spectroscopic analysis

Table 2. Radial velocities, weights and $O-C$ values (in km s^{-1}) for HD 23642, taken from M04. For the SBOP analysis, observations were assigned weights of 2, 1 or 0.5 if their quoted uncertainty was 0.5, 1.0, and 1.5 km s^{-1} respectively.

HJD $- 2\,400\,000$	Primary velocity	$O-C$	Secondary velocity	$O-C$
53039.41273	-85.0 ± 0.5	-0.4	134.5 ± 0.5	0.2
53039.45356	-88.0 ± 0.5	0.2	139.0 ± 0.5	-0.5
53039.50548	-91.0 ± 0.5	0.4	143.0 ± 1.0	-1.0
53043.25049	105.5 ± 1.5	0.4	-136.0 ± 1.0	-2.0
53046.25318	29.5 ± 0.5	0.7	-24.5 ± 1.0	1.6

M04 observed HD 23642 five times with the Élodie échelle spectrograph on the 1.93-m telescope of the Observatoire de Haute-Provence. The derived radial velocities were combined with the spectroscopic observations of Pearce (1957) and Abt (1958), using lower weights for the older data, to calculate a circular spectroscopic orbit.

The low weight – and low precision – of the data of Pearce (1957) and Abt (1958) mean that they contribute little to the accuracy of the spectroscopic orbit. For comparison with the results of M04 we have chosen to derive the orbit using only the five échelle

Table 3. Parameters of the spectroscopic orbit derived for HD 23642.

		Primary	Secondary
Velocity semi-amplitude ($\mathrm{km\,s^{-1}}$)	K	99.10 ± 0.58	140.20 ± 0.57
Systemic velocity ($\mathrm{km\,s^{-1}}$)	V_γ	6.07 ± 0.39	
Mass ratio	q	0.7068 ± 0.0052	
Projected separation ($\mathrm{R_\odot}$)	$a \sin i$	11.636 ± 0.041	
Minimum mass ($\mathrm{M_\odot}$)	$M \sin^3 i$	2.047 ± 0.016	1.447 ± 0.013

velocities for each star. The orbit was computed using SBOP†, with the orbital ephemeris from M04, eccentricity fixed at zero, and equal systemic velocities for both stars. The radial velocities, adopted weights and observed minus calculated (O–C) values are given in Table 2, and the parameters of the spectroscopic orbit are given in Table 3. These parameters, plotted in Fig. 1, are in acceptable agreement with those of Torres (2003) and Abt (1958), although a little different to the values of Pearce (1957) (but see Griffin, 1995).

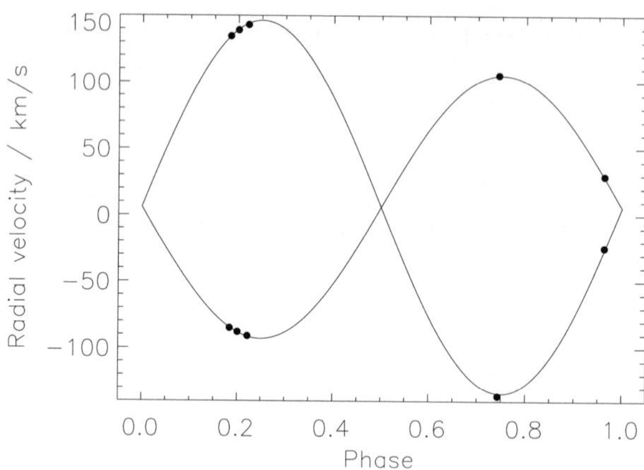

Figure 1. Circular spectroscopic orbit for HD 23642 derived using the Élodie radial velocities.

From comparison between the observed spectra of M04 and synthetic spectra calculated using UCLSYN and ATLAS9 model atmospheres (see Southworth *et al.* 2004a for references), the effective temperatures of the two stars were found to be 9750 ± 250 K and 7600 ± 400 K, in good agreement with the results of M04. The uncertainties in these values include possible systematic errors caused by the slight spectral peculiarity of both components of HD 23642. The rotational velocities of both stars were confirmed to be synchronous with the orbital motion, in agreement with Zahn's (1977) theory of tidal evolution which predicts a synchronisation timescale of 0.5 Myr.

3. Photometric analysis

The B and V light curves contain 492 and 432 individual measurements, respectively, obtained with a 28 cm Schmidt-Cassegrain telescope and photometer by M04. We solved the light curves individually using EBOP‡ (Nelson & Davis 1972; Popper & Etzel 1981).

† Spectroscopic Binary Orbit Program written by Dr. Paul B. Etzel
(`http://mintaka.sdsu.edu/faculty/etzel/`).
‡ Eclipsing Binary Orbit Program written by Dr. Paul B. Etzel
(`http://mintaka.sdsu.edu/faculty/etzel/`).

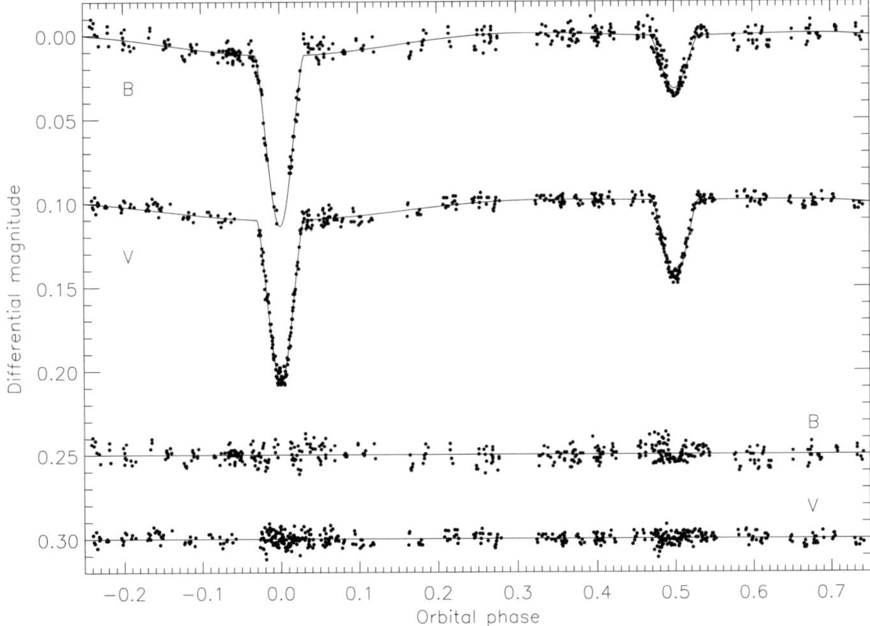

Figure 2. The M04 B and V light curves with our best fitted light curves overplotted. The V light curve is shifted by $+0.1$ mag for clarity. The residuals of the fit are offset by $+0.25$ mag and $+0.30$ mag for the B and V light curves respectively. Note the poor fit around the secondary eclipse in the B light curve due to scattered data.

This is a simple and efficient light curve fitting code in which the discs of the stars are modelled using biaxial ellipsoids. Filter-specific linear limb darkening coefficients were taken from van Hamme (1993), gravity darkening exponents, β_1, were fixed at 1.0 (Claret 1998) and the mass ratio was fixed at the spectroscopic value. The ephemeris given in M04 was used and the orbit was assumed to be circular. As there are no spectral lines due to a third star visible in the high-quality spectra of HD 23642, and third light, L_3, was negligible in initial photometric solutions, we have assumed $L_3 = 0$. However, as this quantity is not very well constrained by the light curves, we have also assessed the effect of assuming $L_3 = 0.05$ (in units of the total brightness of the eclipsing system) on the final result.

Initial light curve solutions using EBOP did not provide a good fit to the observations outside eclipse, so the size of the reflection effect on the secondary star was allowed to vary independently rather than being calculated from the system geometry. The light curves were also solved with the Wilson-Devinney code, using the 1998 version (WD98) in mode 2 and with a detailed treatment of the reflection effect. The B and V light curves were solved independently as before, except for the use of a logarithmic limb darkening law and fixing the stellar albedos at 1.0. The parameters derived from the light curve fits using EBOP and WD98 were very similar except for the ratio of the radii, which is poorly constrained by the light curves (see Fig. 3).

Considering the good agreement between the results of the EBOP and WD98 preliminary light curve fits, we used EBOP for further analysis. This code has two important advantages; firstly, a detailed error analysis is not prohibitively expensive in terms of computer time. Secondly, the philosophy of the EBOP code is to solve for the set of parameters most directly related to the light curve shape, leaving further analysis to the

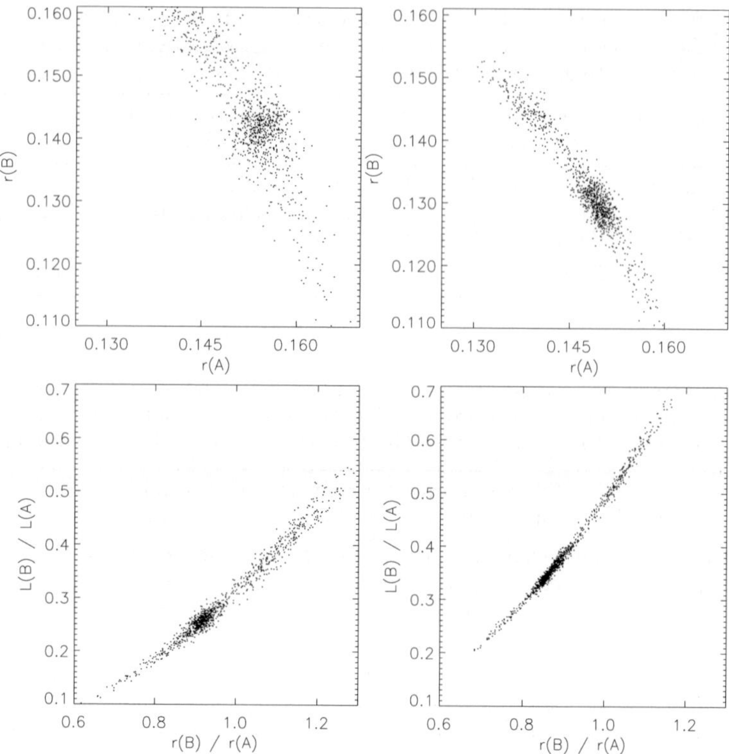

Figure 3. Results of the Monte Carlo analysis for the B (left) and V (right) light curves. The upper panels show primary radius versus secondary radius whilst the lower panels show the ratio of the radii versus the light ratios. All panels show that their parameters are strongly correlated, which is why our adopted solution uses a spectroscopic light ratio to break the degeneracy by allowing the ratio of the radii to be fixed during solution. Note the greater scatter of the synthetic solutions for the B light curve, which is due to the greater observational errors. Only 1000 of the 10 000 points have been plotted in each panel.

researcher. This allows a light curve solution for fixed values of the ratio of the radii and does not require the estimation of an effective temperature before solution.

The B and V light curve fits gave somewhat different results for the parameters in common – for example the sum of the radii, usually very well determined, differs by 5% between the two light curves. Attempts to solve this discrepancy, by manually rejecting discrepant data in the B light curve, were unsuccessful so all photometric data have been retained. This difficulty was not mentioned by M04, who solved the B and V light curves simultaneously using the WD98 code.

Torres (2003) gave a preliminary monochromatic light ratio at 5187 Å of $\frac{l_B}{l_A} = 0.31 \pm 0.03$. The final solution has been obtained by using this light ratio, converted to a V-filter light ratio using synthetic spectra from ATLAS9 model atmospheres, the effective temperatures found in Section 2, the surface gravities of M04, and a V filter transmission function†. The uncertainty of the resulting light ratio, 0.335 ± 0.035, includes possible systematic errors from the use of ATLAS9 model atmospheres. Note that the use of ATLAS9 model atmospheres rather than black bodies has a negligible effect on the resulting light ratio. This light ratio was fixed in the EBOP analysis of the V light curve and used to

---

† The V filter transmission function was taken from the Isaac Newton Group website
`http://www.ing.iac.es/Astronomy/astronomy.html`.

Table 4. Best-fitting parameters for the V light curve found using EBOP and the light ratio of Torres (2003). The individual errors are from Monte Carlo simulations, with a fixed ratio of the radii, around the best-fitting light curves. The last line gives the adopted parameters.

Light ratio	Ratio of the radii, k	Surface brightness ratio, J	Primary stellar radius, r_1 (a)	Secondary stellar radius, r_2 (a)	Orbital inclination $(°)$
0.300	0.8075	0.483 ± 0.014	0.1531 ± 0.0009	0.1237 ± 0.0007	78.04 ± 0.08
0.335	0.8477	0.489 ± 0.015	0.1508 ± 0.0009	0.1278 ± 0.0008	77.90 ± 0.08
0.370	0.8858	0.493 ± 0.015	0.1485 ± 0.0009	0.1315 ± 0.0008	77.79 ± 0.08
0.335	0.8477	0.489 ± 0.021	0.1508 ± 0.0032	0.1278 ± 0.0047	77.90 ± 0.19

Table 5. Best-fitting parameters for the B light curve found using EBOP and the light ratio of Torres (2003). The individual errors are from Monte Carlo simulations, with a fixed ratio of the radii, around the best-fitting light curves. The last line gives the adopted parameters.

Ratio of the radii, k	Light ratio	Surface brightness ratio, J	Primary stellar radius, r_1 (a)	Secondary stellar radius, r_2 (a)	Orbital inclination $(°)$
0.8075	0.190 ± 0.008	0.304 ± 0.014	0.1599 ± 0.0019	0.1291 ± 0.0015	77.47 ± 0.19
0.8477	0.214 ± 0.008	0.309 ± 0.015	0.1580 ± 0.0019	0.1339 ± 0.0016	77.26 ± 0.19
0.8858	0.237 ± 0.009	0.313 ± 0.015	0.1560 ± 0.0019	0.1382 ± 0.0016	77.08 ± 0.19
0.8477	0.214 ± 0.032	0.309 ± 0.020	0.1580 ± 0.0038	0.1339 ± 0.0062	77.26 ± 0.39

determine the ratio of the radii. This ratio of the radii was then used to solve the B light curve. The results are given in Table 4 and Table 5.

The uncertainties in the fitted parameters were estimated using a Monte Carlo method implemented for the study of the high-mass eclipsing binary V453 Cygni (Southworth *et al.* 2004b). This algorithm evaluates the correlations between different light curve parameters around the point of best fit, given the actual phases of observation and observational error. The resulting uncertainties are given in Table 4 and Table 5. The final values of the adjusted photometric parameters are $r_1 = 0.1538 \pm 0.0024$, $r_2 = 0.1300 \pm 0.0037$ and $i = 77°\!.78 \pm 0°\!.17$, where the quantities are the weighted means of the individual determinations from the B and V light curves. The best fit is plotted in Fig. 2.

Fig. 3 shows the results of the Monte Carlo analysis for the B and V light curves using the EBOP code and solving for the parameters including the ratio of the radii. The large parameter correlations illustrate the nature of the light curve and show why the spectroscopic light ratio of Torres (2003) was needed to allow us to fix the ratio of the radii for the final photometric solution. If the light curves are solved without using the light ratio of Torres (2003), the resulting fractional radii are $r_1 = 0.151 \pm 0.005$ and $r_2 = 0.135 \pm 0.007$; they are determined to accuracies no better than 3% and 5% respectively. The χ^2 value of the light curve fits is approximately constant for values of the ratio of the radii between 0.8 and 1.0. The current light curves are not definitive by the criteria of Popper (2000). Our results agree with the results of M04 although our uncertainties are significantly larger than the formal errors quoted by these authors. This is despite the fact that we have incorporated extra information, a spectroscopic light ratio, into our analysis. Formal errors are known to be optimistic (e.g., Popper 1984).

Table 6. Absolute dimensions of the components of HD 23642. The equatorial rotational velocities have been derived assuming that the values of M04 are projected rotational velocities and the orbital inclination derived in Section 3.

	Primary star	Secondary star
Mass ($M_\odot$)	2.193 ± 0.017	1.550 ± 0.014
Radius ($R_\odot$)	1.831 ± 0.030	1.548 ± 0.045
Surface gravity $\log g$ ($\mathrm{cm\,s^{-1}}$)	4.254 ± 0.018	4.249 ± 0.029
Effective temperature (K)	9750 ± 250	7600 ± 400
Equatorial rotational velocity ($\mathrm{km\,s^{-1}}$)	38 ± 1	32 ± 2
Synchronous velocity ($\mathrm{km\,s^{-1}}$)	27.7 ± 0.6	31.8 ± 0.9
B	7.109 ± 0.044	8.782 ± 0.150
V	7.133 ± 0.043	8.320 ± 0.100

4. Absolute dimensions and comparison with stellar models

Consideration of the final results from the photometric analysis (Table 4 and Table 5) and the results of the spectroscopic orbital analysis in Section 2 give the absolute dimensions of the two stars (Table 6). Our results are in reasonable agreement with the results of M04, and although the radius of the secondary star is somewhat larger, the two values are consistent within their uncertainties.

In Fig. 4 the observed properties of HD 23642 have been compared with predictions of the Granada stellar evolutionary models for $Z = 0.01, 0.02$ and 0.03 (Claret 1995, 1997; Claret & Giménez 1995), adopting an age of 125 Myr for the Pleiades (Stauffer, Schultz & Kirkpatrick 1998). The positions of the components of HD 23642 in the mass–radius plane are consistent with the metal abundance of the Pleiades being solar or slightly supersolar. This confirms the atmospheric solar iron abundance determined by Boesgaard & Friel (1990) from high-resolution spectroscopy of F dwarfs, but is in disagreement with suggestions that the 'short' and 'long' Pleiades distances could be reconciled by adopting a low metal abundance for the cluster.

The components of HD 23642 cannot be fitted in the mass–radius plane by models of metal abundance or helium abundance significantly different from the solar abundances; for a solar chemical composition the predictions of the Granada stellar evolutionary models are consistent with the dimensions of HD 23642 for ages between 125 and 175 Myr. Predictions of the Padova (Bressan *et al.* 1993) and Cambridge (Pols *et al.* 1998) stellar models are in agreement with this conclusion, but the Geneva (Schaller *et al.* 1992) evolutionary models predict a smaller radius for the secondary star than observed.

5. The distance to the Pleiades

5.1. *Distance using bolometric corrections*

The traditional method of determining the distance of an eclipsing binary is to calculate the luminosity of each star using the usual formula

$$L = 4\pi\sigma R^2 T_{\mathrm{eff}}^4 \tag{5.1}$$

where R is the stellar radius and σ is the Stefan-Boltzmann constant. Then the absolute bolometric magnitude for each star is found using the definition

$$M_{\mathrm{bol}} = -2.5 \log \frac{L}{L_\odot} + M_{\mathrm{bol}\odot} \tag{5.2}$$

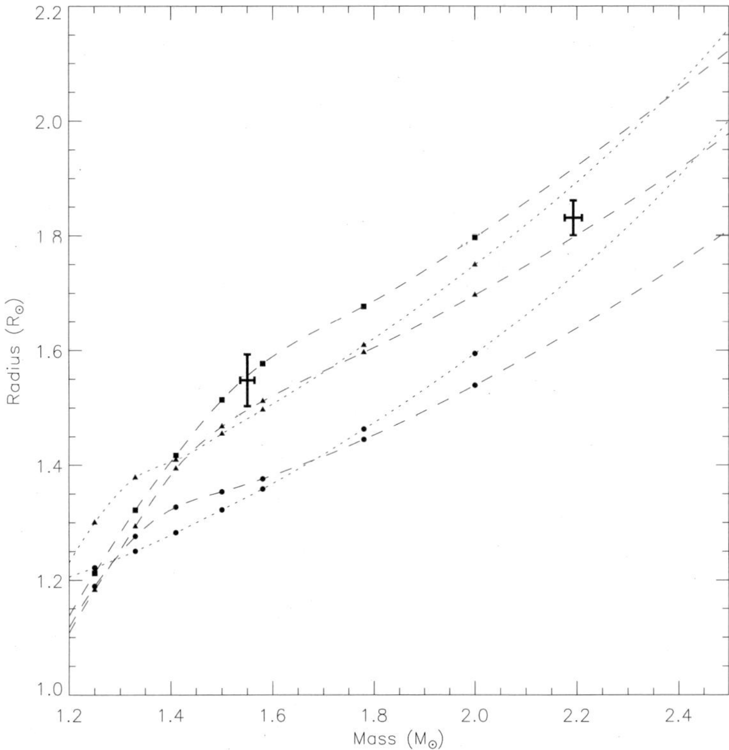

Figure 4. Comparison between the observed properties of HD 23642 and the Granada models for metal abundances of $Z = 0.01$ (circles), $Z = 0.02$ (triangles) and $Z = 0.03$ (squares). Predictions for normal helium abundances are plotted with dashed lines and helium-rich model predictions ($Y = 0.34$ and 0.38 for $Z = 0.01$ and 0.02 respectively) are plotted using dotted lines. An age of 125 Myr was assumed.

The absolute visual magnitude, M_V, of each star is calculated and the distance, d (parsecs), derived using

$$V_0 - M_V = (V - A_V) - (M_{\mathrm{bol}} - BC) = 5 \log d - 5 \qquad (5.3)$$

where V is the apparent visual magnitude of the star, A_V is the total visual extinction, V_0 is the dereddened apparent visual magnitude, and BC is the bolometric correction. This provides a distance to each star or, alternatively, the absolute visual fluxes of the two stars can be summed to find the total absolute visual magnitude of the system.

There are several problems with deriving distances by using bolometric corrections:

(a) The effective temperature scale must be fundamental to avoid systematic errors.

(b) $M_{\mathrm{bol}\odot}$ and $L_\odot$ are not defined to an arbitrary precision.

(c) Bolometric corrections are calculated using stellar models so the final result has a model dependence.

(d) The final distance uncertainty can be very sensitive to the uncertainties in the effective temperatures and must also incorporate uncertainties in the bolometric corrections used.

(e) Consistent values of $M_{\mathrm{bol}\odot}$ and $L_\odot$ must be adopted to avoid systematic errors caused by the use of different zeropoints.

Adopting the astrophysical parameters of the system from Tables 1 and 6, bolometric corrections from Bessell, Castelli & Plez (1998), and uncertainties in bolometric corrections from consideration of other tabulations in the literature, we find a distance to HD 23642 of 138.1 ± 4.7 pc. For the primary component of HD 23642, the uncertainty due to its effective temperature is actually reduced due to the form of the bolometric correction function around 10 000 K. The main contribution to the overall uncertainty (4.5 pc; note that the individual uncertainties are added in quadrature) comes from the uncertainty in the effective temperature of the secondary star. Whilst this is an acceptable result to adopt for the final distance to HD 23642, it is still affected by (probably small) systematic error due to the use of bolometric corrections calculated from theoretical models.

M04 found the distance to HD 23642, from the above method and the absolute bolometric magnitudes of the stars given by the Wilson-Devinney code, to be 131.9 ± 2.1 pc. The bolometric magnitudes given by M04 are consistent with the values $M_{\text{bol}\odot} = 4.77$ and $L_\odot = 3.906 \times 10^{26}$ W, which are adopted in WD98. This $M_{\text{bol}\odot}$ value is different to the standard value of 4.75 (Hilditch 2001) and is also different to the value of 4.74 adopted by Bessell $et\ al.$ (1998), from which M04 have taken their bolometric corrections. If we adopt the temperatures and radii of M04, and the $M_{\text{bol}\odot} = 4.74$ and bolometric corrections given by Bessell $et\ al.$, we find a distance of 134.5 ± 2.5 pc. However, adoption of the bolometric corrections of Flower (1996), and $M_{\text{bol}\odot} = 4.75$ (as Flower does not give the value he adopted), gives a distance of 135.4 ± 2.6 pc, which is 0.9 pc different to the previous value.

5.2. *Distance using surface brightness calibrations*

An alternative to considering the unobserved parts of the stellar spectral energy distributions is to use empirical relations between effective temperature and the visual surface brightness of a star. In this method the angular diameter of the star is compared to its linear diameter, determined from photometric analysis, to find the distance. The main source of uncertainty in this method is usually the light ratio, found from the photometric analysis, needed to calculate the individual magnitudes of the components of the dEB from the apparent magnitude of the system. If the effective temperatures of the two stars are known, this difficulty can be bypassed by comparing the apparent magnitude of the system to the flux-weighted combined angular diameter of the two stars.

The surface brightness of a star is defined to be

$$S_{m_\lambda} = m_\lambda - 5 \log \phi \tag{5.4}$$

where m_λ is the apparent magnitude in filter λ and ϕ is the angular diameter of the star in milli-arcseconds. The zeroth-magnitude angular diameter is defined to be

$$\phi^{(m_\lambda=0)} = \phi \times 10^{\frac{m_\lambda}{5}} = \frac{S_{m_\lambda}}{5} \tag{5.5}$$

(van Belle 1999), i.e., the angular diameter a star would have if its apparent magnitude in the filter m_λ were zero. From consideration of the definitions of surface brightness and angular diameter, it can be shown that the distance to a dEB is given by

$$d = 10^{0.2 m_\lambda} \sqrt{\left[\frac{2R_{\text{A}}}{\phi_{\text{A}}^{(m_\lambda=0)}} \right]^2 + \left[\frac{2R_{\text{B}}}{\phi_{\text{B}}^{(m_\lambda=0)}} \right]^2} \tag{5.6}$$

where, for a distance given in parsecs, the stellar radii R_{A} and R_{B} are given in AU, the zeroth-magnitude angular diameters $\phi_{\text{A}}^{(m_\lambda=0)}$ and $\phi_{\text{B}}^{(m_\lambda=0)}$ are given in arcseconds and λ represents a broad-band filter passband. Calibrations for $\phi^{(m_\lambda=0)}$ are given in terms of effective temperature by Kervella $et\ al.$ (2004, hereafter KTDS04) (where they are denoted using ZMLD_λ) for the broad-band $UBVRIJHKL$ filters. The calibrations with

Table 7. The results and individual error budgets for distance estimates using the effective temperatures and overall apparent magnitudes of the HD 23642 system in the $BVJHK$ filters. All distances are given in parsecs and the total uncertainties are the sums of the individual uncertainties added in quadrature. The observational scatter around the calibrations is denoted as 'cosmic' scatter.

Source of uncertainty	B	V	J	H	K
Effective temperature of the primary star	4.8	3.3	1.7	0.7	0.7
Effective temperature of the secondary star	3.8	3.5	2.1	1.5	1.4
Apparent magnitude of the system	2.1	1.8	2.2	2.1	2.1
Radius of the primary star	1.9	1.7	1.5	1.4	1.4
Radius of the secondary star	0.9	1.0	1.3	1.5	1.5
Intrinsic "cosmic" scatter in the calibration	9.0	8.3	1.7	1.9	1.4
Total uncertainty in distance	11.2	10.0	4.4	3.9	3.6
Distance	142.8	141.4	139.6	138.4	139.1

the least scatter involve the $JHKL$ filters. For example, the calibrations for the V and K filters are

$$\log \phi^{(V=0)} = 3.0415(\log T_{\text{eff}})^2 - 30.9671(\log T_{\text{eff}}) + 53.7010 \qquad \sigma = 5.9\% \qquad (5.7)$$

$$\log \phi^{(K=0)} = 0.8470(\log T_{\text{eff}})^2 - 7.0790(\log T_{\text{eff}}) + 15.2731 \qquad \sigma < 1.0\% \qquad (5.8)$$

For the K filter calibration the scatter is undetectable at a level of 1% so we conservatively adopt a fitting uncertainty of 1.0%. We have applied the calibrations for the B and V filters (using Tycho apparent magnitudes transformed to the Johnson system) and for the JHK filters (using 2MASS data transformed to the SAAO system). As there is no "standard" infrared photometric system, the calibrations of KTDS04 use data from several different $JHKL$ systems, so the systematic uncertainty of not having a standard system is already included in the quoted scatter in the calibrations. For the A stars only, the scatter around the B and V calibrations is much smaller than the overall scatter quoted, so for HD 23642 the B and V distance uncertainties are overestimated by a factor of about two.

The distances found using equation 5.6 and the KTDS04 calibrations are given in Table 7. Note that the uncertainties in the JHK calibration distances are much smaller than in the BV calibration distances. We will adopt the K filter calibration distance of 139.1 ± 3.6 pc as our final distance to HD 23642 and so to the Pleiades. Note that we cannot treat any of the distance estimates investigated above as being independent of each other as they all incorporate an effective temperature of one or both of the stars in HD 23642.

One shortcoming of finding distances using equation 5.6 is that the effective temperature scales used in analysis of the dEB and for the calibration must be the same to avoid systematic errors. This is, however, a more relaxed constraint on the effective temperature scale than that involved in finding distance through absolute bolometric magnitude. We note that our effective temperature uncertainties include contributions due to possible spectral peculiarity and systematic offset relative to the (inhomogeneous) effective temperatures used in the KTDS04 calibration. The uncertainty in distance could be reduced by further observations and estimations of the effective temperatures of the two stars using the same technique as for the stars used to calibrate the surface brightness relations.

A way of avoiding systematic errors in the determination of effective temperatures is to use calibrations between surface brightness and colour indices for the individual stars.

This method is of only minor interest here but gives a distance of 138 ± 19 pc. The large uncertainty in this value is because we only have $B-V$, which is a poor surface brightness indicator due to the small wavelength difference between the B and V filters and because the B filter is sensitive to metallicity due to the effects of line blanketing. Good surface brightness indicators are the $B - L$ and $V - K$ indices (KTDS04; Di Benedetto 1998), where the observed scatter becomes much less than 1%, and the Strömgren c_1 index (Salaris & Groenewegen 2002). This method would also require additional photometry in several filters, including B or V, as well as a significant number of observations – or a complete light curve – in the K or L filters.

6. Conclusion

The distance of the Pleiades open cluster is currently controversial. The 'long' distance of the Pleiades is 132 ± 3 pc and is supported by main sequence fitting, the distance of the astrometric binary Atlas, and by ground-based trigonometrical parallax measurements. The 'short' distance is 120 ± 3 pc and comes from trigonometrical parallaxes observed by the Hipparcos space satellite. It has been suggested that the two distances could be reconciled if the Pleiades cluster is metal-poor, but determinations of the metal abundances of Pleiades F dwarfs suggest that the cluster has a solar iron abundance.

We have reanalysed the observational data on HD 23642 to determine absolute dimensions and effective temperatures. Our uncertainties mainly derive from the adoption of a spectroscopic light ratio to determine an otherwise poorly known ratio of the radii, but Monte Carlo simulations have also been undertaken to estimate the remaining uncertainties in the light curve fits. As the spectroscopic light ratio is preliminary, our results should also be viewed as such.

By comparing the radii of the components of HD 23642 to theoretical evolutionary models we find that the system has a solar or slightly supersolar metal abundance and a normal helium abundance, which removes the possibility that the 'long' and 'short' distance scales could be reconciled by adopting a low metal abundance or high helium abundance for the Pleiades cluster. This result is almost independent of the spectroscopic light ratio as the important quantity, the sum of the radii, is relatively well determined by the light curves.

From the use of bolometric corrections and our own temperatures and radii, we have calculated the distance to HD 23642 to be 138.1 ± 4.7 pc. The error given is a random error, but there is also a small systematic error due to the use of model atmospheres to calculate bolometric corrections and effective temperatures. M04 found a distance of 131.9 ± 2.1 pc using their own astrophysical data. Depending on the bolometric corrections adopted, we find that their temperatures and radii imply a distance to the dEB of 134.5 ± 2.5 pc or 135.4 ± 2.6 pc. This difference is probably due to different assumed values for the solar absolute bolometric magnitude and effective temperature.

We have found the distance to HD 23642 by using the effective temperature – surface brightness calibration of KTDS04. This method avoids the systematic errors due to the use of bolometric corrections and the calculation of luminosities. We find a distance of 139.1 ± 3.6 pc, where the biggest contributor to the uncertainty comes from the apparent magnitude measurement. There may be a slight systematic error in this result due to our method of determining the effective temperatures of HD 23642, but our uncertainties include this and the systematic error is certainly smaller than in the determination of distances by the use of bolometric corrections. An alternative distance of 138 ± 19 pc has been found by the use of the surface brightness – $(B - V)$ calibration of Di Benedetto (1998). This result is entirely empirical but its precision is low. The B filter magnitude is also dependent on the photospheric metal abundance so could be affected by the metallic-lined nature of the secondary star.

Our distance to HD 23642 is slightly larger than the 'long' distance to the Pleiades but is in full agreement with the constraints on the Pleiades distance from study of the astrometric binary Atlas (Pan *et al.* 2004). The distance derived by M04 is also in agreement with the 'long' distance scale. Our distance, and that of M04, is inconsistent with the Hipparcos 'short' distance, and indeed with the Hipparcos parallax distance of the dEB. We also confirm that the Pleiades has a solar metal abundance, removing the possibility of composition effects being used to reconcile the two distances.

Further observations of HD 23642, to determine accurate dimensions, would provide very precise metal abundance and effective temperature measurements for the component stars. This would reduce the uncertainty in its distance and allow further investigation of the system, which is itself an interesting object due to the metallic-lined nature of the secondary star. Further infrared or Strömgren observations, however, would allow the use of entirely empirical surface brightness relations for the calculation of the distance to HD 23642.

Distance estimation using the surface brightness calibrations of KTDS04 appears to have a lot of potential. Distances accurate to 2% could be obtained from definitive analysis of dEBs incorporating infrared JHK photometry. The study of dEBs in open clusters is also a powerful tool for inferring the chemical composition of the clusters.

Acknowledgements

The authors would like to thank Dr. U. Munari for making his data on HD 23642 freely available, and for helpful discussions. This publication makes use of data products from the Two Micron All Sky Survey, which is a joint project of the University of Massachusetts and the Infrared Processing and Analysis Center/California Institute of Technology, funded by the National Aeronautics and Space Administration and the National Science Foundation. JS acknowledges financial support from PPARC in the form of a postgraduate studentship. The authors acknowledge the data analysis facilities provided by the Starlink Project which is run by CCLRC on behalf of PPARC. The following internet-based resources were used in research for this paper: the NASA Astrophysics Data System; the SIMBAD database operated at CDS, Strasbourg, France; the VizieR service operated at CDS, Strasbourg, France; and the arχiv scientific paper preprint service operated by Cornell University.

References

Abt, H. A. 1958 *ApJ* **128**, 139–141.
Abt, H. A. & Levato, H. 1978 *PASP* **90**, 201–203.
Andersen, J. 1991 *A&A Review* **3**, 91–126.
Argelander, F. 1903 *Bonner Durchmusterung des nordlichen Himmels, Eds Marcus and Weber's Verlag, Bonn.*
Bessell, M. S. 2000 *PASP* **112**, 961–965.
Bessell, M. S., Castelli, F. & Plez, B. 1998 *A&A* **333**, 231–250.
Boesgaard, A. M. & Friel, E. D. 1990 *ApJ* **351**, 467–479.
Bressan, A., Fagotto, F., Bertelli, G. & Chiosi, C. 1993 *A&AS* **100**, 647–664.
Cannon, A. J. & Pickering, E. C. 1918 *Annals of Harvard College Obs.* **91**, 1–290.
Castellani, V., Degl'Innocenti, S., Prada Moroni, P. G. & Tordiglione, V. 2002 *MNRAS* **334**, 193–197.
Claret, A. 1995 *A&AS* **109**, 441–446.
Claret, A. 1997 *A&AS* **125**, 439–443.

Claret, A. 1998 *A&AS* **131**, 395–400.

Claret, A. & Giménez, A, 1995 *A&AS* **114**, 549–556.

Crawford, D. L. & Perry, C. L. 1976 *AJ* **81**, 419–426.

Di Benedetto, G. P. 1998 *A&A* **339**, 858–871.

Flower, P. J. 1996 *ApJ* **469**, 355–365.

Gatewood, G., de Jonge, J. K. & Han, I. 2000 *ApJ* **533**, 938–943.

Grenon, M. 1999 *in Garcia Lopez R. J., Rebolo R., Zapaterio Osorio M. R., eds, ASP Conf. Ser. Vol. 223, Proc. 11th Cambridge Workshop on Cool Stars. Astron. Soc. Pac., San Francisco,* p. 359.

Griffin, R. F. 1995 *Journal Royal Ast. So.c. Canada* **89**, 53–62.

Hertzsprung, E. 1947 *Ann. Sterrew. Leiden* **19**, 3.

Hilditch, R. W. 2001 *An Introduction to Close Binary Stars* Cambridge University Press, p. 281.

Johnson, H. L. 1957 *ApJ* **126**, 121–133.

Kervella, P., Thévenin, F., Di Folco, E. & Ségransan, D. 2004 *A&A* **426**, 297–307.

Makarov, V. V. 2002 *AJ* **124**, 3299–3304.

Meynet, G., Mermilliod, J.-C. & Maeder, A. 1993 *A&AS* **98**, 477–504.

Moro, D. & Munari, U. 2000 *A&AS* **147**, 361–628.

Munari, U., Dallaporta, S., Siviero, A., Soubiran, C., Fiorucci, M. & Girard, P. 2004 *A&A* **418**, L31–L34 **(M04)**.

Nelson, B. & Davis, W. D. 1972 *ApJ* **174**, 617–628.

O'Dell, M. A., Hendry, M. A. & Collier Cameron, A. 1994 *MNRAS* **268**, 181–193.

Pan, X., Shao, M. & Kulkarni, S. S. 2004 *Nature* **427**, 326–328.

Pearce, J. A. 1957 *Publ. Dom. Astr Obs.* **10**, 435–445.

Perryman, M. A. C., *et al.* 1997 *A&A* **323**, L49–L52.

Pinsonneault, M. H., Stauffer, J., Soderblom, D. R., King, J. R. & Hanson, R. B. 1998 *ApJ* **504**, 170–191.

Pols, O. R., Schröder, K.-P., Hurley, J. R., Tout, C. A. & Eggleton, P. P. 1998 *MNRAS* **298**, 525–536.

Popper, D. M. 1984 *AJ* **89**, 132–144.

Popper, D. M. 2000 *AJ* **119**, 2391–2402.

Popper, D. M. & Etzel, P. B. 1981 *AJ* **86**, 102–120.

Robichon, N., Arenou, F., Mermilliod, J.-C. & Turon, C. 1999 *A&A* **345**, 471–484.

Salaris, M. & Groenewegen, M. A. T. 2002 *A&A* **381**, 440–445.

Schaller, G., Schaerer, D., Meynet, G. & Maeder, A. 1992 *A&AS* **96**, 269–331.

Soderblom, D. R., King, J. R., Hanson, R. B., Jones, B. F., Fischer, D., Stauffer, J. & Pinsonneault, M. H. 1998 *ApJ* **504**, 192–199.

Southworth, J., Maxted, P. F. L. & Smalley, B. 2004a *MNRAS* **349**, 547–559.

Southworth, J., Maxted, P. F. L. & Smalley, B. 2004b *MNRAS* **351**, 1277–1289.

Stauffer, J. R., Schultz, G. & Kirkpatrick J. D. 1998 *ApJ* **499**, L199–L203.

Stello, D. & Nissen, P. E. 2001 *A&A* **374**, 105–115.

Torres, G. 2003 *IBVS* **5402**.

Vandenberg, D. A. & Bridges, T. J. 1984 *ApJ* **278**, 679–688.

van Belle, G. T. 1999 *PASP* **111**, 1515–1523.

van Hamme, W., 1993, AJ, 106, 2096. 1993 *AJ* **106**, 2096–2117.

van Leeuwen, F. 1999 *A&A* **341**, L71–L74.

Wilson, R. E. & Devinney, E. J. 1971 *ApJ* **166**, 605–619.

Wilson, R. E. 1993 *New Frontiers in Binary Star Research: Pacific Rim Colloquium. Astron. Soc. Pac., San Francisco* , 91–126.

Zahn J.-P. 1977 *A&A* **57**, 383–394.

Discussion

FRITZ BENEDICT: First, I'm very pleased that we agree within our error bars! Secondly, a question: What's the observational evidence for a third light in the system?

JOHN SOUTHWORTH: None. We cannot assume it's not there. It might be; it's just that we can't see it.

FRITZ BENEDICT: So it's just the error bars, basically?

JOHN SOUTHWORTH: It makes a negligible difference, this assumption of 5%. It only adds about a tenth to the uncertainties in the radii. But it's important to take into account, because there's no physical basis to assume that it's not there at all.

DON KURTZ: You think that the primary star is a silicon star. A silicon star in a binary that short is very unusual, but let's assume that it is true. If so, it's magnetic. If it's magnetic, it's got spots, and I notice you particularly have a problem fitting your secondary eclipse. A spot might handle that problem, so that's a little bit of extra you can throw in, at least at secondary eclipse time.

JOHN SOUTHWORTH: I do not agree with that entirely. Firstly, it may have a slightly enhanced silicon abundance. The spectral classification is quite old and we've not confirmed it so far, because we've only been at this data for about a month. Secondly, if I go back to the fitting of our secondary eclipse, the best fit does not appear to pass through the locus of the points. However, if you look at the residuals, you'll find it's not the problem with the model; I believe that's a problem with the photometry. I can improve the fit by rejecting data selectively. I'm not very keen on doing that. I should also point out that Munari *et al.* did not have this problem because they solved the data simultaneously, and the V light curve is a lot better defined than the B light curve, so it forced the results of the B light curve to where you think the data should be. I'm not convinced that's a problem with the model.

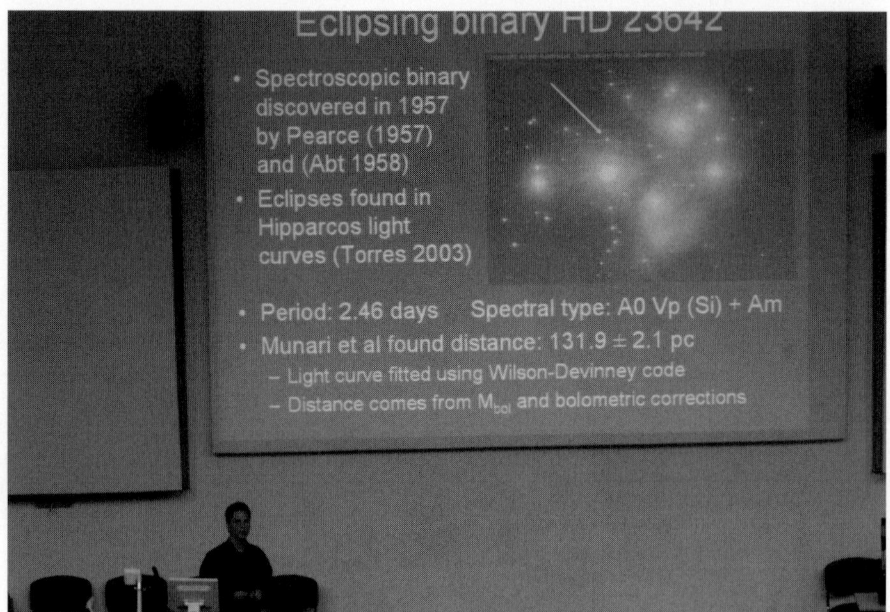

John Southworth presenting his paper

Fritz Benedict, Ricky Smart and Dimitri Pourbaix

Transits of Venus: New Views of the Solar System and Galaxy
Proceedings IAU Colloquium No. 196, 2004
D.W. Kurtz, ed.

© 2004 International Astronomical Union
doi:10.1017/S1743921305001547

Chromatic effects in Hipparcos parallaxes and implications for distance scale

Dimitri Pourbaix†

Institute of Astronomy and Astrophysics, Université libre de Bruxelles CP 226, B-1050
Brussels, Belgium   email: pourbaix@astro.ulb.ac.be

and

Department of Astrophysical Sciences, Princeton University, Princeton NJ 08543-1001, USA

Abstract. The change in the measured position due to the chromatic effect in the optical system is usually so small that it can be neglected. However, in the case of the Hipparcos mission, this effect is of the same magnitude as the targeted precision; it is therefore mandatory to correct for it. The different assumptions adopted to build up such corrections in the Hipparcos framework are presented. Some of their limitations and how they can be overcome to improve the results even today are described. The chromaticity mainly affects red stars and has its strongest impact on the astrometry of very red and highly variable stars such as Miras. Nevertheless, the implications on distance scale determination are rather small.

1. Introduction

The Hipparcos Catalogue (ESA 1997) contains parallaxes of nearly 120 000 stars brighter than $V \approx 12.4$, up to roughly 1 kpc away. In order to achieve a very high precision (and hopefully accuracy) in the derived quantities, several corrections were applied to the raw observations, including some to account for the chromaticity effects.

Owing to the pressure of releasing the catalogues, some simplifying assumptions were adopted in the reductions. For instance, a fixed color and, therefore, a constant chromatic correction was assumed for each individual star. Though sound for most stars, this assumption apparently is not correct, for instance, in the case of red long period variable stars such as Miras, LPVs (Mira-like, SR, ...) and may result in biased parallaxes.

In this contribution, we first note that the number of binaries in the Hipparcos catalogue is correlated with the color of the object – the redder the color the higher the likelihood to be a binary. We describe the way the chromaticity was originally corrected for in the Hipparcos framework and show that the connection between the color and binarity most likely is an artifact of the original assumption about the fixed color of stars. We then focus on how those initial results can benefit from ground-based photometric data and how, once combined together, they improve the accuracy of parallaxes, hence, the distance scale for long-period red variable stars.

2. Duplicity and color

Wielen (1996, 1997) suggested that unresolved binaries with one variable component would exhibit a motion of the photocenter correlated with the overall brightness change. An annex of the Hipparcos Catalogue, namely the DMSA/V, is dedicated to such objects called VIM, for Variability-Induced Movers. For all of them, it was assumed that the

† Research Associate, FNRS, Belgium

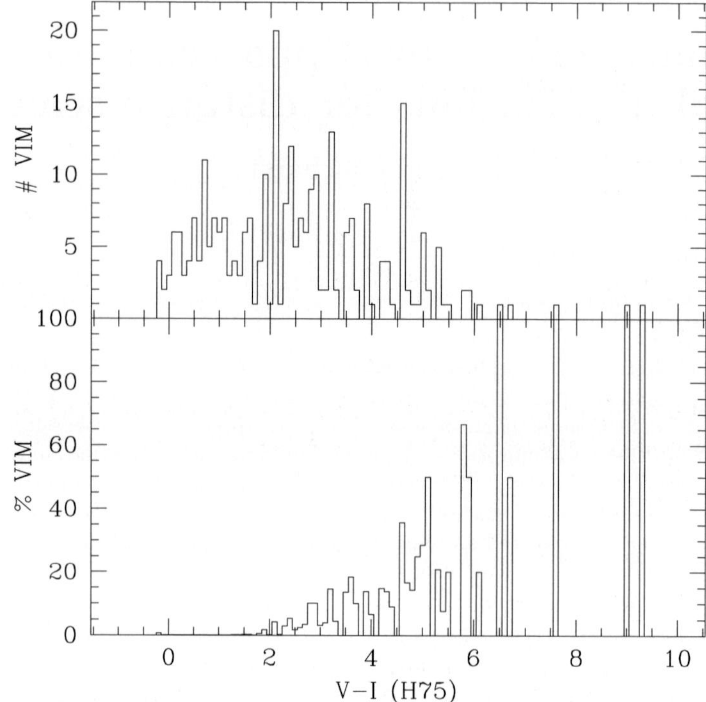

Figure 1. Distribution of the $V - I$ color of VIM. The top panel represents the actual number of VIM whereas the bottom panel gives the distribution of the percentage of VIM among all Hipparcos stars of similar color.

orbital period was long enough for the relative motion of one component with respect to the other to be negligible over the mission lifetime.

The position of the photocenter (ξ, η) is thus given by:

$$\xi = \alpha_0^* + \mu_{\alpha^*}(t - t_0) + P_\alpha \varpi + \left(10^{0.4(Hp_{tot}(t) - Hp_{ref})} - 1\right) D_{\alpha^*},$$

$$\eta = \delta_0 + \mu_\delta(t - t_0) + P_\delta \varpi + \left(10^{0.4(Hp_{tot}(t) - Hp_{ref})} - 1\right) D_\delta$$

where $\alpha_0^*, \delta_0, \varpi, \mu_{\alpha^*}, \mu_\delta$ are the five fundamental astrometric parameters (position, parallax and proper motion), P_α, P_δ are the parallactic factors, $Hp_{tot}(t)$ and Hp_{ref} are respectively the total Hp magnitude at time t and a reference magnitude. D_{α^*} and D_δ are the components of the direction of the secondary with respect to the primary star. A VIM object thus requires a 7-parameter model where D_{α^*} and D_δ are obtained together with $\alpha_0^*, \delta_0, \varpi, \mu_{\alpha^*}$ and μ_δ.

Though this model is legitimate and one expects such binaries to be present in a large stellar sample such as the Hipparcos catalogue, the color distribution of those VIMs are rather unexpected, as shown in Fig. 1. The percentage of VIMs is clearly color dependent: the redder the star, the higher the percentage. If VIMs are indeed binaries, that would mean that red stars are more likely to be binaries which no other investigation has shown. A similar correlation between the color and binary percentage was also noticed by Platais *et al.* (2003) on a larger sample of Hipparcos stars.

Rather than questioning the binary distribution/likelihood over the Hertzsprung-Russell diagram, it might be worth looking for some neglected instrumental effects or biases arising from the data reduction. For instance, what role does the color of an object play in the reduction, what were the assumptions?

3. Hipparcos chromaticity correction

Like any optical device with refracting elements, the Hipparcos satellite was subject to chromatic effects, i.e. light rays at different wavelengths follow different paths and therefore hit slightly different parts of the detector. While such an offset is so small that it can be safely ignored in most applications, it had to be accounted for and corrected in order to reach the very high precision aimed at by Hipparcos.

One can assume that the offset in position (for Hipparcos – along the scanning direction) is proportional to the offset in color:

$$\text{Offset} = p_0(C_* - C_{\text{Ref}}) \tag{3.1}$$

where C_* denotes the color of the object and C_{Ref} is the reference color for which zero chromatic displacement is adopted. In the case of Hipparcos, a term was added to account for aging of the instrument, i.e. a gradual evolution of detector's response function over the lifetime of the mission:

$$\text{Offset} = p_0(C_* - C_{\text{Ref}}) + p_1(C_* - C_{\text{Ref}})(t - 1991.25). \tag{3.2}$$

Since Hipparcos did not carry an on-board multi-band photometer, the ground-based colors were adopted. Initially built upon $B - V$ colors, the $V - I_C$ color index was finally adopted as a more reliable color indicator for the Hipparcos sample. While both data reduction consortia (FAST and NDAC) used $V - I_C$, they had their own chromaticity correction expressions. For instance, FAST had an additional quadratic term in Eq. 3.2 whereas NDAC calibrated the correction a function of effective wavelength which was used in lieu of the $V - I$ color index.

However, the $V - I_C$ color for all $\sim$120 000 stars was not more readily available than the $B - V$ color. The adopted values were either directly measured from ground (2929 stars) or derived from other colors, e.g. $B - V$. Nineteen different methods for such a derivation were used depending on the spectral type and luminosity class of stars. Thus, 97% of the chromaticity corrections are based on the empirically estimated $V - I_C$ colors. And yet, the worst is still to come.

As illustrated in Fig. 2, a change in magnitude often yields a change in color. Although the amplitude of the latter is not as large as of the former, it can still cause a shift of several milli-arc seconds (mas) across the detector. If this variable shift is unaccounted for, the position of the photocenter is apparently correlated with the brightness. This is a typical case of VIM but without a real secondary. It is thus likely that a fraction of the VIMs, especially the red ones, are just an artifact of the reduction pipeline and should rather be called Color-Induced Movers. In order to assess this alternative explanation, one needs to apply a better chromaticity correction due to the change in color and then decide whether the VIM model is still worth adopting. While removing the original chromaticity correction and applying a new one based on a variable color are rather straightforward (at least for the second part), deriving an instantaneous color is much more difficult as we see in the following section.

4. Deriving $V - I_C$ for red Hipparcos stars

How do we obtain $V - I_C$ color for each individual Hipparcos observation despite the fact that no on-board instrument measured that color and no parallel ground-based

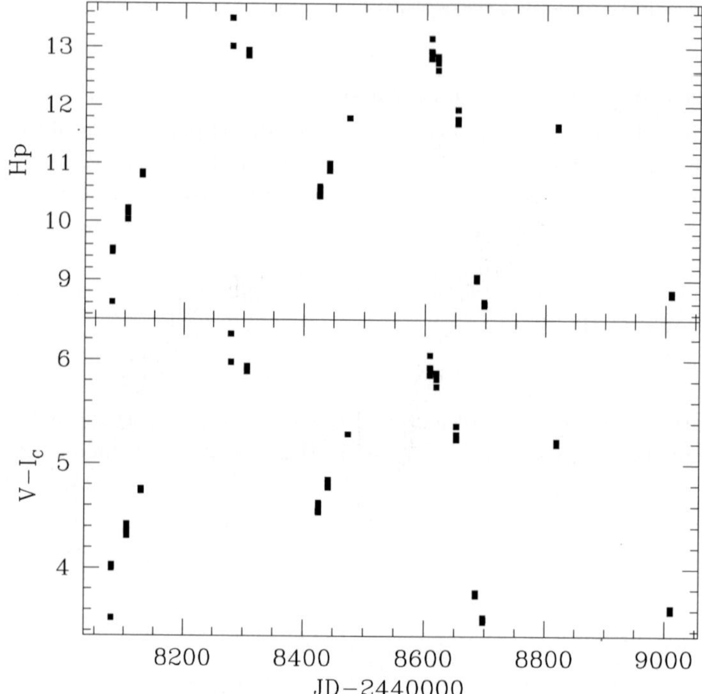

Figure 2. Variability in magnitude and color.

observing campaign was set up for that purpose? We here outline how the instantaneous Cousins $V - I$ were "restored" from a combination of the ground-based and Hipparcos/ Tycho observations (Platais et al. 2003).

First, we collected all the existing $V - I$ for a large sample of M, S, and C spectral type stars. Some of these $V - I$ were already published while other were measured for this purpose. Whenever necessary, those colors were changed into the natural Cousins system using color-color transformation listed in Volume 1 of the Hipparcos catalogue. The result was a set of 567 $V - I_C$ color indices. Although this sample already makes it possible to assess the accuracy of the adopted Hipparcos $V - I_C$, it does not address the variability of this color index over the mission lifetime.

It was noticed that, for reasonably high fluxes a new instantaneous color index for red stars, $Hp - V_{T2}$, is strongly correlated with Hp, and can be well-modeled for each such star with a straight line.

$$Hp = \frac{b_0 + V_{T2}}{1 - b_1}.$$

Those lines thus express Hp in terms of V_{T2}. Volume 1 of the Hipparcos catalogue also lists transformations from $V - I_C$ to V_T (measured as V_{T2}). Combining all these results, Platais et al. (2003) therefore established a color-color transformation from $V - I_C$ to $Hp - V_{T2}$. It is represented by two fourth-order polynomials (Fig. 3) depending on the spectral type of the stars.

Such a tight relationship now makes it possible to derive the $V - I_C$ color for any epoch. Using this relationship for a given red star, a set of Hipparcos and Tycho photometric

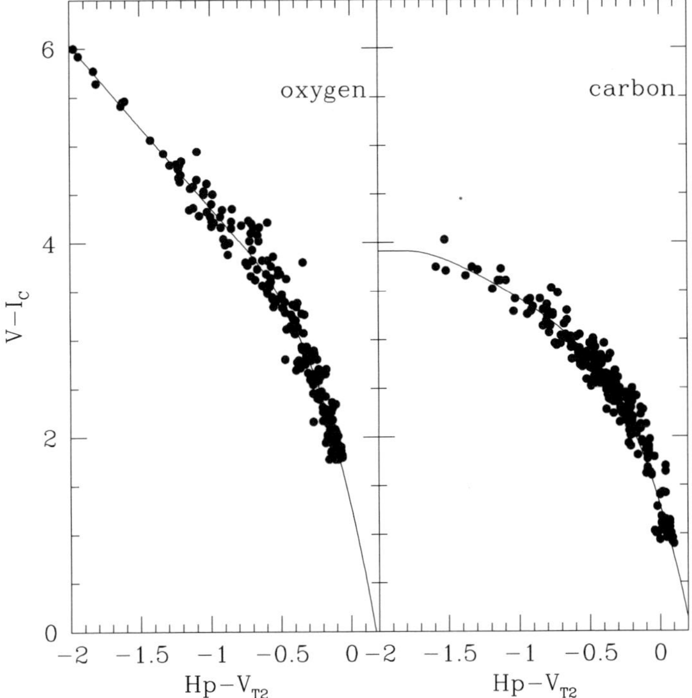

Figure 3. Color-color diagram of oxygen and carbon long period variable stars.

observations themselves allow us to *predict* what was the corresponding instantaneous $V - I_C$ color index at each epoch of Intermediate Astrometric Data.

5. Revised model and parallaxes

Once the observations are properly corrected for chromaticity thanks to the significantly improved colors, one can evaluate how the number of VIMs changes and, if the adopted model is different, how this affects the parallax. According to what is stated in the previous section, most VIMs are actually color-induced movers and, therefore, should disappear when correct colors are adopted.

The evolution of the distribution of most VIMs is given in Fig. 4. The first thing to notice is that the number of VIMs falls by 20% if, instead of using FAST astrometric data (as originally done), one takes advantage of the NDAC part as well (upper left corner). Only 188 VIMs are carbon or oxygen LPVs for which epoch $V - I_C$ were derived. Using FAST and NDAC data, 150 VIMs survive (upper right corner). So, even without changing anything to the chromaticity correction, just by adding the NDAC data, 21% of the VIM get already discarded. The lower panels of Fig. 4 present the results when epoch $V - I_C$ are used. The number of VIMs is given for two values of the statistical confidence (10% which correspond to the original one and 0.27%). The left panel uses the original $V - I$ on its x-axis whereas the right panel uses the median of the epoch $V - I_C$ instead. Clearly, the number of VIMs drops (by at least 47%) and the red tail of the distribution simply disappears.

Among the original VIMs, using epoch $V - I_C$ often leads to the replacement of the VIM model by the single star model, thus potentially affecting the parallax, especially

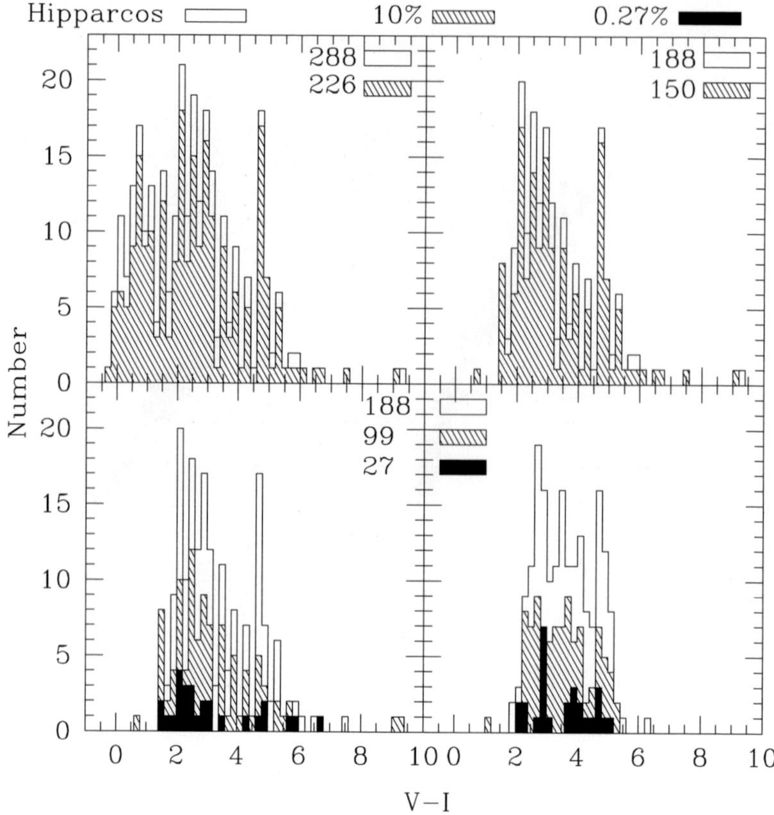

Figure 4. Evolution of the distribution of VIMs in terms of $V - I_C$

for those variable stars with periods close to one year. Even for the objects originally processed as single stars, the revised chromaticity correction potentially changes the astrometric solution and its uncertainty.

Once compared with independent determinations of the distance of LPVs (e.g., Mennessier & Luri 2001), some of the newly derived parallaxes are in much better agreement than were the original ones. For example, the parallax of V1943 Cyg changes from 1.82 ± 1.82 mas to 5.02 ± 0.97 mas – versus 4.7 mas (Mennessier & Luri 2001; Neugebauer & Leighton 1969); R Hya from 1.62 ± 2.43 mas to 8.44 ± 1.00 mas – versus 7.14 mas (Whitelock et al. 2000).

One way of extending the Hipparcos results beyond 1 kpc consists in using its parallaxes to derive the Period-Luminosity relation of Mira-like variables. Adopting the slope obtained in the LMC (Feast et al. 1989), van Leeuwen et al. (1997) derived the zero-point of the relation based upon the Hipparcos parallaxes of 16 pre-selected Mira variables and obtained:

$$M_K = -3.47 \log P(\text{days}) + 0.94 \pm 0.18, \tag{5.1}$$

where M_K denotes the absolute K magnitude and P the period of variability. From a sample of 255 Mira-like variables and still adopting the slope from Feast et al. (1989), Whitelock & Feast (2000) derived:

$$M_K = -3.47 \log P(\text{days}) + 0.84 \pm 0.14. \tag{5.2}$$

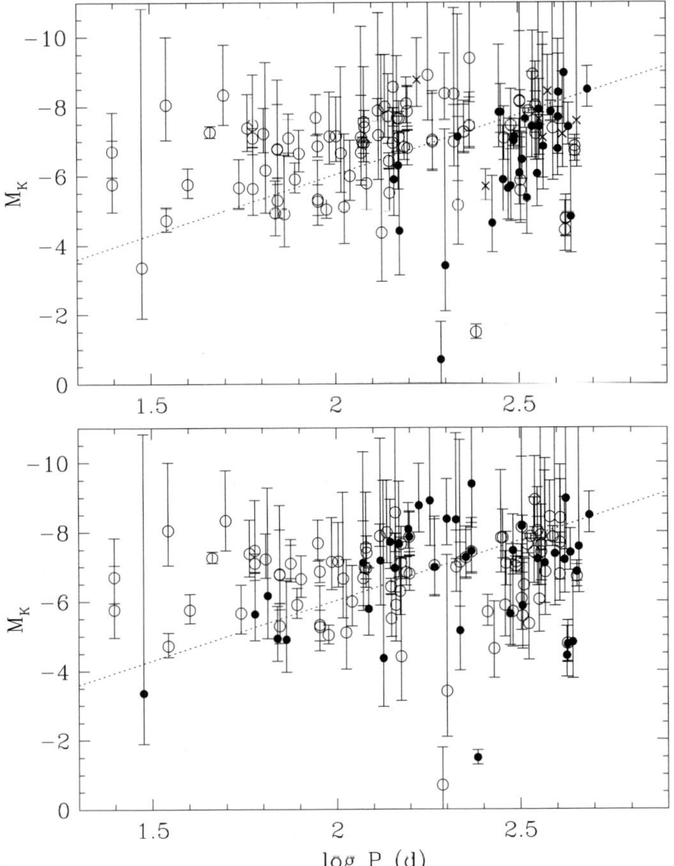

Figure 5. Distribution of LPVs in the $M_K - \log P$ plane. Upper panel: stars are sorted by variable type – filled circles = Miras; open circles = SR/SRb; crosses = SRa. Lower panel: stars are sorted by chemical type – filled symbols = carbon stars; open symbols = oxygen stars; the dashed line is the P-L relationship of Eq. (5.2).

All previous derivations assume the empirical LMC slope, not really knowing whether it also applies to Galactic Miras. In order to establish the soundness of that approach, Knapp et al. (2003) derived both the slope and the zero-point from the Hipparcos data revised with the new chromaticity correction. Fitting

$$\varpi = P^{\frac{a}{5}} 10^{\frac{b}{5}} 10^{\frac{10-K+A_K}{5}} \tag{5.3}$$

instead of

$$M_K = a' \log P(\text{days}) + b' \tag{5.4}$$

because of the Gaussian error on ϖ only, they obtain

$$M_K = -3.39(\pm 0.47) \, \log P(\text{days}) + 0.95(\pm 3.01). \tag{5.5}$$

A_K denotes the interstellar absorption in the K band. Although the parallaxes are accurate enough to yield a relation consistent with the previous ones, the uncertainty on the parallaxes is still way too large to precisely derive both the slope and the zero-point.

6. Conclusions

We have shown that a simplistic assumption about fixed colors of many red variable stars over the Hipparcos mission yielded some questionable parallaxes together with a strong correlation between the color and binary percentage. A nice feature of the Hipparcos catalogue is the availability of the low level data which make it possible to improve the original data processing upon the availability of new data.

Although ESA had not foreseen that the chromaticity correction could be revised as well, Platais et al. (2003) nevertheless succeeded in taking advantage of ground-based $V-I_C$ colors to improve the correction of the chromatic effects. Thanks to these improved colors, the number of stars seemingly best described as VIM vanished quite substantially. The (default) single star model not only fits the data at a satisfactory level but, in some cases, it also improves the parallax accuracy.

Regardless of which model presently fits the observations, the revised parallaxes can be used to derive the parameters of the period-luminosity relationship for long-period variables. The LMC slope is confirmed (unlike other investigations where that slope is simply assumed) but those parallaxes are neither accurate enough nor precise enough to reliably adjust both the slope and the zero-point.

Acknowledgements

This research was supported in part by ESA/PRODEX C15152/01/NL/SFe(IC) and by NASA via grant NAG5-11094.

References

ESA 1997, The Hipparcos and Tycho Catalogues. ESA SP-1200

Fabricius C., Makarov V. V., 2000, A&A, 356, 141

Feast M. W., Glass I. S., Whitelock P. A., Catchpole R. M., 1989, MNRAS, 241, 375

Høg E., Fabricius C., Makarov V. V., Urban S., Corbin T., Wycoff G., Bastian U., Schwekendiek P., Wicenec A., 2000, A&A, 355, L27

Knapp G. R., Pourbaix D., Platais I., Jorissen A., 2003, A&A, 403, 993

Mennessier M. O., Luri X., 2001, A&A, 380, 198

Neugebauer G., Leighton R. B., 1969, Two-Micro Sky Survey Catalogue. NASA SP-3047

Platais I., Pourbaix D., Jorissen A., Makarov V. V., Berdnikov L. N., Samus N. N., Lloyd Evans T., Lebzelter T., Sperauskas J., 2003, A&A, 397, 997

Pourbaix D., Platais I., Detournay S., Jorissen A., Knapp G., Makarov V. V., 2003, A&A, 399, 1167

van Leeuwen F., Feast M. W., Whitelock P. A., Yudin B., 1997, MNRAS, 287, 955

Whitelock P., Feast M., 2000, MNRAS, 319, 759

Whitelock P., Marang F., Feast M., 2000, MNRAS, 319, 728

Wielen R., 1996, A&A, 314, 679

Wielen R., 1997, A&A, 325, 367

Discussion

FLOOR VAN LEEUWEN: I would like to make a few clarifications. In the science team for Hipparcos, when we were approaching the publication of the data it was decided that it was better to keep $V - I$ constant than to try to put in a variable $V - I$ which was uncertain. So we kept everything constant with the idea that it could be corrected more easily from that value than from a variable input. We didn't have the time to derive the epoch $V - I$ photometry that you have been deriving. So that was the best we could do. This is not a question of a mistake. This is a question of a purposeful best comprise. That's why we provide both the $V - I$ as used in the reduction, and what was afterwards thought of as the best $V - I$; you find both values there. Secondly, the work that you've been doing, it is worth noticing, is all incorporated in the new reduction [see van Leeuwen, these proceedings]. So the $V - I$ variability is all already incorporated in the new reduction, and doing a very good job. I can see that all the calibrations that concerned the colour are a lot better than they have been before. Because $V - I$ not only works through what you have been looking at, but it also works at a much deeper level – where the data are not available for correction.

MIKHAIL MAROV: You introduced the covariance matrix and it's very powerful for the data processing. How well did the data fit along the main diagonal and how large is the scatter? Are the scattered points meaningful, or just noise?

DIMITRI POURBAIX: I only mentioned one covariance matrix, the one given in the Hipparcos Catalogue, so that's something I can't change. The only thing I can change in that relation is the p, the Δp. So, for instance, if I take the right ascension, I know the value published for the right ascension. I will try another one; the difference between the two will give me the Δp_1. So that's the only place where I have room for change. I can't change the V there. The covariance matrix is given from the reduction; so it's just the weight of all the observations. I have no room for improvement there.

MIKHAIL MAROV: So, it has nothing to do with the multi-band photometry?

DIMITRI POURBAIX: No.

FRITZ BENEDICT: You threw out about 90% of those VIMs as possible binaries, and you did mention that the navy was going after them. Did their statistics agree with yours? Did they get about a 10% hit rate while looking at them?

DIMITRI POURBAIX: The problem is, in their case, they confirm a binary if they see two components. If they don't see two components, they can't say anything. They can just say, "no detection"; that's all. What we did was to reduce the effort in looking for new binaries.

RICKY SMART: You got a calibration for $H - V$ for Miras. Did you apply it to everything?

DIMITRI POURBAIX: No. Only to Miras.

RICKY SMART: So all your VIMs are Miras?

DIMITRI POURBAIX: Most of them, yes. For the VIMs we look only for Miras, and that's indeed why you expect most of them, since you need a big change in brightness in order to see a displacement.

RICKY SMART: Why did you use the $H - V$, which are quite close bands, instead of the $B - V$ from Tycho?

DIMITRI POURBAIX: That's where we got the maximum flux.

FLOOR VEN LEEUWEN: The main benefit here is that the H-band is actually extending quite nicely towards the red. The difference between the V and the H band is the red wing.

Transits of Venus: New Views of the Solar System and Galaxy
Proceedings IAU Colloquium No. 196, 2004
D.W. Kurtz, ed.

© 2004 International Astronomical Union
doi:10.1017/S1743921305001559

The use of eclipses in the evaluation of absolute stellar information

Edwin Budding[1]†, Denis Sullivan[2] and Michael Rhodes[3]

[1]Carter National Observatory of New Zealand, PO Box 2909, Wellington, New Zealand
email: ebudding@comu.edu.tr

[2]Victoria University of Wellington, Wellington, New Zealand

[3]Brigham Young University, Provo, Utah, USA.

Abstract. Eclipsing binary light curves have provided the 'royal road' (Russell 1948) to fundamental astrophysical information on stars. If radial velocities of the components and a reliable colour/temperature/flux relation are available, parallaxes may be determined and compared with direct measures, e.g. by Hipparcos. Accuracies of existing measures are considered and aspects of the development of this and other methods of distance determination reviewed. Roles for multiwavelength techniques (e.g. VLBA, broad-spectrum photometry) are noted. The recovery of information on new planets orbiting remote stars by transit phenomena will be looked into within this context.

1. Introduction

The fitting of physical models to photometric data – particularly, in the stellar context, sets of relative flux measures at discrete phases (light curves) – concerns how information in the data maps into the parameters: sizes, masses, luminosities, and notably, for this meeting, *distances*, that scientists are seeking. Henry Norris Russell (1948) said that eclipsing binary stars offer a 'royal road' for such purposes, and although this route to parallax determinations has been increasingly recognized, particularly since the work of Lacy (1978), it may not be as direct to this particular destination as to some others.

2. Photometric calibration

The 31.572126 magnitudes by which the Sun's apparent magnitude would increase if moved to a distance of 10 pc carries the same accuracy as that of the A.U. However, the actual value of $m_v(\odot) = -26.75$ is still hardly known to better than four decimal figures for various technical reasons. These include bringing a body as unique as the Sun, from the terrestrial point of view, on to the same scale of measurement as applies to other stars, and the lengthy, interdependent framework of comparison measurements needed to build up a particular magnitude scale or system. Still, four decimal digits is a relatively high accuracy in the context of stellar parallaxes.

At 10 pc, the Sun's apparent magnitude would also be its absolute magnitude (4.82), which, folded with a standard photopic transmission curve, comes to 2.94×10^{-8} lux or $4.32 \times 10^{-11}\,\mathrm{W\,m^{-2}}$. Comparison of normalized transmission curves for standard optical filters then leads to the zero point of, say, Johnson's V scale as $3.08 \times 10^{-9}\,\mathrm{W\,m^{-2}}$.

† Present address: Physics Dept., 18 March University of Canakkale, TR 17020, Turkey.

Received fluxes f can be directly measured by a normal astronomical photometer, generally as a number of photon-events per second, and, especially if the photometer behaves linearly and the zero point of the relevant scale is known, such counts can be directly transferred to watts per square meter. Typically, after taking account of various absorption effects and response efficiencies, when using a regular telescope set up for photometry, measured optical fluxes are on the order of $\sim 10^{-10}$ to $10^{-12}\,\mathrm{W\,m^{-2}}$ for primary standard stars (although not normally given in such units). But notice that these fluxes, if transmitted across empty space, are simply related to surface fluxes F at the photospheres of the measured stars by

$$f = FR^2/r^2, \tag{2.1}$$

where R is the radius of the measured star and r is its distance. Hence, the parallax, $1/r$ (pc) $= 4.433 \times 10^7 \sqrt{f/F}/R_\odot$, is determined if we can obtain the ratio of measured to surface flux and the radius of the star, here expressed in solar units.

The most direct way to obtain surface fluxes is to use Eq. (2.1) again, but with the ratio R/r as half the angular diameter, combining a measurement of that with f. Angular diameters may be found by interferometry, or sometimes, at least for some zodaical stars, by a rather special kind of eclipse, namely occultations by the hard edge of the Moon, particularly in IR light. The derivation of surface fluxes, or corresponding temperatures, has become a focus for such angular diameter and apparent magnitude measurement. If stellar parallaxes become known, as they notably have been in the wake of the Hipparcos programme (ESA 1997), corresponding radii are determinable.

The known behaviour of continuum radiation from models of normal photospheres allows us to expect a fairly simple connection between surface flux and colour. Empirical forms have received attention, particularly since the work of Barnes & Evans (1976), and if such a form can be supported, for example, through angular diameter and apparent magnitude checks, it can be applied to parallax determinations when the radii of the stars are independently known. This is the case for eclipsing binary stars that have good quality light and radial velocity curves and it forms the basis of the 'eclipse method'.

Let us first rewrite (2.1) for the parallax Π, following usual terminology (e.g. Popper 1998):

$$\log \Pi = 7.450 - \log R - 0.2 m_V - 2F_V' \tag{2.2}$$

where $F_V' = 0.25 \log F_V$, F_V being the surface flux as in (2.1) in the Johnson V system. This expression is related to the effective surface temperature T_e by

$$F_V' = \log T_e + 0.1\mathrm{BC}, \tag{2.3}$$

where BC is the bolometric correction $m_{bol} - m_V$, connecting visual and total-radiation magnitude scales.

The dependence of flux on colour, or form of $F_V' = F(B-V)$, found by Popper (1998) is well represented by the following:

$$F_V'(\text{early}) = 3.957 - 0.311(B-V) + 0.586(B-V)^2 - 5.713(B-V)^3, \tag{2.4}$$

$(-0.2 < B - V \leqslant 0.1)$; and

$$F_V'(\text{late}) = 3.958 - 0.361(B-V) + 0.103(B-V)^2 - 0.076(B-V)^3 \tag{2.5}$$

$(0.1 < B - V \leqslant 1.2)$.

3. Eclipsing binary stars

Russell's original light curve 'solution' procedure lacked the sophistication of present-day modelling, although a couple of points may be worth noting. Thus, (i) the introduction of more elaborate theoretical models, by itself, does not deal with the underlying issues of curve-fitting adequacy, uniqueness and determinacy and (ii) Russell and his students were concerned with the application of the results of light curve analysis to stars in general. Emphasis in more recent work often concerns the special peculiarities of close binaries themselves. Interactive evolution, for example, has engendered its own detailed research and the development of relevant physics. But stellar models suitable for general parameter evaluation (including parallaxes), to data-limited accuracies, of detached pairs were available in the wake of Kopal's (1959) 'classical' monograph.

Kopal's approach utilized well-known spherical harmonic expansions for the component interactions. These are formally integrable and thus able to characterize light variations to the same order of approximation as the distortion of figure and in a comparable number of manageable terms. Kopal (1959) also dealt with surfaces of zero velocity and limiting dynamic stability within close binaries, pointing out the relatively tractable form of the 'Roche' approximation, when internal density distributions are neglected. This has proved useful in photometric studies of close binaries. Depending on quadrature accuracy, when components become relatively very close, with radii expressed as a fraction of the orbital separation $R/A >\sim 0.35$, representations of surfaces based on this approximation may become closer to actual physical distortions than those using series approximations with finite density distribution that stop after terms in $(R/A)^5$ (as in Kopal's original presentation). While numerical integration approaches are much more demanding in computer time, that is becoming increasingly less of an issue. User-friendly software packages, such as Bradstreet's *Binary Maker* (based on the approach of Wilson & Devinney (W-D) 1971), allow such models to be easily compared with data or those of other approaches.

Close binary system models for the light variations require the separate specification of around 16 independent physical quantities (cf. Budding 1993). The effects these parameters have on the light curve shape are, however, sometimes quite similar to each other, so that, in a deterministic sense, they become interdependent. A typical close system light curve would be insufficient, by itself, to furnish all physical parameters, although one could normally expect to make fair estimates for most, if not all, of the more basic geometric ones, using reasonable assumptions about the others.

It is important to remember this interdependence of parameters in light curve analysis. The relative radius of the primary component r_1, for example, i.e. the mean spherical radius divided by the mean orbital separation, is derived from eclipse photometry. While published findings often quote values to three or more significant decimal digits, a two-digit value is already rather optimistic for the likely information content of many light curves, if we admit to the correlated effects of uncertainty in all the other parameters. The extent of disagreement between different sources for even the more well-known examples suggests two significant figures is, conservatively, realistic, though a third digit is usually included.

4. Practical tests

The application of eclipsing binary data to the direct checking of basic stellar parameters was the subject of a number of seminal publications of D. M. Popper. The review already cited (Popper 1998) examined the properties of 14 well-studied systems within

125 pc with low scatter Hipparcos parallaxes. The thrust of that paper was towards establishing the flux-colour relation for Main Sequence (MS)-like stars and the most commonly used filters (V and B), for which purpose the choice of relatively near stars rendered insignificant the potential complication of interstellar absorption. The procedure can then be inverted by using (2.2) to test the agreement of individual parallaxes. We show the results in Fig. 1.

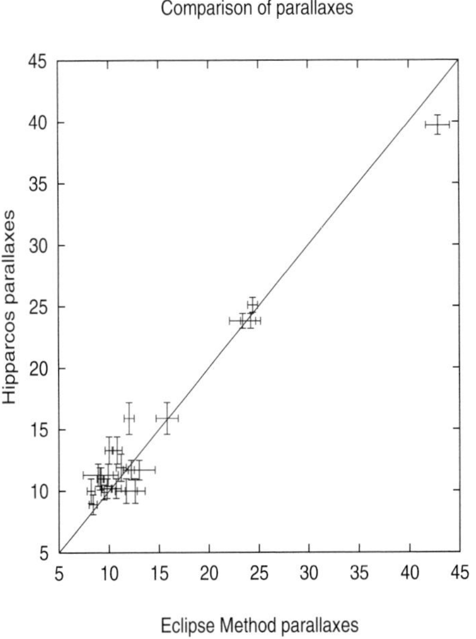

Figure 1. Comparison of Hipparcos parallaxes (in mas) with those from eclipsing binaries containing stars of known magnitudes, sizes and surface fluxes derived from their $B - V$ colours using Popper's (1998) data.

The high-parallax star in the upper right of Fig. 1, playing an important role in establishing the correlation across the range of values, though with some apparent departure from prediction, is β Aur. This is one of the closest of a small number of eclipsing binaries with measured angular diameters, light and radial velocity curve solutions, and direct measurements of parallax. Nordström & Johansen (1994) provided a detailed review of this system, in which they gave a parallax of 40.3 ± 0.4 mas, confirmed by interferometry and cross-checked by several other methods. This is within its own error limit of the Hipparcos parallax (39.7 ± 0.8 mas). Nordström & Johansen give radii of primary and secondary components as 2.77 and 2.63 $R_\odot$, to a claimed 1% accuracy. The angular diameters of the two stars, from these values, are 1.04 and 0.99 mas, i.e. somewhat less than the cited average measurement of 1.05 mas, which yields an average temperature of 9120 K. The decreased average radius implies a slightly hotter average, in essential agreement with the 9280 K average from Nordström & Johansen.

The V magnitudes of the two components, using Nordström & Johansen's adopted luminosity ratio and the SIMBAD listed combined value, are 2.587 and 2.713. Corresponding $B - V$ values are 0.061 and 0.092. Substitution in the above equations then yields 40.9 and 42.8 mas for the parallaxes. These may be compared with a similar treatment by Semeniuk (2000), who used Lacy's (1979) adopted form of $F'_V = F(V - R)$, and derived a mean parallax of 46.6 mas. With the Popper flux-colour relation, the primary appears almost within the error limit of the Hipparcos parallax, though the secondary is distinctly less in agreement. But it is worth noting, as did Semeniuk (2000), that β Aur has eclipses that are only 0.1 mag deep and for which, therefore, we should not expect parameters to have the same determinacy as more typical examples. We note that slightly larger radii, particularly for the secondary, would reduce the disparities, both for the angular diameter measurement and the flux-dependent parallax.

A number of points lower down the correlation appear rather discordant in the sense of the photometric parallaxes being larger than Hipparcos values. These turn out to be the 'RS CVn' binaries SZ Psc, LX Per, RT And and UV Psc. Light curves of the latter two were intensively studied by a group at the University of New Mexico at Albuquerque, using the 'spot-cleaning' technique of Budding & Zeilik (1987). These systems contain stars that are quite similar to the Sun, except that they are rotating $\sim 30 - 40$ times faster. Since, from (2.2), parallaxes can be increased by decreasing radii, it is tempting to think that these may have been overestimated in the analyses. Even if we push the radii down to their reasonably lowest limits, however, the parallax disparity remains, and we should look towards the more influential flux term to seek an explanation. In fact, as Popper (1998) noticed, the existence of dark photospheric spots would reduce the *mean* surface flux, while not changing the apparent colour. This can be easily checked from the apparent temperatures that go with the angular diameters: only $\sim$5100 K for the G5 primary of UV Psc, for example. This is an interesting point in relation to the spot-cleaning. That technique dealt only with the differential, cyclically varying component of maculation and not directly with its steady, phase-independent component. That component would reduce mean flux estimates of stars like RT And and UV Psc, even with cleaned light curves, below what could be estimated from the flux-colour relation of a spot-free photosphere. The extent of the parallax disparity suggests the steady component of the darkening is quite appreciable: up to 30% of the total luminosity would be affected, or $\sim$3 times greater than typical 'spot-wave' components.

Semeniuk (2000) considered 19 stars from a homogeneous sample of 47 eclipsing binaries studied by Lacy (1979) that includes 8 of the binaries in Popper's (1998) review. These stars all had relatively low Hipparcos errors, usually of around 10% or less. We show a similar sample in Fig. 2, that includes 7 of the Semeniuk stars not in Popper's list, a further 2 stars (YZ Cas and KW Hya) for which very good eclipsing binary data exist (cf. Malkov 1993; http://simbad.u-strasbg.fr) and also the binary HD 209458 (see Section 6, below). We have also placed the Sun, as if seen at the distance of 25 pc, in the same figure. The uncertainty of the photometric parallax is here reflecting uncertainties of the solar magnitudes in the BV system and the derived stellar flux calibration.

Semeniuk (2000) concluded that a higher accuracy of distance modulus was possible, using such well-known binaries, than would follow from the Hipparcos parallaxes, provided certain stars were excluded. As with Popper, she found the RS CVn stars systematically low in mean surface flux values. She also proposed rejecting the Algol binaries β Per and δ Lib from her high-accuracy list. Although these stars have experienced interactive evolution histories, their primaries, at least, show photospheric conditions resembling those of normal MS stars. The results of applying new solution data indicate a satisfactory accord with Hipparcos parallaxes and this is potentially advantageous for reasons considered below. Details on all 24 reference binaries considered in this paper are presented in Table 1.

5. Algols and checks on distances

Algols are the most commonly observed type of eclipsing binary. The reason is the selection effect of a relatively deep primary minimum, caused by the primary's occultation

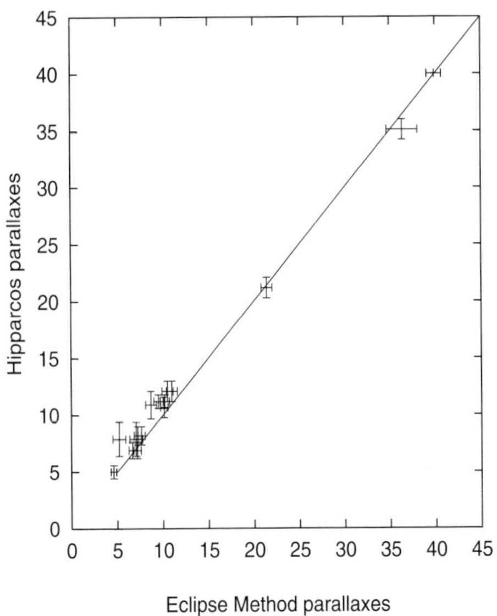

Figure 2. Comparison of Hipparcos parallaxes with 10 eclipsing binaries, mainly from Semeniuk's (2000) list.

by a cool, subgiant secondary. As well as favouring their easy recognition, this circumstance facilitates their discovery in surveys of remote systems, such as the neighbouring galaxies. Another interesting consequence of the constitution of many of these binaries is that the rapidly rotating secondaries show 'activity' characteristics in their convective envelopes. This means that, among other things, we can expect Algols to be microwave sources, and existing data indeed show radio characteristics similar to those of RS CVn stars (Slee *et al.* 1987). Hence, this opens up Algols to the positional finesse of modern high-precision radio-interferometric techniques. Mutel *et al.*'s (1998) VLBI map showing coronal features around the secondary of β Per was also giving support to less-direct positional information from the optical domain on the system's geometry and parallax (cf. also Lestrade *et al.* 1999). It would be very interesting to check if such features, resolved for β Per, could be made out for some other nearby Algols.

Another, still relatively near, Algol is δ Lib: a close binary made up of A0V + K0IV stars that has been the subject of a recent intensive study (Budding *et al.* 2004a) combining photometric, spectroscopic and radio information. Like β Per, δ Lib seems to be in a gravitationally bound system with 3, or more stars. Suspicions of such arrangements often come from data on the times of minimum light, which are again facilitated by deep eclipses. For δ Lib, more precise times of minima could check on a tentative 13 min amplitude oscillation indicated in recent photoelectric eclipse times. This would allow estimates of the semi-major axis, period and eccentricity parameters of a third-body orbit. Detailed light curve analysis, with relatively high determinacy, of wide-spectrum photometry including the infra-red light curves of Lazaro *et al.* (2002), gave persuasive evidence of a late A or early F type companion having a feasible (MS) mass of $\sim$1.4 M$_\odot$.

Extra stars in a light source produce complications for the Hipparcos reduction procedures, so various issues are interrelated for binaries with wider orbit companions and it is worthwhile to recheck the astrometry. Worek (2001) already postulated a low mass companion for δ Lib to account for irregularities in the primary's radial velocity

Table 1. Parameters of 24 binaries (and Sun) used to compare Eclipse Method and Hipparcos parallaxes.

Star	R	V	F_V'	Π_{EM}	err.	Π_{Hip}	err.
ζ Phe A	2.85	4.221	4.032	12.2	0.3	11.7	0.8
ζ Phe B	1.85	5.591	3.975	13.0	1.6	11.7	0.8
β Aur	2.49	2.643	3.947	42.9	1.2	39.7	0.8
WW Aur	1.88	6.603	3.903	11.2	0.5	11.9	1.1
RR Lyn A	2.50	5.928	3.883	12.6	1.0	10.0	1.0
RR Lyn B	1.93	6.878	3.860	11.7	1.1	10.0	1.0
ZZ Boo	2.22	7.543	3.835	8.4	0.4	8.9	0.8
HS Hya A	1.28	8.791	3.816	9.1	0.3	11.0	0.9
HS Hya B	1.22	8.961	3.806	9.2	0.3	11.0	0.9
SZ Psc A	1.40	8.746	3.809	8.6	1.5	11.3	0.9
SZ Psc B	4.95	7.566	3.624	9.9	1.0	11.3	0.9
Z Her A	1.85	7.664	3.809	10.7	0.5	10.2	0.8
Z Her B	2.75	8.664	3.643	9.8	0.5	10.2	0.8
V1143 Cyg	1.34	6.613	3.806	24.4	0.5	25.1	0.6
LX Per A	1.64	8.913	3.787	7.5	0.4	10.0	1.0
LX Per B	3.05	8.833	3.661	7.5	0.4	10.0	1.0
UX Men	1.31	8.973	3.781	9.5	0.3	9.9	0.6
RT And A	1.24	9.045	3.778	9.8	0.4	13.3	1.1
RT And B	0.87	11.325	3.605	10.9	0.5	13.3	1.1
UV Psc A	1.11	9.291	3.743	11.5	0.5	15.9	1.3
UV Psc B	0.83	10.571	3.608	15.8	1.1	15.9	1.3
AR Lac B	1.52	7.121	3.730	24.2	1.0	23.8	0.6
AR Lac B	2.81	6.621	3.654	23.4	1.3	23.8	0.6
AR Aur A	1.78	6.851	3.973	7.7	0.4	8.2	0.8
AR Aur B	1.82	6.941	3.972	7.2	0.4	8.2	0.8
EI Cep A	2.90	8.113	3.851	4.6	0.3	5.0	0.6
EI Cep B	2.33	8.686	3.841	4.6	0.3	5.0	0.6
V624 Her A	3.03	6.629	3.893	7.2	0.4	6.9	0.7
V624 Her B	3.23	7.429	3.816	6.7	0.4	6.9	0.7
UV Leo	1.10	9.733	3.752	8.8	0.6	10.9	1.2
BH Vir A	1.12	9.868	3.772	7.0	0.7	7.9	1.5
BH Vir B	1.06	11.678	3.686	4.0	0.7	7.9	1.5
YZ Cas A	2.53	5.747	3.944	11.0	0.6	11.2	0.6
YZ Cas B	1.35	8.357	3.835	10.5	0.6	11.2	0.6
KW Hya A	2.13	6.352	3.906	11.0	0.6	12.1	0.9
KW Hya B	1.48	7.882	3.841	10.5	0.6	12.1	0.9
HD 209458 A	1.15	7.653	3.764	21.4	0.6	21.2	0.9
β Per A	3.09	2.206	3.979	36.4	1.7	35.1	0.9
δ Lib A	3.28	5.070	3.957	10.1	0.4	10.7	0.9
Sun (25 pc)	1.000	6.820	3.746	39.3	0.8	(40.0)	

measurements. Worek had a relatively short period for the third orbit at 2.762 yr, which can be checked from the Hipparcos observations that covered a $\sim$4-yr time interval. Trials of this orbit have failed to produce an improvement in Hipparcos residuals, however, while there are other reasons to prefer a different one, for example, the new photometric solution indicates a significantly more massive and brighter companion star than Worek's. If we accept Worek's γ-velocities as of the right order, there would have to be an order of magnitude increase in the period of the third orbit to increase the mass-function to an appropriate value. A longer period, while able to reconcile the mass with the scale of radial velocity changes, would allow Hipparcos to fail to detect significant changes of

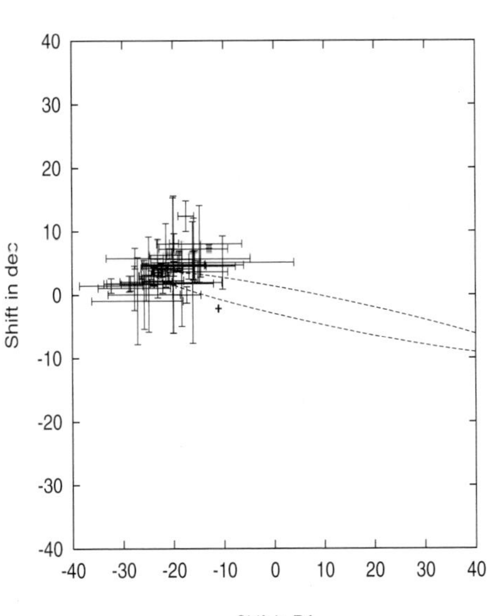

Figure 3. Plot of Hipparcos data trial fit to an orbit of δ Lib about a barycentre with a third body, corresponding to parameters given in Table 2.

Table 2. Tentative parameters of the wide orbit of δ Lib corresponding to the fitting of O – C residuals together with those of the Hipparcos abscissal measures.

Parameter	Value	Parameter	Value
P	23.5 y	Ep.	2448347.75
a	39.3 mas	e	0.49
i	80°.3	ω	317°
Ω	265°	$\Delta\alpha$	17.1 mas
m_c	0.48	$\Delta\delta$	8.8 mas

position, particularly if the third star was close to its apse at the time of observation. A longer period ($\sim$23.5 yr) is in keeping with the O – C oscillation mentioned above.

In Fig. 3, we show the result of a trial fitting in which the period, projected semi-major axis, eccentricity and longitude values were taken from the provisional O – C analysis, and inclination and nodal angle values were optimized from the Hipparcos astrometry, also allowing for a mean shift in the centroid of the orbit from the position given for δ Lib in the Hipparcos catalogue. The corresponding parameters are listed in Table 2. This approach is essentially similar to that of Ribas *et al.* (2001) for R CMa. The s.d. error of an individual residual is 1.8 mas in this fitting, which is close to typical Hipparcos solutions for fixed stars, so that this result compares with the no-orbit solution of the Hipparcos catalogue. The third body mass that would go with the light-time effect mass function, however, would be only 0.48 M$_\odot$ with such a high inclination, and this is inconsistent with the light curve model. Other areas of uncertainty about this have been noted by Budding *et al.* (2004a). VLBI radio observations could help probe this subject through its high accuracy positional information, as with the parallel case of β Per (Lestrade *et al.* 1999).

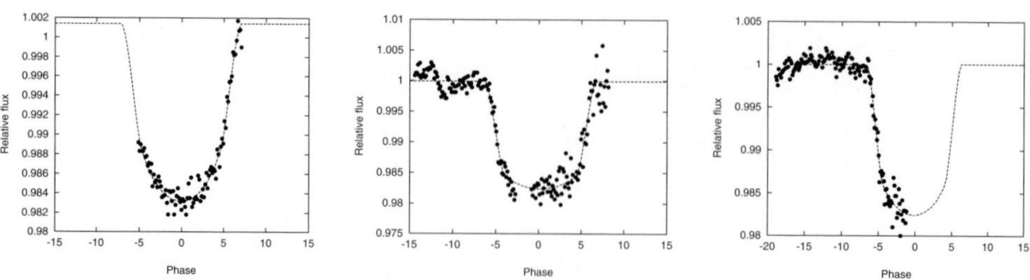

Figure 4. B light curves of three transits across the disk of HD 209458 by its planetary companion as observed by Sullivan & Sullivan (2003). NB: the axes, scaled according to the data distribution of each set, are different for the three transits.

6. Binaries containing very low mass components

The circumstances of this meeting invite consideration of the cosmic frequency of planetary transits. The possibility that they may be more detectable than generally supposed was made by Demircan (private communication), noting the example of the Trojan asteroids in the solar system. The vanishing azimuthal gravity derivative in the restricted two-body problem at the Lagrangian equilateral points $L_{4,5}$ allows matter accumulation, and, noticeably, this should be in the same plane as the two main bodies. This implies that sufficiently careful monitoring of classical eclipsing binaries around phases $\pm 60°$ could reveal the existence of Trojan planets. A planet comparable to Venus would produce a light diminution of only 1 mmag. Current plans for satellite-based photometry, however, are specifying 10^{-4} mag as a precision guideline (Luri *et al.* 2002; Giménez 2003). An interesting case of what a light curve of such an occurrence may resemble comes from the example of HD 209458. In Fig. 4, data of the 'planetary' transits in this system observed from Hawaii in 1999 and 2000 are presented.

Information on the circumstances of these observations was given by Jha *et al.* (2000). The general subject has been reviewed by Perryman (2000) and Black & Stepinski (2003), while Sullivan & Sullivan (2003) referred to http://www.obspm.fr/encycl/encycl.html as a useful website for up-to-date information. HD 209458 was one of a small number of candidate stars showing spectral evidence of low mass companions with reasonably short periods, some proportion of which would be likely to show eclipses. Slight light losses, interpreted as the transits of a planetary sized body, were first observed in September 1999 (Charbonneau *et al.* 2000) and these were confirmed from the Fairborn Observatory in November 1999 (Henry *et al.* 2000) a few days before our data was gathered at Mauna Kea.

The data shown in Fig. 4 were analysed using the CURVEFIT package, most recently reported on by Budding *et al.* (2004b). Speed of evaluation of an algebraic form of fitting function is an advantage when exploring a wide range of parameter space and evaluating the parameter error matrix. That this matrix corresponds, geometrically, to a closed ellipsoid is sufficient and necessary for formal determinacy of the underlying model. Examination of the error matrix is necessary in order to find out that the parametrization neither surpasses nor under-utilizes the information content of the data. Its evaluation is a key feature in this approach (cf. Banks & Budding 1990). Each data set was fitted separately and the number of parameters to be optimized appropriately chosen, with initial guidelines coming from the papers of Henry *et al.* (2000) and Sullivan & Sullivan (2003). We give the curve-fitting details in Table 3. Combining these relative parameters with the absolute data on the primary of HD 209458 given in Table 2 and using also the calibration equations (2.2) and (2.3) we derive the absolute parameters given in Table 4. It may be noted that these values are within the error limits of those given by Brown

Table 3. HD 209458 curve-fit details: optimal parameters and errors.

Parameter	15/11/99		12/11/00		19/11/00	
U	1.0012	0.0009	1.0002	0.0005	1.0003	0.0005
T_0	1497.7993	0.001	1860.8471	0.0002	1867.8965	
r_1	0.1165		0.1159	0.007	0.1161	
r_2	0.0140	0.0006	0.0149	0.0003	0.0147	0.0011
i	86.9	0.4	86.1	0.7	86.1	0.9
u	0.77	0.10	0.38	0.07	0.57	0.1
$\Delta l'$	0.001		0.0018		0.001	
χ^2/ν	0.74		0.97		1.03	

Table 4. Adopted Absolute Parameters for HD 209458.

Parameter	Value	p.e.
Period	3.52474 d	0.00001 d
Epoch (HJD)	2451497.7993	
A	9.837	0.04 $R_\odot$
R_1	1.14	0.01 $R_\odot$
R_2	1.40	0.03 R_{Jup}
i	86°.4	0.3°
u	0.57	0.14
M_1	1.03 $M_\odot$	*
M_2	0.62 M_{Jup}	*
V_{Abs}	4.31	
V	7.653	*
$B - V$	0.594	*
T	5920 K	
Dist.	46.7	0.8 pc

* These values come from Henry *et al.* 2000

et al. (2001), while the mean limb-darkening coefficient u is also within its error limit of modern theoretical estimates (e.g. 0.71 from van Hamme 1993).

7. Concluding remarks

Semeniuk (2000), following a Barnes-Evans flux calibration, pointed out that parallaxes obtained by the eclipse method could have an accuracy surpassing that of the Hipparcos programme if care was taken about the selection of suitable binaries. We have confirmed that the method can be extended, using the calibrations of Popper (1998), to similar MS-like stars, but making use of the UBV system in which large numbers of objects have reliable measurement records. We propose also that the primaries of classical Algol systems can be included in the procedure, provided care is taken in establishing their radii and fluxes. Of course, only one star in $\sim$1000 shows clear eclipses, but the fact that the effect can be discerned, in principle, out to as far as stars can be individually monitored gives the method a cosmic significance not shared by astrometric work of the present generation.

Eclipse data present other factors of relevant interest. For example, careful photometry over a wide spectral range, can allow detailed analysis to reveal the presence of companion objects to the eclipsing pair. This was strongly indicated in the case of δ Lib from $UBVJHK$ light curve fittings. The timing of eclipses can also indicate the presence of physical companions and even lead to estimates of a number of key parameters of the third orbit. In the case of nearby objects, Hipparcos observations can be reanalysed, using these parameters, to check on orbital solutions in order to find the remaining two elements not given by O − C data. Since the absolute orbital scale can then be known, as well as its apparent size on the sky, the parallax directly follows.

Further checks of distances are possible when the stars happen to be radio sources, as was clearly shown by Lestrade *et al.* (1999) for Algol, and indirectly shown in the resolved map of the Algol source produced by Mutel *et al.* (1998). Slee *et al.* (1987) listed a few dozen stars that may fall into the category of having VLBI-determinable parallaxes.

Analysis of the B observations of HD 209458 was carried out to elicit parameter resolution for ground-based photometry, but of higher accuracy than typical. It is worth noting that submillimag accuracy is achievable even with a 0.6-m telescope and 2 min integrations from a ground-based site, provided that site is suitably located: in this case at an altitude of $\sim$4000 m. On this basis, hour-long integrations with a $>$1-m telescope from similar locations should allow μmag accuracies to be approachable for brighter stars. HD 209458 turns out to be high up on the list of nearby eclipsing binaries. In the well-studied lists from Popper (1998) and Semeniuk (2000), it comes in fifth, after the very well-known stars β Aur, β Per, V1143 Cyg and AR Lac, and may be the nearest simple binary with complete eclipses.† This point alone seems a strong indication of the likely high relative frequency of such systems cosmically.

Taking these last two points together, a project to check for less than Jupiter-sized bodies accompanying the brightest stars seems feasible even with ground-based instruments. One possible initial set of targets could come from checking, with near μmag accuracy, the 60° phase regions of the nearest eclipsing binaries. From one point of view, it may be unfortunate that a relatively high proportion are those nearby sunlike binaries, such as UV Psc, RT And and BH Vir, whose light curves are known to be complicated by activity effects. On the other hand, this could simply give an extra reason motivating a prolonged photometric survey. But if complication-free binaries are to be preferred, after V1143 Cyg, whose eclipses are still too shallow to ensure a suitable degree of coplanarity, probably the best candidate would be the 'textbook' example YZ Cas. Otherwise, candidates should probably come from the more general arguments proposed by Henry *et al.* (2000) and others.

Acknowledgements

E.B. would like to acknowledge useful comments and support from Prof. O. Demircan and the Physics Dept., 18th March University of Canakkale, Turkey.

References

Banks, T. & Budding, E. 1990 *Ap&SS*, 167, 221.
Barnes, T.G. & Evans, D.S. 1976, *MNRAS*, 174, 489.
Black, D.C., & Stepinski, T.F. 2003, *Proc. 8th IAU Asian-Pac. Reg. Meet.*, eds. S. Ikeuchi *et al.*, *Astron. Soc. Pacific*, 77.

† We are grateful to J. Southworth for drawing attention to the case of α CrB. Tomkin & Popper's (1986) solution, for the primary at least, yields a parallax 42.2 $\pm$ 0.3 mas, very close to that of β Aur. Its Hipparcos parallax 43.7 $\pm$ 0.8 mas indeed surpasses the latter. The completely eclipsing model used by Tomkin & Popper discarded a relatively large number of eclipse observations, however, and the resulting model for the (non-MS) secondary produces discordant results (e.g. a photometric parallax of 49$\pm$3 mas). It thus seems safer to follow Tomkin & Popper's suggestion of waiting for the difficult, but worthy, task of producing definitive light curves to be completed before including α CrB among the reference stars.

Brown, T.M., Charbonneau, D., Gilliland, R.L., Noyes, R.W. & Burrows, A. 2001, *ApJ*, 552, 699.

Budding, E. 1993, *Introduction to Astronomical Photometry*, Cambridge Univ. Press, Cambridge

Budding, E. & Zeilik, M. 1987, *ApJ*, 319, 827.

Budding, E. *et al.* 2004a, *Ap&SS*, in press.

Budding, E., Rhodes, M., Priestley, J. & Zeilik, M. 2004b, *Astron. Nachr.*, 325, 433.

Charbonneau, D., Brown, T.M., Latham, D.W., & Mayor, M. 2000, *ApJ*, 529, L45.

ESA 1997, *Tycho Catalogue* (ESA SP-1200), 542.

Giménez, A. 2003, *Publ. Canakkale Onsekiz Mart Univ. (Astrophys.)*, 3, 19.

Henry, G.W., Marcy, G.W., Butler, R.P., & Vogt, S.S. 2000, *ApJ*, 529, L41.

Jha, S., Charbonneau, D., Garnavich, P.M., Sullivan, D.J., Sullivan, T., Brown, T.M. & Tonry, J.L. 2000, *ApJ*, 540, L45.

Kopal, Z. 1959, *Close Binary Systems*, Chapman & Hall, London & New York.

Lacy, C.H. 1978, *Ph.D. Thesis*, Univ. Texas, Austin.

Lacy, C.H. 1979, *ApJ*, 228, 817.

Lazaro, C., Arevalo, M.J. & Claret, A. 2002, *MNRAS*, 334, 542.

Lestrade, J-F. *et al.* 1999, *A&A*, 344, 1014.

Luri, X., Figueras, F. & Torra, J. 2002, *EAS Publ. Ser.*, eds. O. Bienaymé and C. Turon, EDP Sciences, 2, 171.

Malkov, O.Y. 1993, *Bull. Inf. Centre Donn. Stellaires*, 42, 27.

Mutel, R.L., Molnar, L.A., Waltman, E.B. & Ghigo, F.D. 1998, *ApJ*, 507, 371.

Nordström, B. & Johansen, K.T. 1994, *A&A*, 291, 777.

Perryman, M.A.C. 2000, *Rep. Prog. Phys.*, 63, 1209.

Popper, D.M. 1998, *PASP*, 110, 919.

Ribas, I., Arenou, F, & Guinan, E.F. 2002, *AJ*, 123, 2033.

Russell, H.N. 1948, *Centennial Symposia* ed. C. Payne-Gaposchkin, Harvard University Press, Cambridge, Mass.

Semeniuk, I. 2000, *Acta. Astron.*, 50, 381.

Slee, O.B., Nelson, G.J., Stewart, R. T., Wright, A.E., Innis, J.L., Ryan, S.G., & Vaughan, A.E. 1987, *MNRAS*, 229, 659.

Sullivan, D.J. & Sullivan, T. 2003, *Bullic. Astron.*, 12, 145.

Tomkin, J. & Popper, D.M. 1998, *AJ*, 91, 1428.

van Hamme, W. 1993, *AJ*, 106, 2096.

Wilson, R.E. and Devinney, E.J. 1971, *ApJ*, 166, 605.

Worek, T.F. 2001, *PASP*, 113, 964.

Discussion

DIMITRI POURBAIX: In your data for the δ Librae system you have looked for a third body over a period of 23 years. Have you checked that the Hipparcos proper motion is discrepant with respect to the Tycho-2 proper motion? Because if the period is indeed that long with respect to the length of the mission, then you should see a departure of the two proper motions. One based on just 3 years and one based on 100 years.

ED BUDDING: Yes, I'm aware of that technique. This approach has come from the study of Ribas *et al.* [2002, AJ, 123, 2033] on R CMa. The great emphasis in that paper is on the apparent shift in position. They had to sum the proper motions down to just 4 individual points in order to get error bars that gave a significant effect. Given that you have only got at most a 4-year time interval in a 23-year orbit, the change of the speed on the sky is going to be a bit small. You can certainly get the mean value from a century of observation. The very long-term observations – going back a 100, 150 years – have appreciable errors on the milli-arcsecond scale.

DIMITRI POURBAIX: During the 3 years you will see essentially the orbiter motion, so what is quoted as the proper motion in the Hipparcos catalogue is likely to be confused with the orbiter motion. Whereas in the 100 year period of the proper-motion-based Washington catalogue, you would get just a proper motion, so you should see a departure of the Hipparcos proper motion with respect to that proper motion.

ED BUDDING: I'm not sure how much that departure differs within the scale of the error bars involved. I have not checked this in detail, as we wanted to concentrate first on the possible positional orbit. We've found one, but I think there could be more. This isn't necessarily a unique solution for this star.

DIMITRI POURBAIX: No, but if you take the Tycho-2 proper motion, that would give you two parameters, thus giving more chance to detect your actual orbiter motion. So you would reduce the number of consistent orbiter solutions.

JOHN SOUTHWORTH: About HD 209458 being near the star with complete eclipses: α CrB shows complete eclipses – about 0.05 mag deep. It might be closer, but it is too bright for modern telescopes, so good luck observing it.

ED BUDDING: It wasn't on my list. The four which I found that were nearer were β Aur, Algol, AR Lac and one other, V(some-big-number) Cyg. Those came from Popper's list and Semeniuk's list.

Transits of Venus: New Views of the Solar System and Galaxy
Proceedings IAU Colloquium No. 196, 2004
D.W. Kurtz, ed.

© 2004 International Astronomical Union
doi:10.1017/S1743921305001560

High precision pulsar astrometry and its applications

Walter Brisken

National Radio Astronomy Observatory, PO Box O, Socorro, NM 87801, USA

Abstract. Very long baseline interferometry (VLBI) is a radio astronomy technique that allows imaging and astrometry at milliarcsecond or better resolution. The last decade has seen improvements in VLBI techniques and equipment which have enabled precision astrometry of pulsars. Most notably, VLBI arrays have become more sensitive and calibration has become more accurate. Pulsar astrometry using VLBI can readily yield a proper motion and in many cases a trigonometric parallax as well. These observables have have numerous applications which justify the difficulty of the measurements. The goals of pulsar astrometry include: 1. determination of the distribution of pulsar birth velocities, a critical probe of supernova asymmetry; 2. improvement in understanding of the ISM through observations of pulsar bow shocks, interstellar scattering, and timing dispersion measures; and 3. understanding characteristics of individual pulsars though knowledge of their distance and velocity. I will discuss these science cases, some recent results, and prospects for better astrometry in the near future.

1. Introduction

Neutron stars are extreme astrophysical objects exhibiting mass densities greater than that of atomic nuclei and magnetic fields approaching the electron-positron pair production limit making them distant particle physics laboratories. When these high density objects are in close orbits with other stars, especially other neutron stars, they become fantastic laboratories for testing theories of gravity, such as General Relativity. The high velocities and spins imparted to neutron stars provide fossil clues about their origin. Distances and velocities of these objects figure crucially in many neutron star research areas, including all of the above.

Most known neutron stars are radio-emitting pulsars owing to their regular pulses that provide unique signatures for detection. Normal isolated pulsars have spin periods usually between 20 ms and 2 s. Recycled pulsars, those that have undergone 'spin-up' by accretion of matter from a companion star, have spin periods as short as 1.6 ms. This paper will concentrate mainly on normal pulsars.

2. Pulsar timing

Accurately measuring the arrival times of pulses can yield a wealth of information about a pulsar. A measure of the period P and its period derivative $\dot{P}$ provides estimates for the age, $\tau_{\mathrm{char}} = P/2\dot{P}$ and magnetic field, $B \sim 3.2 \times 10^{19}\sqrt{P\dot{P}}$ (in Gaussian units). The changing pathlength caused by Earth orbiting the sun imparts a one year period modulation of magnitude $\sim 500\,\mathrm{s}\cos\beta$ to the timing measurements. The magnitude of the modulation reveals the ecliptic latitude β of the pulsar and the phase provides its ecliptic longitude. A pulsar's proper motion may be derived by observing the change in these parameters over a very long period (possibly decades). Timing noise limits the precision of timing-derived proper motions to a few milliarcseconds (mas) per year for normal pulsars (cf. Hobbs et al. 2004). The much older recycled pulsars have very little timing noise and hence can provide astrometry that is up to 100 times better, which in some cases can even measure the curvature of the pulse wavefront providing timing parallaxes.

Pulse arrival times are different at different observing frequencies. This is due to propagation through ionized interstellar medium (ISM) which imparts a delay of 4.12×10^3 s $\frac{\text{DM}}{\nu^2}$ at a frequency ν in MHz for a dispersion measure of DM measured in parsecs cm^{-3}. The DM is the integrated column density of electrons along the line of site. A simple model of the electron distribution within the Galaxy (Taylor & Cordes 1993; Cordes & Lazio 2002) provides a method to estimate the distance to a pulsar based on its readily measured DM. Pulsars with known distances are crucial in determining the details of the electron density models, making independent distance measurements important. Based on parallax measurements of nearby pulsars, this model is thought to produce distances accurate to about 40% (Brisken et al. 2003b). Distances to pulsars that are far from the galactic plane ($|z| > 150$ pc) are even less accurate.

3. VLBI

The desire for angular resolution at radio wavelengths that is comparable with that of optical instruments has led to the development of large radio interferometers such as the Westerbork Radio Synthesis Telescope, the Australia Telescope Compact Array, and the Very Large Array with baselines up to 30 km, providing arcsecond resolution at frequencies as low as 1.5 GHz. Nothing fundamentally prevents even larger arrays from producing measurements at even higher resolution. The practical problems of transporting data and synchronizing clocks at distant stations have led to what is now known as Very Long Baseline Interferometry. VLBI has become a mature science with NRAO† Very Long Baseline Array (VLBA) operating as a dedicated VLBI array and other consortia of radio telescopes scheduling several VLBI sessions per year. Below is a very brief description of some of the standard methods used to calibrate and image radio interferometry data. These basic techniques are described in much more detail in *Synthesis Imaging in Radio Astronomy II* (1999).

3.1. *Visibilities and imaging*

The fundamental measurement made by a radio interferometer is the fringe visibility, a complex number produced by combining two antennas' voltage sampled data streams, E_i and E_j at a correlator and integrating the correlation function over a time T,

$$V_{ij} = \langle E_i(t) E_j^*(t) \rangle_T . \tag{3.1}$$

An N element array produces $N(N-1)/2$ independent visibility measurements for each integration time. Most modern correlators are spectral line correlators, concurrently producing visibilities for each of several frequency channels. The projected baseline vector, (u, v), is recorded with each visibility measurement for use in imaging and astrometry.

The visibility measured by a pair of antennas is the measurement of a single spatial frequency component of the brightness distribution,

$$V_\nu(u, v) = \int \int \frac{I_\nu(l, m)}{\sqrt{1 - l^2 - m^2}} e^{-2\pi i \nu (ul + vm)/c} dl\, dm, \tag{3.2}$$

where (l, m) are the direction cosines parameterizing the sky distribution, and c is the

† The National Radio Astronomy Observatory is a facility of the National Science Foundation operated under cooperative agreement by Associated Universities Inc.

speed of light. A collection of visibility measurements can be inverse Fourier transformed to reveal a 'dirty' image of the sky in the direction of interest. The point spread function is readily determined by Fourier Transforming the (u, v) plane sampling. The 'Clean' algorithm can then be used to deconvolve the dirty image, restoring the brightness distribution, $I_\nu(l, m)$. For this inversion to be faithful, an appropriate sampling of the visibility function, $V_\nu(u, v)$ must be made. Earth's rotation causes each pair of antennas to trace elliptical paths in the (u, v) plane.

3.2. *Phase-referencing*

VLBI data is notoriously difficult to calibrate properly, though better hardware, improved algorithms, and increased bandwidth has over the years have simplified its use greatly, especially for imaging experiments. Astrometry remains quite tricky, especially at frequencies below 2 GHz where the ionosphere can cause rapid phase fluctuations and at frequencies greater than 15 GHz where the troposphere causes similar grief.

The simplest form of calibration, phase-referencing, involves nodding the antennas between a bright point-like calibrator source of accurately known position and the potentially much weaker target source. The nodding time depends on frequency but ranges usually between 30 s and 5 min. The visibility corresponding to a single point source at location (l, m) relative to the center of the field of view is

$$V_\nu^{\text{point}}(u, v) = \frac{S_\nu}{\sqrt{1 - l^2 - m^2}} e^{-2\pi i \nu (ul + vm)/c}, \tag{3.3}$$

Where S_ν is the source flux density. Calibration involves determining in a least-squares sense the *station-based* complex gain factors, g_i that minimize

$$\chi^2 = \sum_{i,j} \left| g_i g_j^* V_{ij}^{\text{observed}} - V_{ij}^{\text{model}} \right|^2 \tag{3.4}$$

by observing a calibrator source with known V_{ij}. Calibration proceeds by interpolating the complex gains derived from the calibrator observations and applying them to the target observations. The amplitudes are almost always more stable than the phases in VLBI observations, and it is the phases that contain the astrometric information, so the remainder of this section will focus on phase calibration.

3.3. *In-beam calibration*

Phase-referencing leaves uncorrected spatial gradients in the phase front caused by the non-uniform ionosphere and troposphere. The calibration is improved by observing a calibrator source as near the target as possible. At the lower frequencies, where the primary beam of the individual telescopes is large enough, one can often find with the aid of Very Large Array imaging compact sources within the same primary beam as the target. These sources can be much weaker (10 mJy rather than about 80 mJy at 1.5 GHz) since the bulk of the phase variation can be removed by nodding to a brighter, more distant calibrator. This technique was first used on a pulsar in the parallax measurement of B0919+06 by Chatterjee et al. (2001).

3.4. *Ionosphere calibration*

At frequencies below about 5 GHz, the phase fluctuations are dominated by the ionosphere. The phase of a visibility of a point source at location (l, m) on a baseline (u, v) at frequency ν is

$$\phi = \frac{\nu}{c} (u \, l + v \, m) + \epsilon(\alpha, \delta, t, \nu) + \text{noise}. \tag{3.5}$$

The error term, ϵ has contributions from the troposphere, the ionosphere, and the instrument,

$$\epsilon(\alpha,\delta,t,\nu) = \underbrace{A(\alpha,\delta,t)\cdot\nu}_{\text{Troposphere}} + \underbrace{B(\alpha,\delta,t)\cdot\nu^{-1}}_{\text{Ionosphere}} + \underbrace{C(\alpha,\delta,t,\nu)}_{\text{Instrument}}. \tag{3.6}$$

The contribution from the instrument is very stable for modern antenna arrays and can usually be considered constant. Since the phase error due to the ionosphere has a different frequency dependence ($\propto \nu^{-1}$) than that of the troposphere or any other pure delay ($\propto \nu$), it can be measured and removed given a strong enough target source whose observations span a large enough frequency range. Note that the geometric term that is being determined has a frequency dependence identical to that of a pure delay making a similar calibration for the troposphere not possible. A similar technique can be used by simultaneously observing at 2.3 and 8.4 GHz. Pulsars are too weak to be observed at these frequencies, but there is enough bandwidth to span in the 1.2 to 1.7 GHz band. This technique was developed for measuring the parallax to pulsar B0950+08 (Brisken et al. 2000), see Fig. 1 and has been extended to determine parallaxes for 8 other pulsars (Brisken et al. 2002).

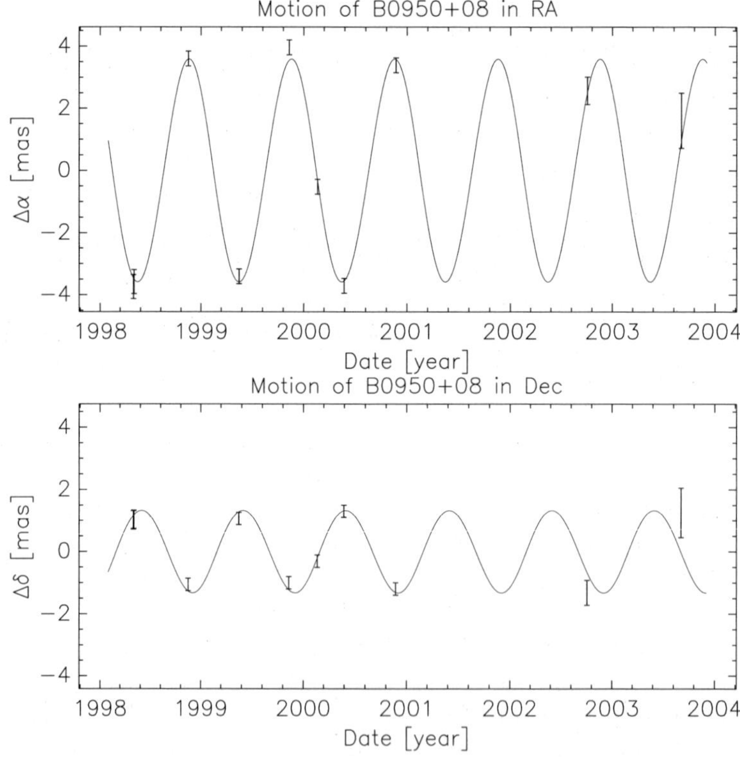

Figure 1. The parallax fit of Pulsar B0950+08 using ionosphere calibration (Brisken et al. 2000). The proper motion and parallax values fit for this pulsar are $\mu_\alpha = -2.07\pm0.07\,\mathrm{mas\,yr^{-1}}$, $\mu_\delta = 29.37\pm0.05\,\mathrm{mas\,yr^{-1}}$, and $\pi = 3.81\pm0.07\,\mathrm{mas}$ respectively. Note that the fit proper motion has been subtracted to show more clearly the parallax.

4. The VLBA

The VLBA consists of ten identical 25-m radio telescopes operating between 327 MHz and 90 GHz. The maximum baseline is 8600 km, yielding 5×11 milliarcsecond (mas) resolution at 1.5 GHz for sources at moderate ($\sim 20°$) declination. It nominally observes over a 32 MHz bandwidth, but can observe at twice or four times that bandwidth at a correspondingly reduced duty cycle. Because of its year-round availability, simple scheduling, flexible tuning capabilities and reliable operation it has been the instrument of choice for pulsar VLBI observations.

4.1. *VLBA pulsar observations*

Pulsars have very steep spectra, $S_\nu \propto \nu^{-\alpha}$ for $1 < \alpha < 2.5$ usually. Thus pulsar observations are usually made at ~ 1.5 GHz or sometimes 5 GHz as a compromise between source brightness and instrument resolution. Scattering, especially near the Galactic center, can be a problem at frequencies below 5 GHz. As observing bandwidths increase, the observing frequency will likely increase as calibration is easiest around 8 GHz. The VLBA correlator in Socorro, NM has a pulsar gate which allows a factor of 3 to 5 increase in signal-to-noise ratio by blanking the signal inputs to the correlator during the off-pulse portions of each pulsar rotation.

5. Applications of pulsar astrometry

The following sections describe some applications of pulsar astrometry and some recent results.

5.1. *Pulsar velocity distribution*

Pulsars have typical velocities of 100 to 300 km s^{-1}, far faster than the class of stars thought to be their progenitors. It is fairly clear based on this simple observation and on positional properties of pulsar-supernova remnant associations that supernovae impart significant momentum to the pulsars they form. Determining the distribution of pulsar birth velocities would improve our understanding of the core collapse process.

Sample biases greatly challenge any effort to measure the distribution of velocities. Many proper motion studies are biased toward the brighter, easier to study pulsars, biasing toward the slower moving pulsars that loiter in the solar neighborhood. The more distant pulsars tend to be traveling more along the line of sight than in the plane of the sky. Since only the transverse velocity is amenable to observation, this tends to lower the observed velocity. Also pulsars with large z velocities tend to leave the sample volume, whereas pulsars with velocities parallel to the plane of the Galaxy tend to enter and leave the sample volume at almost the same rate. Dispersion measure based distances are inaccurate for pulsars with $|z| > 200$ pc since they are above the bulk of the Galactic electron content.

Brisken et al. (2003b) measured proper motions of twenty eight pulsars using the VLA. Wide-field images made at 1.4 GHz typically contain about 15 compact ($< 1"$) background objects that were used to define a coordinate grid on the sky near the pulsar. The pulsar's motion against this grid of points can then be measured to a precision of roughly 1 to 10 mas yr^{-1}, depending on the brightness of the pulsar and somewhat on its location in the sky. These results were combined with six additional pulsars with high quality VLBI proper motions and parallaxes in a velocity distribution analysis. In order to minimize the bias related to pulsars with large z velocities, only the velocity component parallel to the plane of the Galaxy was used. This also avoids the need to disentangle acceleration due to the gravitational potential of the Galaxy from the velocity.

Several model forms were fit to the observed proper motion and distance estimates. The best fit among those tested was a velocity distribution with two normally distributed components. The values obtained compared reasonably with an independent method that used forward evolution of a sample of pulsars to determine the model (see Fig. 2).

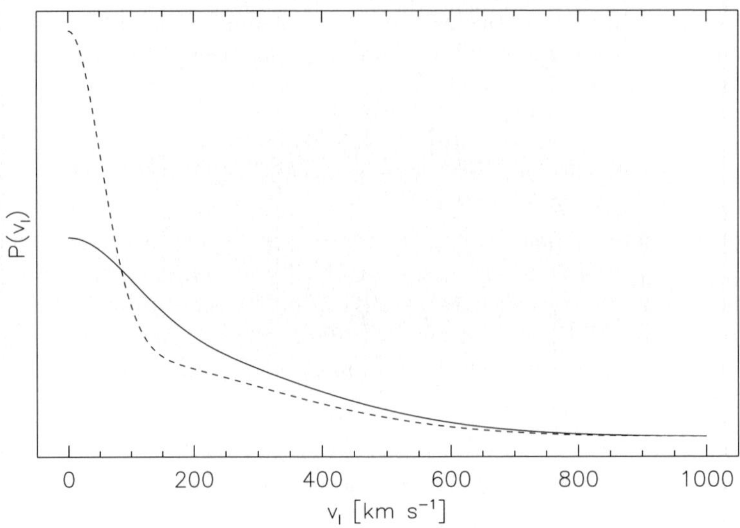

Figure 2. The best fit velocity distribution from Brisken et al. (2003b) is shown by the solid curve. In this distribution, 20% of pulsars are contained in a normal distribution with a width of $99\,\mathrm{km\,s^{-1}}$ and the remainder in a normal distribution of width $294\,\mathrm{km\,s^{-1}}$. This is compared to the fit of Arzoumanian et al. (2002) shown dashed. They found 40% in a $52\,\mathrm{km\,s^{-1}}$ distribution and 60% in a $289\,\mathrm{km\,s^{-1}}$ distribution.

5.2. *Galactic electron distribution*

The ionized interstellar medium retards the arrival of radio waves by an amount Δt given by

$$\Delta t = \frac{e^2}{2\pi m_e c \nu^2} \int_0^d n_e dl, \tag{5.1}$$

where m_e and e are the mass and charge of an electron, ν is the frequency of observation, and n_e is the local electron density. The total integrated column density of electrons toward a pulsar is called the dispersion measure (DM) and is directly observable with multi-frequency timing observations. Thus given a model for the electron distribution within the Galaxy, $n_e(d, l, b)$ and the DM, one can determine the pulsar's distance. Using limited pulsar distance data, the distribution of pulsar dispersion measures and other physical constraints, Taylor & Cordes (1993) published a model of the Galactic electron distribution that included spiral arms.

Recent VLBI parallaxes, as well as parallaxes derived from the timing of stable millisecond pulsars, a large number of new pulsar discoveries, more and improved scattering measurements as well as a handful of other constraints have yielded a newer model called NE2001 (Cordes and Lazio 2001). Ongoing VLBI parallax measurements will likely again double the number of direct parallax distances and improve the model details out to greater distances. See Fig. 3.

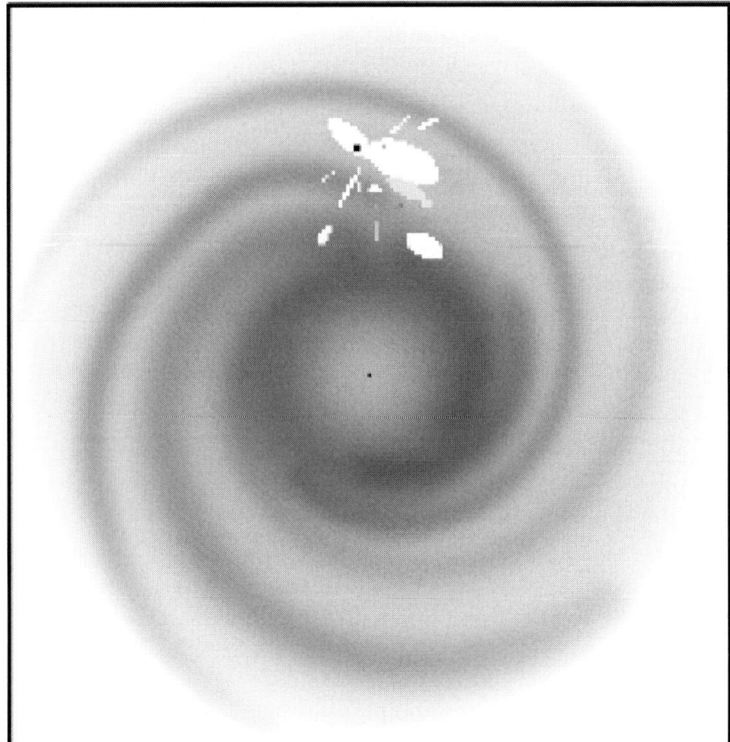

Figure 3. The Galactic electron density plotted over an area of $30 \times 30\,\mathrm{kpc}^2$. The greyscale levels are logarithmic. Around the solar neighborhood, small structures have been modeled such as the Gum Nebula and Vela SNR. Also some local low density regions have been included in the model. Figure is from Cordes & Lazio (2002).

5.3. *Pulsar / supernova remnant association*

Pulsars and supernova remnants provide complementary information about supernova explosions when two such objects can be connected. The SNR provides information about the ISM local to the supernova and can in some cases be used to obtain distance and age information. The pulsar's proper motion can be used to determine the center-point of the explosion. Pulsar astrometry has recently firmly established three pulsar/SNR associations. Since the discovery of Pulsar B1951+32 its association to SNR CTB80 has been assumed based on the apparent interaction of the two objects (see Fig. 4). A recent proper motion by Migliazzo et al. (2002) has confirmed the predicted direction of motion of the pulsar. Using long term timing data for pulsar J0538+2817, its association with SNR S147 has been strengthened (Kramer et al. 2004). This association has led to a measurement of this pulsar's initial spin period of 139 ms based on its age and proper motion.

The distance to pulsar B0656+14 was sought because of the possibility of a radius measurement for this object (see section 5.5). This pulsar lies near the center of the Monogem ring, a large X-ray bright ring that was confirmed to be a SNR by Plucinsky et al. (1996). Models place the ring at a distance of about 300 pc. The association of these two objects was not initially considered due to the discrepant distances when using the pulsar's DM as a proxy for its distance. A five-epoch VLBA observation using a bright in-beam reference source yielded a distance of 288^{+33}_{-27} pc (Brisken et al. 2003a). Thorsett

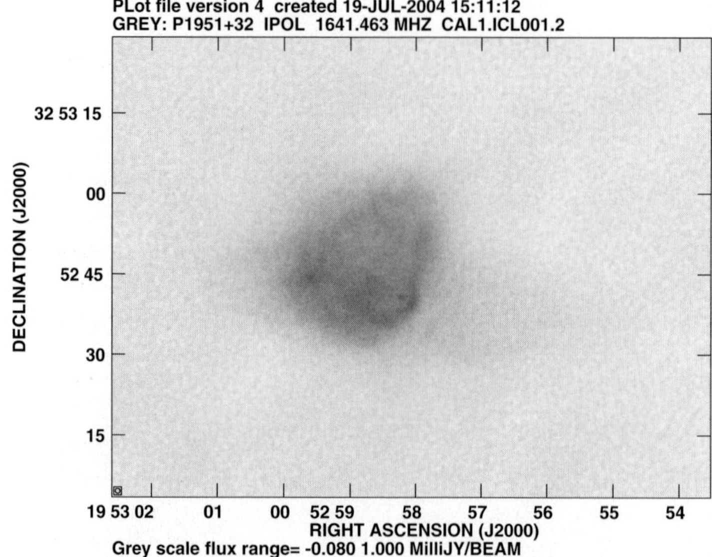

Figure 4. The central region of SNR CTB80. Pulsar B1951+32 (point source near lower right of the central nebula) was found to be moving toward the arc-like structure right of the pulsar. The nature of this arc is still uncertain. Image courtesy of Ben Zeiger, made from archival VLA data.

et al. (2003) revived the association hypothesis and with this provided an age for the Monogem ring.

5.4. *Core collapse*

Supernova core collapse leaves its signature in the large velocities and the spins of the pulsars they produce. The misalignment angle of a pulsar's spin and velocity vectors probes the process by which a neutron star is kicked. Most models that produce the significant velocities that are observed require the collapse to be asymmetric. Spruit & Phinney (1998) postulated that this asymmetry would cause correlation between spin and velocity that could be observed. With the advent of high-resolution X-ray imaging from space, the spin axes of young pulsars can be probed by examining the symmetry of the X-ray wind nebula surrounding some pulsars. This involves fitting observed X-ray data to models consisting of one or two equatorial disks and/or polar jets (Romani & Ng 2003; Ng & Romani 2004; see Fig. 5). Combined with high precision astrometry of the pulsar, this idea is now being tested. The spin-velocity misalignment angle is plotted against measured initial spin period for 5 pulsars in Fig. 6. These suggest that the timescale over which rotational averaging aligns the two vectors is about 2 to 3 s.

5.5. *Neutron star properties*

The determination of some neutron star properties, such as luminosity, depend directly on the distance to the pulsar. Pulsar B0656+14 is an X-ray bright source consisting of a two-component blackbody and a power-law spectrum. The two blackbody components are interpreted as those from the neutron star surface and the hotter polar caps; the power-law is likely magnetospheric in origin. Recent fits for the surface temperature by Marshall & Shultz (2002) suggest a surface temperature of $T_\infty = 8.0 \pm 0.3 \times 10^5$ K. A distance measurement would allow a direct radius measurement given the well determined blackbody measurements. The distance of 288 pc measured with the VLBA implies a

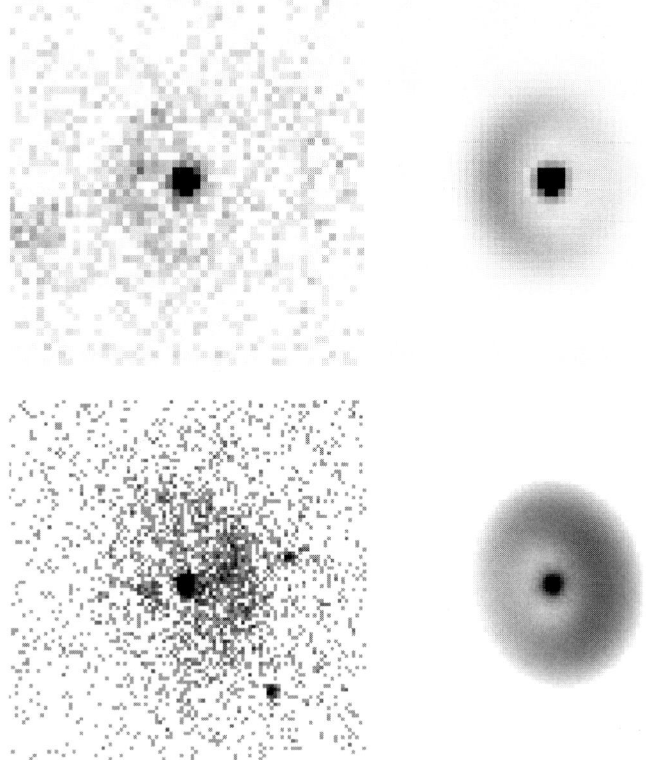

Figure 5. Top: a cleaned Chandra X-ray Observatory image of Pulsar J1930+1852 and its best fit one-disk model. Bottom: The same for Pulsar J2229+6114. Images courtesy of Roger Romani.

blackbody radius in the range 6.9 to 8.5 km, depending on the details of the modeling (Brisken et al. 2003a). Since this radius is not supported by any reasonable neutron star equation of state, this measurement was turned around and used as a probe of the neutron star atmosphere, which distorts the blackbody spectrum. A fit by Anderson et al. (1993) using a magnetized hydrogen atmosphere model of Shibunov et al. (1993), places the radius at $R_\infty = 13$ km given the new distance measurement, which is consistent with the canonical neutron star radius. Given uncertainties in the modeling, any radius between about 13 and 20 km is allowed by the observations.

Acknowledgements

I'd like to thank Don Kurtz and Gordan Bromage for organizing such an interesting conference that addressed a wide range of interests and for providing such a unique opportunity to view the Transit of Venus in historical context.

References

Anderson, S. B., Córdova, F. A., Pavlov, G. G., Robinson, C. R. & Thompson, R. J. Jr. 1993, ApJ, 414, 867
Arzoumanian, Z., Chernoff, D. F., & Cordes, J. M. 2002, ApJ, 568, 289
Brisken, W. F., Benson, J. M., Beasley, A. J., Fomalont, E. B., Goss, W. M. & Thorsett, S. E. 2000, ApJ, 541, 959

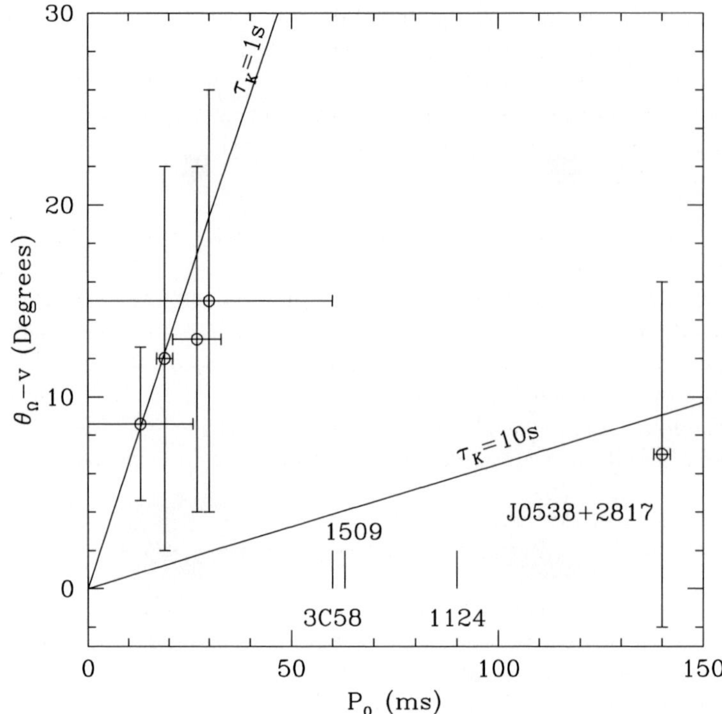

Figure 6. Spin-velocity angle misalignment plotted against pulsar initial spin period for 5 pulsars. These data suggest a 2- to 3-s kick duration.

Brisken, W. F., Benson, J. M., Goss, W. M., & Thorsett, S. E. 2002, ApJ, 571, 906

Brisken, W. F., Thorsett, S. E., Golden, A., & Goss, W. M. 2003a, ApJ, 593L, 89

Brisken, W. F., Fruchter, A. S., Goss, W. M., Herrnstein, R. M., & Thorsett, S. E. 2003b, AJ, 126, 3090

Chatterjee, S., Cordes, J. M., Lazio, T. J. W., Goss, W. M., Fomalont, E. B. & Benson, J. M. 2001, ApJ, 550, 287

Cordes, J.M. & Lazio, T. J. W. 2002, astro-ph/0207156

Hobbs, G., Lyne, A. & Kramer, M. 2004, IAU Symposium 218, 431

Kramer, M., Lyne A. G., Hobbs, G., Löhmer, O., Carr, P., Jordan, C., & Wolszczan, A. 2004, ApJ, 593L, 31

Marshall, H. L. & Shultz, N. S., 2002, ApJ, 574, 377

Migliazzo, J. M., Gaensler, B. M., Backer, D. C., Stappers, B. W., van der Swaluw, E. & Strom, R. G. 2002, ApJL, 567, L141

Ng, C. Y. & Romani, R. W. 2004, ApJ, 601, 479

Plucinsky, P. P., Snowden, S. L., Aschenback, B., Egger, R., Edgar, R. J., & McCammon, D. 1996, ApJ, 463, 224

Romani, R. W. & Ng, C. Y. 2003, ApJ, 585L, 41

Shibunov, Y. A., Zavlin, V. E., Pavlov, G. G., Ventura, J. & Potehkin, A. 1993, in Isolated Pulsars, eds. Van Riper, K. A., Epstein, R. I., & Ho, C., Cambridge University Press, 174

Spruit, H. C. & Phinney, E. S. 1998, Nature, 393, 139

Synthesis Imaging in Radio Astronomy II, eds. Taylor, G. B., Carilli, C. L. & Perley, R. A. 1999, ASP conf. series 180

Taylor, J. H. & Cordes, J. M. 1993, ApJ, 411, 674

Thorsett, S. E., Benjamin, R. A., Brisken, W. F., Golden, A., & Goss, W. M. 2003, ApJ, 592L, 71

Discussion

MIKHAIL MAROV: You're completely right saying that the ionospheric, and partially tropospheric, effects crucially influence the accuracy of data collected, and you also mention that GPS suffers very much from that. It is a most important component influencing the data accuracy. So GPS people are now eliminating, or at least decreasing, the effect using a multiple – say double or triple – frequency approach. Will the same idea improve the data accuracy of VLBI?.

WALTER BRISKEN: That's right. We are doing this in two stages: The first uses the published maps from GPS observations to calibrate out the bulk of the delay, while leaving us just differential uncorrected data. The other is that geodesists and other astrometrists have been using dual frequency observations for many years. Usually at S-band at 2.4 GHz and X-band at 8 GHz; this has worked very well. A technique I developed from my thesis is to use multiple frequencies within just the L-band at 1.4 GHz and 1.7 GHz; the signature of the ionosphere is so strong at these lower frequencies that just within that 300 MHz we can already measure and remove the effect of the ionosphere.

MIHAIL MORAV: Is it then possible to avoid drifting anomalies in the ionosphere, or some kind of irregularities stirred up by solar activity, especially sudden disturbances?

WALTER BRISKEN: Yes, we have actually observed these MSTIDs – medium scale travelling ionospheric disturbances (as they are called). If you map the magnitude of the differential ionosphere on 30-second timescales, you see many degrees of phase change. So we do see those, and we are in some cases able to correct for them.

CORYN BAILER-JONES: I'd heard that somebody had been using VLBI of radio stars to look for exo-planets. Do you know anything about this?

WALTER BRISKEN: I have not heard about that observation. I assume it's using the wobble technique. I suppose if you are able to get it at 8 GHz where the ionosphere is less problematic and your resolution is five times higher, then you should easily be able to get down to 50 micro-arc seconds resolution. Recently someone demonstrated 10 micro-arc second astrometry in the measurement of the speed of gravity. It's something you may have heard about where Jupiter passed very near a quasar. They wanted to look for the apparent motion of the quasar as a result of Jupiter's gravity influence. So 10 micro-arc seconds is currently the state of the art and I assume that people have started to use this to look for exo-planets. But it's not a very easy technique – there's a lot of work involved – so it's probably not done on more than 10 objects or so at a time.

DON KURTZ: Walter, if I can go back to Monday when Myles and Lena told us we know the astronomical unit to a precision of metres: Let's pretend that we didn't have their technique. We know that ten years ago planets were artificially discovered around a pulsar because the barycentre correction was wrong. How precisely could you determine the AU with your parallaxes?

WALTER BRISKEN: That's a good question. Unfortunately, none of the pulsars that have timing distances also have VLBI distances, but I guess those are probing the same AU anyway. I'm not actually sure what you would do.

DON KURTZ: If the astronomical unit were wrong – if you had the barycentre in the wrong place   you would see an annual motion in your positions for the pulsar. So, at what level would you detect that, was the question. For the public that might be an interesting number to know. I was very intrigued when you said that you can pick up the angular momentum change in the Earth when you release water from a dam. At what level are you timing the Earth's period with the VLBA?

WALTER BRISKEN: The Earth orientation service does make daily, maybe hourly measurements of the orientation of the Earth and it shows up in this data. Maybe Myles Standish knows more?

DON KURTZ: Myles, do you know the Earth's period to micro-second precision, nano-second precision?

MYLES STANDISH: [response unrecorded]

DON KURTZ: Then how can you detect the change in the Earth's rotation rate with just the release of water from a dam?

MYLES STANDISH: I don't know. I didn't say that!

[general laughter]

DON KURTZ: I think that it's a change at the level about a millisecond for El Niño – but that's from massive flows in the Pacific – so it's an amazing claim.

WALTER BRISKEN: I think it's micro-seconds accuracy. We can get the axis of the Earth in the reference frame of the ICRF to the order of 100 or maybe 10 micro-arcseconds, and if it's an effect that integrates with time such as a rotation rate – then observations over a period of time may show this effect.

Walter Brisken observing the transit

Transits of Venus: New Views of the Solar System and Galaxy
Proceedings IAU Colloquium No. 196, 2004
D. W. Kurtz, ed.

© 2004 International Astronomical Union
doi:10.1017/S1743921305001572

Statistical calibrations of trigonometric parallaxes

T. Tsujimoto[1], Y. Yamada[2] and N. Gouda[1]

[1]National Astronomical Observatory, Mitaka-shi, Tokyo 181-8588, Japan
email: taku.tsujimoto@nao.ac.jp, naoteru.gouda@nao.ac.jp

[2]Department of Physics, Kyoto University, Kyoto 606-8502, Japan
email: yamada@scphys.kyoto-u.ac.jp

Abstract. We examine statistical methods for calibrating trigonometric parallaxes to retrieve the absolute magnitudes of stars, using Monte Carlo simulations. Here we consider the case of the zero-point of the period-luminosity relation for Cepheid variables. The method originally proposed by Ratnatunga & Casertano was revisited by introducing a *realistic* density distribution of sample stars belonging to the catalogue through prior calculations of the photometric distance for each star. It is found that our method gives an unbiased estimate, regardless of any dispersions in their absolute magnitude. We further investigate the reliability of results which depend on the accuracy of parallax. Our finding is that the accuracy ($\sim$1 mas) of Hipparcos parallaxes is not enough to obtain a reliable result due to a large variation among different ensembles of stars. More precise determination of parallaxes to an accuracy of 200 μas at least, which will be easily realized by the ongoing astrometric space satellites, will give a precise zero-point together with a dispersion in absolute magnitude.

1. Introduction

The Hipparcos satellite has brought us a new era for the distance determination using the trigonometric parallaxes, π, of stars. Before Hipparcos, the ground-based observations offered π only for stars within a few tens pc. We have now reached the stage where we can obtain π for stars with their distances extending to $\sim$12 kpc. As a result, the Hipparcos catalogue involves many valuable stars, in which Cepheid variables and RR Lyrae stars which serve as primary distance indicators are included.

It is well known that Cepheids obey the period-luminosity (PL) relation. The zero-point for PL relation gives a distance modulus $\mu_{\rm LMC}$ to the Large Magellanic Cloud (LMC), which is no doubt an important step for determining the distances to distant galaxies and thus for the Hubble constant (H_0). Instead of the indirect calibrations of the zero-points so far, Hipparcos has enabled us to derive them directly from π of Galactic Cepheids for the first time. However, in fact, direct calibrations have confronted a serious problem which comes from the fact that almost all of π for these stars are measured with very large errors, including negative parallaxes. In such cases, the distance d for each star cannot be calculated from $d(\rm pc) = 1/\pi$. We therefore need to deduce the physical quantities (e.g., the zero-points) statistically from an ensemble of Hipparcos parallaxes (e.g., Smith 1987, 1988).

To retrieve statistically an *unbiased* estimate is indeed a difficult task (e.g., Smith 2003). Many studies have been done since Roman (1952) and Jung (1971) (see the review by Arenou & Luri 1999). Essentially, two different approaches have been investigated since then, and both were applied to the calibrations with Hipparcos parallaxes. Ratnatunga &

Casertano (1991) have proposed a maximum likelihood method (hereafter, ML method) for correcting the notable Lutz-Kelker bias (Lutz & Kelker 1973), allowing full use of low-accuracy and negative parallaxes. Tsujimoto, Miyamato & Yoshii (1998) applied their method to Hipparcos RR Lyraes, and found the absolute magnitude $M_V(\mathrm{RR})$ of RR Lyraes at $[\mathrm{Fe/H}] = -1.6$ is 0.59, which corresponds to $\mu_{\mathrm{LMC}} = 18.41\,\mathrm{mag}$ with the data of LMC RR Lyraes by Walker (1992), given the slope 0.20 of the $M_V(\mathrm{RR}) - [\mathrm{Fe/H}]$ relation. Arenou et al. (1995) also investigated extensively the statistical properties of the errors of Hipparcos parallaxes by a similar algorithm to the ML method.

Feast & Catchpole (1997) have proposed another method, the so-called reduced parallax method (hereafter, FC method), to estimate the zero-point for the Cepheid PL relation from the weighted mean of the formula free from biases such as the Lutz-Kelker one. Using Hipparcos Cepheids, they found a 0.2 mag brighter zero-point than the previous value (Laney & Stobie 1994), which results in $\mu_{\mathrm{LMC}} = 18.70\,\mathrm{mag}$. This value gives a upper bound for μ_{LMC} among numerous determinations of the distance to the LMC (Gibson 2000), and it seems a bit too large (e.g., Freedman et al. 2001). However, Pont (1999) concluded this method to be the most rigorous one, and Lanoix, Paturel & Garnier (1999) further confirmed it by Monte Carlo simulations. In any case, even confined to the determinations of μ_{LMC} from Hipparcos parallaxes, there exists a large uncertainty in μ_{LMC}, which is one of the largest remaining uncertainties in the overall error budget for the determination of H_0.

In this paper, we perform Monte Carlo simulations as done by Lanoix, Paturel & Garnier (1999) to analyze the zero-point for Cepheid PL relation with the ML method. The ML method allows us to incorporate the density distribution of stars into the model arbitrarily. Without knowing the zero-point of the Cepheid luminosity in advance, *relative* approximate distances of individual Cepheids can be deduced from their photometric information assuming an arbitrary zero-point. Then from these relative distances, a realistic density distribution of stars belonging to the catalogue can be obtained. We show that the ML method combined with a density distribution of stars thus obtained leads to an unbiased estimate of the zero-point for the Cepheid PL relation, regardless of any values for the dispersion of the absolute magnitude. In fact, the intrinsic dispersion for Cepheids is estimated to be $\sim 0.2\,\mathrm{mag}$ (e.g., Ngeow & Kanbur 2004). However, it could possibly be broadened by a large reddening (Luri et al. 1998).

Furthermore, the dependence of the reliability of the results on the accuracy of parallaxes is investigated. We will show how not only the absolute magnitude precision, but also its dispersion precision will improve according to the parallax accuracy. These results are also compared with those obtained by the FC method. The precise determination of biases and variances involved in the final results for the statistical calibrations is certainly demanded for not only the further study using Hipparcos data, but also the future work using the highly-precise astrometric data to be obtained by the ongoing space satellite projects such as Gaia (Perryman 2002) and JASMINE (Japan Astrometry Satellite Mission for INfrared Exploration, Gouda et al. 2003; Gouda et al., these proceedings).

2. The method

2.1. *Construction of pseudo catalogues*

We make pseudo catalogues based on the Hipparcos data for Cepheids. To investigate the dependence of the end results on the accuracy of parallaxes, we prepare two kinds of pseudo catalogues: One is totally based on the Hipparcos data, which is identical to simulated catalogues made by Lanoix, Paturel & Garnier (1999); another is constructed with some changes to be done in order to be compatible with forthcoming catalogues with a high accuracy of parallax determination. Here we follow the way of DIVA, which was an astrometric satellite project once planned to be launched by the German Space Agency.

The absolute magnitude of Cepheids is expressed as $M_{V0} = \delta \log P + \rho$. An absolute magnitude M_{V0} given here is the value of the center of distribution with intrinsic magnitude dispersion. We assume that the slope δ of the PL relation is $\delta = -2.77$ (e.g., Madore & Freeman 1991). Using our pseudo catalogues, we will investigate how precisely the zero-point ρ can be determined by the improved ML method and its dependence on the accuracy of parallax. For individual stars in our pseudo catalog, the distance r, the logarithm of the period $\log P$, the absolute magnitudes in the V band M_V and the B band M_B, the apparent B and V magnitudes m_v and m_b, and the observed parallax π and its error σ_π are given by the following procedure.

Tentative stellar distribution

The spatial distribution of stars is first assumed to be uniform and then expressed as $n(r) \propto r^2$ for $0 < r < r_{\max}$. Here we introduce the parameter $r_{\max}$ corresponding to the maximum distance where stars are distributed. We adopt different values of $r_{\max}$ for two kinds of catalogues as shown in Table 1. The final stellar distribution will be changed so as to be compatible with the actual distribution of the observed stars.

Table 1. List of parameters and their input values.

parameter	
r_{max} (kpc)	2.1/8.4
$\langle \log P \rangle$	0.8554
$\sigma_{\log P}$	0.2865
δ	-2.77
ρ	-1.33
N_{star}	250/15700
N_{try}	100
σ_{m0}	0.21
σ_{mobs}	0.005

Distribution of $\log P$

Then, we give the value of $\log P$ to each star, following the distribution of $\log P$ expressed by the log-normal form; $P(\log P) = G(\log P, \langle \log P \rangle, \sigma_{\log P})$, where the notation $G(x, \langle x \rangle, \sigma)$ indicates a Gaussian distribution function of x with an average $\langle x \rangle$ and dispersion σ;

$$G(x, \langle x \rangle, \sigma) = \frac{1}{\sqrt{2\Pi}\sigma} \exp\left[-\frac{1}{2}\left(\frac{x - \langle x \rangle}{\sigma}\right)^2\right]. \tag{2.1}$$

Here Π denotes a circular constant. In our analysis, the details of this distribution form do not affect the results. Adopted values of $\langle \log P \rangle$ and $\sigma_{\log P}$ are listed in Table 1, together with other input values of the parameters.

Absolute magnitude

Since the value of $\log P$ is given to each star, we can give individual absolute magnitudes. It is assumed that absolute magnitude M_V in V band has a Gaussian distribution with mean M_{V0} and dispersion σ_{m0}, following a formula $P(M_V) = G(M_V, M_{V0}, \sigma_{m0})$. Using the PL relation, individual absolute magnitudes in V band are expressed as $M_V = \delta_V \log P + \rho_V + \Delta_1$, where Δ_1 is a deviation from M_{V0}. Its mean and dispersion

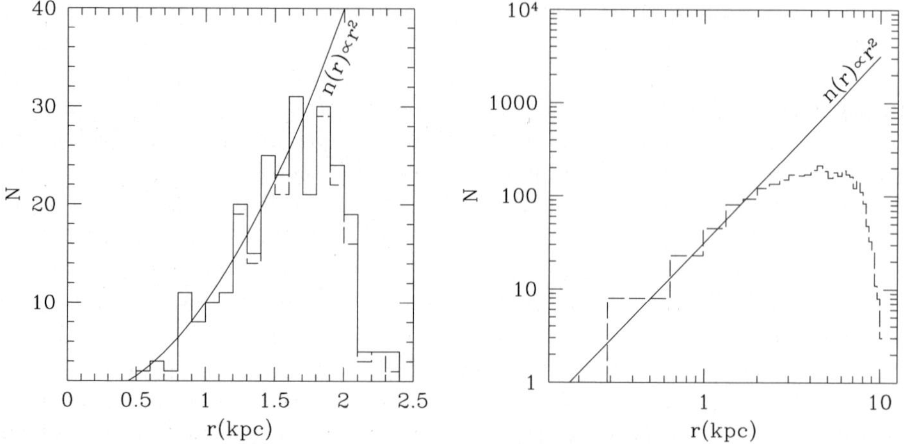

Figure 1. *left panel*: The approximate distribution of Hipparcos Cepheids deduced by giving the photometric distance to each star with the help of the PL relation (solid histogram) together with one in a pseudo catalogue that we made (dashed histogram). The solid line denotes a uniform distribution. *right panel*: The stellar distribution in a pseudo catalogue with an accuracy of parallax $\sim$200 μas (dashed histogram) compared with a uniform distribution (solid line).

are $\langle \Delta_1 \rangle = 0$ and $\sigma(\Delta_1) = \sigma_{m0}$. In the same manner, absolute magnitude M_B in B band can be expressed as $M_B = \delta_B \log P + \rho_B + \alpha \Delta_1 + \Delta_2$.

We analyze the observed PL and period-color relations in the LMC (Laney & Stobie 1994), and obtain 0.322, 0.236, and 0.0974 for dispersions of B, V, and $B - V$, which lead to the values; $\sigma(\Delta_1) = 0.236, \sigma(\Delta_2) = 0.0521$, and $\alpha = 1.35$. Although the value of $\sigma(\Delta_2)$ is small, it results in a rather large effective intrinsic dispersion as mentioned in the next section.

Apparent magnitude

In order to define apparent magnitude for each star, it is necessary to estimate the extinction A_V. As shown by Lanoix, Paturel & Garnier (1999), the color excesses $E(B - V)$ for Hipparcos Cepheids are uniformly distributed against the distance r in the range of $0.05 < E(B-V)/r < 0.5$. We take such a uniform distribution of $E(B-V)$. The relation between A_V and $E(B - V)$ is then used, i.e., $A_V = R_V E(B - V), A_B = R_B E(B - V)$ with $R_V = 3.3$ and $R_B = 4.3$. Actual apparent magnitudes are given by the formulae: $P(m_V) = G(m_V, M_V + 5 \log r + 10 + A_V, \sigma_{mobs})$ and $P(m_B) = G(m_B, M_B + 5 \log r + 10 + A_B, \sigma_{mobs})$, where σ_{mobs} means observational error, and the value is set to be 0.005.

Parallax and its error

It is assumed that the observed parallaxes π_o are distributed following the formula, $P(\pi_o) = G(\pi, 1/r, \sigma_\pi)$. The observational error of parallaxes, σ_π, depends on an apparent magnitude m_V. For the pseudo catalogues with the Hipparcos accuracy, we give σ_π to each star to reproduce the observed $m_V - \sigma_\pi$ distribution for the Hipparcos Cepheids. For cases with high accuracy of parallaxes, we assume the following formulae for reference, which are in good agreement with the case of DIVA.

$$\sigma_\pi = \begin{cases} \sigma_0 & m_V < 8 \\ \sigma_0 \times 10^{0.0146(V-8)^2 + 0.00036(V-8)^3} & 8 < m_V < 18 \\ 6.6225\sigma_0 10^{0.4(V-15.5)} & m_V > 18 \end{cases} \qquad (2.2)$$

Selection of stars – reconstruction of the stellar distribution –

The actual stellar distribution in the catalogue should deviate from an uniform one mainly due to the Malmquist bias. The approximate distribution of Hipparcos Cepheids can be deduced by giving the photometric distance to each star with the help of the Cepheid PL relation. Given $\rho = -1.33$ and $\delta = -2.77$ together with the apparent magnitudes rectified with absorption, we obtain the distribution shown in the left panel of Fig. 1 (solid histogram), which is indeed different from the first assumed $n(r) \propto r^2$ (solid line) at larger distance. According to the derived distribution, we should reconstruct the stellar distribution for the pseudo catalogues, which is realized by the following procedure introducing the Malmquist bias adopted by Lanoix, Paturel & Garnier (1999). To select stars belonging to the final pseudo catalogues, we generate uniform random numbers in the range of $0 \leqslant t \leqslant 1$, and calculate t_0 from $t_0 = \frac{1}{1+\exp[\gamma(V-V_{\mathrm{lim}})]}$. When the value t of the obtained random number is smaller than t_0, we use a star for a member of a sample, regarding the star as observed. Otherwise, we throw the star away from the sample. In our calculation, $\gamma = 1$ is adopted. In the Hipparcos case, we adopt $V_{\mathrm{lim}} = 12.5$ and $V_{\mathrm{lim}} = 15.5$ in higher accuracy cases. We use the congruent algorithm of 48-bit alignment in order to generate a uniform random number.

The resultant distributions are shown by dashed histograms in Fig. 1 for two kinds of pseudo catalogues. As shown in the left panel of Fig. 1, the final distribution is in good accord with the one for Hipparcos Cepheids (solid histogram).

2.2. *A maximum likelihood method*

Here we analyze our pseudo catalogues by the maximum likelihood method. This method was first proposed by Ratnatunga & Casertano (1991) to obtain the period-color relation for disk dwarf stars, and further applied to the metallicity-luminosity relation for Hipparcos RR Lyrae stars by Tsujimoto, Miyamato & Yoshii (1998). The merit of this method to be noted is a simultaneous determination of the zero-point of a linear-relation such as the Cepheid PL relation and the intrinsic dispersion of absolute magnitude.

In this method, the values of parameters to be determined are obtained as those which maximize the likelihood function $\mathcal{L}$, i.e., $\mathcal{L}(\rho, \sigma_m) = \sum^{N_{star}} \ln[P(\pi_o|\rho, \sigma_m)]$. The probability distribution P can be expressed as

$$P(\pi_o|\rho, \sigma_m) \propto \int_{-\infty}^{\infty} d\Delta_V \; p(\pi_o|\pi)m(\Delta_V, \sigma_m)\nu(r)10^{0.6\Delta_V} \tag{2.3}$$

where

$$p(\pi_o|\pi) = \begin{cases} G(\pi_o, \pi, \sigma_\pi) & \pi_o \in (\pi_-, \pi_+) \\ 0 & \text{otherwise} \end{cases}$$

$$m(\Delta_V, \sigma_m) = G(\Delta_V, 0, \sigma_m)$$

$$\Delta_V = \rho + \delta \log P + 10 + 5\log\frac{1}{\pi} - m_V$$

Δ_V means a difference between the true absolute magnitude and the calibrated absolute magnitude, and $\nu(r)$ is the number density of stars at the position of each star. The range of allowable parallax for the analysis is denoted by (π_-, π_+). Here we take $\pm\infty$ for these limits. A probability function $P(\pi_o|\rho, \sigma_m)$ is normalized by π_o.

The term $\nu(r)10^{0.6\Delta_V} d\Delta_V$ corresponds to the number of stars per unit solid angle between r and $r+dr$, and thus equivalent to $\nu(r)r^2 dr$. The number density $\nu(r)$ is assumed to be constant in the original method of Ratnatunga & Casertano (1991). However, as shown in the left panel of Fig. 1, the distribution of Hipparcos Cepheids deviates from a uniform distribution. Furthermore, in the pseudo catalogues with higher accuracies

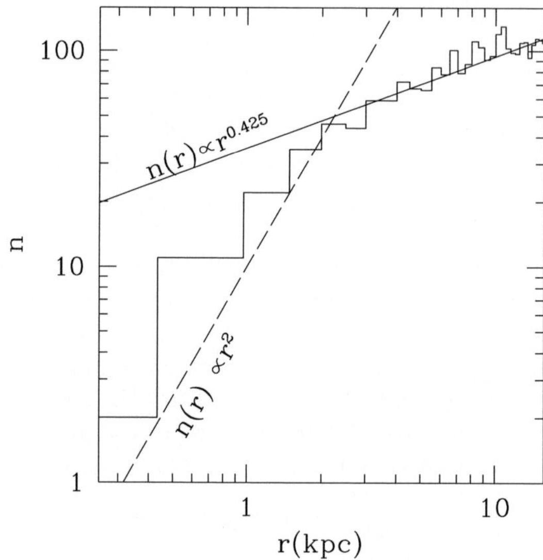

Figure 2. The stellar distribution of one sample in the pseudo catalogues with an accuracy of parallax $\sim 200\,\mu$as. The solid line is obtained by the power-law fitting at larger distance. The dashed line denotes a uniform distribution.

of parallaxes, the deviation from a uniform distribution is significant, as shown in the right panel of Fig. 1. Therefore we introduce a stellar distribution close to the actual one, assuming the power-law form of $\nu(r) \propto r^{-\alpha}$. We perform power-law fitting for each catalogue, and obtain different values of the power index. Fig. 2 shows one sample with $n \propto r^{0.425}$, which gives $\nu(r) \propto r^{-1.575}$ from the relation $n(r) = \nu(r)r^2 = r^{2-\alpha}$. As in this case, we tried to fit the distribution at larger distances such as $r > 2\,$kpc because most of stars are resided in this area.

3. Results

For one hundred pseudo catalogues with an accuracy of parallax corresponding to Hipparcos, the resultant distribution of ρ and σ_m is shown in the left figure on the left panel of Fig. 3. The dashed lines denote the input values, i.e., $\rho = -1.32$ and $\sigma_m = 0.2$. It is found that there is no bias for the returned values of ρ, regardless of the returned values of σ_m, though there exists a large scatter in the values of ρ, ranging over $-1.7 < \rho < -1$. This means that the accuracy (~ 1 mas) of Hipparcos parallaxes is not enough to obtain a reliable result for the zero-point for Cepheid PL relation. Besides, we cannot expect to obtain the right σ_m. For comparison, the results calculated with the ML method in which a stellar distribution is assumed to be uniform are shown in the right figure on the left panel of Fig. 3. For the cases which return higher values of σ_m than the true (input) value, biases are apparently seen in the returned values of ρ. These biases appeare in the opposite sense to the Malmquist bias which yields a bias toward a brighter magnitude. This is caused by an overcorrection of the Malmquist bias for the assumed uniform density distribution of stars.

The results of one hundred pseudo catalogues with an accuracy of parallax corresponding to DIVA, i.e., $200\,\mu$as are shown on the right panel of Fig. 3. Here we perform calculations for three input values of σ_m, i.e., $\sigma_m = 0.2$, 0.5, and 0.7 denoted by three

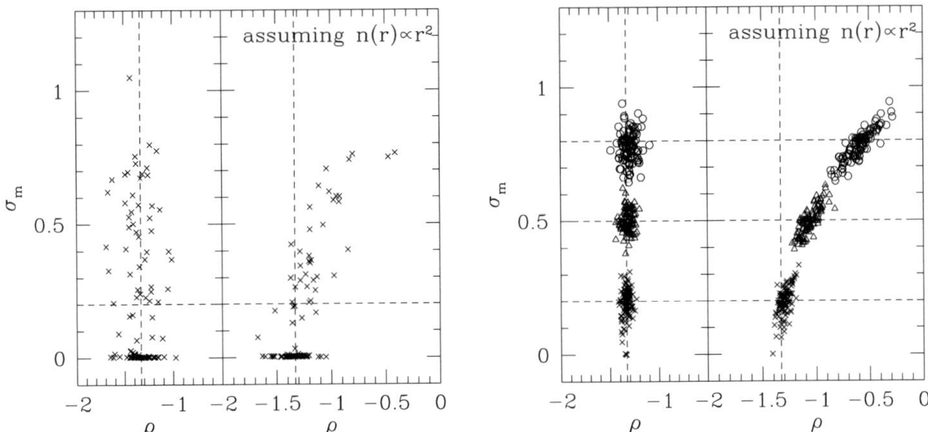

Figure 3. *left panel*: A left figure shows a distribution map of one hundred trial results for the case of the accuracy equivalent to Hipparcos parallaxes. The dashed lines correspond to the input values of ρ and σ_m. A right figure shows the results obtained by the original ML method in which a uniform stellar distribution is assumed. *right panel*: The distribution maps for one hundred pseudo catalogues with an accuracy of parallax $\sim 200\,\mu$as for three cases of $\sigma_m = 0.2$, 0.5 and 0.7.

dashed lines. Although the values of $\sigma_m = 0.5$ and 0.7 are much larger than an intrinsic dispersion of M_V, they could possibly be due to the error in the value of the reddening. Furthermore, it should be remarked that a large value of $\sigma_m \sim 0.7$ is suggested by the statistical calibration of parallaxes and proper motions for Hipparcos Cepheids (Luri et al. 1998). For all cases, our modified ML method gives unbiased estimates of ρ with a small scatter (the left figure). However, this is not the case for the ML method with a assumption of a uniform stellar distribution (the right figure). The biases become larger according to larger input σ_m.

The left panel of Fig. 4 clearly demonstrates that the area of the distribution of (ρ, σ_m) becomes smaller according to higher accuracy of parallaxes. Providing that an accuracy of parallax better than $\sim 50\,\mu$as is reached, we will be able to obtain the precise combination of ρ and σ_m. The right panel of Fig. 4 shows the returned values of ρ with 1σ errors as a function of an accuracy of parallax (crosses with solid bars) together with the results obtained by FC method (open triangles with dashed bars). Interestingly, in the case of the FC method, the bias always remains regardless of the accuracy of parallax, though the application of the FC method to cases with high accuracies such as $10\,\mu$as is not appropriate. This bias might be ascribed to an effective intrinsic dispersion $\sigma_{\rm eff}$ of the absolute magnitude. In fact, the FC method gives a reliable value of the zero-point as long as $\sigma_{\rm eff}$ can be reduced to a small value such as $0 - 0.1$. It was then proposed by Feast & Catchpole (1997) that such a small $\sigma_{\rm eff}$ could be realized if we use a color-period relation to estimate a reddening, and combine it with the luminosity-color-period relation. However, we obtain

$$\sigma_{\rm eff} \sim \sqrt{((\alpha - 1)R_V - 1)^2 \sigma(\Delta_1)^2 + R_V^2 \sigma(\Delta_2)^2} \sim 0.176, \tag{3.1}$$

using the values already obtained, $\sigma(\Delta_1) = 0.236, \sigma(\Delta_2) = 0.0521$, and $\alpha - 1.35$. This relatively large $\sigma_{\rm eff}$ is a result of non-zero $\sigma(\Delta_2)$, while Feast & Catchpole (1997) assumed $\sigma(\Delta_2) - 0$.

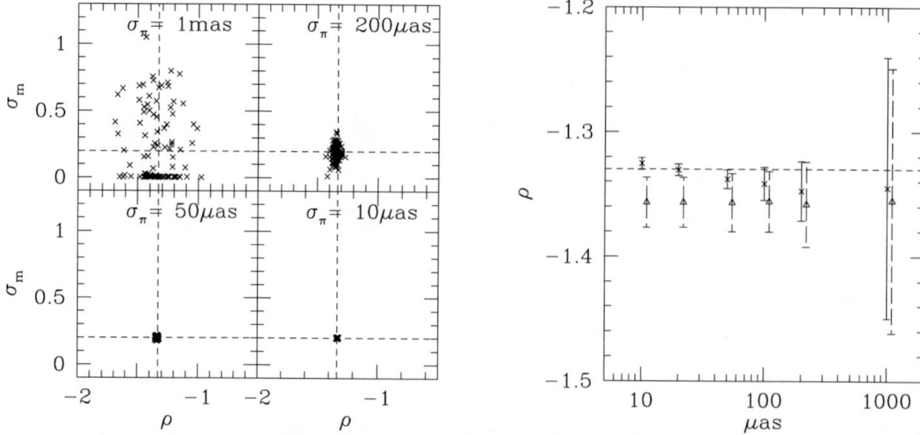

Figure 4. *left panel:* Distribution maps of one hundred trial results in the ρ-σ_m diagram for four cases with accuracies of $1\,\mathrm{mas}$, $200\,\mu\mathrm{as}$, $50\,\mu\mathrm{as}$, and $10\,\mu\mathrm{as}$ for parallax. *right panel:* The returned values of ρ as a function of an accuracy of parallax calculated by two methods, i.e., the modified ML method (crosses with solid bars) and the FC method (open triangles with dashed bars).

4. Conclusions

Here we present a maximum likelihood method in which a *realistic* density distribution of sample stars to be analyzed is introduced. Prior information on the stellar density is obtained through calculation of the photometric distance for each star instead of use of the Galactic model as done by Arenou et al. (1995). Our method gives an unbiased estimate of the absolute magnitude of stars, regardless of any dispersions in their absolute magnitude. However, an accuracy of $200\,\mu\mathrm{as}$, or better, for parallaxes is required to obtain a reliable result from only one ensemble of stars. If not, as in the case of Hipparcos, we cannot get a definitive conclusion due to a large variation in the estimate. Furthermore, an accuracy more than $\sim 50\,\mu\mathrm{as}$ promises to give a precise value of the dispersion in absolute magnitude together with its mean value.

Acknowledgements

This work is supported by a Grant-in-Aid for the 21st Century COE Center for Diversity and Universality in Physics.

References

Arenou, F., Lindegren, L., Froeschlé, M., Gómez, A. Z., Turon, C., Perryman, M. A. C. & Wielen, R. 1995 *A&A* 304, 52–60.

Arenou, F. & Luri, X. 1999 in *ASP Conf. Ser. 167, Harmonizing cosmic distance scale in a post-Hipparcos era* (ed. D. Egret & A. Heck), pp. 13–32. ASP.

Feast, M. W. & Catchpole, R. M. 1997 *MNRAS* 286, L1–L5.

Freedman, W. L., *et al.* 2001 *ApJ* 553, 47–72.

Gibson, B. K. 2000 *Mem. Soc. Astron. Ital.* 71, 693–700.

Gouda, N. *et al.* 2003 *SPIE*, vol. 4850, pp. 1161–1168.

Jung, J. 1971 *A&A* 11, 351–358.

Laney, C. D. & Stobie, R. S. 1994 *MNRAS* 266, 441–454.

Lanoix, P., Paturel, G. & Garnier, R. 1999 *MNRAS* 308, 969–978.

Luri, X., Gómez, A. E., Torra, J., Figueras, F. & Mennessier, M. O. 1998 *A&A* 335, L81–L84.

Lutz, T. E. & Kelker, D. H. 1973 *PASP* 85, 573–578.

Madore, B. F. & Freeman, W. 1991 *PASP* 103, 933–957.

Ngeow, C.-C. & Kanbur, S. M. 2004 *MNRAS* 349, 1130–1136.

Perryman, M. A. C. 2002 *Astrophys. and Space Sci.* 280, 1–10.

Pont, F. 1999, in ASP Conf. Ser. 167, Harmonizing cosmic distance scale in a post-Hipparcos era (ed. D. Egret & A. Heck), pp. 113–128. ASP.

Ratnatunga, K. U. & Casertano, S. 1991 *AJ* 101, 1075–1088.

Roman, N. 1952 *ApJ* 116, 122–143.

Smith, H. 1987 *A&A* 171, 336–347.

Smith, H. 1988 *A&A* 198, 365–369.

Smith, H. 2003 *MNRAS* 338, 891–902.

Tsujimoto, T., Miyamoto, M. & Yoshii, Y. 1998 *ApJ* 492, L79–L82.

Walker, A. R. 1992 *ApJ* 390, L81–L84.

Seiji Ueda, Takuji Tsujimoto (centre), Naoteru Gouda, Taihei Yano, Vladimir Elkin

Transits of Venus: New Views of the Solar System and Galaxy
Proceedings IAU Colloquium No. 196, 2004
D.W. Kurtz, eds.

© 2004 International Astronomical Union
doi:10.1017/S1743921305001584

Parallaxes of L and T dwarfs

R. L. Smart[1], C. A. L. Bailer-Jones[2] and H. R. A. Jones[3]

[1]Osservatorio Astronomico di Torino – INAF, Strada Osservatorio 20, Pino Torinese,
TO 10025, Italy
email: smart@to.astro.it

[2]Max-Planck-Institut für Astronomie, Königstuhl 17, D-69117 Heidelberg, Germany
email: calj@mpia-hd.mpg.de

[3] School of Physics Astronomy and Mathematics, University of Hertfordshire, College Lane,
Hatfield AL10 9AB, UK
email: hraj@star.herts.ac.uk

Abstract. We discuss a new program to measure the parallaxes of a number of L and T dwarfs, objects that bridge the gap between M dwarfs and planets. This pilot project tests the feasibility of using large telescopes with infrared detectors to determine parallaxes at the level of milli-arcseconds (mas). First results show that we are able to achieve the required centroiding precision and simulations indicate that when the final observations come in we should be able to achieve our goal of parallaxes with accuracies of 2 mas. The main problems will be focal plane astrometric distortions and stability.

1. Introduction

T and L dwarfs are ultra-cool objects cooler than M dwarfs that bridge the gap between stellar and substellar objects, also known as brown dwarfs. They have spectra dominated by molecular absorption due to water, methane and pressure-induced molecular hydrogen. Methane, which first appears prominently in T dwarfs, is expected to remain an important atmospheric constituent down to the temperature of Jupiter ($\sim 125\,\mathrm{K}$), where it is also prominent in the infrared spectrum. This means these objects provide an important link to extrasolar giant planets. Indeed, nearly ten years ago the announcements of the discoveries of the first brown dwarf and extrasolar planet were made at the same conference. Since then many new discoveries have been made, so the prototype of the T class, the companion to the nearby M dwarf star Gl229 (Nakajima et al. 1995), has been supplemented by the discovery of more than 300 L and T dwarfs. These come primarily from the Sloan Digital Sky Survey (Strauss et al. 1999; Tsvetanov et al. 2000; Leggett et al. 2002; Geballe et al. 2002; Knapp et al. 2004) and from 2MASS (Burgasser et al. 1999, 2000, 2002a, 2002b, 2003a, 2003b).

Model atmosphere analyses indicate temperatures of $2500\,\mathrm{K}$ for L dwarfs down to $750\,\mathrm{K}$ for lower T dwarfs, although significant uncertainties remain. Absolute luminosities are the most direct route toward an empirical temperature scale for ultra-cool dwarfs (e.g., Vrba et al. 2004). In this paper we discuss an ongoing program to measure the parallaxes and hence distances and absolute luminosities of nine T dwarfs and two L dwarfs.

2. Observational program

In Table 1 we list the targets under observation along with their nominal 2MASS J magnitudes (Cutri et al. 2003), spectral types in the Kirkpatrick et al. (1999) and Burgasser et al. (2002) systems, number of observations and range of observational epoch

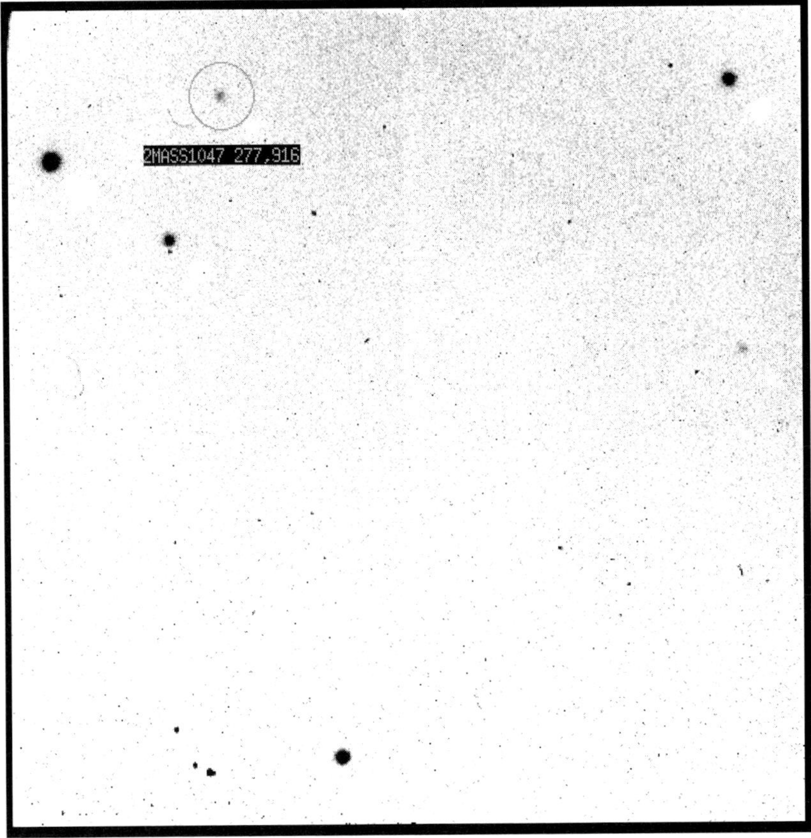

Figure 1. An example field for the object 2MASS1047. The target has a large offset from the center of the CCD to enable sufficient reference stars.

in months. For these faint magnitudes a 3-m class telescope is required, and we have successfully obtained time on the infrared camera, Omega Cass, of the 3.5-m telescope at Calar Alto.

In Fig. 1 we show the field of 2MASS1047. An empty field like this is typical for this program. It forces us to offset the target star from the center in order to maximise

Table 1. Target list

Target	RA	Dec	J	Sp	Number of epochs	Epoch span months
2MASS1021	10:21:09	-03:04:20	16.3	T3	7	11
2MASS1047	10:47:53	+21:24:23	15.8	T6.5	7	11
2MASS1217	12:17:11	-03:11:13	15.9	T7.5	7	13
2MASS1145	11:45:57	+23:17:29	15.4	L1	7	11
2MASS1225	12:25:54	-27:39:47	15.2	T6	6	11
2MASS1237	12:37:39	65:26:15	15.9	T6.5	8	13
SDSS1254	12:54:53	-01:22:47	14.7	T2	8	13
SDSS1346	13:46:46	-00:31:50	15.9	T6	8	13
GL570D	14:57:15	-21:21:50	15.3	T8	6	13
2MASS1507	15:07:47	-16:27:38	12.8	L5	5	13
SDSS1624	16:24:14	00:29:16	15.5	T6	5	13

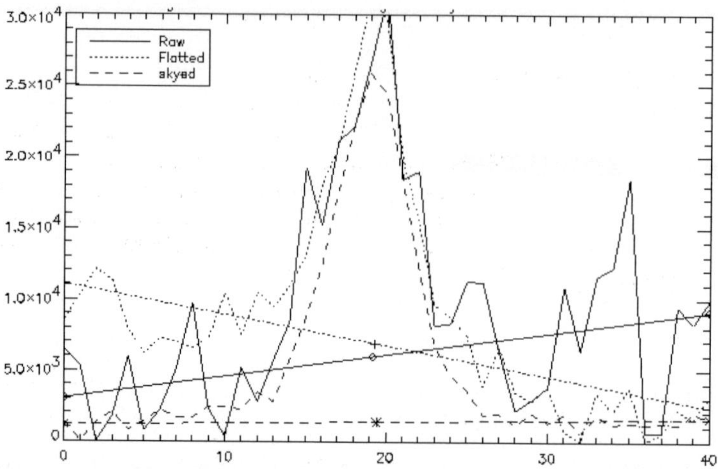

Figure 2. Normalized marginal distributions of a bright star in three phases of calibration as indicated in the legend.

the number of reference stars, and also limits the size of any dither pattern we can employ.

The determination of a parallax in the infrared follows the same process as that of the optical. See, for example, Smart et al. 2003, however, for a number of reasons it should be more precise:

1) The refraction of the atmosphere is lower and varies more slowly so the effects of differential refraction are smaller and probably negligible.

2) The PSFs have smaller FWHM, so in theory can be centred better.

3) The exposure times are in general shorter (of the order of seconds as opposed to minutes), so more observations can be made leading to a higher redundancy.

4) The sky background is always higher, but observations can be made during the period between astronomical and civil twilight leading to observations at marginally higher parallax factors and the possibility to observe when optical observations are not possible.

However, most of these benefits are outweighed by the noisier background and the subsequent necessity to carry out sky subtraction.

In Fig. 2 we show the normalized marginal distributions of a relatively bright star in three phases of calibration: the raw profile, the flattened profile and the sky subtracted profile. To obtain our precision goal we are required to center to at least 0.05 pixels. The centroids from these three distributions vary by over 0.3 pixels using standard weighted moments and 0.2 pixels using 2 dimensional Gaussians.

A number of experiments are being carried out to find the best way to centroid these images. Applying offset images for sky subtraction is the normal procedure, and, as such, we flattened all frames, made 10″ dithered images and sky subtracted using adjacent frames. In these cleaned images we found positions of all stars using two dimensional Gaussian fitting and then aligned them using a linear transform. From these multiple observations of each star we found positional standard deviations. In Fig. 3 we show two comparisons, one of observations in the same positions (i.e. no offsets) and one of all observations (i.e. including dithered images). The larger standard deviations in the comparison of offset images is indicative of a variable astrometric distortion map, or perhaps because of the lower number of stars in common. Further work is under way to understand these differences.

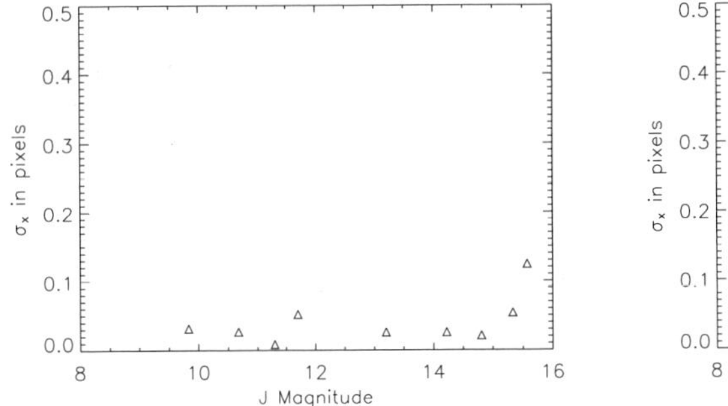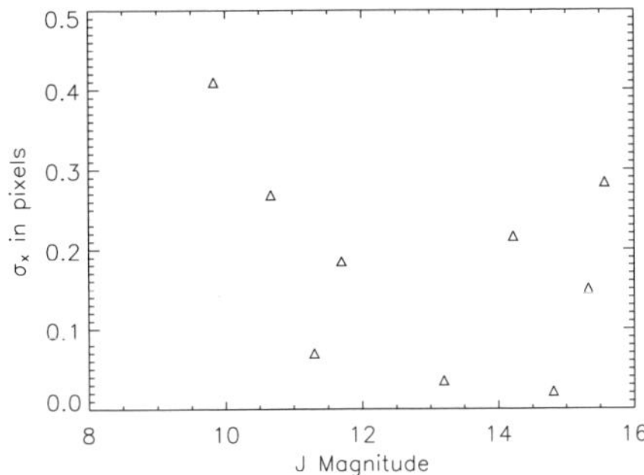

Figure 3. Centroiding sigmas for consecutive observations of the field 2MASS1047. Left panel: detector position always the same. Right panel: dithered positions.

Given that we are able to centroid to within 0.05 pixels we have used simulations (Smart et al. 2001) to see the expected precision with various observational strategies. The best approach is to have a season of intensive observations to sample the parallax ellipse, and then carry out only nominal observations in the following years to sample the proper motion movement. This means the precision of the final parallax, given a fixed number of observations, is a maximum when they are they are distributed in the manner indicated in Fig. 4. This simulation with a centroiding error of 0.05 pixels predicted a final parallax precision of 2 mas.

3. Conclusions

In Fig. 5 we plot absolute magnitudes derived from all published T dwarf distances – 30 distances for 23 objects. The overall trend of a hump in the middle of the range can be clearly seen. However, the details are lost in the noise which is mainly due to inaccurate parallaxes. The results of this project will improve this picture and allow us

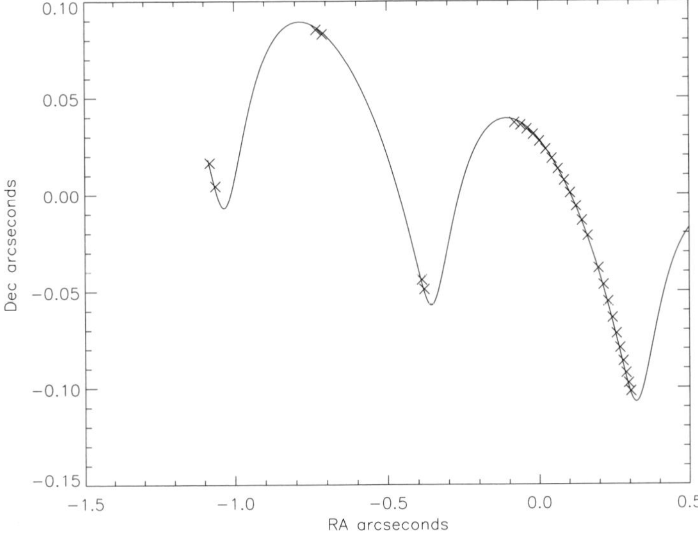

Figure 4. Simulation of the expected motion and observations for a star at RA 12 h and Dec 17°, distance 10 pc and proper motion $1'' \mathrm{yr}^{-1}$.

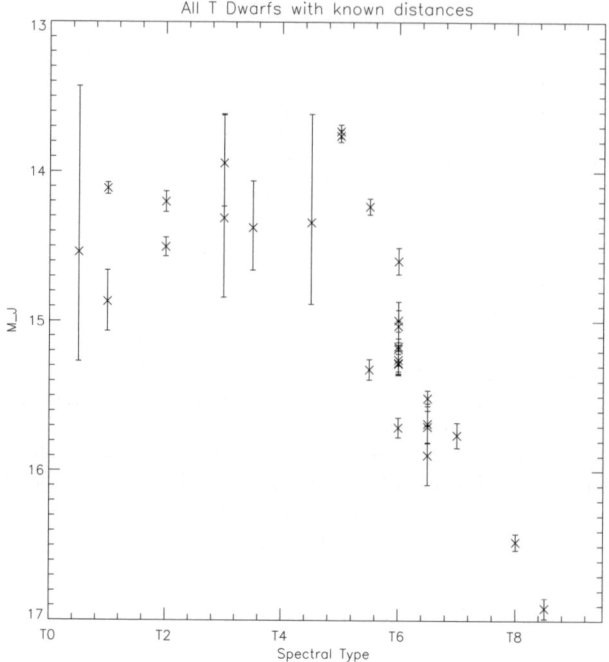

Figure 5. Absolute J magnitudes derived from all known T dwarf distances. 2MASS J magnitudes and spectral types taken from Burgasser (ref).

to distinguish between competing scenarios: cloud disruption (Burgasser et al. 2002a) or a thin dust cloud deep in the photosphere (Tsuji 2002).

The work on T dwarfs has, however, only just begun. While these results will aid the understanding of such features as the L-T transition and the T hump, as more objects are found more issues will present themselves. For example, the differences between T dwarf spectra appear to be relatively subtle. The T8 dwarf Gl570D (Burgasser et al. 2000) must be around 250 K cooler than Gl229B, yet their spectra are similar and bear more resemblance to the reflectance spectra of Jupiter (e.g. **?**) than to L dwarfs (Geballe et al. 2002). Indeed, spectral types may have as much to do with cloud properties as with effective temperatures. Thus absolute luminosities are a key ingredient to a better understanding of T dwarf spectra.

It is also important to point out that for these objects we would be mistaken to wait for the results of the planned space astrometric missions. Gaia (Perryman et al. 2001) will probably not observe any of the T dwarfs, and not many of the L dwarfs, as they are too faint in the magnitude band of the astrometric field. JASMINE (Yamada et al., these proceedings) will observe some of the brighter L and T dwarfs in the Galactic plane, but the number and hence range of properties will be limited. It appears that only dedicated programs such as the present one will provide distances for future objects discovered in the large surveys such as 2MASS, SDSS and UKIDSS.

Acknowledgements

Based on observations collected at the Centro Astronómico Hispano Alemán (CAHA) at Calar Alto, operated jointly by the Max-Planck Institut für Astronomie and the Instituto de Astrofsica de Andaluca (CSIC).

References

Burgasser, A. J., Kirkpatrick, J. D., Brown, M. E., et al. 1999, *ApJ Lett.* **522**, L65

Burgasser, A. J., Kirkpatrick, J. D., Cutri, R. M., et al. 2000, *ApJ Lett.* **531**, L57

Burgasser, A. J., Kirkpatrick, J. D., Brown, M. E., et al. 2002a, *ApJ* **564**, 421

Burgasser, A. J., Liebert, J., Kirkpatrick, J. D., & Gizis, J. E. 2002b, *AJ* **123**, 2744

Burgasser, A. J., Marley, M. S., Ackerman, A. S., et al. 2002c, *ApJ Lett.* **571**, L151

Burgasser, A. J., Kirkpatrick, J. D., Burrows, A., et al. 2003a, *ApJ* **592**, 1186

Burgasser, A. J., Kirkpatrick, J. D., Liebert, J., & Burrows, A. 2003b, *ApJ* **594**, 510

Cutri, R. M., Skrutskie, M. F., van Dyk, S., et al. 2003, *VizieR Online Data Catalog* **2246**, 0

Geballe, T. R., Knapp, G. R., Leggett, S. K., et al. 2002, *ApJ* **564**, 466

Kirkpatrick, J. D., Reid, I. N., Liebert, J., et al. 1999, *ApJ* **519**, 802

Knapp, G. R., Leggett, S. K., Fan, X., et al. 2004, *AJ* **127**, 3553

Leggett, S. K., Golimowski, D. A., Fan, X., et al. 2002 *ApJ* **564**, 452

Nakajima, T., Oppenheimer, B. R., Kulkarni, S. R., Golimowski, D. A., Matthews, K., & Durrance, S. T. 1995, *Nature* **378**, 463

Perryman, M. A. C., de Boer, K. S., Gilmore, G., et al. 2001, *A&A* **369**, 339

Smart, R. et al.: 2001, in *ASP Conf. Ser. 232: The New Era of Wide Field Astronomy*, p. 335

Smart, R. L., Lattanzi, M. G., Bucciarelli, B., et al. 2003, *A&A* **404**, 317

Strauss, M. A., Fan, X., Gunn, J. E., et al. 1999, *ApJ Lett.* **522**, L61

Tsuji, T. 2002, *ApJ* **575**, 264

Tsvetanov, Z. I., Golimowski, D. A., Zheng, W., et al. 2000 *ApJ Lett.* **531**, L61

Vrba, F. J., Henden, A. A., Luginbuhl, C. B., et al. 2004, *AJ* **127**, 2948

David Sellers, Walter Brisken and Coryn Bailer-Jones

Roger Clowes and Dave Monet

Jacqueline Mitton, Barbara Hassall and Simon Mitton

Part 6

NEW VIEWS OF THE GALAXY: FUTURE SPACE AND GROUND-BASED PROGRAMMES

Hideyuki Kobayashi and Taihei Yano observe the transit

Coryn Bailer-Jones

Transits of Venus: New Views of the Solar System and Galaxy
Proceedings IAU Colloquium No. 196, 2004
D.W. Kurtz, ed.

© 2004 International Astronomical Union
doi:10.1017/S1743921305001596

Microarcsecond astrometry with Gaia: the solar system, the Galaxy and beyond

Coryn A.L. Bailer-Jones

Max-Planck-Institut für Astronomie, Königstuhl 17, 69117 Heidelberg, Germany
email: calj@mpia-hd.mpg.de

Abstract. Gaia is an all-sky, high precision astrometric and photometric satellite of the European Space Agency (ESA) due for launch in 2010. Its primary mission is to study the composition, formation and evolution of our Galaxy. Over the course of its five-year mission, Gaia will measure parallaxes and proper motions of every object in the sky brighter than visual magnitude 20, amounting to a billion stars, galaxies, quasars and solar system objects. It will achieve an astrometric accuracy of $10\,\mu$as at $V = 15$ – corresponding to a distance accuracy of 1% at 1 kpc – and $200\,\mu$as at $V = 20$. With Gaia, tens of millions of stars will have their distances measured to a few percent or better. This is an improvement over Hipparcos by several orders of magnitude in the number of objects, accuracy and limiting magnitude. Gaia will also be equipped with a radial velocity spectrograph, thus providing six-dimensional phase space information for sources brighter than $V \sim 17$. To characterize the objects (which are detected in real time, thus dispensing with the need for an input catalogue), each object is observed in 15 medium and broad photometric bands with an onboard CCD camera. With these capabilities, Gaia will make significant advances in a wide range of astrophysical topics. In addition to producing a detailed kinematical map of stellar populations across our Galaxy, Gaia will also study stellar structure and evolution, discover and characterise thousands of exoplanetary systems (extending down to about ten Earth masses for the nearest systems) and make accurate tests of General Relativity on large scales, to mention just some areas. I give an overview of the mission, its operating principles and its expected scientific contributions. For the latter I provide a quick look in five areas on increasing scale size in the universe: the solar system, exosolar planets, stellar clusters and associations, Galactic structure and extragalactic astronomy.

1. Introduction

Distance measurement has been historically one of the most fundamental challenges in astronomy. To measure cosmic distances from the Earth's surface we must generally rely on parallaxes, i.e. the apparent change in position of an object relative to some other object (or reference frame) brought about by a known displacement of the observer. Astronomical distance measurement is hard because these displacements (e.g. the motion of the Earth around the Sun) are small compared to the distances we want to measure (e.g. the distance to the Galactic centre). That is we need to measure very small angular changes in position.† The lack of accurate distances has been, and continues to be, one of the most significant limitations in studying the universe.

The topic of this conference – the transit of Venus across the Sun – is central to this issue, because the attempts to observe it were a milestone in observational astronomy and in our potential to measure distances. Following first Halley's and then Deslise's outline of a method to measure the solar parallax (and hence the Astronomical Unit,

---

† This problem has been captured more elegantly by Douglas Adams (1978): "Space is big. Really big. You won't believe how vastly hugely mindbogglingly big it is. I mean you may think it's a long way down the road to the chemist, but that's just peanuts compared to space."

Table 1. Comparison of the capabilities of Gaia with its predecessor Hipparcos.

Quantity	Hipparcos	Gaia
Magnitude limit	$V = 12.4$	$G = 20$
Completeness limit	$V = 7.3 - 9.0$	$G = 20$
Number of sources	120 000	26 million to $G = 15$
		250 million to $G = 18$
		1000 million to $G = 20$
Number of quasars	none	$0.5 - 1$ million
Number of galaxies	none	$1 - 10$ million
Target selection	input catalogue	onboard; magnitude limited
Astrometric accuracy	$\sim$1000 μas	$2 - 3\,\mu$as at $G < 10$
		$5 - 15\,\mu$as at $G = 15$
		$40 - 200\,\mu$as at $G = 20$
Broad band photometry	$2\ (B_\mathrm{T}, V_\mathrm{T})$	$4 - 5$ bands
Medium band photometry	none	$10 - 12$ bands
Spectroscopy	none	$R = 11\,500\ (848 - 874\,\mathrm{nm})$
Radial velocities	none	$\sigma = 1 - 10\,\mathrm{km\,s^{-1}}$ to $G = 17 - 18$

AU) from timing a Venus transit, numerous expeditions were mounted to observe the four transits occurring during the 18th and 19th centuries. While the degree of accuracy and consistency of these methods in determining the AU were not as high as hoped or expected (for reasons discussed elsewhere in this volume), they nonetheless made a vital contribution. The timing of transits, for example, was an ingenious addition to our otherwise limited way of measuring cosmic distances.

Now, some 370 years after the first transit observations of Mercury and Venus in the 1630s, distance measurement in the universe remains a fundamental issue in astronomy. It is vital for understanding the structure and evolution of stars, the formation and composition of our Galaxy and ultimately for tracing the origin of the universe. Almost all aspects of astrophysics rely to some degree on accurate distances, and this ultimately relies on astrometry: the measurement of positions over time and derivations of parallaxes and proper motions from them.

The Gaia mission will mark a significant step forward in astrometry. Following in the wake of the very successful Hipparcos mission (ESA 1997), Gaia will extend Hipparcos' capabilities by several orders of magnitudes and through this make significant breakthroughs in numerous astrophysical topics. Table 1 gives a brief comparison of the main capabilities of Gaia compared to Hipparcos.

In this contribution I summarise the main aspects of the satellite, its mission and its observational principles, and provide a sample of expected science contributions. For further information the reader is referred to the concept study report (ESA 2000) and its summary (Perryman 2001). I will say little about the radial velocity spectrograph on board Gaia: for this see the contribution in this volume by Mark Cropper.

2. Global astrometry

Astrometry is the practice of accurately measuring the positions of objects on the celestial sphere. At any instant the position of a celestial object is given by two coordinates, e.g. Right Ascension and Declination. By measuring positions repeatedly over time, we can measure parallax (due to the cyclic motion of the observing platform, e.g. the Earth about the Sun) and proper motions (linear motions through space in some reference frame). Combined with the radial velocity, these yield six astrometric

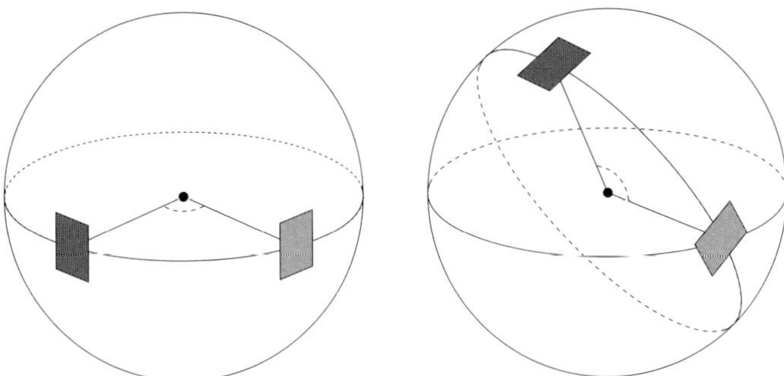

Figure 1. The key to global astrometry is to measure positions simultaneously in two fields separated by a large, fixed basic angle. By repeating this for a given field at different orientations (to give different reference stars) we can, in principle, determine absolute parallaxes.

parameters comprising three position co-ordinates and their first order time derivatives (three velocity co-ordinates). †

Astrometry performed from the ground is done so over a narrow field. That is, stellar positions can only be measured accurately relative to nearby stars, typically within the field-of-view of the telescope. Thus the derived parallaxes (in particular) are only relative: all stars in the same direction share a common parallactic effect due to the motion of the Earth about the Sun. To perform *global astrometry* over the entire celestial sphere we must measure the positions of stars relative to other stars separated by a large angle on the sky, such that they have a different parallactic effect. Repeating this (for a given field) for many different positions angles ('reference fields'), and then measuring the positions of star in *those* reference fields with respect to yet other stars separated by large angles, we can eventually build up an entire grid of relative position measurements over the celestial sphere (Fig. 1). Note that we only accurately measure positions in one dimension, i.e. along the great circle arc connecting the two fields. Two dimensional positions are obtained from the fact that we measure along great circles inclined at a range of orientations. We know that there are 2π radians in any great circle so we can use this 'closure condition' to derive absolute positions of the objects. The choice of axes (e.g. RA and Dec) and zero point (first point of Aries) are then a matter of convention. In practice, the reduction of the data involves a large iterative solution to simultaneously derive the five astrometric parameters of a large number of stars.

The key to global astrometry is therefore to measure simultaneously the positions of stars in two widely separated fields-of-view and to observe stars spread over the entire celestial sphere several times every year for several years. (The last condition is imposed by the need to lift the degeneracy between parallax and proper motion.) To achieve this in practice we must observe from space (also to overcome complex refraction problems in the Earth's atmosphere).

This principle was used by the Hipparcos satellite, the first (and so far only) mission to perform global astrometry (Perryman et al. 1989; Kovalevsky 1995).

The *scanning law* describes how the satellite observes the sky, i.e. how the two fields shown in Fig. 1 move with time. With Gaia, it essentially consists of a three-axis motion. First, Gaia rotates about its spin axis with a period of 6 hr. The two viewing directions

---

† Deviations from this six parameter model are important for unresolved binary stars, which will show imposed Keplerian motions. Moreover, higher order time derivatives (e.g. accelerations) can in principle also be measured from an astrometric time series, and in fact will be important with Gaia for some bright, nearby stars (the *perspective acceleration*).

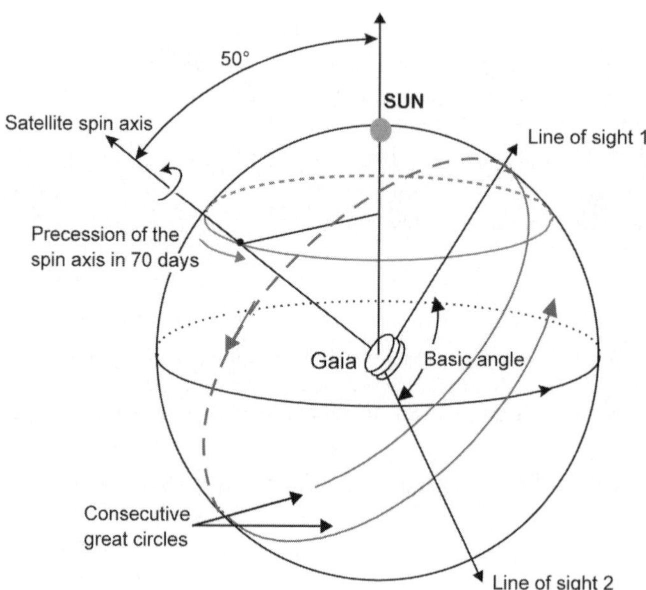

Figure 2. The Gaia measurement principle and scanning law. Gaia simultaneously observes in two viewing direction separated by a constant *basic angle*. Gaia continuously rotates about an axis perpendicular to these two viewing directions with a period of 6 hr. This axis precesses in such a way that it maintains a constant angle with respect to the Sun of 50°. The precession period is 70 d, so that every 6 hr Gaia traces a quasi-great circle across the sky. From the combination of these two motions Gaia can observe the whole sky. © ESA

lie in a plane perpendicular to this axis (Fig. 2). To ensure it can observe the whole sky, this spin axis simultaneously precesses (with a period of 70 d) such that it maintains a constant angle of 50° with respect to the Sun (the *Sun aspect angle*, SAA – the Sun is chosen for reasons which are described below). As it rotates and precesses, Gaia traces quasi great circles across the sky. Finally, the satellite orbits the Sun with a period of 1 yr. The paths traced out by this somewhat complex motion are shown in Fig. 3.

The angle between the two viewing directions – the *basic angle* – is fixed. It is not necessary to know this angle exactly (it can be derived as part of the reduction given the close condition) but it is essential that it remain fixed, or at least that small variations of it be measured. Specifically, to reach microarcsecond precision, the basic angle must be constant to a few μas over the 6-hr spin period. This places extremely stringent requirements on the thermal and mechanical stability of the satellite. For example, thermal gradients across the optical bench must be kept below 25 μK. Mechanical drags (e.g. from the Earth's tenuous atmosphere) or thermal loads (e.g. from passing into the Earth's shadow) could make achieving this almost impossible. For this reason, Gaia will be placed in orbit about the Earth-Sun L2 Lagrange point, situated about 1.5 million km from the Earth (four times the Earth-Moon distance) in the antisolar direction.† As the Earth and Sun now lie in the same direction from Gaia, fixed in its rotating reference frame, it precesses about this axis in order to avoid ever observing the Earth or Sun. With SAA = 50° Gaia never observes closer than 90° − 50° = 40° from the Earth/Sun. Furthermore, with a flat sun shield deployed perpendicular to its spin axis (Fig. 4), this

† The L2 point itself in unstable so Gaia performs Lissajous orbits about it. The orbit is such that Gaia does not pass into the Earth's shadow during its five years of science operations.

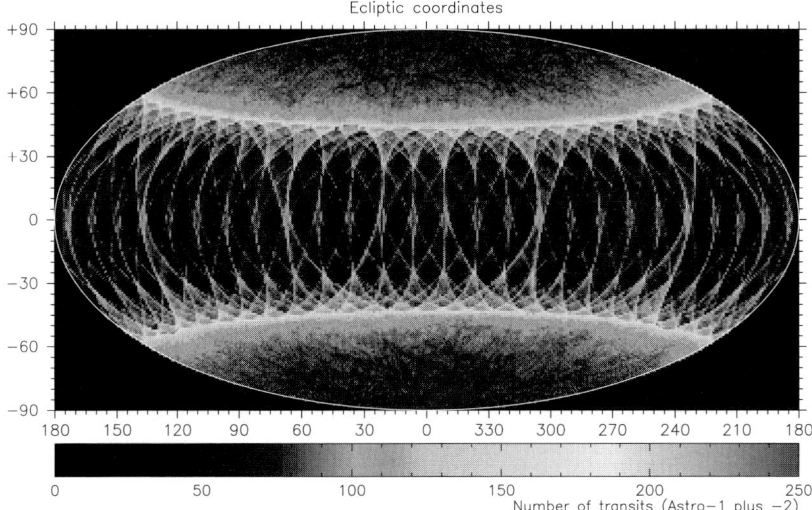

Figure 3. Sky coverage by Gaia. The two-axis motion of Gaia (rotation and precession) shown in Fig. 2 results in a non-uniform sky coverage. The coverage depends on the specific scanning law parameters, the size of the field-of-view and the duration of the mission. This shows the number of observations at each point on the sky for the nominal Gaia mission (5 yr), plotted in ecliptic co-ordinates. © J. de Bruijne / ESA

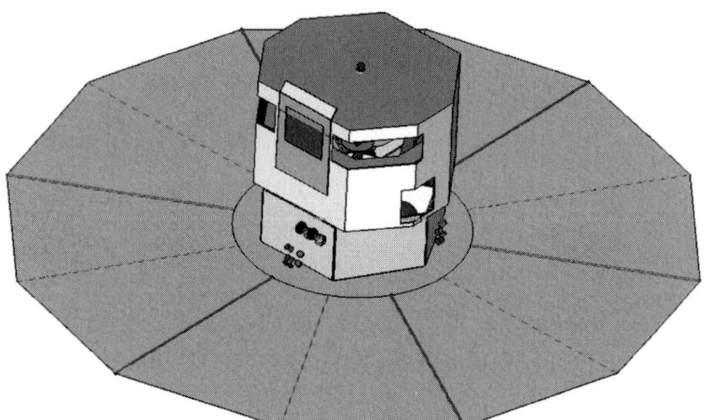

Figure 4. The Gaia satellite and sun shield. The payload module is 3.1-m high and the sun shield has a diameter of 11 m. Between the two entrance slits for the two astrometric fields at the top is the astrometric focal plane. © Astrium

arrangement also ensures that the solar flux on the spacecraft is constant as it precesses and orbits the Sun.

3. Gaia satellite

The scientific payload of Gaia is built around a rigid optical bench (Fig. 5). Onto this are mounted two astrometric telescopes which image the two viewing directions onto a common focal plane (together these form the *Astro* instrument). This focal plane consists of a very large number of CCD detectors: there are some 170 arranged into a rectangular grid. Each CCD contains 4500×1966 pixels with a pixel scale of $44\,\mathrm{mas} \times 133\,\mathrm{mas}$. As the satellite rotates, stars enter the focal plane from the left and drift across it at a rate of

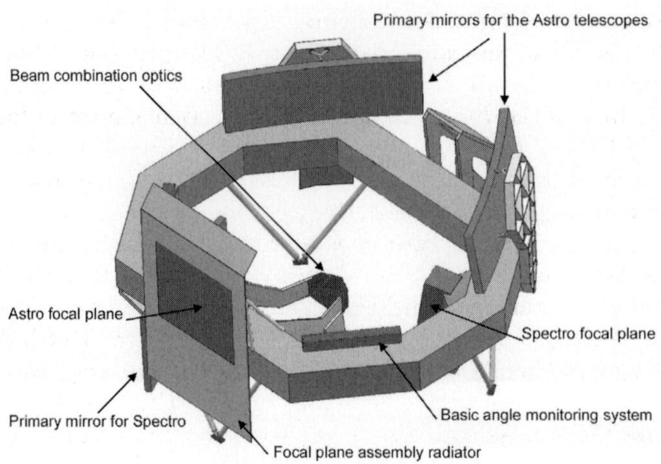

Figure 5. The Gaia payload. The two astrometric primary mirrors are $1.4\,\mathrm{m} \times 0.5\,\mathrm{m}$ in size.
© Astrium

$1°\,\mathrm{min}^{-1}$, crossing the focal plane in about $55\,\mathrm{s}$. The CCDs are clocked in time-delayed integration (TDI) mode at the same rate to track the stars. The reason for having a large number of CCDs in the along-scan direction is to acquire a high signal-to-noise ratio (SNR) in the stars' point spread function (PSF). This is the key ingredient to precise centroiding of the PSF and hence precise angular measurements in the along scan (quasi-great circle) direction.

Gaia detects objects in real time. There is no input catalogue telling it approximately where the target objects are, as was the case with Hipparcos. With Gaia, anything with a point source magnitude of $G < 20$ is detected in real time (the G band is defined below). This is done by the *star mapper* CCDs at the leading edge of the focal plane: an initial strip of CCDs does the initial detection, and another the subsequent confirmation and rejection of cosmic rays etc. There is a separate strip of star mappers for each field-of-view, masked such that each can only see one field-of-view. The reason for these star mappers is two-fold. First, there exists no sufficiently accurate catalogue complete to $G = 20$ over the whole sky at the required spatial resolution which could act as an input catalogue. Second, there is not enough bandwidth capacity in the Gaia antenna to transmit the entire sky to the ground. To do this we would require a data rate of around $1\,\mathrm{Gb\,s}^{-1}$, whereas Gaia is limited to a few $\mathrm{Mb\,s}^{-1}$, i.e. more than a hundred times lower.†
Thus by using the star mapper detections, Gaia electronically allocates CCD windows (of size $265\,\mathrm{mas} \times 1590\,\mathrm{mas}$) to stars and transmits only these parts of the focal plane to the ground.

The CCDs in the main part of the astrometric focal plane are unfiltered to maximise the number of photons collected. The resulting photometric passband is determined by the CCD quantum efficiency curve and the reflectivity of the optics and is named the G band. To maximise sensitivity for the typical star (which is cool and/or reddened) the

---

† The reason for this limited capacity is that Gaia must use an electronically steerable array rather than a mechanically movable one, which would permit a larger data rate. A mechanically movable antenna is incompatible with the very severe requirements on mechanical stability onboard the satellite, which is essential for high accuracy astrometry.

mirrors are silver coated. The net passband therefore lies between 400 and 1000 nm. At the high level of centroiding precision required (ca. 1 part in 1000), chromatic effects in the optics with this broad passband are significant. The position of the centroid of the stellar PSF depends on the colour of the star and this must therefore be corrected for. This is the main task of the *Broad Band Photometer* (BBP), a set of four or five filters attached to the CCDs at the trailing edge of the focal plane. (A different colour filter is attached to each of the final strips of CCDs so that all five colours are determined essentially simultaneously with the astrometry.)

The spectroscopic instrument, *Spectro*, is a separate telescope mounted on the same optical bench as Astro (see Fig. 5). It is a 3-mirror telescope with a single viewing direction. At the focal plane, part of the field is intercepted by a dichroic mirror. This reflects the red part of the light into the *Radial Velocity Spectrograph* (RVS), described in the article by Mark Cropper in this volume. The blue transmitted light – as well as the white light in the other part of the field – arrives at another CCD focal plane, the *Medium Band Photometer* (MBP). As with Astro, this consists of an array of individual CCDs, although not as many and with a larger pixel scale than in the case of Astro. This CCD focal plane operates in the same way as the astrometric CCDs, namely clocked in TDI mode to track stars as the satellite rotates and also with an independent set of star mappers. Now, however, a different medium band filter is attached to each strip of CCDs. As a star traverses the focal plane we therefore sample its spectral energy distribution in some $10 - 12$ passbands. The photometric system and number of filters is not finalised, but it will cover most of the wavelength region between 250 and 1000 nm. The purpose of MBP is to classify the objects observed and determine the physical parameters (temperature, metallicity, surface gravity etc.) for the stars. Techniques for doing this are discussed by Bailer-Jones (2002, 2003).

4. Astrometric accuracy

The design of the Gaia satellite, in particular the size of the mirrors and focal plane, its scanning law and mission duration, are driven primarily by the astrometric accuracy we want to achieve. These in turn are dictated by the science goals, discussed in detail in the Gaia Concept and Technology Study Report (ESA 2000). Essentially, to be able to answer the main questions about the three-dimensional structure of the Galaxy and to establish the kinematics of tracer stars to sufficiently large distances and with sufficient accuracy, there are two main requirements. The first is to achieve an end-of-mission astrometric accuracy of $10\,\mu$as at $G = 15$. The second is for the survey to extend to $G = 20$ and still retain good accuracy.

To first order, the uncertainty in an astrometric parameter (position, parallax or proper motion), σ_a, is given by $\sigma_a \sim 1/\sqrt{N}$, where N is the number of photons collected (Lindegren 1978; ESA 2000). Large N is achieved with a large aperture, a large field-of-view (so that a given star will be observed more often with a given scanning law), a large focal plane (so that each star is observed for longer during a transit) and a long mission life (so each star is observed many times).

Denoting the uncertainty in a parallax measurement with $\delta\varpi$, the corresponding uncertainty in the distance, d, is $\delta d \sim d^2 \delta\varpi$ (for $\delta\varpi \ll 1$). Thus for a given parallax accuracy, the fractional distance accuracy decreases linearly with increasing distance. For a given type of star (fixed intrinsic luminosity), $N \sim 1/d^2$, and hence $\delta d \sim d^3$. Roughly speaking, given the 5-yr mission of Gaia, the astrometric and parallax accuracy are the same as each other (e.g. $10\,\mu$as) and the same as the proper motion accuracy expressed per year ($10\,\mu$as yr^{-1}). From this we can see how the distance accuracy varies with distance

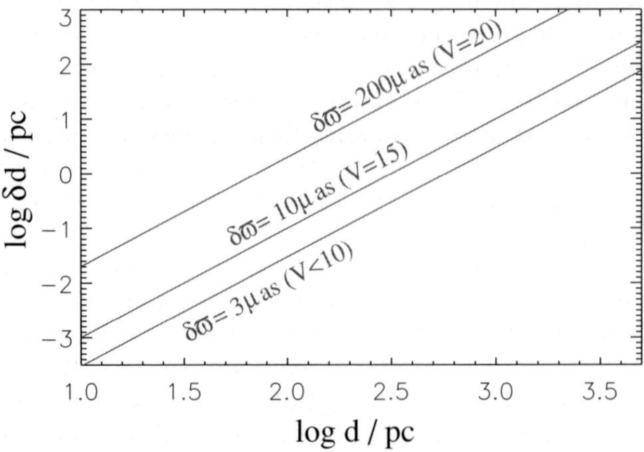

Figure 6. The Gaia distance accuracy at three magnitudes. To first order the uncertainty in the distance, d, is $\delta d \sim d^2 \delta\varpi$, where the parallax accuracy, $\delta\varpi$, depends only on the apparent brightness (to first order) as $\delta\varpi \sim 1/\sqrt{N}$ (N is the number of photons collected).

Table 2. Predictions (from ESA 2000) of the distance and tangential velocity accuracy which Gaia will achieve. It may be useful to recall than at a distance of 200 pc, a velocity of $1\,\mathrm{km\,s}^{-1}$ gives rise to a proper motion of $1\,\mathrm{mas\,yr}^{-1}$.

distance accuracy	*tangential velocity accuracy*
0.2% at $d = 200\,\mathrm{pc}$ at $V = 15$	$0.1\,\mathrm{km\,s}^{-1}$ at $d = 2\,\mathrm{kpc}$ at $V = 15$
1% at $d = 1\,\mathrm{kpc}$ at $V = 15$	$1\,\mathrm{km\,s}^{-1}$ at $d = 20\,\mathrm{kpc}$ at $V = 15$
20% at $d = 1\,\mathrm{kpc}$ at $V = 20$	$1\,\mathrm{km\,s}^{-1}$ at $d = 1\,\mathrm{kpc}$ at $V = 20$
accuracies achieved	*accuracies achieved*
$< 0.1\%$ for 700 000 stars	$0.5\,\mathrm{km\,s}^{-1}$ for 44 million stars
$< 1\%$ for 21 million stars	$1\,\mathrm{km\,s}^{-1}$ for 85 million stars
$< 10\%$ for 220 million stars	$5\,\mathrm{km\,s}^{-1}$ for 300 million stars

(Fig. 6). Combining these estimates with Galaxy models we can determine the number of objects which obtain various accuracy levels (Table 2). We see immediately the vast numbers of objects for which very accurate distances are determined, e.g. some 20 million with distances better than 1% (which includes stars down to $G = 20$ and with distances of up to a few kpc). In comparison, Hipparcos only obtained this distance accuracy for stars brighter than $V = 10$ which were nearer than 10 pc, a mere handful of objects.

5. Scientific contributions

A brief overview of Gaia's capabilities and likely contributions in five scientific fields is now presented. For more information and references see the Gaia Concept and Technology Study Report (ESA 2000) and the summary by Perryman et al. (2001), from which much of this section has been drawn. For more up-to-date information see the Gaia website `http://www.rssd.esa.int/gaia`

5.1. *Solar system*

Due to its high proper motion sensitivity, Gaia will detect numerous minor planets in our solar system. In particular, it is expected that it will detect some $10^5 - 10^6$ new objects in the main asteroid belt, compared to fewer than 10^5 currently known. The detailed statistics of orbital elements so derived can be used to investigate the formation of the solar system. Knowing their distances, the apparent magnitudes of asteroids can be used to derive their albedos. The medium and broad band photometry from Gaia will be used to classify asteroids and thus derive their chemical compositions. Furthermore, it is expected that during the lifetime of Gaia some 100 close mutual encounters will be observed and their open (hyperbolic) orbits traced: from this asteroid masses can be determined. This would increase the number of directly (i.e. dynamically) determined asteroid masses from the present number of about 20. Combined with size measurements from stellar occultations, densities may be derived, thus providing insight into their chemical composition and thus that of the primordial solar environment.

Because Gaia will be able to observe nearer to the Sun (within $40°$) and with much higher accuracy than is possible from the ground, it will make important contributions toward the study of Near Earth Objects (NEOs). Some of these asteroids, such as those of the Apollo and Athen groups, cross the Earth's orbit and so are potentially at risk of collision. As Gaia performs real-time onboard detection, it will detect and observe such objects many times over the course of its mission. It should be able to discover and determine accurate orbits for all NEOs with diameters larger than a few hundred metres.

The Trojans are asteroids situated around the Lagrange L_4 and L_5 points of planets, where stable orbits can be maintained for hundreds of millions of years. Planetary perturbations and mutual close encounters mean that they can have orbits elongated around the L_4 and L_5 points of their 'host' planet. Trojans are known for the outer planets, but in the inner solar systems the only known ones are two Martian Trojans. It is still not known whether the Trojans were captured or formed in situ, or even whether they have similar compositions to the main belt asteroids. Due to the large area it covers, Gaia will undertake a systematic survey for the Trojans of all planets, including, for the first time, for Venus.

Gaia will just about be sensitive enough to detect the very faint, red population in the Edgeworth-Kuiper Belt (EKB). The EKB, situated around $40\,\mathrm{AU}$ from the Sun, is a population of icy bodies which appear to be a remnant of the formation of the solar system. Pluto is its largest member. An improved census of the EKB members, their orbits and intrinsic properties will provide further insight into the formation of planetary systems. Although Gaia's sensitivity will not permit detections of large numbers of these very faint, cold objects, it is predicted to discover some 50 new objects. Of these, $5-10$ are expected to be binaries.

5.2. *Exosolar planets*

Through its high precision, multi-epoch astrometry Gaia will be able to detect the positional 'wobble' of a star due to the gravitational force of a companion. Not only will this allow Gaia to detect hundreds of millions of visually unresolved binary stars, but the sensitivity is high enough to detect motions caused by planetary mass companions. To date, most exoplanets have been detected by the radial velocity method, i.e. the variable Doppler shift of spectral lines induced by relative motions, although a few have been detected by astrometry (see the article by F. Benedict in this volume) and via photometric detection of transits. The astrometric amplitude of the motion of a star of mass M_s due a planet of mass M_p orbiting at a distance a is

$$\alpha/'' = \left(\frac{M_p}{M_s}\right)\left(\frac{a/\mathrm{AU}}{d/\mathrm{pc}}\right)$$

where d is the distance to the system from the Earth. For example, the 47 Ursa Majoris system is predicted to give an amplitude of $360\,\mu$as. The astrometric method is most sensitive to large or long period orbits, which is the opposite of the radial velocity (RV) technique. Therefore the techniques are complementary. But because the astrometric signature is observed in a two-dimensional plane we can determine the inclination of the Keplerian orbit. This allows a determination of the mass of the planet without the $\sin i$ ambiguity inherent to the RV technique (in which we only observe the projected orbital radial velocities).†

Gaia will be most useful for finding planets around stars with $V < 13$, with good capabilities out to about $d = 200\,$pc. This covers a vast number of stars – around $100\,000$, far more than have been observed with current RV planet surveys – over a wide range of spectral types. With a mission of 5 yr, Gaia will be able to determine orbits for planets with periods of up to about 10 yr. Extrapolating from current known systems (and only a limited part of the orbital parameter space has been explored), it is expected that Gaia will detect some 5000 new planetary systems (compared to some 120 known as of mid-2004) and determine accurate orbital elements for $1000 - 2000$ of them. From the above equation, we see that for a given size of orbit and mass of star, the minimum mass planet which Gaia can detect decreases linearly with distance. For the nearest stars – closer than 10 pc – Gaia will be able to detect planets with 3% of the mass of Jupiter, or 10 Earth masses.

This extensive exoplanetary survey will provide detailed statistics on the types of planetary systems which stars of different mass, metallicity, and age do and do not host. This will mark a major contribution to understanding the formation and evolution of planetary systems.

Gaia should also be able to detect some planets when they transit across their host stars and thus cause a dimming of the integrated light received by Gaia. Of course this only occurs for near edge-on orbits (i.e. $i \simeq 90°$). As Jupiter has a diameter about ten times smaller than the Sun, its transit would cause the Sun to dim by 1%. To detect this with a signal-to-noise of 10 in the G band requires a photometric precision of 1 mmag, easily achieved in single epoch photometry for stars brighter than about $G = 13 - 14$. Despite the relatively poor time sampling for transits, Gaia observes millions of such stars. Predictions have been made that of order 6000 planets around F–K type stars with semi-major axes in the range $0 - 2\,$AU will be detected this way. Incidentally, and in keeping with the theme of this conference, a transit of Venus across the Sun (radius ratio of 1/115) observed from a large distance (not from the Earth!) causes a maximum dimming of 0.08 mmag. This is at the level at which Gaia photometry will probably be dominated by systematic errors, so is unlikely to be detectable even for the brightest stars.

5.3. *Star formation and stellar clusters*

Most star formation occurs in cold, dense, dark molecular clouds. As gas is converted into stars via gravitational collapse (and quite possibly via the dissipation of supersonic turbulence), the cloud emerges from being an embedded cluster to an optically visible young star-forming region or open cluster, such as the dense Orion Nebula Cluster or the less dense low-mass star-forming region Taurus-Auriga (with ages of a few million years).

---

† With both the astrometric and RV techniques we can only determine the mass of the planet (or rather $m \sin i$ for the RV method) if we assume a mass for the star: just astrometry or RV alone does not allow us to determine the individual masses of unresolved systems. However, in many cases, the stellar mass can be determined with reasonable accuracy from Gaia's photometry.

Such clusters evolve dynamically: internal encounters will slowly evaporate the lower mass members and interactions with the Galactic tidal field and passing interstellar clouds will disrupt all but the densest clusters. Surveys of stellar clusters show a marked drop off in numbers beyond a few hundred million years in age. Thus is appears that almost all clusters will disperse into the Galactic field population over a timescale somewhat less than 1 Gyr.

By studying the structure and content of clusters – in particular the initial mass function and binarity – at different ages and abundances we therefore study the recent star formation history of the Galaxy.

Gaia will make some of its most fundamental contributions in this area. First, by measuring very accurate distances for individual stars in clusters, it will allow an accurate determination of stellar luminosities across a wide range of evolutionary phases, thus making vital tests of theories of stellar structure and evolution. For the nearest clusters, such as the Hyades, Gaia will determine the IMF (corrected for binarity) down into the brown dwarf regime and down to a solar mass for a cluster as far away as 3 kpc. By accurately determining the distances to subgiants, the position of the Main Sequence turn-off can be determined to much higher precision than is presently possible, resulting in a significant improvement in age determinations through isochrone fitting.

Of the 1000 clusters currently known (mostly within 2 kpc), distances are only known to about half of them, of which many are still highly uncertain, by factors of two or more. According to WEBDA there are some 70 clusters within 500 pc, of which 20 are closer than 200 pc. For all of these, Gaia will determine distances to *individual* members brighter than $V = 15$ to better than 0.5%. This corresponds to a depth precision of 2.5 pc at 500 pc or better for nearer cluster and/or brighter stars. These data will provide a three dimensional map of clusters and their mass distribution. Combined with accurate three dimensional velocities from the radial velocities and proper motions, we will be able to study the dynamical evolution of clusters across a range of characteristics in much more detail than has previously been possible. Recalling that Gaia's distance accuracy, δd, depends on distance, d, as $\delta d \sim d^2$ for a star of given apparent magnitude, then at a distance of 100 pc the depth uncertainty is only 0.1% or 0.1 pc. For a given type of star, i.e. of given intrinsic luminosity, $\delta d \sim d^3$. Thus out to 200 pc, Gaia will measure the distances to all giants and all dwarfs earlier than about K5–M0V to better than 1 part in 250. In terms of kinematics, all these stars will have their tangential velocities determined to better than $1\,\mathrm{km\,s^{-1}}$. Extending the range to 500 pc, then giant star kinematics are still measured with this precision or better, as are dwarfs earlier than mid A types. Armed with this information we can investigate mass segregation, ejection and the the dispersion of clusters.

The power of Gaia comes in the vast number of stars it observes over the whole sky with a well-defined selection function. By searching for clusterings in the multidimensional space formed by the 3D spatial co-ordinates, the 3D velocity co-ordinates, as well as "astrophysical co-ordinates" ($\mathrm{T_{eff}}$, [M/H] etc. obtained from Gaia photometry) we can use the Gaia archive to detect new clusters, associations and moving groups out to several kpc. This will provide a much more complete, systematic and reliable survey for stellar clusters than has hitherto been possible.

5.4. *Galactic structure*

Gaia's combination of 6D phase space information and astrophysical parameter determination make it a unique instrument for determining the large scale structure of our Galaxy. It is here that its faint limiting magnitude, all-sky coverage and determination of stellar properties (especially metallicity) are particularly important. We can, for example, investigate the age–metallicity relation (using long-lived K and M dwarfs) with a much larger sample than has been possible to date. A dynamical determination of the mass of the disk and its dark matter content will likewise be possible, as is a determination of the Galactic rotation curves out to large Galactocentric distances.

Within the Galactic disk, Gaia will provide an accurate mapping of star forming regions and spiral arms. Interstellar extinction will ultimately limit how far Gaia can see in the Galactic plane, but by identifying and measuring parallaxes of very luminous Cepheid variables, for instance, we can map out structure to large distances.

Moving to the halo, the Gaia database will allow us to search for stellar clusterings in the Galactic halo which may be hallmarks of merger events. In addition to the spatial overdensity searches generally carried out with existing star-count data sets, we can apply multidimensional cluster search algorithms to look for intrinsic patterns in a space–velocity–astrophysical parameter space. This would be relevant for finding halo streams which may not be spatially distinct but nonetheless share a common orbit and/or star formation epoch. Even at $10\,\mathrm{kpc}$, Gaia can determine tangential velocities to $0.5\,\mathrm{km\,s^{-1}}$ for stars as faint as $G = 15$ (i.e. with absolute magnitudes less than zero, such as OB main-sequence stars and K giants). In the outer halo ($> 20\,\mathrm{kpc}$ from the Galactic Centre), most stars will have relatively uncertain parallaxes (parallax errors of greater than 20% for $G > 15$). However, photometric 'parallaxes' may then provide a reasonable distance estimate from which proper motions may still be converted to tangential velocities of useful precision. Moreover, as intervening foreground objects will have much larger (and accurate) parallaxes, these foreground 'contaminants' can be distinguished from the background tracers giants which may then be used as kinematic and density tracers in the halo.

5.5. *Extragalactic*

Beyond our Galaxy, the parallaxes of individual objects will be negligible. By averaging over a population, useful results may be obtained. For example, the Large Magellanic Cloud (LMC) has a parallax of about $20\,\mu\mathrm{as}$. But by averaging over a suitable set of members, a geometric distance to the LMC with an accuracy of about 1% should be obtainable.

In studying local group galaxies, Gaia proper motions will permit a better distinction between stars in our Galaxy and those in external galaxies. With a reliable selection, mean distances and motions of local group galaxies can be determined. By determining 3D orbits within the local group out to $1 - 2\,\mathrm{Mpc}$ (which includes some 20 galaxies), it may be possible to probe fluctuations in the initial density distribution of matter on cosmological scales.

An important aspect of Gaia is its direct optical observations of quasars. From these, Gaia astrometry will be tied to a quasi-inertial reference frame with an accuracy of better than $1\,\mu\mathrm{as\,yr^{-1}}$. Down to its limiting magnitude of $G = 20$, it is predicted that Gaia will detect around half a million quasars. These will be observed in all of Gaia's 15 photometric bands at some 100 epochs from which the classes of quasars and their variability may be studied. Interestingly, Gaia will be sensitive to the motion of the solar system barycentre about the Galactic Centre. This motion induces an aberration in the position of distant objects such as quasars. With the Sun moving at $220\,\mathrm{km\,s^{-1}}$ at a distance of $8.5\,\mathrm{kpc}$ about the Galactic Centre with a period of $250\,\mathrm{Myr}$, quasars will show an apparent proper motion of about $4\,\mu\mathrm{as\,yr^{-1}}$.

Gaia performs real-time onboard detection of everything with a point source magnitude brighter than $G = 20$. Thus transient events, in particular supernovae, will be detectable. Based on the current known supernova rate, Gaia will catch about 50 supernovae *per day*. A real-time classification and alerts system is envisaged for such events, from which rapid follow up by other observatories will be possible.

Finally, because Gaia will measure distances accurately to very large numbers of different types of stars, including RR Lyraes and Cepheids, it will permit a much better calibration of primary distance indicators than has hitherto been possible. This will provide a geometrically calibrated distance scale which is independent of CMB or cosmological models.

6. Current status and data analysis

At the time of writing (mid-2004) Gaia is planned for launch in 2011. Present efforts focus on several essential technology developments on the industrial side, related, for example, to the CCDs, SiC mirrors and FEEP thrusters. On the scientific side, the major on-going effort is preparation for the data analysis.

Gaia will produce data at a rate of about $1\,\mathrm{Mb\,s^{-1}}$ for 5 yr, producing a total of some 100 TB raw data. This quantity of data is not actually that large. The challenging issue with the Gaia data processing it the complexity of the data. Due to the continuous scanning of the satellite over a period of 5 yr, the astrometric data stream essentially consists of a strip of focal plane data 2.6 million degrees in length. Each object typically appears 100 times in this data strip. Thus the data reduction involves object matching in the 7000 or so different quasi-great circle scans making up this strip. The relative one-dimensional positions of a few hundred million stars must then be determined from which the five basic astrometric co-ordinates of each of these stars are solved for in the so-called *Global Iterative Solution* (GIS). This process must simultaneously solve for the attitude of the satellite and the various instrument calibration parameters. Numerous additional astrometric reduction tasks much be included, such as solving for higher order astrometric terms of binary star systems. Furthermore, the radial velocity data and photometric data must also be included, often simultaneously. For example, the colour of the star must be included in the astrometric reduction, as must the radial velocity for nearby, high proper motion stars (as their parallaxes and proper motions will not be constant with time).

The data analysis with Gaia is therefore a major challenge. A data reduction prototype, GDAAS, has been running for a few years which is studying these problems. Beyond this, efforts are also ongoing to ensure that the resulting Gaia database can be properly exploited for subsequent scientific work. These include experiments in data mining and work on theoretical models against which Galactic structure and kinematic data may be compared.

Acknowledgements

Gaia is a project of the European Space Agency (ESA) with guidance from the scientific community organised into several working groups under the auspices of the Gaia Science Team (GST). It is through the efforts of these people, ESA staff and our industrial partners that Gaia has reached its present advanced stage of development. Much of this article has drawn from the Gaia Concept and Technology Study Report (ESA 2000), complied by the former Gaia Science Advisory Group, from which more detailed information and references can be obtained.

References

Adams D. 1978 *The Hitchhikers guide to the Galaxy*, (radio series), BBC.
Bailer-Jones C A L 2002 *Ap&SS* **280**, 21–29.

Bailer-Jones C.A.L. 2003 In *GAIA spectroscopy, science and technology* (ed. U. Munari). ASP Conf. Ser. 298, pp. 199–208. Astron. Soc. Pacific.

ESA 1997 *Hipparcos Venice '97* ESA-SP-402.

ESA 2000 *GAIA: Composition, formation and evolution of the Galaxy* ESA-SCI(2000)4

Lindegren L. 1978 In *Modern Astrometry* (ed. F.V. Prochazka, R.H. Tucker). IAU Colloquium no. 48, pp. 197–217. Institute of Astronomy (University Observatory), Vienna.

Kovalevsky J. 1995 *Modern astrometry* Springer, Heidelberg.

Perryman M.A.C. & Hassan H. et al. 1989 *The Hipparcos mission. Pre-launch status* ESA SP-111.

Perryman M.A.C. et al. 2001 *A&A* **369**, 339–363.

Discussion

FRITZ BENEDICT: I'm interested in the things that are transient, like a supernova or a transit. Is the procession rate so slow that you come back in six hours and look at the same object again?

CORYN BAILER-JONES: Yes, with the two fields of view, depending on exactly where the first object passes the focal plain. Because if it's not too much at the edge, then the second field-of-view will get it, and may even get it on the next rotation. I think it's possible to get it up to three or four times; if it's situated right on the top of the focal plain you can actually get it several times in the same orbit, but the sampling is very non-uniform over the sky.

DAVE MONET: Very nice talk, thank you. I would argue that we have, in fact, looked and catalogued the sky down to at least 7 times your radial velocity limits, so there is something that looks like an input catalogue, at least for counting stars. A question: How do you extend the dynamic range with normal CCDs and A-to-D converters? If you have a decent response at 20th magnitude, it's something like 5 magnitudes and things begin saturating. So how do you go from 15th to 10th to 5th and fill in the bright end of the catalogue?

CORYN BAILER-JONES: Let me just answer your first point. Of course we have catalogues of objects down to 20th magnitude, clearly. The problem partly is the sampling. With the pixel scales on Gaia you would need to have a catalogue down to the less-than-0.1 arc-second scale. To answer your question in terms of saturation: It's still very much open how bright we can go with Gaia, where the saturation limit sets in. There are various ways you can do this: for example, using gates on CCDs to actually limit the integration time of the CCD. Another option that has been suggested is that you saturate the star, but because the PSF has wings, it may be possible to calibrate the wings of the PSF. Obviously, you have to control carefully the charge, the blooming and the leakage. There are on-going studies about this.

BARBARA BROMAGE: You were talking about the complexity and the size of the data handling problems: Are you working along side the ASTROGRID group in planning how you are going to deal with this, or are you thinking of putting in a bid for funding the development of the data handling yourself?

CORYN BAILER-JONES: We are currently running a fairly big data processing study called GDAAS, which is looking in some detail of the core data processing tasks; this is in collaboration with a software company in Spain. We are also looking at GRID technologies at the moment, but our approach is not to rely too much on where things like GRID maybe going in terms of doing the processing. One of our real headaches is to work out the size of the problem, the structure of the problem and how we are going to distribute the problem. The goal is that the data processing – like Hipparcos – will be done by a dedicated team from the scientific community. I think nobody is suggesting that you should leave this job for scientists on their own. We are good at some things, but we are terrible at doing other things. So there's going to have to be quite a serious input from professional software engineers. We are in discussion with other groups that have had to deal with these kind of large problems – in particular at CERN, and also with the Sloan [Digital Sky Survey] – because they have had these kind of problems to deal with. The plan is that there will be an announcement of opportunity for the scientific community to get involved and do this data processing. That could be next year, or soon thereafter.

Coryn Bailer-Jones waits for a look at the transit

Transits of Venus: New Views of the Solar System and Galaxy
Proceedings IAU Colloquium No. 196, 2004
D.W. Kurtz, ed.

© 2004 International Astronomical Union
doi:10.1017/S1743921305001602

Radial Velocities with Gaia

Mark Cropper[1], David Katz[2], Ulisse Munari[3], Tomaz Zwitter[4] and Andrew Holland[5]

[1]Mullard Space Science Laboratory, University College London, Holmbury St Mary, Dorking, Surrey RH5 6NT, UK email: msc@mssl.ucl.ac.uk

[2]Observatoire de Paris, GEPI, 5 Place Jules Janssen, F-92195 Meudon, France

[3]Osservatorio Astronomico di Padova INAF, sede di Asiago, 36012 Asiago (VI), Italy

[4]University of Ljubljana, Department of Physics, Jadranska 19, 1000 Ljubljana, Slovenia

[5]Centre for Electronic Imaging, School of Engineering and Design, Brunel University, Uxbridge, UB8 3PH, UK

Abstract. Gaia is a mission to map the positions and determine the velocities of 1 billion stars in the Galaxy, in order to determine how it was formed and how it has evolved. Spectroscopy provides one component of the space velocities for 200 million objects, and, in addition, astrophysical information such as temperature and metallicity for the brighter 40 million stars in a large unbiased Galactic sample. Together the instruments on Gaia will transform our knowledge of the Galaxy. We describe here a sample of the science that will be done with Gaia concentrating on aspects in which the spectroscopy plays a central role.

1. Introduction

Gaia is an ESA Cornerstone science mission due for launch in 2011. Gaia's role is to determine how the Galaxy was born and how it has evolved, and it will do this by measuring the positions and space velocities of $\sim 10^9$ stars. The positions will be obtained by precise astrometry at the level of $10\,\mu$arcsec (for stars of $V = 15$), with the distances following though parallax measurements. Transverse space velocities are determined from proper motions throughout the six-year mission, while the radial velocities are obtained via a dedicated Radial Velocity Spectrometer (RVS) with a resolving power of 11 500 and working in a band from $8480 - 8740\,\text{Å}$ around the Ca triplet. Gaia will also have a photometric capability, allowing luminosities to be determined, and, together with the RVS, will provide detailed physical characteristics (temperature, metallicity, gravity, etc.) of each object observed. It should be noted that owing to the scanning of the satellite, the observations are multi-epoch, providing information on variability and binarity. Astrometric and photometric measurements extend to beyond $V = 20$, while spectroscopic measurements reach $V = 17$. The completeness and scale of this dataset will be unprecedented, with selection functions that are clear and simply treated.

The scientific harvest of Gaia will be far-reaching, extending beyond the Galaxy to the local group, to the discovery of several supernovae and 250 QSOs per day, many of them lensed. It will find thousands of new Jupiter-mass planets, and within our Solar system, will detect large numbers of Kuiper-belt objects and asteroids. Gaia will impact profoundly on fundamental physics, for example providing tight constraints on the distribution and characteristics of Dark Matter, on the Modified Newtonian Dynamics (MOND) theories and on changes in the gravitational constant. An overview of the mission, including the science case, is available from Perryman et al. (2001), while up-to-date information can be obtained from the project website at http://www.rssd.esa.int/GAIA/.

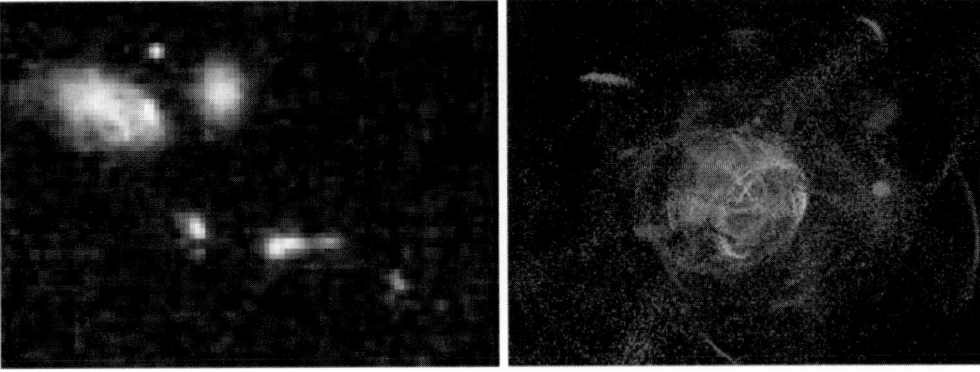

Figure 1. Left: Detail from the Hubble Space Telescope Ultra-Deep Field. Image courtesy of NASA. Right: A simulation depicting the formation of a halo by the accretion of 50 dwarf galaxies at various times during the last 10 Gyr. Image by Paul Harding, available from http://burro.astr.cwru.edu/heather/spag.html (Harding et al. 2001)

Gaia builds on the legacy of ESA's successful Hipparcos mission, which pioneered space-based astrometry. The scientific harvest from Hipparcos was diverse (see, for example, Perryman 1997). However, Hipparcos lacked a facility to measure the radial motions, with implications that were appreciated at the time. One example was due to Blaauw (1988):

> "...there are, of course many things left that one would like to see done. One of them, perhaps the most urgent, is in the field of radial velocities...there is no really dedicated Hipparcos radial velocity programme...may we be sure that, when the Hipparcos proper motions become available, we will not be faced with a deplorable lack of data on the third component of stellar motions?"

Since then the absence of radial velocities for the Hipparcos data has been noted several times (e.g. Binney et al. 1997). While a large dataset of some 13 500 velocities has recently become available (Nordström et al. 2004), it should be noted that this is still only 10% of the Hipparcos dataset, and the situation is expected to remain unsatisfactory. The Nordström et al. (2004) catalog contains F and G dwarf stars to $V \sim 10$ with the peak in the distribution at $V \sim 8.5$.

It was understood early on in the conceptual design of Gaia (Favata & Perryman 1997) that, given the scale of the output catalog, the measurement of radial velocities would need to be obtained in parallel with the astrometric measurements. The mission was therefore designed with the RVS as an essential component of the instrument suite. This paper provides a brief overview of the science return from the RVS. More complete expositions can be obtained from Katz et al. (2004) and Wilkinson et al. (2004), with a wider view available from the Monte Rosa Conference proceedings (e.g. Munari 2003a).

2. RVS Science Topics

The highest priority science areas for RVS include the perspective acceleration correction (essential for astrometric measurements), thin disk, thick disk and halo kinematics, and the determination of the Galactic gravitational potential. Other science goals include mapping reddening, spiral arm and open cluster kinematics, binary systems, chemical abundance gradients, and the Galaxy disk warp.

RVS will achieve $< 1\,\mathrm{km\,s^{-1}}$ velocity accuracy for $\sim 4 \times 10^7$ stars, and $< 15\,\mathrm{km\,s^{-1}}$ accuracy for $\sim 10^8$ stars.

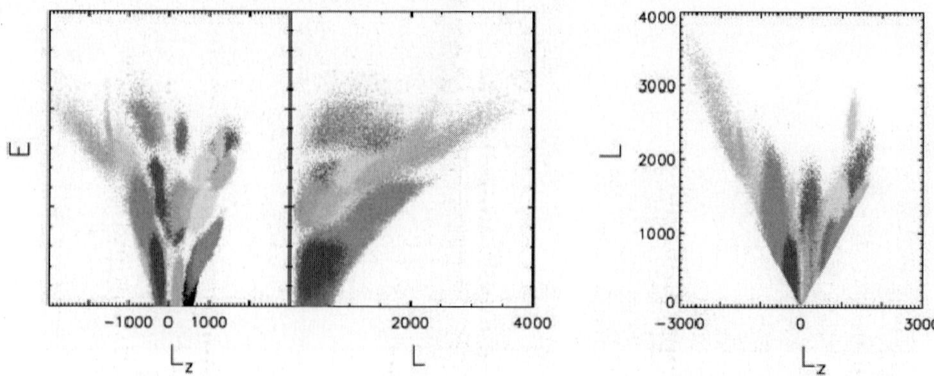

Figure 2. The distribution of particles within Galactic halo in the integrals of motion space after 12 Gyr. Particles denote bright giants ($M_V = 1$) with color coding representing different accreted satellites. From Helmi & de Zeeuw (2000).

This section examines some of the areas of Gaia science in which spectroscopy plays a central role.

2.1. Halo Streams

The Milky Way is the only place where *direct* tests of galaxy formation scenarios can be made. Gaia's main quest is to understand how the Galaxy was formed and how it has evolved. This has been a long-standing issue in astronomy since the early seminal work in this area (such as Kapteyn 1922). For some time the accepted model was that of Eggen, Lynden-Bell & Sandage (1962), in which the Galaxy formed in a monolithic collapse, with the halo as the oldest component, and the flattened disk resulting from conserved angular momentum as the youngest. More recent work has emphasised the importance of hierarchical merging of smaller systems with the Galaxy, perhaps responsible for generating the thick disk (Quinn, Hernquist & Fullagar 1993). Recent images from the HST Deep Field and Ultra-Deep Field (Fig. 1, left) have shown how irregular most galaxies are at early epochs. This results (at least partly) from a combination of the relaxation of the dark matter halo and relatively active merging, together with the effects of early star formation activity.

Because of the long timescale for interactions of the halo stars, the best place to search for evidence of the merger history of the Galaxy is in the halo, since this is where relics would retain their identity longest. This led several groups to examine the possibilities from a theoretical point of view (Johnston, Hernquist & Bolte 1996; Helmi & White 1999). Fig. 1 (right) shows a simulation in which the outer halo is composed of a number of different streams, with many older streams widely spread throughout the halo. Helmi & de Zeeuw (2000) addressed this issue from the point of view of the integrals of motion (energy, momentum) and found that these quantities for individual satellites retained their identity despite the spatial dispersion of their stars (see Fig. 2). This raised the possibility of tracing up to 80% of the merger history of the Galaxy (i.e. to $z < 4$). At the same time progress was made observationally in identifying the disruption of satellites, for example in the dSgr galaxy where this process has proceeded towards almost complete disruption (Ibata, Gilmore & Irwin 1994).

For this work to identify the various merger components in the merger scenarios, the integrals of motion (energy, momentum) must be derived from 6-dimensional position and velocity information. These motions are derived mainly from the kinematics of halo

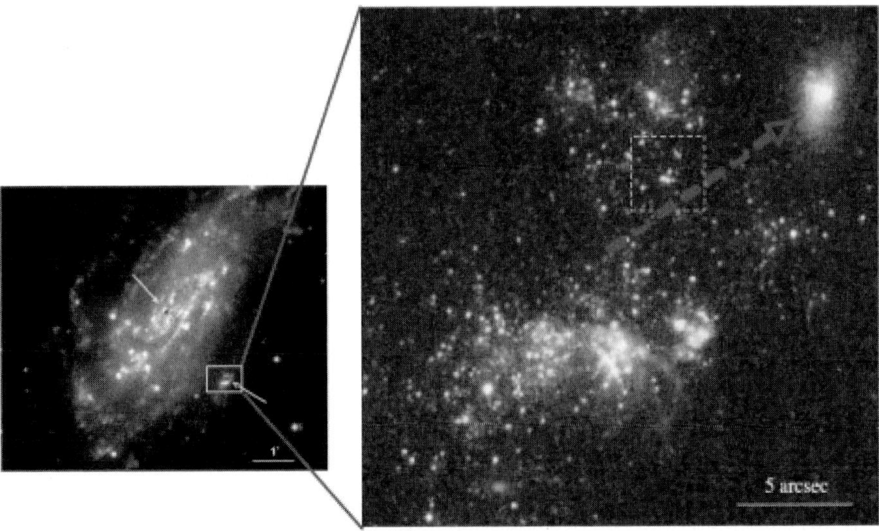

Figure 3. Star formation in the outskirts of NGC 4559. On the left is a UV image from the Optical Monitor on XMM-Newton while the enlarged image on the right is a 3-colour HST image. Soria et al. (2004) suggest this region is caused by the passing of a dwarf spheroidal galaxy (orange, upper right) through the outer disk of NGC 4558. From Soria et al. (2004).

stars with $V > 16$, and it is necessary to reach $V > 17$ to generate a sufficient sample. This is a significant driver for RVS performance.

2.2. *Solar System Origins and Evolution*

Despite the greater loss of information from kinematic motions within the disk compared to the halo arising from the greater number of interactions (mostly with spiral arms and molecular clouds), because of the accuracy and completeness of the astrometric and velocity measurements, Gaia will provide information on the origin of the Solar system and its subsequent evolution. It should be possible to distinguish among different scenarios, for example, whether the Sun formed within our main Galaxy, or whether it was incorporated as the result of a merger, as currently suggested for Arcturus (Navarro et al. 2004). Alternatively, because the extent to which star formation is triggered by mergers has become increasingly evident, for example in the Antennae galaxies, or NGC 4559 (Fig. 3), was the Solar system formed as a result of such a merger or interaction?

With regards to the more recent evolution of the Solar system and life within it, it was possible with Hipparcos data to reconstruct the motions of stars in the Solar vicinity over the past few Myr to examine which have passed sufficiently close to disturb the Oort Cloud, which is thought to extend a significant part of the distance to the nearest star (Algol passed within 2.5 pc of the Sun 7 Myr ago). For Gaia the full space motions will be known with much greater accuracy, so that ancient star encounters can be examined over geological timescales > 200 Myr. This will make it possible to test suggestions that extinctions in the number of species of life have been caused by cometary impacts resulting from disturbances of the Oort Cloud. It will also be possible to predict future nemeses.

2.3. *Normal Stars*

"Normal" stars, whether single or in resolved binaries, will form the bulk of the Gaia harvest. This will be a huge homogeneous sample for which astrophysical parameters

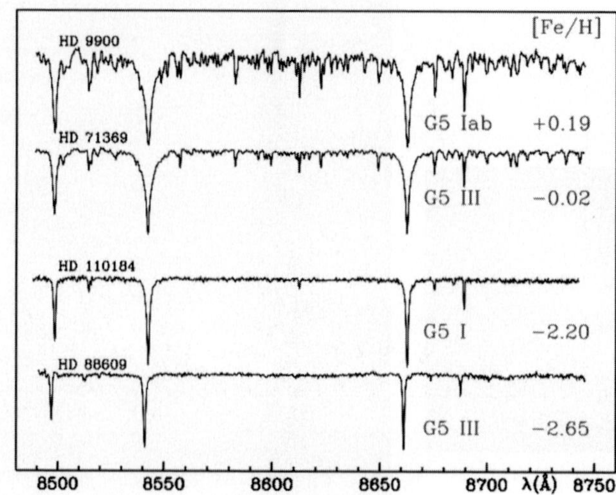

Figure 4. A metallicity sequence for spectra of G5 giant and supergiant stars. From Munari (2003a).

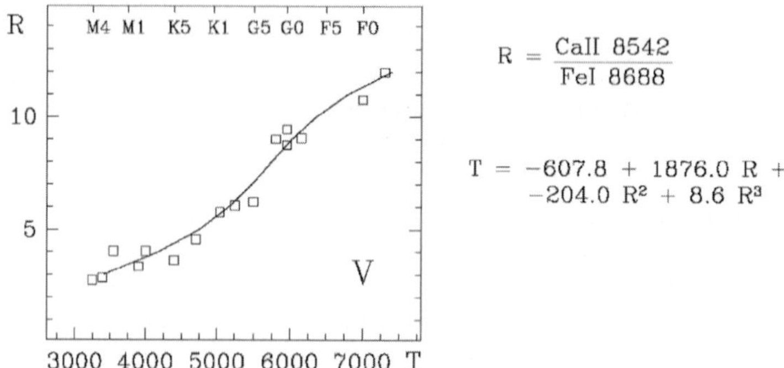

$$R = \frac{\mathrm{CaII}\ 8542}{\mathrm{FeI}\ 8688}$$

$$T = -607.8 + 1876.0\ R + -204.0\ R^2 + 8.6\ R^3$$

Figure 5. A calibration of the ratio between Ca II 8542Å and Fe I 8688Å line equivalent widths for dwarf stars as a function of temperature. From Boschi et al. (2003)

such as effective temperature, metallicity, gravity and rotation can be measured. From these can be inferred the masses, luminosities and ages of the stars, and when combined with the kinematics, this will form the basis for the work on the chemical evolution of the Galaxy, on the IMF and star formation rates and on feedback processes. Being a spectrograph, RVS is an essential component in this, uniquely providing the detailed elemental abundances and rotations, and independent measures of T_{eff} and gravity. An example is shown in Fig. 4, where a sequence in metallicity [Fe/H] in giants and supergiants is evident from the strength of the absorption lines of elements other than Ca (note that the Ca triplet is a strong spectral feature, and thus provides a good velocity marker even in extremely metal-poor stars). Fig. 5 shows an example calibration of the ratio of the Ca II 8542Å and Fe I 8688Å lines with temperature.

Such detailed astrophysical information will be obtained for the brighter stars within the sample $V < 15$. This is still a sample an order of magnitude larger than that available from any foreseen spectral surveys.

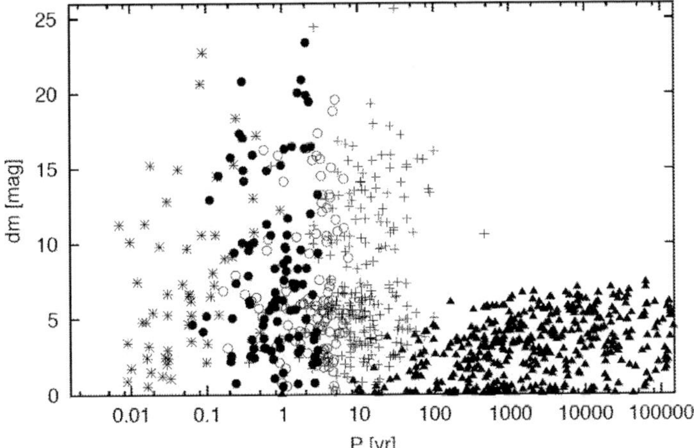

Figure 6. The properties for binaries as a function of period (years) and visual magnitude difference of the stellar components, for $V < 15$. Each mark represents 5000 systems. Binaries detected by spectroscopy alone are designated by $*$, while those detected by both astrometry and spectroscopy are designated by $\bullet$. Other symbols indicate detection by astrometry alone. From Söderhjelm (2003).

2.4. *Binaries*

Perhaps up to 75% of stars in the Galaxy exist as binaries, but the real fraction is still not well known (e.g. Duchêne 1999). The spectrum of separations and mass ratios is also not well characterised, and is affected strongly by selection effects. Binaries are important because they are (except in exceptional circumstances) coeval pairs. They are thus direct tests of stellar evolution, and they offer the opportunity of direct mass and radius measurements, essential inputs to our understanding of stellar structure. Examining the binary population within different Galactic constituents (halo, thick disk, thin disk) can determine how stellar parameters depend on mass, age and chemical abundance. Moreover, a significant fraction of the angular momentum in the initial star forming cloud is locked up in binary and multiple systems, so it is impossible to examine the nature of star formation in any detail without including in the models the appropriate binary fraction and associated mass ratios and orbital characteristics.

The presence of binaries has a number of implications for Gaia. Astrometric measurements will be affected by orbital motion, so that the astrometric solutions need to include the simultaneous derivation of the binary parameters, as for Volume 10 of the Hipparcos catalog (Perryman et al. 1997). In many cases the signature of binarity will be subtle, for example in close systems, or very wide systems or where mass ratios are large; in some cases the binary will contain variable stars, which could result in photocentre changes, and thus errors in the parallax determination (Pourbaix, these proceedings). On the other hand, the combination of astrometric, spectroscopic and photometric instruments on Gaia will be orders of magnitude more powerful than any current facilities in detecting binary systems, and in determining their characteristics.

Fig. 6 shows how the whole range of orbital periods is covered with Gaia. Close pairs with periods from less than a day are detected in the RVS, while wide pairs with periods up to $\sim 10^6$ yr are detected from astrometric motions. Binaries with periods from a few days to ~ 5 yr are detected with both techniques.

Spectroscopic double-lined eclipsing binaries will be most important for constraining stellar parameters. Currently only a few dozen of these are known, but $> 100\,000$ are

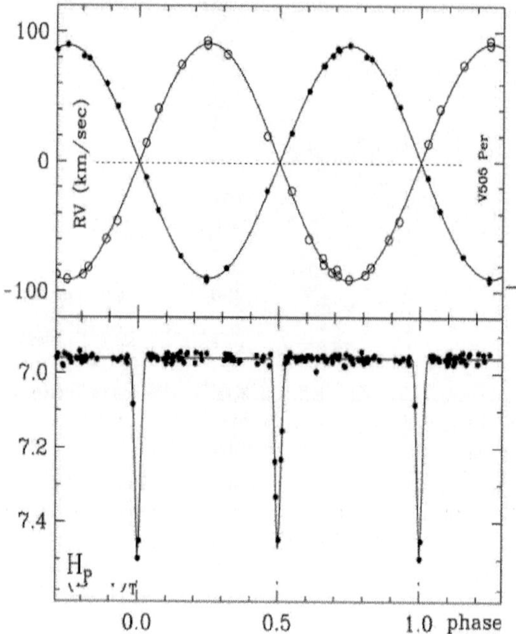

Figure 7. V505 Per radial velocities (top) and Hipparcos photometry (bottom) for this double-lined eclipsing binary. The data have been folded in phase. From Munari et al. (2001).

expected to be discovered for $V < 15$, mostly with the RVS. Fig. 7 shows an example of ground-based spectroscopic observations of V505 Per, found to be eclipsing from Hipparcos photometry. Gaia will also provide astrometric measurements. Complete solutions will be obtained for masses, radii, luminosities, and semi-major axes. Some 1% of the sample, > 1000 binaries, will have exceptionally precise measurements and will form a reference sample for stellar parameter determination.

2.5. *Non-normal Stars*

A significant fraction of the Gaia sample will be of "peculiar" stars of various types. These continue to be very important in elucidating particular aspects of astrophysics and include stars with coronal emission, winds, accreting systems, magnetically interacting systems, eruptive, pulsating and oscillating systems, systems in particular stages of evolution (some very short), and so on. Even for relatively rare systems, the sample sizes of such systems will be transformed by Gaia.

This group is too diverse to provide anything more than a few examples in the RVS spectral range. Fig. 8 (left) shows examples of Carbon star spectra, in which the Ca triplet is completely dominated by other absorptions. Fig. 8 (right) shows spectra of halo symbiotics, in which subtle signatures of wind emission are evident – note the relative velocity shifts between these halo objects.

Fig. 9 shows spectra of nova-like Cataclysmic Variables in the RVS spectral range. Broad Paschen lines in emission are evident from the accretion disk, together with narrower components from the heated secondary star. The variation of these lines with orbital period allows the velocity structure and hence tidal effects in the disk to be determined; the multi-epoch nature of the Gaia survey is an important ingredient in this case.

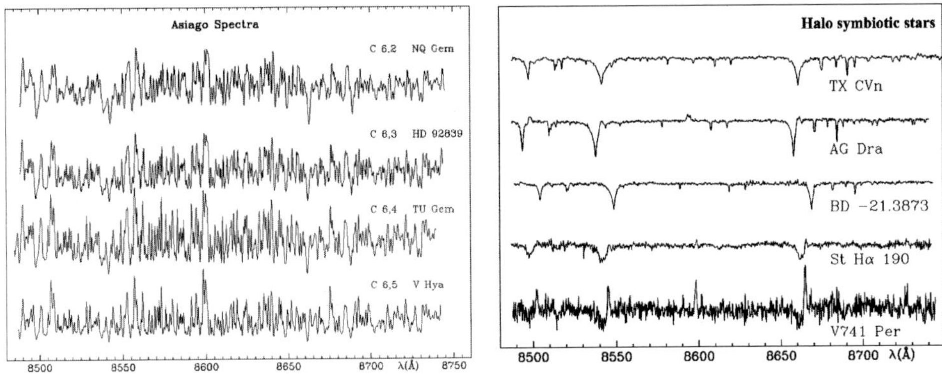

Figure 8. Left: Spectra of four Carbon stars in the RVS spectral range. Right: Halo symbiotic stars, some showing P Cygni profiles as a signature of wind emission in those systems. Both figures from Munari (2003b)

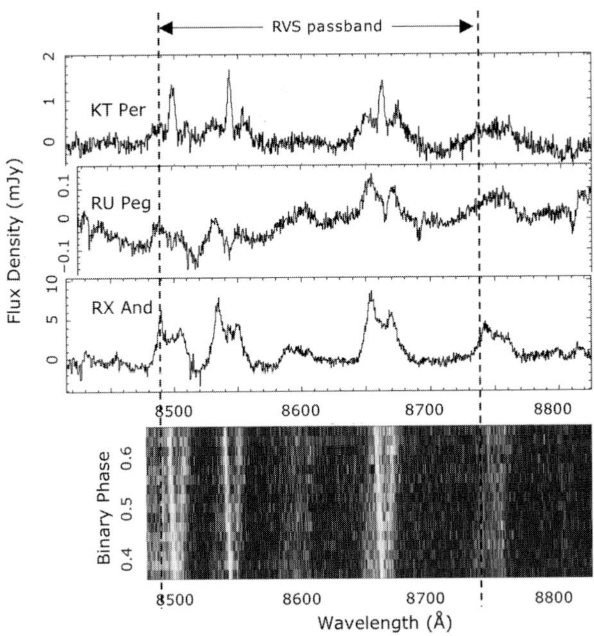

Figure 9. Spectra of the nova-like Cataclysmic Variables KT Per, RU Peg and RX And in the RVS spectral range. The plot at the bottom shows the variation of the spectral features as a function of time in the case of RX And. From Cropper & Marsh 2003.

3. RVS Characteristics

RVS consists of an Offner-type spectrograph fed by a specific telescope on Gaia. The telescope also provides images for the Medium Band Photometer (see Katz et al. 2004). This optical system provides good image quality with minimal distortion over the field of view. Gaia scans the sky continuously, so images drift over the focal plane, with the detector clocking synchronised to the scan rate. Distortion is important to minimise in a scanning instrument, since it translates directly to image degradation.

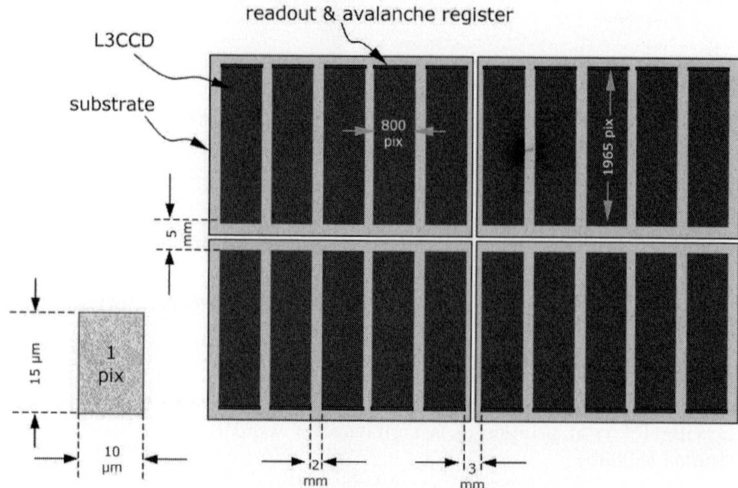

Figure 10. The layout of the RVS focal plane. This consists of 4 sets of 5 L3CCDs, with each set on a single substrate. The avalanche registers providing the gain in the devices follow the readout registers, and are located at the top and bottom of the focal plane.

The RVS focal plane consists of 20 CCDs of dimension 1965 × 800. The pixels are rectangular, with dimensions $10 \times 15\,\mu$m – see Fig. 10. The CCDs are of a particular type, L3CCDs (low-light-level CCDs) which use avalanche gain in a multiplication register to amplify the signal before it is read out (Jerram et al. 2001). This effectively reduces the readout noise in the device, at the cost of some additional statistical noise but which does not dominate for fainter signals. The use of a number of CCDs allows the images to be remapped in software to correct for optical distortion, and the across-scan motion of the spectra (perpendicular to the scan and dispersion direction) to be corrected before the data from the individual CCDs are added together for telemetering.

RVS is a slitless spectrograph. The data format will be dispersed spectra from CCD detectors with spatial width ~ 2 pixels and spectral extension over ~ 690 pixels, set by the resolution requirement and the spectral bandpass. Radial velocities are then obtained by cross-correlation with a template. Most stars are detected at the limits of visibility and will require co-adding of all ~ 100 epochs of exposure scans. A simulation of the images from the RVS is shown in Fig. 11.

Acknowledgements

We are grateful to ESA for partially funding this work; also to the national funding agencies in the UK, France, Italy and Slovenia. We acknowledge the fruitful interactions with the Alenia/Alcatel and Astrium industry teams.

References

Binney, J. J., Dehnen, W., Houk, N., Murray, C. A., Penston, M. J. 1997, *Proceedings of the ESA Symposium 'Hipparcos – Venice 97'* ESA SP-402, p473

Blaauw, A. 1988, *Proc. Colloq. Sitges, Hipparcos Scientific Aspects of the Input Catalogue Preparation II* ed. Torra J., Turon C. p490. Comissió Interdepartamental de Recerca i Innovació Tecnològica, CIRIT, Generalitat de Catalunya, Barcelona

Boschi, F., Munari, U., Sordo, R., Marrese, P. M., 2003, In *Symbiotic Stars Probing Stellar Evolution* ed. R. L. M. Corradi, R. Mikolajewska & T. J. Mahoney. ASP Conference Proceedings, vol. 303, p535. Astronomical Society of the Pacific.

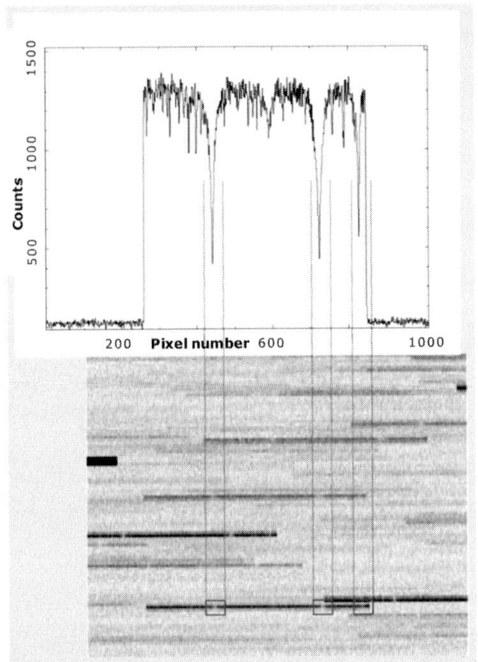

Figure 11. Bottom: a small section of simulated data from the RVS. The Ca absorption lines can be seen in the dispersed spectra. Top: an example of a spectrum extracted from the 2-dimensional data. The edges of the instrument wavelength bandpass are artificially sharp in this simulation.

Cropper, M., Marsh, T. 2003, In *Gaia Spectroscopy: Science and Technology* ed. U. Munari. ASP Conference Proceedings, vol. 298, p407. Astronomical Society of the Pacific.

Duchêne, G. 1999, *A&A* 341, 547

Eggen, O. J., Lynden-Bell, D., Sandage, A. R. 1962, *ApJ* 136, 748

Favata, F., Perryman, M. A. C. 1997, In *Proceedings of the ESA Symposium 'Hipparcos – Venice 97'* ESA SP-402, p.771

Hardfing, P., Morrison, H. L., Olszewski, E. W., Arabadjis, J., Mateo, M., Dohm-Palmer, R. C., Freeman, K. C., Norris, J. E. 2001, *AJ* 122, 1397

Helmi, A., de Zeeuw, P. T., 2000, *MNRAS* 319, 657

Helmi, A., White, S. D. M. 1999, *MNRAS* 307, 495

Ibata, R., Gilmore G., Irwin, M. 1994, *Nature* 370, 194

Jerram, P., Pool, P., Bell, R., Burt, D. J., Bowring, S., Spencer, S., Hazelwood, M., Moody, I., Catlett, N., Heyes, P. S. 2001, *SPIE* 4306, 178

Johnston, K. V., Hernquist, L., Bolte, M. 1996, *ApJ* 465, 278

Kapteyn, J. C. 1922, *ApJ* 55, 302

Katz, D., Munari, U., Cropper, M., and 47 co-authors 2004, *MNRAS*, in press

Munari, U. 2003a, In *Gaia Spectroscopy: Science and Technology* ed. U. Munari. ASP Conference Proceedings, vol. 298, p51. Astronomical Society of the Pacific.

Munari, U. 2003b, In *Gaia Spectroscopy: Science and Technology* ed. U. Munari. ASP Conference Proceedings, vol. 298, p227. Astronomical Society of the Pacific.

Munari, U., Tomov, T., Zwitter, T., Milone, E. F., Kallrath, J., Marrese, P. M., Boschi, F., Prsa, A., Tomasella, L., Moro, D. 2001, *A&A* 378, 477

Navarro, J. F., Helmi, A., Freeman, K. C., 2004, *ApJ* 601, L43

Nordström, B., Mayor, M., Andersen, J., Holmberg, J., Pont, F., Jørgensen, B. R., Olsen, E. H., Udry, S., Mowlavi, N. 2004, *A&A* 418, 989

Perryman, M. A. C. 1997, In *Proceedings of the ESA Symposium 'Hipparcos – Venice 97'* ESA SP-402

Perryman, M. A. C., Lindegren, L., Kovalevsky, J. and 16 coauthors 1997, *A&A* 323, L49

Perryman, M. A. C., deBoer, K. S., Gilmore, G., Høg, E., Lattanzi, M. G., Lindegren, L., Luri, X., Mignard, F., Pace, O., deZeeuw, P. T. 2001, *A&A* 369, 339

Quinn, P. J., Hernquist, L., Fullagar, D. P. 1993, *ApJ* 403, 74

Söderheljm, S., 2003, In *Gaia Spectroscopy: Science and Technology* ed. U. Munari. ASP Conference Proceedings, vol. 298, p351. Astronomical Society of the Pacific.

Soria, R., Cropper, M., Pakull, M., Mushotzky, R., Wu, K. 2004, *astro-ph/0409568*

Wilkinson, M. I., Vallenari, A., Turon, C. and 35 co-authors 2004, *MNRAS* submitted

Discussion

DAVE MONET: Is the spectral range 8500 Å to 8800 Å with a resolution of about 1.5 Å?

MARK CROPPER: The resolving power is 11 000 and it's over a 25-nm range at 850 nm. This range has also been subsequently adopted for the RAVE programme.

JOHN SOUTHWORTH: If you are getting velocities to $1\,\mathrm{km\,s^{-1}}$, you will need a good wavelength calibration. How do you do this?

MARK CROPPER: You do. There are radial velocity standards, and Gaia is continually sweeping across these. There are not very many, but you are able to go to the primary references. The CORAVEL velocities are very good, so they are coming in to this range. But we've realised that the very best way for us to do this is through asteroids. You know the position very accurately, and you know the solar spectrum, so you can predict the radial velocity zero-point from asteroids to $\mathrm{m\,s^{-1}}$. That easily good enough for reference and there are plenty of them – we'll see several a day and that keeps everything calibrated.

Transits of Venus: New Views of the Solar System and Galaxy
Proceedings IAU Colloquium No. 196, 2004
D.W. Kurtz, ed.

© 2004 International Astronomical Union
doi:10.1017/S1743921305001614

JASMINE: Japan Astrometry Satellite Mission for INfrared Exploration

N. Gouda[1], T. Yano[1], Y. Kobayashi[1], Y. Yamada[2], T. Tsujimoto[1],
T. Nakajima[1], M. Suganuma[1], H. Matsuhara[3], S. Ueda[4],
and the JASMINE Working Group

[1]National Astronomical Observatory of Japan, Mitaka, Tokyo, 181-8588, Japan
email: naoteru.gouda@nao.ac.jp, yano.t@nao.ac.jp,
yuki@merope.mtk.nao.ac.jp, taku.tsujimoto@nao.ac.jp, tadashi@dodgers.mtk.nao.ac.jp,
suganuma@merope.mtk.nao.ac.jp

[2]Department of Physics, Kyoto University, Kyoto, 606-8502, Japan
email:yamada@scphys.kyoto-u.ac.jp

[3]Institute of Space and Astronautical Science, Japan Aerospace Exploration Agency (JAXA),
Sagamihara, Kanagawa 229-8510, Japan
email:maruma@ir.isas.ac.jp

[4]The Graduate University for Advanced Studies, Mitaka, Tokyo, 181-8588, Japan
email:seiji.ueda@nao.ac.jp

Abstract. We introduce a Japanese plan for infrared (z-band: $0.9\,\mu$m) space astrometry (the JASMINE-project). It will measure parallaxes, positions with the accuracy of $10\,\mu$as and proper motions with the accuracy of $10\,\mu$as/yr for stars brighter than z$\sim$14. JASMINE can observe about 10^8 stars belonging to the disk and bulge components of our Galaxy which are hidden by interstellar dust extinction in optical bands. The number of stars with $\sigma_\pi/\pi < 0.1$ in the direction of the Galactic central bulge is about 10^3 times larger than those observed in optical bands, where π is a parallax and σ_π is an error of the parallax. The main objective of JASMINE is to provide very useful and important astrometric parameters for studying fundamental structures and evolution of the disk and bulge components of the Milky Way Galaxy. Furthermore, the astrometric parameters given by JASMINE will give us exact absolute luminosities and motions of many stars in the bulge and the disk far away from us, so it will promote the study of stellar physics. The information of infrared astrometry that JASMINE will provide is very useful also for investigating stars in star formation regions, gravitational lens effects due to disk stars, extra-solar planets, etc. JASMINE will be launched around 2014 and a candidate for the orbit is a Lissajous orbit around the Sun-Earth L2 point with about a 5-yr mission life. We adopt a 3-mirror optical system (modified Korsch system) with a primary mirror of $\sim$1.5-m diameter in an instrument design of JASMINE. A beam combiner should be used for performance of the global astrometry as used in the Hipparcos satellite. On the astro-focal plane, we put about 100 new-type CCDs for the z-band in which TDI mode (drift scan mode) can be operated. The effective field of view is 0.23 square degrees. The consideration of overall system (bus) design is now going on in cooperation with the Japan Aerospace Exploration Agency (JAXA). Furthermore, we introduce the Nano-JASMINE project which uses a nano-satellite with a size of about $20\,$cm^3 and a weight of a few kg. The objective of Nano-JASMINE is verification of the observing strategy adopted in JASMINE and examination of some important technical issues for the JASMINE project. It will be launched around 2006.

---

1. Introduction

Astrometric measurements provide the most fundamental parameters of stars such as the absolute trigonometric parallaxes (i.e. distance), positions on the celestial sphere and proper motions, which are the backbone in many branches of astronomy and astrophysics. So the improvement of astrometric parameters enables advances across numerous branches of astronomy and astrophysics. The Hipparcos satellite was launched to measure the astrometric parameters without certain limitations (atmospheric turbulence and refraction, etc.) in the ground-based measurements. Hipparcos provided remarkably interesting results in many fields of astronomy.

The success of Hipparcos has triggered several proposals for astrometric satellites that would observe more stars with better accuracies than those in Hipparcos. We need better astrometry because a drastic increase in the accuracy and the number of parallaxes and proper motions will result in remarkable advances in kinematics and dynamics of the Galaxy and furthermore in fields of stellar evolution, extra-solar planets and the extragalactic distance scale.

If we have parallaxes with errors larger than about 10%, we would have some biases in deriving distances by the parallaxes and so we could not determine the distances with enough accuracy. The accuracy of the parallaxes in Hipparcos is about 1 mas, thus we cannot accurately evaluate the distances of the stars which are about $100\,\mathrm{pc}$ or more distant from us using the parallaxes given by Hipparcos. But we require the accurate distances of stars which are at least around $10\,\mathrm{kpc}$ distant from us in order to investigate the bulge component and almost the inner disk structure of the Galaxy. Hence we need a level of $10\,\mu\mathrm{as}$ accuracy in parallax. Proposed astrometric satellites perform astrometric measurement with this level of accuracy.

The proposed space projects of optical astrometry are Gaia and SIM. Gaia is a survey type astrometric mission for all-sky survey, which is proposed by ESA. Gaia aims at measuring positions and parallaxes accurate to $10\,\mu\mathrm{as}$ at $V = 15$ with proper motion errors $10\,\mu\mathrm{as/yr}$. The limiting magnitude is $V = 20$. SIM is a NASA's mission using optical interferometer with a 10-m baseline. SIM is not a sky-scanning, but rather a pointing instrument which is set on selected targets of about 10^4 stars brighter than $V = 20$. SIM has the capability to measure absolute parallaxes accurate to $4\,\mu\mathrm{as}$. It should be noted that both Gaia and SIM observe stars in optical bands. In Japan we have two projects which perform the global astrometry. Those are VERA and JASMINE which measure astrometric parameters in radio and infrared bands, respectively. They have advantages in observing stars on the Galactic plane which are hidden by the interstellar dust in optical bands. VERA is a VLBI system dedicated to differential VLBI to measure the astrometric parameters of about 1000 masers with $10\,\mu\mathrm{as}$ level accuracy. VERA will achieve the targeted accuracy of the parallax only in a few years and determine accurately some fundamental parameters of the Galaxy such as the distance to the Galactic center.

JASMINE is an infrared (z-band: $0.9\,\mu\mathrm{m}$) astrometric mission designed to perform an astrometric survey on the Galactic plane, determining positions and parallaxes accurate to $10\,\mu\mathrm{as}$ for stars brighter than z= 14, with proper motion errors of $\sim 10\,\mu\mathrm{as/yr}$. JASMINE will observe about 10^8 stars around the bulge and the disk of the Galaxy; VERA does not observe stars themselves but maser sources and the number of targets is only 1000. We expect that JASMINE will extend the area of studies on Galactic structure and stellar populations after the success of VERA.

In this paper, we introduce the outline of the JASMINE project. §2 is devoted to the scientific objectives of JASMINE. We briefly review a mission design, an instrument design and a spacecraft system in §3, §4 and §5, respectively. In §6 a management plan of JASMINE is briefly mentioned. Finally, §7 is a summary.

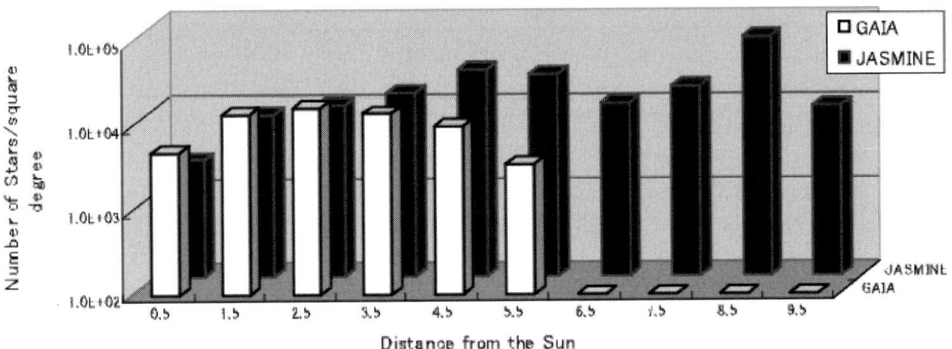

Figure 1. Number of stars (per square degree) measured with $\sigma/\pi \leqslant 0.1$ toward the direction of $\ell = 0°$ and $b = 1°$.

2. Scientific objectives

As mentioned in §1, JASMINE will provide the astrometric parameters to promote studies in many branches of astronomy and astrophysics. One of the most important scientific objectives among them is the formation, evolution and structure of the Milky Way Galaxy. The quantitative analysis of the Galaxy needs distances, 2-dimensional (or even better, 3-dimensional) motions of stars in the Galaxy. Especially, most of the stars, and almost all of the interesting dynamics, are found at low Galactic latitudes in crowded fields. So there is a great requirement to measure accurate astrometric parameters of stars in the fields at low Galactic latitudes. On the other hand, the light in optical bands from the stars at low Galactic latitudes is effectively absorbed by interstellar dust. This extinction effect decreases both the number of observable stars and the accuracy of the astrometric parameters. So we need to survey the Galactic plane to measure the astrometric parameters in near infrared bands which penetrate the obscuring dust. Then, at first, we will explain the necessity of infrared astrometry based on the quantitative analysis using a Galaxy model. After that, we will mention a few important topics of science which can be expected to make remarkable progress by JASMINE.

2.1. *Advantages of infrared astrometry*

JASMINE performs unique astrometric measurements in the infrared band (z-band: $0.9\,\mu$m) in order to get the accurate astrometric parameters of many stars on the Galactic plane. The galactic bulge is an important scientific target as explained in §2.2. Then we focus on the galactic central bulge as an example for showing an advantage of infrared astrometry. Interstellar dust effectively absorbs visible light and a large amount of dust is located in the direction of the galactic bulge. However, there are a few windows where the amount of dust is rather smaller than that in other central regions, so we can observe many stars through those windows. A famous window is Baade's window ($\ell = 0°, b = -3.9°$) with an area of $\sim$1 square degree. Gaia will detect about 8000 stars with parallax accuracies of $\sigma/\pi \leqslant 0.1$ (Vallenari et al. 1999). Here π is the parallax of a star and σ the error of that parallax. If $\sigma/\pi \leqslant 0.1$, then we do not need worry about errors due to biases which arise in converting parallaxes to distances. However, Gaia will not detect a large number of stars with such good parallax accuracies toward directions other then the central bulge.

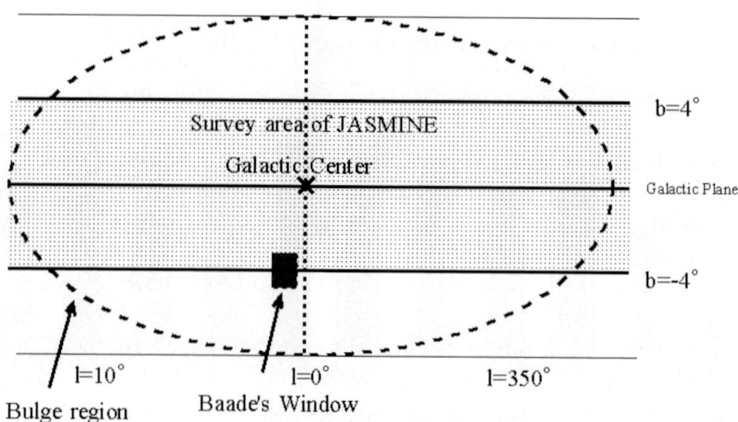

Figure 2. Survey area of JASMINE around the bulge and Baade's window.

For example, Fig. 1 shows the expected numbers of stars per square degree to be observed with $\sigma/\pi \leqslant 0.1$, estimated using our Galactic model. This model is based on the "sky" model developed by Cohen and his collaborators (Wainscoat et al. 1992; Cohen 1994; Cohen, Sasseen & Bowyer 1994; Cohen 1995). The center of the field of view is pointed toward Galactic longitude $\ell = 0°$ and Galactic latitude $b = 1°$. The horizontal line represents the distance from us. The Galactic center is assumed to be at 8.5 kpc. The black histogram shows the number of the stars evaluated for z-band observations of JASMINE while the white histogram shows those for V-band observations with parallax accuracies (e.g. 10 μas accuracy at $V = 15$) of Gaia design. We can see from Fig. 2 that the number of stars observed in the z-band is much larger than that observed in the V-band at distances of more than a few kpc from the sun on the Galactic plane. JASMINE can detect about 7.3×10^5 stars of the bulge within its survey area $(|b| \leqslant 4.0°, \ell = 0° - 360°)$, while Gaia will detect about 400 stars of the bulge in the same area, excluding Baade's window. Then JASMINE will measure the distances and proper motions of many stars of the central bulge with high accuracies over the large survey area within the inner parts of the bulge at lower galactic latitudes than the latitude of Baade's window (see Fig. 2). This is the most important advantage of infrared space astrometry. We hope that JASMINE can be complementary to Gaia for the survey of the bulge.

We think that some may worry about the confusion limit. That is, that we cannot accurately determine the position of stars fainter than the confusion limiting magnitude due to contamination in crowded regions. We estimated the confusion limit magnitude in the survey area of JASMINE using our Galactic model and found that the minimum magnitude of the confusion limit, which is achieved around the center of the Galaxy, is z= 18. This value is above the limiting magnitude of JASMINE ($\sim$17), so we need not worry about confusion in JASMINE.

2.2. Structure, formation and evolution of the Galaxy

Dynamical analysis, derived primarily from accurate distances and kinematics data is the key to understand the structure, formation and evolution of the Galaxy. The Galaxy is believed to be representative of the giant spiral galaxies which dominate the luminosity of the Universe. Then a well studied template in the Galaxy underpins analysis of unresolved galaxies.

The proper motions and distances obtained by JASMINE will give the distribution of matter, high-luminosity and dark, in the bulge and disk of the Galaxy. We will have an accurate 3-dimensional map of significant portions of the Galaxy to understand the structure. Size, shape and kinematic properties of the different components of the Galaxy such as the bulge, spiral arms, thin and thick disk, and warp are pending problems in the Galaxy. These problems will be resolved by JASMINE. For example, the formation, evolution and dynamics of the bulge is an unresolved interesting problem. JASMINE will provide important information on distances and transverse motions of stellar populations in the bulge to resolve the problem using other information on radial velocities, metallicities and ages of the stars given by other facilities.

Many questions about shape, dynamical structure and formation history of the galactic bulge still remain open. For example, the shape of the bulge is still not clear while there is substantial evidence for a bar structure. Several models for the density distribution in the bulge have been proposed over the years from spherical to triaxial systems. So to know exact distances of stars in the bulge is the key to clarify the shape of the bulge. Furthermore, investigation of the dynamics in the bulge derived from distances and kinematics is the key not only to resolve the formation and evolution of the bulge, but also to study the dynamical structure such as the character of the steady state mechanism of relaxation process in self-gravitating, many-body systems.

The origin of spiral arms is also an interesting problem. We can determine whether spiral arms are due to density waves, or not, using our analysis technique (Yano, Chiba & Gouda 2002) with JASMINE data. This technique is based on the kinematic analysis of the Galactic disk stars to clarify whether the internal motions of the stellar system in spiral arms follow those expected in the density wave theory. The method uses the linear relation between the phases of spatial positions and those of epicyclic motions of stars, as predicted by the theory.

As mentioned before, the Galaxy is believed to be typical of the mean population in the universe. So detailed analysis of the Galaxy can contribute to the general study of galaxy formation and evolution in the early universe. That is, the Galaxy should retain a fairly direct fossil record of the conditions at the time of Galaxy formation in the early universe. Information such as 3-dimensional positions and motions, and absolute luminosities of stars is necessary to resolve the structure and evolution of the Galaxy. The astrometric parameters of many stars in the bulge and the disk obtained by JASMINE give information on Galactic formation. On the other hand, Gaia will accurately provide the structure of the halo and the disk within a few kpc from us. Then JASMINE can be complementary to Gaia.

2.3. *Stellar evolution*

All the models of internal structure of stars are very strongly constrained by models built for the Sun. But the characteristics of the stars especially in the Galactic bulge are not known well. The extension to other types of stars would demand more accurate information on the characteristics of many different types of stars. This requires accurate distances of stars in the bulge and JASMINE will provide those. The extensive amount of data of extreme accuracy given by JASMINE will stimulate a revolution in the exploration of stellar formation for many types of stars crowded at the Galactic plane. Furthermore, exact luminosities calibrated by accurate trigonometric parallaxes are based on the distance indicator.

2.4. *Gravitational lensing effect*

The gravitational lensing effect causes the luminosity of a star to change with time. In fact, this phenomenon has been observed in the Galaxy by the MACHO project on disk stars. But information about a lens is restricted because mass, motion of the lens, the

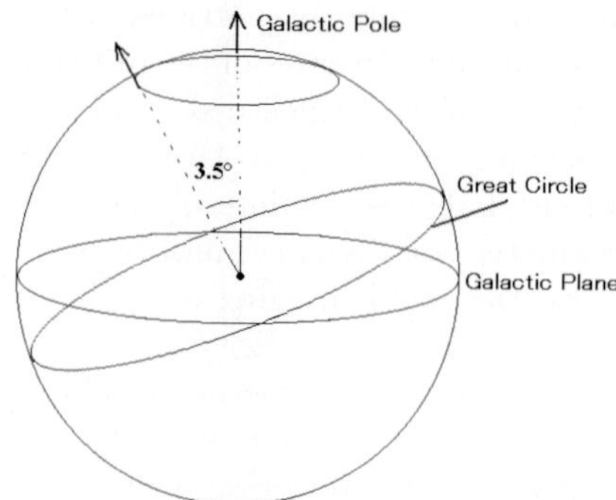

Figure 3. The spin axis of the JASMINE spacecraft is aligned 3.5° from the Galactic pole – the spacecraft line with a rotation period of about 5 hr.

distance to the lensed star, and the position of the lens on the celestial sphere cannot be separately resolved only using the light curve of the lens. On the other hand, the lensing effect results in a elliptical motion of the lensed star. Highly accurate astrometry can measure this motion and this information can resolve the degeneracy of the characteristics of the lens, such as mass. JASMINE will observe a large number of stars in the Galactic plane. The fraction of lensing events is about 10^{-3}, so about 10^4 lensing events might be expected in JASMINE. The information will be useful to resolve the mass and kinematics of lensing objects.

2.5. *Extra-solar planets*

Astrometric measurement of a star allow the detection of extra-solar planets due to nonlinear motion of the star with planets and is complementary to radial velocity measurement of the star. JASMINE might detect extra-solar planets of stars covered by a dust layer or bright stars in the z-band.

2.6. *Reference frame*

JASMINE will provide positions and proper motions with high accuracies in the z-band. The combination of JASMINE data with other data observed by Gaia and radio catalogues will result in an accurate reference frame in the z-band. This reference frame will be a useful catalogue for many purposes in astronomy.

2.7. *Other scientific objectives*

The data of JASMINE will promote some other important investigations on stellar physics of binary stars, variable stars and novae. Moreover, we can test general relativity using measurements of the gravitational lensing effect with high accuracies.

3. Mission design

3.1. *Observing strategy*

A possible candidate for the orbit of JASMINE is a Lissajous orbit around the Sun-Earth Lagrange point L2, because this region provides a very stable thermal environment, amongst other advantages. The JASMINE spacecraft rotates slowly with a period of about 5 hr with a perpendicular rotation axis to look in the directions of two fields of view. The CCDs in the telescope focal plane are clocked in time-delayed integration (TDI) mode so that the accumulated charge packets track star images as they sweep across the

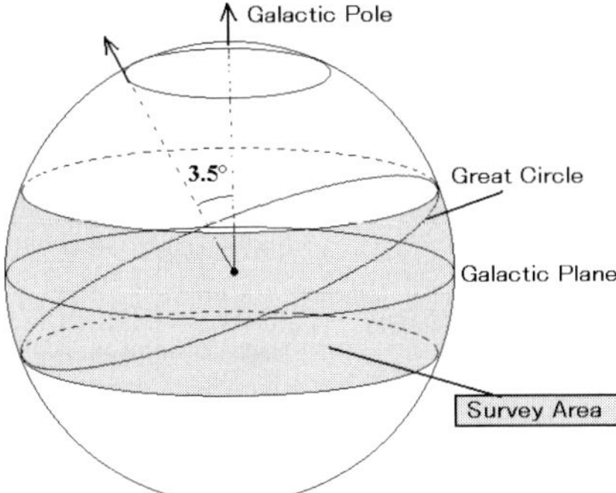

Figure 4. The JASMINE spacecraft precesses around the Galactic pole with a period of about 37 d and the sky area around the Galactic plane is scanned with this precession.

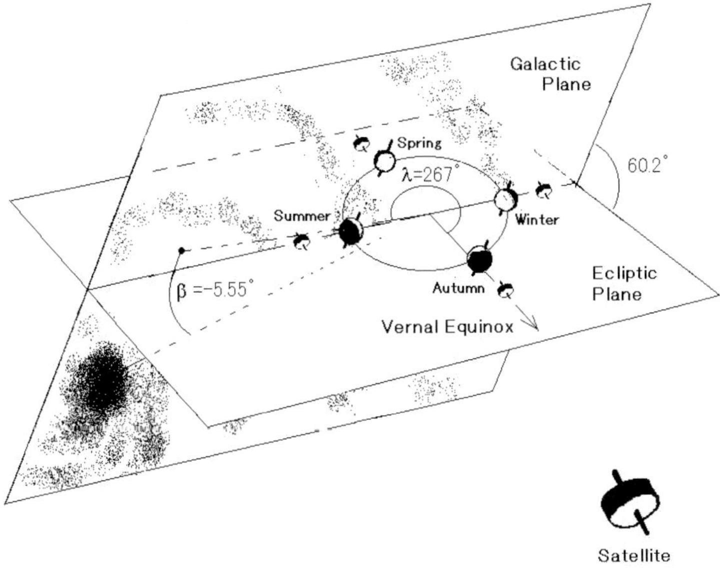

Figure 5. The Galactic plane and the Ecliptic plane.

CCD. TDI mode makes it possible to decrease the effect of readout noises on the signal of star images. The rotation axis of the JASMINE spacecraft will be aligned 3.5° from the spacecraft – Galactic pole line as shown in Fig. 3. The precession of the JASMINE spacecraft will be forced and JASMINE can survey the Galactic plane with a region of 360° (along Galactic longitude)×7° (along the Galactic latitude) with a precession period of about 37 d as shown in Fig. 4. The mission life will be 5 yr.

As seen from Fig. 5, the direction toward the sun from the spacecraft overlaps with that of the Milky Way in two quarters of a year (summer and winter seasons). In these seasons, the spin axis of the JASMINE spacecraft is changed to be almost perpendicular to the Galactic pole – spacecraft line and then JASMINE observes toward the Halo regions instead of the Milky Way. The observing mode for the Milky Way is then restricted to a half of the total mission life.

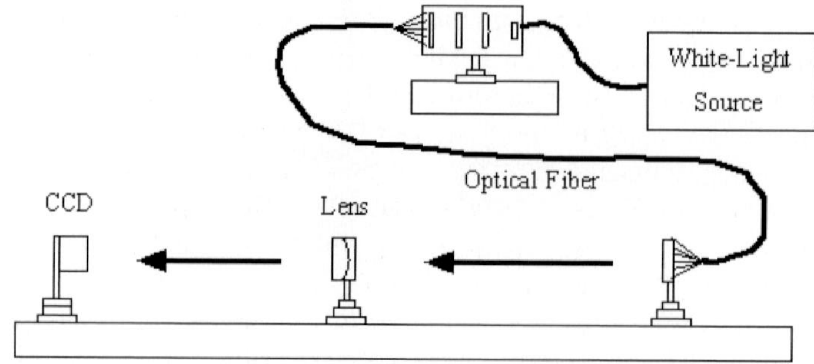

Figure 6. Centroiding Experiment Layout

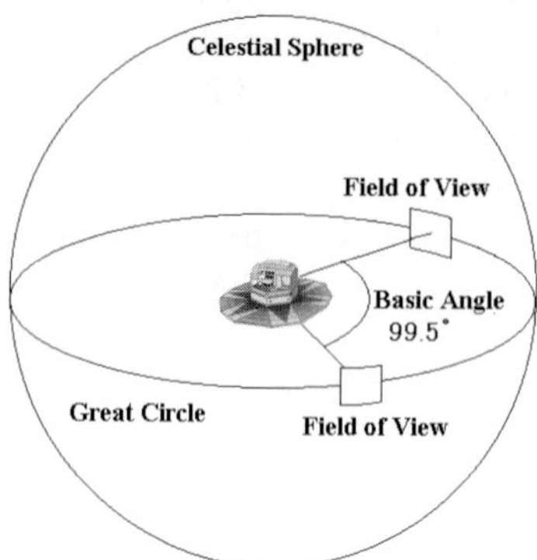

Figure 7. The two fields of view are perpendicular to the rotation axis and are separated by a $99°\!.5$ basic angle.

3.2. *Astrometric data reduction*

The data analysis for JASMINE will proceed as follows:

(i) A star image is sampled in a binned window with 5 pixels (along the scan direction) $\times$ 9 pixels (along the cross-scan direction). The centroid of the star image in the scan direction is required to reach an accuracy of about 1/140 pixel for JASMINE. We perform the measurements of centroids using our algorithm which takes a weighted mean of photon numbers with correction factors of some errors (Yano et al. 2004).

We need demonstrate that the required single look centroiding precision can be achieved in an simulated operating environment using a CCD. We have carried out experiments on the ground measuring centroids of artificial star images on a CCD for test of our algorithm and verification of technical issues (Yano et al. 2004). These experiments have been done in collaboration with ILOM (In-situ Lunar Orientation Measurement)-project team at the National Astronomical Observatory of Japan. The schematic layout of our centroiding experiment is shown in Fig. 6. It consists of a star field projector which produces point spread functions and a CCD focal plane on a precision linear stage to simulate the linear sweep of the star field across the CCD array. We have found that the accuracy of the centroids reaches about 1/300 pixel while the TDI mode has not been operated yet. The experiments are going on in operating the TDI mode.

(ii) As JASMINE rotates, the field of view will map out a spiral band on the sky as described in the previous subsection. The next stage of data reduction is to determine the relative positions of the centered images of stars along the observing spiral. JASMINE observes two fields of view separated by 99°.5 simultaneously as shown in Fig. 7. The separation of view is referred to as the basic angle. This limits the growth of errors in relative star positions over large angles. The relative positions of stars are determined by the angular velocity $\omega(t)$ of the spacecraft and transit times of the centroid images. A Fourier expansion of $\omega(t)$ will be integrated over the time for a star to cross both fields of view. The Fourier coefficients are obtained by equating such integration to the basic angle. This use of the two fields of view is taken from the Hipparcos design.

(iii) The next stage is the determination of individual stars' astrometric parameters using the relative star positions on many spiral bands: The observing spiral bands are then combined to form a single global system in this stage. The next procedure is to determine individual stars' astrometric parameters (position, proper motion, and parallax), observing-spiral origins and orientations, local parameters of the instruments and the spacecraft, and global parameters. The local parameters are included in models which describe thermal and/or machinery time variations with *short* periods, of some instruments and the spacecraft system while the global ones represent time variations with *long* periods. A system of observation equations can be formed from the data. The astrometric parameters and both local and global parameters are determined using least squares fits. The entire process is iterated until corrections to the parameters converge. Residuals will be examined for signs of nonlinear proper motion, which would indicate the presence of a nearby gravitating body or a gravitational lensing effect.

4. Instrument design

The JASMINE instrument uses a beam combiner to observe two fields of view simultaneously. The JASMINE beam combiner consists of two flats that feed a common telescope with fields of view separated by the basic angle of 99°.5. The value of the basic angle should avoid unwanted correlations of measurements on successive scans and so its value should not be an integer divisor of 360°. These values are ideally determined by limits of certain Fibonacci series. The basic angle of about 99°.5 is the limit of a Fibonacci series in which the first and the second term are $2\pi/3$ and $2\pi/4$, respectively.

The two fields of view are fed into a common telescope with a 50-m focal length and a circular primary mirror with a 1.5-m diameter. The JASMINE telescope has three mirrors (modified Korsch system) with 4 folding flats to fit the back focal length into the available volume (Fig. 8; see Yano et al., these proceedings). The acceptable mass budget at launch requires that the weight of the telescope should be as little as possible. Furthermore, it is desirable that the telescope has low thermal expansion, high stiffness and high thermal conductivity. Therefore we need to develop suitable new materials with these desired characteristics. One candidate is the use of a new, high-strength, reaction-sintered silicon carbide (RS-SiC) which is now being developed at a Japanese company. We are now trying to make a prototype telescope with a 5-cm diameter primary mirror using this RS-SiC.

The telescope provides a flat image plane consisting of an array of large format CCDs. A total of 98 4k×2k CCDs with 15 μm square pixels are read out in TDI mode to transfer the charge across the devices at the same rate that the images are moving due to the spacecraft rotation. JASMINE observes in the z-band so we need CCDs whose sensitivity is very high around the z-band. Thus we are developing a new type of CCD in collaboration with a Japanese company. This is a back-illuminated, fully-depleted CCD image sensor. The quantum efficiency of the CCD will be about 90% at the z band.

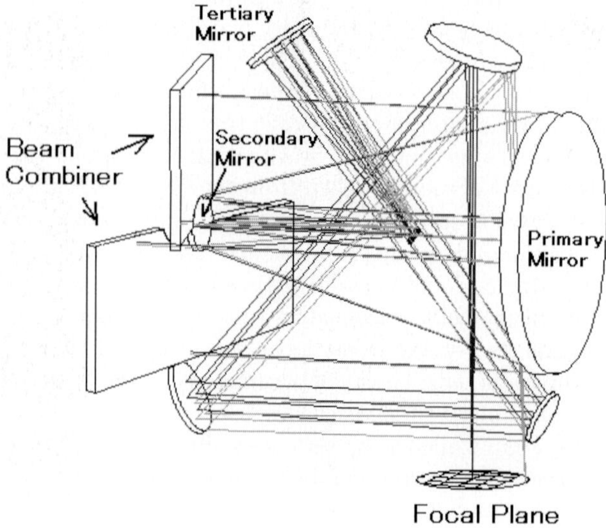

Figure 8. Optics of JASMINE telescope.

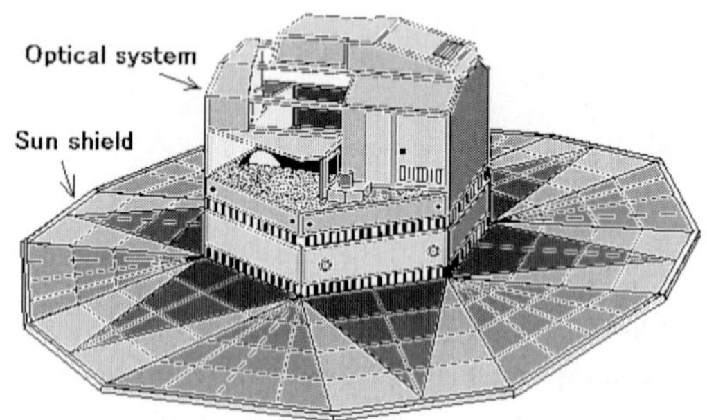

Figure 9. Schematic figure of the JASMINE spacecraft.

5. Spacecraft system

The spacecraft subsystems provide all the necessary support to the payload instruments. The JASMINE spacecraft system has been investigated in collaboration with Japan Aerospace Exploration Agency (JAXA).

A Lissajous orbit around the Sun-Earth Lagrange point L2 is a preferred option because this region provides a very stable thermal environment, minimization of eclipses, and other advantages. The launch strategy is based on a dual launch with H-IIA rocket of JAXA. The mission lifetime will be 5 yr.

The JASMINE spacecraft rotates slowly and precesses, as mentioned in §3. A 3-axis stabilization will be carried out during the observation phase. The attitude control system must meet stringent requirements on the instrument line-of-sight stability, as well as on the spin-axis pointing and rate measurements during the operation mode. The relative pointing error of 60 mas is required during 3.5 s in order to avoid blurring during each sub-field of view integration time of 3.5 s. The absolute pointing error is about 3 arcmin in order that a coming spiral band after one revolution can overlap along the cross-scan direction at least by 1/8 region of the spiral band observed just before.

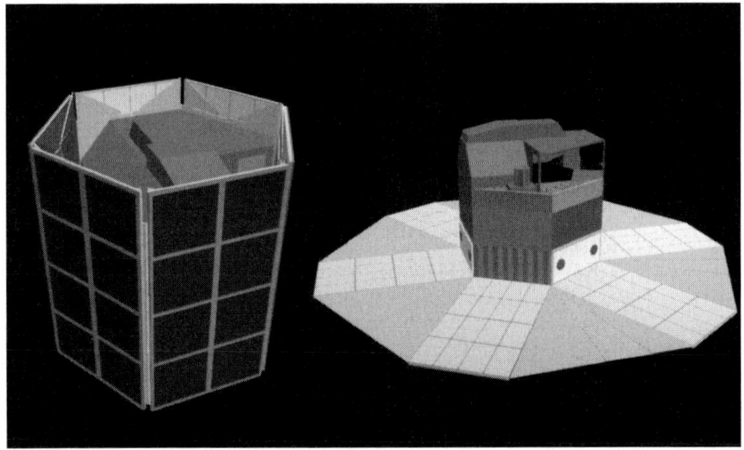

Figure 10. Sunshield of JASMINE. The left figure shows the satellite at launch and the right one shows the satellite in orbit.

Opto-mechanical high stability of the payload is required in the JASMINE spacecraft. Especially high stability of the basic angle of the beam combiner is required over the satellite revolution period (5 hr). The short-term basic angle stability over 5 hr is the only critical parameter so far identified which cannot be properly calibrated by on-ground data processing. A basic angle stability of $10\,\mu$as rms should be attainable. A basic angle variation of $10\,\mu$as rms corresponds to a thermal gradient variation of ~ 1 mK at the beam combiner. We are examining methods of thermal control in the JASMINE spacecraft. If such stability is not attainable, we should measure the variation of the basic angle with an accuracy of $10\,\mu$as. We are investigating measurement devices such as a wave-front sensor.

The Sun shield for JASMINE is a large shield used to protect the JASMINE payload and service module from direct sun illumination. Due to its large size, it is necessary to fold it for launch, to comply with the volume of the launcher fairing. In Fig. 10 the schematic spacecraft with the sunshield is shown.

The total science raw data rate is about 0.42 Mbps on the sky average. If the possible time duration of the downlink is 8 hr, the telemetry rate is about 1.3 Mbps. We suppose that an electrically scanned phased array antenna is preferred to avoid turbulence to the attitude control of the spacecraft. But the acceptable maximum telemetry rate of the phased array antenna is about 1 Mbps. So we need reduce the data rate such that it becomes compatible with the actual telemetry link. We are considering some methods of data compression.

We have other technical problems beside those described above. The investigations are going on in collaboration with JAXA. Furthermore, we plan a Nano-JASMINE project whose objective is the verification of observing strategy and examinations of some technical issues in JASMINE. Nano-JASMINE uses a nano-satellite whose size and weight are about $20\,\mathrm{cm}^3$ and a few kg, respectively. The definition of a "nano-satellite" is that the range of its weight is between 1 kg and 10 kg.

The optics of Nano-JASMINE are similar to that of JASMINE and the size of the telescope is reduced to the 5 cm diameter of a primary mirror with a focal length of about 1.7 m. We put one CCD with $1\mathrm{k} \times 1\mathrm{k}$ pixels on the focal plane. The candidate orbit is a sun-synchronous orbit. The detailed objective of Nano-JASMINE is the verification of the observing strategy adopted in JASMINE such as a great circle reduction. Furthermore,

Table 1. Summary of the instrument parameters

Optics design	Korsch System (3 mirrors)
Aperture size	1.5 m
Focal length	50.0 m
pixel size	15 μm
pixel on sky	61.9 mas
Array size	6 cm $\times$ 3 cm
Pixels per detector	4096 $\times$ 2048
Number of detectors	98 (7 $\times$ 14)
Basic Angle	99$.\!\!^\circ$5

Table 2. Summary of the scanning law

Mission Time	5 yr
Rotation Period	5.0 hr
Precession Period	36.9 d
Rotation Axis	around the Galactic Pole
Launcher	H-II A
Orbit	Lissajous orbit around the Sun-Earth L2 point

we will examine the operation of the TDI mode on the new type of CCD, damages due to radiation on the CCD, on-board processing, thermal variation of the basic angle, and so on. The cost of Nano-JASMINE is low and it will be launched about 2006. The development of the spacecraft is going on in collaboration with Prof. Nakasuka and his group at the University of Tokyo. His group successfully launched a nano-satellite whose name is Cube-Sat(XI-IV) in 2003 June.

6. Management of JASMINE

The establishment of the JASMINE working group at JAXA was approved last year by a science committee of ISAS (Institute of Space and Astronautical Science) of JAXA. The working group includes many scientists and engineers who are investigating the basic design of JASMINE and technical problems. The working group aims at proposal of the JASMINE mission to JAXA to get an approval for launch in 4 or 5 years.

7. Summary

JASMINE will measure parallaxes, proper motions and positions of about 10^8 stars mainly within the central bulge and disk components of the Galaxy. JASMINE aims at high precision astrometry of 10 μas for stars brighter than z= 14 in the z-band. The primary scientific targets of JASMINE are to clarify the structure and evolution of the bulge and the disk. The instrument parameters and scanning law of JASMINE are summarized in Table 1 and Table 2, respectively.

Jasmine is a name of a plant. The flower language of jasmine is elegance. Thus we greatly hope that JASMINE will be *elegantly* successful in the future. (Please refer also to the JASMINE web page: http://www.jasmine-galaxy.org/index.html)

Acknowledgements

We would like to acknowledge ILOM-project team (N. Kawano, H. Hanada, K. Araki, H. Noda, S. Tsuruta, K. Asari and S. Tazawa) for the collaboration in the CCD centroiding experiment. We would like to thank Y. Kawakatsu, A. Noda, A. Tsuiki, M. Utashima, A. Ogawa, N. Sakou, H. Ueda for their collaboration in the investigations on the JASMINE spacecraft system. Furthermore, we would like to thank S. Nakasuka and members of his laboratory. This work has been supported in part by the Grant-in-Aid for the Scientific Research Funds (15340066) and Toray Science Foundation.

References

Cohen, M. 1994 *ApJ* **107**, 582–593.

Cohen, M. 1995 *ApJ* **444**, 874–878.

Cohen, M., Sasseen, T. P. & Bowyer, S. 1994 *ApJ* **427**, 848–856.

Vallenari, G, Bertelli, G., Bressan, A. & Chiosi, C. 1999 *Baltic Astronomy* **8**, 159–170.

Wainscoat, R.J., Cohen, M., Volk, K., Walker, H.J. & Schwartz, D.E. 1992 *ApJ Suppl.* **83**, 111–146.

Yano, T, Chiba, M. & Gouda, N. 2002 *A&A* **389**, 143–148.

Yano, T, Gouda, N., Kobayashi, Y. Tsujimoto, T., Nakajima, T., Hanada, H. Kan-ya, Y., Yamada, Y., Araki, H., Tazawa, S., Asari, K. Tsuruta, S. Kawano, N. 2004 *PASP*, in press.

Discussion

Coryn Bailer-Jones: I'm slightly concerned about your scanning law because your satellite is keeping a constant axis pointing in the galactic plane. This means, as it orbits the sun, some aspect angle will change enormously. Doesn't this mean that sometimes you are actually illuminating payload and not being blocked by the sunshield, so you get thermal gradients on the satellite? This means that for the some of the orbit around the sun, the sun shield will be pointing away from the sun not towards the sun.

Naoteru Gouda: Yes. In the summer and winter seasons the payload is illuminated. We can observe in the autumn season and the spring season – for only half of a year.

Floor van Leeuwen: The way that you described the scanning strategy you will be observing virtually in the ecliptic plain at some point. This means you are going to get problems with scattered light inside the instrument from the sun. This is not what you want. It looks, from both Coryn's and my point of view, that the scanning strategy that you are proposing for your mission is the opposite of what you want in terms of thermal stability, for which you would like to have a constant fixed solar aspect angle. I don't understand this strategy.

Naoteru Gouda: Yes, I agree. We are now investigating the method.

Dave Monet: Overlapping fields in the galactic plane – in particular, in Baade's window – prevent severe image crowding. Are you adopting special tactics for dealing with centroiding in such crowded fields?

Naoteru Gouda: Yes, there are many stars and some problems with contamination. We have investigated the contamination; of course, the centre we cannot see, but excluding the galactic centre the computed limit is about $z = 18$. Then it is safe for JASMINE.

Transits of Venus: New Views of the Solar System and Galaxy
Proceedings IAU Colloquium No. 196, 2004
D.W. Kurtz, ed.

© 2004 International Astronomical Union
doi:10.1017/S1743921305001626

Overall design of JASMINE

**Yoshiyuki Yamada[1], Naoteru Gouda[2], Takuji Tsujimoto[2],
Yukiyasu Kobayashi[2], Tadashi Nakajima[2], Hideo Matsuhara[3],
Taihei Yano[2], Seiji Ueda[4], Masahiro Suganuma[2],
and the JASMINE Working Group**

[1]Department of Physics, Kyoto University, Kyoto, 606-8502, JAPAN
email: yamada@amesh.org

[2]National Astronomical Observatory of JAPAN, National Institutes of Natural Sciences,
Mitaka, Tokyo, 181-8588, JAPAN
email:naotcru.gouda@nao.ac.jp,taku.tsujimoto@nao.ac.jp,
yuki@merope.mtk.nao.ac.jp,tadashi@dodgers.mtk.nao.ac.jp,
t.yano@nao.ac.jp,suganuma@merope.mtk.nao.ac.jp

[3]Institute of Space and Astronautical Science, Japan Aerospace Exploration Agency(JAXA),
Sagamihara, Kanagawa, 229-8510, JAPAN
email:maruma@ir.isas.ac.jp

[4]The Graduate University for Advanced Studies, Mitaka, Tokyo, 181-8588, JAPAN
email:seiji.ueda@nao.ac.jp

Abstract. JASMINE (Japan Astrometry Satellite Mission for INfrared Exploration) is a mission to determine positions and parallaxes accurate to $\sim 10\,\mu$arcsec, with proper motion errors $\sim 10\,\mu$arcsec yr^{-1} for Galactic stars observed in the z-band ($0.9\,\mu$m). In this article, we report some technical investigations.

1. Introduction

JASMINE (Japan Astrometry Satellite Mission for INfrared Exploration) is a mission to determine positions and parallaxes accurate to $\sim 10\,\mu$arcsec, with proper motion errors $\sim 10\,\mu$arcsec yr^{-1} for Galactic stars observed in z-band ($0.9\,\mu$m). The JASMINE spacecraft rotates slowly with a rotation axis aligned $3°\!.5$ from the Galactic pole-spacecraft line to perform a continuously scanning observation. This precession makes the JASMINE spacecraft scan around the Galactic plane with a $360° \times 7°$ area.

The JASMINE instrument uses a beam combiner to observe two fields of view separated by $99°\!.5$ simultaneously. This makes it possible to perform global measurements to get absolute parallaxes. The two fields of view are fed into a common telescope whose detailed design is described by Yano et al. (these proceedings). The present mission and instrument design are summarized in Table 1.

JASMINE will be launched around 2014 and a candidate for the orbit will be a Lissajous orbit around the Sun-Earth L2 point with a 5-yr mission lifetime. The stability requirement for rotation of the satellite is about $6 \times 10^{-6}\,\mathrm{deg\,s^{-1}}$ ($t < 1\,\mathrm{s}$ where t is the sampling rate). The stability requirement of the relative angle of a beam combiner is less than $10\,\mu$as during the period of the spacecraft rotation. These requirements are very important technical issues for achievement of the required astrometric accuracies. Detailed consideration of the overall system design continues in cooperation with JAXA (Japan Aerospace Exploration Agency).

In Section 2 we show the method for calculating JASMINE specifications, and Section 3 contains discussions of several technical developments. In this paper, we limit ourselves to

Table 1. Overview of system design

Mission Parameter	
Mission Time	5 yr
Rotation Period	~5.0 hr
Precission Period	28.6 d
Rotation Axis	around the Galactic Pole

Optical System Parameter	
Aperture size	~1.5 m
Focal length	~50.0 m
pixel size	15 μm
pixel on the sky	~61.9 mas
Basic angle	99°.5

technical items which other papers about JASMINE in these proceedings presentation do not cover. We are constructing integrated simulation software for the JASMINE mission design (JASMINE Simulator; see Ueda et al., these proceedings). Optical system design and centroiding algorithms are introduced by Yano et al. (these proceedings)and for JASMINE scientific objectives, see Gouda et al. (these proceedings).

2. Specification

Specifications for JASMINE are calculated as follows. From centroiding experiments (see Yano et al., these proceedings), the size of the point spread function in focal length is chosen as $N_{PSF} = 2$ pixels.

$$\frac{\lambda}{D}\frac{f}{w} = N_{PSF} = 2,$$

where λ, D, f, and w are wavelength, diameter of primary mirror, focal length of the optics, and pixel size, respectively. The angular sizes of the fields of view for one detector, α_s and α_c, are

$$\alpha_{s/c} = \frac{N_{pix-s/c}w}{f} = \frac{N_{pix-s/c}}{N_{PSF}}\frac{\lambda}{D},$$

where indices s and c are scanned and cross-scanned direction, respectively.

The number of photons $N_{\rm ph}$ for achieving the required accuracy for JASMINE ($\sigma = 10\,\mu$arcsec rms) is calculated by an equation

$$\sigma = c\frac{\lambda}{D\sqrt{N_{\rm ph}}} = 10\,\mu\mathrm{arcsec}$$

where c is constant (assumed to be 0.5).

The number of photons observed per unit time F_m is denoted as

$$F_m = A\varepsilon E_0 10^{-0.4m}/h,$$

where A, E_0, and h are the area of primary mirror, energy flux of 0 magnitude stars, and the Planck constant, respectively. The value ε represents total efficiency factor and multiples of four factors: (1) The average detector efficiency is now 0.7 but will be 0.9 within a few years; (2) The efficiency of the optics system is estimated as $(0.97)^8 \sim 0.8$; (3) The number ratio of photons within the window of a stellar image(5×9 pixels) and total photons in our optical system $R_w = 0.63$; (4) The band resolution $\Delta\lambda/\lambda \sim 0.22$.

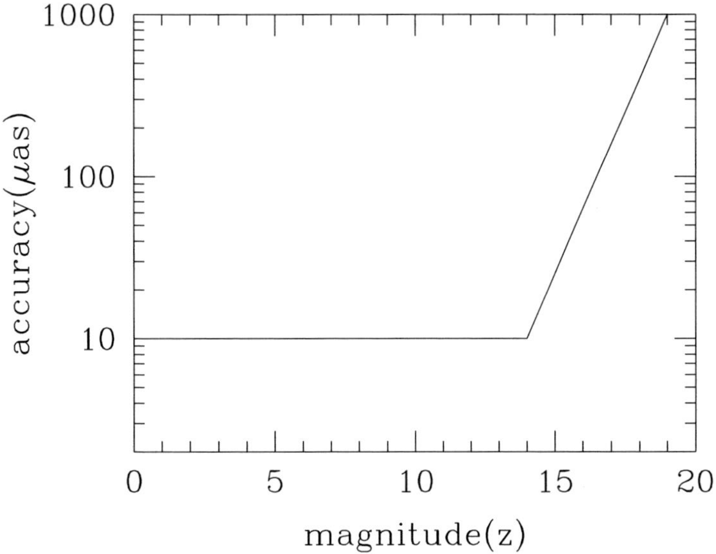

Figure 1. Position accuracy of JASMINE observations.

Using the above estimates, we get $\varepsilon \sim 0.1$. The effective area of our optical system is

$$A = \frac{\pi}{4 N_{BC}}(D^2 - D_{in}^2),$$

where $D \sim 1.5\,\mathrm{m}$ and $D_{in} \sim 0.53\,\mathrm{m}$. The value N_{BC} is the number of the beam combiner. The required integration time for determining the position of one $z = m\,(= 14)\,\mathrm{mag}$ object with an accuracy of $10\mu\mathrm{as}$ is

$$t_{int} = \frac{N_{ph}}{F_m}.$$

In our PSF, the ratio of the photons dropped into the central pixel and the total photons is $R_c = 0.109$. We introduce a new constraint that an object of $z > ms\,(= 9.8)$ does not saturate,

$$R_c F_{ms} t_{int} = N_e,$$

where N_e is full well capacity. For parameters of satellite motion, the periods of spin and precession are denoted as T and T_{pole}, respectively.

$$T = \frac{2\pi}{\alpha_s}t$$

and

$$T_{pole} = \frac{2\pi \sin i}{S \alpha_c n_c}T,$$

where $i = 3°\!.5$ is the inclination of the precession axis, n_c is the number of detectors along a cross scan, S is ratio for the FOV change by precession. We introduce two other parameters, the duty-cycle of observation, β, and the mission time, T_{mis}. The number of observations per year, X, is denoted as

$$X = \beta \frac{1\,\mathrm{yr}}{T_{pole}} N_{BC} n_s \frac{2}{S}.$$

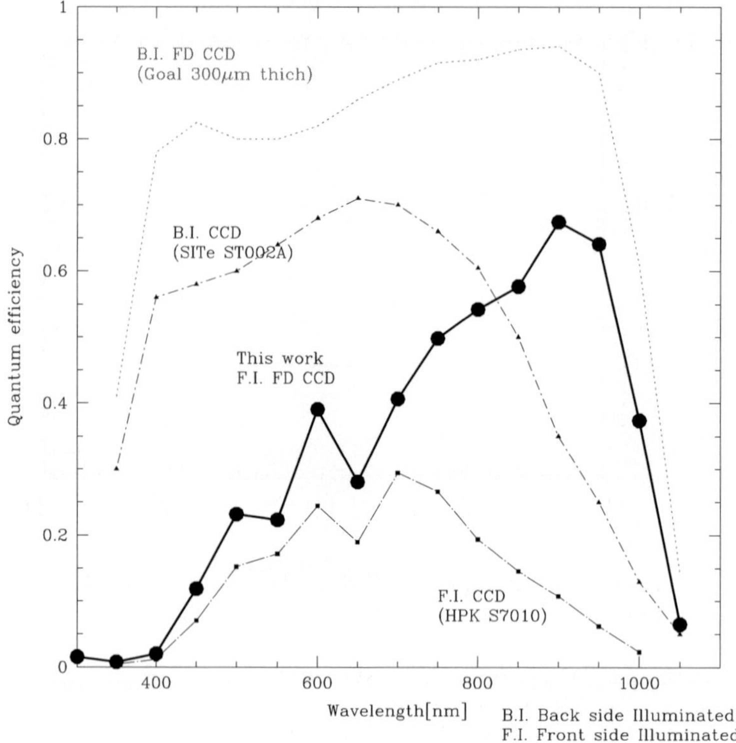

Figure 2. Quantum efficiency a prototype of the CCD under development.

On the other hand, the condition for the integration time is denoted as

$$XT_{mis}t = t_{int}.$$

Combining the above conditions, the total required number of detectors is

$$n_c n_s \sim 88 \left(\frac{c}{0.5}\right)^2 \left(\frac{\sigma}{10\mu as}\right)^{-2} \left(\frac{A}{1.55m^2}\right)^{-1} \left(\frac{\varepsilon}{0.1}\right)^{-1} \left(\frac{\beta}{0.4}\right)^{-1} \left(\frac{T_{mis}}{5yr}\right)^{-1}$$
$$\times \left(\frac{N_{pix-s}}{4096}\right)^{-1} \left(\frac{N_{pix-c}}{2048}\right)^{-1} \left(\frac{\sin i}{\sin 3.5deg}\right).$$

3. Key Technologies

The study objective is to identify technology activities that shall be performed before the Project Implementation Phase, aiming at reducing the project implementation costs. Some of studies are shown in other presentations in these proceedings: Optics design and centroidings (Yano et al., these proceedings); system simulation and software activities (Ueda et al., these proceedings).

3.1. *Detector*

On the focal plane, about 7×14 CCDs are assembled and they measure the stellar images in time-delayed integration (TDI) mode. We are developing a new type of CCD, a back-illuminated fully-depleted CCD for infrared astronomy. The quantum efficiency of a prototype of the CCD is shown in Fig 2 as a solid line. The efficiency at $0.9\,\mu$m

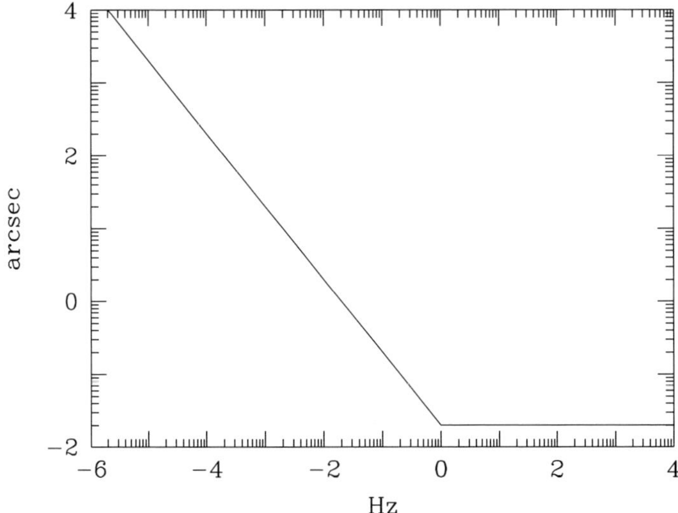

Figure 3. The requirement of satellite attitude control.

will be 90% in the near future. Impurity gathering techniques during the processing are employed to move impurities from the active sensor regions.

3.2. *Attitude requirement*

In our observations, the quality of the final scientific output is strongly dependent on the centroiding accuracy. In our simulation of centroiding algorithms, the satellite attitude is required to be controlled within 1 pixel for both scan and cross-scan directions The requirement is shown in Fig. 3.

3.3. *On board data handling*

On-board object detection will ensure that variable stars, supernovae, transient sources, micro-lensed events, and minor planets will all be detected and catalogued to a faint limit. Each astrometric field comprises an astrometric sky mapper (ASM) and the astrometric field (AF). The sky mapper system provides an on-board capability for star detection and selection and for the star position and satellite scan-speed measurement. From the shape of the point spread function, timing of TDI operation is synchronized with satellite spin, and the spin axis is aligned with the detector implementation.

For each detected star, a window is selected around the star to optimise the complex trade-off between minimising read-out noise, maximising scientific return, and demanding feasible communications.

3.4. *Monitoring the basic angle*

JASMINE will measure the speed and position of a billion stars with extremely high precision ($10\,\mu$as accuracy). The lines-of-sight of the two telescopes are separated by an angle (called the "basic angle") of about $99.^{\circ}5$. The basic angle stability should be within $10\,\mu$as rms over the satellite revolution period of 5 hr, or should be at least monitored with this accuracy.

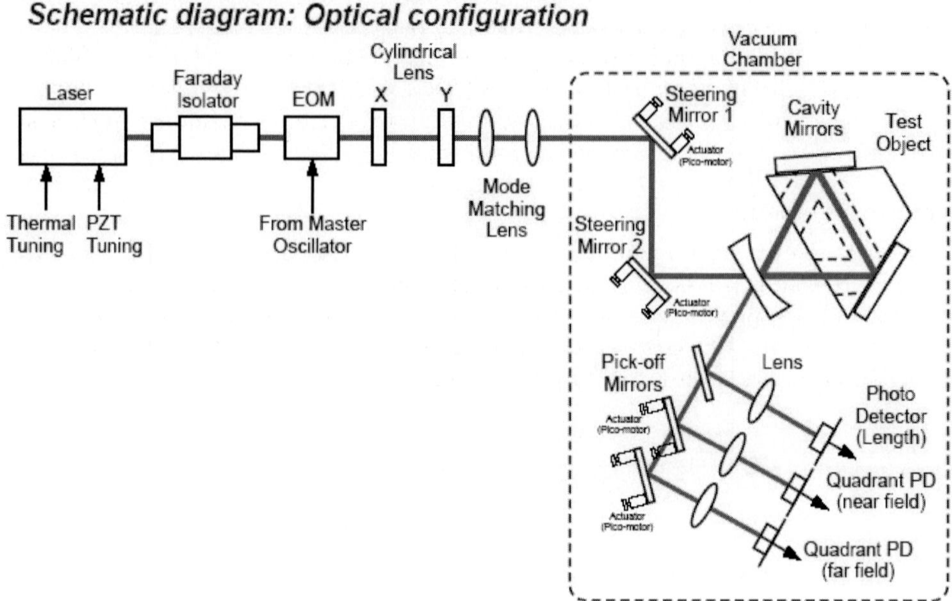

Figure 4. The basic angle monitor.

A technique of monitoring angle is developing in gravitational wave detection projects. The optimum localisation accuracy of the fringe pattern equals:

$$\alpha = \frac{\lambda}{2\pi B \sqrt{N}}$$

where λ, B, N, and α are the wavelength, the length of the baseline, total number of detected photons, and the accuracy, respectively.

We have started ground-based experiments of basic angle monitor. The design of the instruments are shown in Fig. 4.

3.5. *Telecommunication*

The observed number of stars ($z < 17$ mag) is estimated to be 3.2×10^4 per square degree. If we use all data within the stellar window, $5 \times 9 \times 16$ bits are needed for each star. Considering the JASMINE specification parameters, 9.1 Mbps are required for data transfer. We assumed that telecommunication is enabled during 8 hr/day. For transferring such large amount of data, a big parabolic antenna is needed, which may affect the attitude stability.

For telemetry data for faint stars, pixels in the cross-scan direction are integrated and a 5×20-bit stream is used. The advantage of this procedure is not only that it reduces data rates, but also decreases read-out noise. By doing this, data rates are reduced to 1.3 Mbps. Applying an appropriate loss-less data compression method, a phased array antenna can be applied for our telecommunication system. For data compression, Karhunen Loeve transformation and Rice compression may be applied.

4. Conclusions

JASMINE is a unique space astrometric mission using the infrared band. JASMINE will deliver huge scientific impact across the whole of astrophysics, especially the structure and evolution of the Galaxy. But JASMINE will observe only the Galactic disk and bulge. Maintaining close contact between the various proposed space astrometry missions is strongly required for understanding the whole of the Galactic structure. Since FY 2003, we have started collaboration with NASDA, ISAS (both institutes are unified to JAXA), Tokyo University, and the Gravitational wave detection team in NAOJ. System investigation has progressed very much. Until 2009, a mission proposal will be performed.

Acknowledgements

We would like to acknowledge Y. Kawakatsu, A. Noda, A. Tsuiki, M. Utashima, A. Ogawa, N. Sakou, H. Ueda for their collaboration in the investigations on the JASMINE spacecraft system. Furthermore we would like to tank S. Kawamura and K. Arai, who are members of TAMA-300 group. This work has been supported in part by the Grant-in-Aid for the Scientific Research Funds (15340066) and Toray Science Foundation.

Transits of Venus: New Views of the Solar System and Galaxy
Proceedings IAU Colloquium No. 196, 2004
D.W. Kurtz, ed.

© 2004 International Astronomical Union
doi:10.1017/S1743921305001638

The optical system for JASMINE and the CCD centroiding experiment

Taihei Yano[1], Naoteru Gouda[1], Yukiyasu Kobayashi[1], Takuji Tsujimoto[1], Tadashi Nakajima[1], Hideo Hanada[1], Yoshiyuki Yamada[2], Hiroshi Araki[3], Seiichi Tazawa[3], Kazuyoshi Asari[3], Seiitsu Tsuruta[3], Nobuyuki Kawano[3] and Naruhisa Takato[4]

[1]National Astronomical Observatory of Japan, Mitaka, Tokyo 181-8588, Japan
email: yano.t@nao.ac.jp, naoteru.gouda@nao.ac.jp, yuki@merope.mtk.nao.ac.jp,
taku.tsujimoto@nao.ac.jp, tadashi@dodgers.mtk.nao.ac.jp, hanada@miz.nao.ac.jp

[2]Graduate School of Science, Kyoto University, Sakyo-ku, Kyoto 606-8502, Japan
email: yamada@scphys.kyoto-u.ac.jp

[3]National Astronomical Observatory of Japan, Mizusawa, Iwate 023-0861, Japan
email: arakih@miz.nao.ac.jp, tazawa@miz.nao.ac.jp, asa@miz.nao.ac.jp,tsuruta@miz.nao.ac.jp,
kawano@miz.nao.ac.jp

[4]Subaru Telescope, National Astronomical Observatory of Japan, 650 Noth A'ohoku Place,
Hilo, HI 96720, USA Hawaii
email: takato@subaru.naoj.org

Abstract. We have investigated the optical design for the Japan astrometry satellite mission (JASMINE). In order to accomplish measurements of astrometric parameters with high accuracy, optics with a long focal length and a wide focal plane for astrometry are required. In 1977 Korsch proposed a three mirror system with a long focal length and a wide focal plane. The Korsch system is one of the convincing models. However, the center of the field is totally vignetted because of the fold mirror. Therefore we consider an improved Korsch system in which the center of the field is not vignetted. Finally, we obtain the diffraction limited optical design with small distortion. Our project needs a common astrometric technique to obtain precise positions of star images on solid state detectors to accomplish its objectives. In order to determine the centers of stars, an image of the point source must be focused onto the CCD array with a spread of a few pixels. The distribution of photons (photoelectrons) over a set of pixels enables us to estimate positions of stars with sub-pixel accuracy. We modify the algorithm to estimate the real positions of stars from the photon weighted mean, which was originally developed by the FAME (Full-Sky Astrometric Mapping Explorer) group. Finally, we obtain the results from the experiment that the accuracy of estimation of distance between two stars has a variance of about 1/300 pixel; that is, the error for one measurement is about 1/300 pixel, which is almost an ideal result given by Poisson photon noise. We also investigate the accuracy of estimation of positions with a different size of PSF. In this case also, we find that the accuracy of estimation has a variance of about 1/300 pixel.

1. Optics for JASMINE

We have investigated the optical design for the Japan astrometry satellite mission (JASMINE). JASMINE will measure parallaxes, positions and proper motions of stars in our Galaxy with a precision of $10\,\mu$arcsec in order to study the fundamental structure and evolution of the disk and the bulge components of the Milky Way Galaxy. In order to accomplish such measurements with high accuracy, optics with a long focal length and

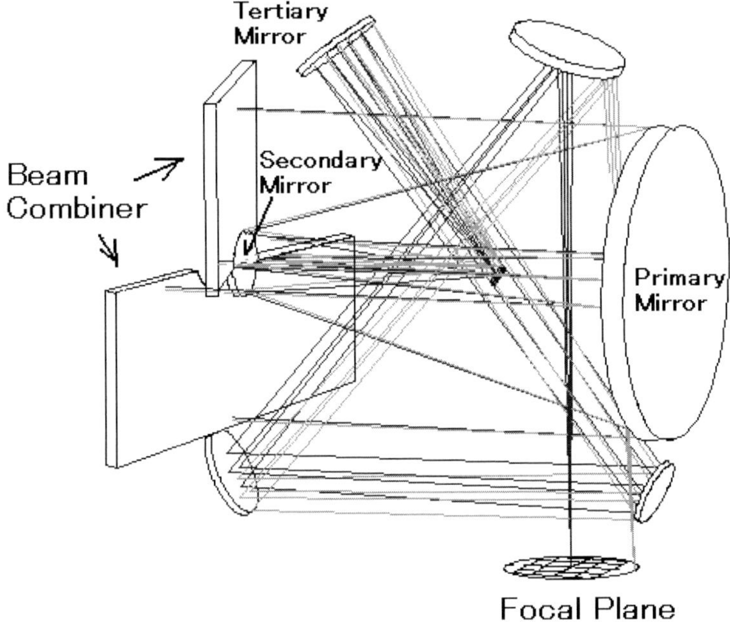

Figure 1. A schematic of the JASMINE optics

a wide focal plane are required. Numerous modern telescope objectives are of Ritchy-Chretien form, e.g., the Hubble space telescope, Subaru, and so on. A Ritchy-Chretien system has two mirrors, both of which are hyperboloids. This system is corrected for spherical aberration and coma, leaving astigmatism and strong field curvature uncorrected. This strong field curvature is hard to correct in such an optics. In 1977 Korsch proposed a three mirror system with a long focal length and a wide focal plane. The Korsch system is one of the convincing models. However, the center of the field is totally vignetted because of the fold mirror. Therefore we consider the improved Korsch system in which the center of the field is not vignetted. A schematic of the optics is shown in Fig. 1.

The aperture size of the optics is 1.5 m, and its focal length, f, is 50 m in order to accomplish $f\lambda/Dw = 2$, where λ, D, and w is the wavelength, the aperture size, and pixel size, respectively. The size of the detector for z-band is 6 cm × 3 cm with 4096 × 2048 pixels. The pixel size is 15 μm which corresponds to 61.9 milliarcsec (mas) on sky. These

Table 1. Summary of the instrument parameters

Optics design	Korsch System (3 mirrors)
Aperture size	1.5 m
Focal length	50 m
pixel size	15μm
pixel on sky	61.9 mas
Array size	6 cm × 3 cm
Pixels per detector	4096 × 2048
Number of detectors	98 (7 × 14)
Basic Angle	99°.5

Table 2. Surface data summary

Surf	Type	Radius	Thickness	Glass	Diameter	Conic
OBJ	STANDARD	Infinity	Infinity		0	0
1	STANDARD	Infinity	2325		1530.435	0
STO	STANDARD	-5400	-2175	MIRROR	1500.682	-0.9861102
3	STANDARD	-1750.010	1312.5	MIRROR	322.9818	-5.174046
4	COORDBRK	-	0		-	-
5	STANDARD	Infinity	0	MIRROR	107.043	0
6	COORDBRK	-	-1350		-	-
7	STANDARD	2378.875	1350	MIRROR	450.1022	-0.7149477
8	STANDARD	Infinity	1275		302.0378	0
9	COORDBRK	-	0		-	-
10	STANDARD	Infinity	0	MIRROR	318.6309	0
11	COORDBRK	-	-2175		-	-
12	COORDBRK	-	0		-	-
13	STANDARD	Infinity	0	MIRROR	438.1686	0
14	COORDBRK	-	2725		-	-
15	COORDBRK	-	0		-	-
16	STANDARD	Infinity	0	MIRROR	568.9629	0
17	COORDBRK	-	-2475		-	-
IMA	STANDARD	-14002.99			654.6421	0

parameters are summarized in Table 1. We also show the summary of the surface data in JASMINE optics in Table 2.

The image quality of the field of view has been analyzed. The spot diagram is shown in Fig. 2. This result shows that the field of view with diffraction limited image is achieved. Furthermore, the field distortion is also investigated. The value of distortion is 0.02% at maximum. This is enough small for our astrometry mission.

Finally we obtain the diffraction limited optical design with small distortion.

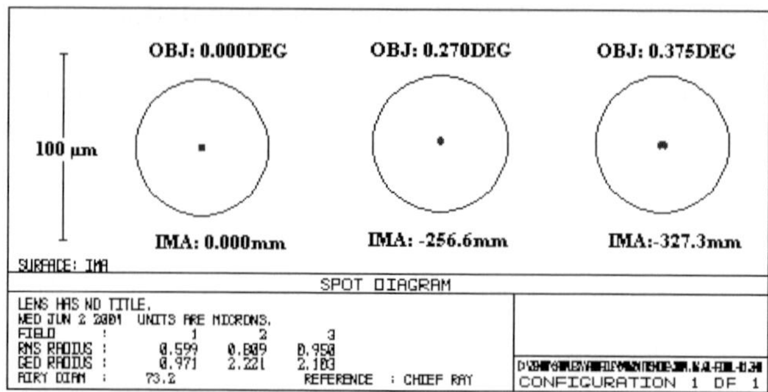

Figure 2. A spot diagram showing that a diffraction limited image is achieved.

2. CCD Centroiding Experiment

Measuring the centroiding of stars is one of the most important problems for astrometry. We examined the accuracy of the centroid of stars and obtained the relative distance of stars to an accuracy of 1/300 pixel (Yano et al. 2004); that is almost ideal. Our experimental method is shown below.

2.1. *Algorithm*

In order to estimate the precise distance of two point sources in image frames to sub-pixel accuracy, the following algorithm is proposed. Here, we show the algorithm used in this experiment. Before the analysis, each image frame is bias subtracted and flat fielded. First, we pick two stars to measure the distance. Next we seek the pixel in which a number of photons is maximum in each star. Then we pick up a square subset of 5×5 pixels around the peak pixel of each star image. Accordingly, the number of photons is the maximum value at the center of pixels in both two stars. Only the pixel values of the two subsets are used to measure the distance of the two stars. We calculate the photon weighted mean of each star by the following equation.

$$\begin{pmatrix} x_c \\ y_c \end{pmatrix} = \frac{1}{\sum_i \sum_j N_{ij}} \begin{pmatrix} \sum_i \sum_j N_{ij} i \\ \sum_i \sum_j N_{ij} j \end{pmatrix},$$

where N_{ij} is the number of photons at the position (i,j). The photon weighted means (x_c, y_c) derived by the above equation are different from the real positions (x_a, y_a). Here, we assume that the difference between the photon weighted mean and the real position is proportional to the deviation of the photon weighted mean from the center of the pixel.

$$x_a - x_c = k x_c, \tag{2.1}$$

where k is a coefficient for the correction of the position of a star. This assumption was originally adopted by the FAME group for their laboratory experiment (Triebes et al. 1999, 2000). However, our treatment of the coefficient k is different from that of the FAME group. Triebes et. al. regarded k as a single parameter across the image frame, but we allow k to be specific to each star. This way, we take into account the variation of the shape of PSF. We calculate the parameters, k, by using the least squares method. Then we obtain the real position x_a from the estimated parameter, k. The algorithm, used in this experiment, is very useful. The reason is as follows. First, it is easy to calculate the photon weighted mean from the data. Second, we need not assume the shape of the PSF. We note that the shape of the PSF is assumed from the estimated parameter, k, implicitly.

Below we show the above algorithm explicitly. We define the positions of star1 and star2 as x_{a1} and x_{a2}, respectively.

$$x_{a1} = x_{c1} + k_1 x_{c1} \tag{2.2}$$
$$x_{a2} = x_{c2} + k_2 x_{c2} \tag{2.3}$$

where, x_c is the photon weighting mean of a star.

Here we define a function I as

$$I = 0 (x_{c2} > x_{c1})$$
$$I = 1 (x_{c2} < x_{c1}). \tag{2.4}$$

The relative distance of the two stars $|\delta x_a|$ is

$$\begin{aligned} |\delta x_a| &= x_{a2} - x_{a1} + I \\ &= x_{c2} - x_{c1} + k_2 x_{c2} - k_1 x_{c1} + I \\ &= (1 + k_2)(x_{c2} - x_{c1}) + (k_2 - k_1)x_{c1} + I \\ &\equiv \alpha \Delta + \beta x_{c1} + I \end{aligned} \tag{2.5}$$

where, $\alpha = (1 + k_2)$, $\beta = (k_2 - k_1)$, and $\Delta = (x_{c2} - x_{c1})$.

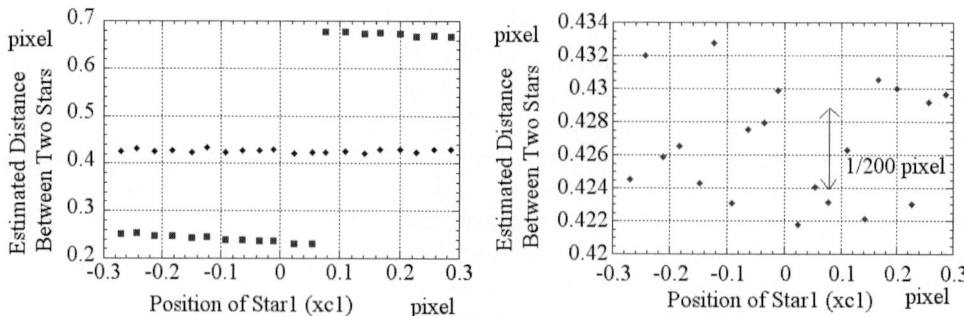

Figure 3. (Left panel) Relative distance between two stars against the photon weighted mean of star1. The squares indicate only the distances between photon weighted means of two stars, that is, no correction is performed. On the other hand, the diamonds indicate the estimated distances by linear correction of the photon weighted mean described here. In this experiment, $f\lambda/Dw$ is equal to about 3. (Right panel) Same with the diamonds in the left panel.

We wish to derive the values of the parameters k_1 and k_2, with which the above equation is satisfied with the smallest error. In other words, we use a least squares method. So, we define S as

$$S = \sum (\alpha\Delta + \beta x_{c1} + \gamma + I)^2, \tag{2.6}$$

where $\gamma = -|\delta x_a|$. The derivative of S by each parameter is equal to zero. Then the following relations are satisfied.

$$\frac{\partial S}{\partial \alpha} = \alpha \sum \Delta^2 + \beta \sum \Delta x_{c1} + \gamma \sum \Delta + \sum \Delta I = 0, \tag{2.7}$$

$$\frac{\partial S}{\partial \beta} = \alpha \sum \Delta x_{c1} + \beta \sum x_{c1}^2 + \gamma \sum x_{c1} + \sum x_{c1} I = 0, \tag{2.8}$$

$$\frac{\partial S}{\partial \gamma} = \alpha \sum \Delta + \beta \sum x_{c1} + \gamma \sum 1 + \sum I = 0. \tag{2.9}$$

From the above relations, we obtain the positions of two stars, x_{a1} and x_{a2} from the estimated parameters. Finally, we note again that the above least squares method is easy to calculate without explicit assumption of the PSF. Furthermore, it is an advantage for calculating thousands of stars because of its simplicity.

2.2. *Experimental results*

We have taken twenty image frames by sliding the CCD array. The interval of sliding is $1\,\mu\text{m}$; that is, twenty steps correspond to 1 pixel. From these twenty image frames, we estimate the distance of two stars, using the algorithm shown in the previous section. An image of the point spread function (PSF) of a star is focused onto the CCD array with a spread of about three pixels, using a lens with focal length of $200\,\text{mm}$. The distance between the image field and the lens is about $570\,\text{mm}$, and the length between the lens and the CCD camera is about $310\,\text{mm}$. For a light source of the simulated stars, white light is used. The results for our experiment are shown in Fig. 3. The abscissa indicates the photon-weighted mean of star1. The ordinate indicates the distance between two stars.

Here, we note that the integer part of the distance is eliminated. Accordingly, the distance has a value between 0 and 1. The squares show only the separations between photon-weighted means of two stars, that is, no correction is performed. On the other hand, the diamonds are the estimated distances between two stars by using the algorithm.

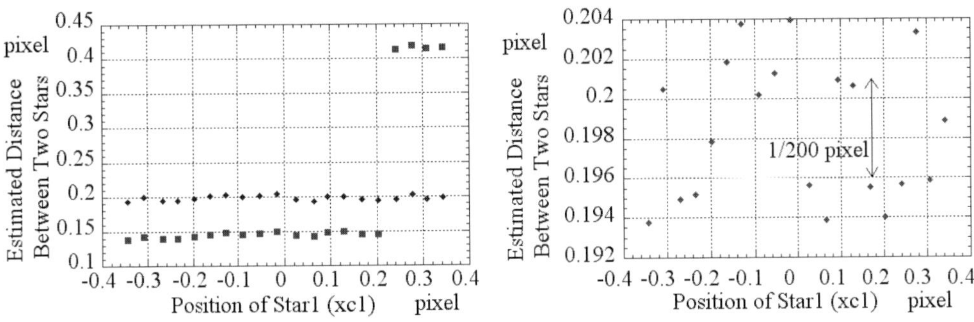

Figure 4. Same as Fig. 3, but $f\lambda/Dw = 1$.

Distances of two stars with no correction, that is, difference of photon-weighted means of two stars, are distributed to two groups, one is a value around 0.24, and the other, around 0.68. On the other hand, estimated distances using the linear correction, all the measurements are the values of around 0.43. Then the accuracy of estimation becomes exceedingly high. These estimated distances (the diamonds) are shown again in the right panel of Fig. 3. As we can see, the variance of the estimated distances of two stars is about 1/300 pixel; that is, the error of the estimation is 1/300 pixel for one measurement, which is almost the ideal one given by the Poisson noise of photons.

Next, we investigate accuracy of estimation with a different size of PSF, using a lens with focal length of 100 mm. The distance between the image field and the lens is about 730 mm, and the length between the lens and the CCD camera is about 110 mm. In this case, the image size of the PSF is about 1 pixel. The results are shown in Fig. 4. The squares show only the separations between photon-weighted means of two stars. On the other hand, the diamonds are the estimated distances between two stars, which are shown again in the right panel of Fig. 4. In this case also, the accuracy of estimation is about 1/300 pixel.

Here we consider the reason that the photon weighted mean differences between star1 and star2 are discontinuous. We pick up a square subset around the peak pixel of star image in this experiment. When the peak pixel moves to the adjoining pixel by sliding the CCD array, the region of a square subset changes. Accordingly, the calculated photon weighted mean jumps discontinuously. The analytical value of this discontinuity is $\frac{k_2}{k_2+1}$. This analytical form shows that the small discontinuity represents the small k_2. The value of the discontinuity in the first case of experiments is about $0.68 - 0.23 \simeq 0.45$. In this experiment, the estimated parameters α, β, γ, k_1, and k_2 are 1.815, 0.112, -0.427, 0.703, and 0.815, respectively. The error in the experiment is about 3.3×10^{-3}. The value of k_2 is consistent with the discontinuity of 0.45. In the second case, the discontinuity in Fig. 5 is about $0.41 - 0.14 \simeq 0.27$. In this case, these parameters, α, β, γ, k_1, and k_2 are 1.363, -1.80×10^{-2}, -0.198, 0.381, and 0.363, respectively. This discontinuity consistent with the value of k_2. Comparing these two cases, the discontinuity in Fig. 3 is larger than that in Fig. 4. This is because the value of k_2 in the first case is larger than that in the second case.

For comparison, we estimate the distance of the two stars with the algorithm in which we use a common parameter k for the two stars, that is, $k_1 = k_2$ is satisfied. This algorithm is essentially the same with that in FAME. In this case we find that the accuracy of estimation is about 1/100 pixel, which is worse than the above results obtained from our algorithm.

3. Concluding remarks

The three mirror optical design for the Japan astrometry satellite mission has been studied. In order to accomplish measurements of astrometric parameters with high accuracy, optics with a long focal length and a wide focal plane for astrometry are required. Korsch proposed a three mirror system with a long focal length and a wide focal plane in 1977. The Korsch system is one of the convincing models. However, the center of the field is totally vignetted because of the fold mirror. Therefore we consider the improved Korsch system in which the center of the field is not vignetted. Finally We obtain the diffraction limited optical design with small distortion(0.02%).

We have experimented with the measurement of centers of star images on a CCD for investigating the accuracy of finding the positions of stars, using the algorithm for estimating the positions of stars from the photon weighted means of stars. Then we obtain the results from the experiment that the accuracy of estimation of distance between two stars has a variance of about 1/300 pixel; that is, the error for one measurement is about 1/300 pixel, which is almost an ideal result given by Poisson noise of photons. We also investigate the accuracy of estimation of positions with a different size of PSF. In this case also, we find that the accuracy of estimation has a variance of about 1/300 pixel.

In our experiment, the separation of two stars is measured by using the image of the field on the CCD array. Precisely speaking, the lens system distorts an image of a star field, and then the estimated distance of stars includes the error by the distortion. Therefore we must correct the distortion in order to estimate the real separations of stars. Development of the algorithm which corrects the distortion of the image by the lens system is needed. These are the future work of this experiment.

References

Triebes, K., Gilliam, L., Harris, F., Hilby, T., Horner, S., Monet, D., Perkins, P., and Vassar, R. 1999, Bull. AAS, Vol. 31, p.1505

Triebes, K. J., Gilliam, L., Hilby, T., Horner, S. D., Perkins, P., Vassar, R. H., Harris, F. H., & Monet, D. G. 2000, in *UV, Optical, and IR Space Telescopes and Instruments*, James B. Breckinridge and Peter Jakobsen, eds., Proc. SPIE, Vol. 4013, p. 482–492.

Yano, T., Gouda, N., Kobayashi, Y., Tsujimoto, T., Nakajima, T., Hanada, H., Kan-ya, Y., Yamada, Y., Araki, H., Tazawa, S., Tsuruta, S., Kawano, N. 2004, *PASP*, 116, 667–673.

Transits of Venus: New Views of the Solar System and Galaxy
Proceedings IAU Colloquium No. 196, 2004
D.W. Kurtz, ed.

© 2004 International Astronomical Union
doi:10.1017/S174392130500164X

JASMINE simulator

Seiji Ueda[1], Yoshiyuki Yamada[2], Takashi Kuwabara[3], Naoteru Gouda[4], Takuji Tsujimoto[4], Yukiyasu Kobayashi[4], Tadashi Nakajima[4], Hideo Matsuhara[5], Taihei Yano[4], Masahiro Suganuma[4], and the JASMINE Working Group

[1]The Graduate University for Advanced Studies, Mitaka, Tokyo, 181-8588, Japan
email: seiji.ueda@nao.ac.jp

[2]Department of Physics, Kyoto University, Kyoto, 606-8502, Japan
email: yamada@amesh.org

[3]Department of Physics, Niigata University, Niigata, 950-2181, Japan
email: takashi@astro.sc.niigata-u.ac.jp

[4]National Astronomical Observatory Japan, National Institutes of Natural Sciences, Mitaka,
Tokyo 181-8588 Japan
email: naoteru.gouda@nao.ac.jp, taku.tsujimoto@nao.ac.jp, yuki@merope.mtk.nao.ac.jp,
tadashi@dodgers.mtk.nao.ac.jp, t.yano@nao.ac.jp, suganuma@merope.mtk.nao.ac.jp

[5]Institute of Space and Astronautical Science, Japan Aerospace Exploration Agency,
Sagamihara, Kanagawa, 229-8510 Japan
email: maruma@ir.isas.ac.jp

Abstract. We explain simulation tools in the JASMINE project (JASMINE simulator). The JASMINE project stands at the stage where its basic design will be determined in a few years. Therefore it is very important to simulate the data stream generated by astrometric fields at JASMINE in order to support investigations of error budgets, sampling strategy, data compression, data analysis, scientific performances, etc. We find that new software technologies, such as Object Oriented (OO) methodologies, are ideal tools for the simulation system of JASMINE (the JASMINE simulator). In this article, we explain the framework of the JASMINE simulator.

1. Introduction

The JASMINE project stands at the stage where its basic design will be determined in a few years. Thus it is very important to simulate the data stream generated by astrometric fields at JASMINE in order to support investigations of error budgets, sampling strategy, scientific performances, etc. The simulation system should include all components in JASMINE as "objects" in order to resolve all issues which can be expected beforehand and make it easy to cope with some unexpected problems which might occur during the mission of JASMINE. Components includes not only models of concrete materials such as scientific instruments and satellite bus system, but also abstract ones such as observation methods, orbits, data transfer, algorithms of data analysis, etc. Furthermore, many researchers in various fields, including engineering of the bus system, should participate in developing the simulator. Then the simulator includes many objects and furthermore, flexibility and interaction are needed to cope with the objects. On the other hand, calibrations must be protected from unintentional modification for each model of the objects. Thus we conclude that the Object Oriented methodologies are ideal for the demands described above in the simulator. High maintainability and reusability

484 S. Ueda *et al.*

of the object oriented approach is very useful in making the simulator system. Thus we are now constructing the JASMINE simulator using techniques developed in the field of information science.

In §2, requirements of JASMINE simulator are discussed; §3 contains some examples of JASMINE system investigations with the simulator software; §4 is devoted to the future schedule of software development.

In this paper we will make reference to terms with specific meanings within a object-oriented programming context. These are:

class: encapsulates common behaviour of a group of objects;

attributes: data member of a class;

methods: functions which may be performed on instances of a class;

object: an instance of a class;

abstract class: a class from which no instances may be created;

inheritance: a way to form new classes or objects using predefined objects or classes where new ones simply take over old ones' implementations and characteristics.

2. Specification requirement of framework

Implementation of the simulator is carried out in two steps. In the first step, we construct or choose the framework of the software. The second step is building software which are JASMINE specific under the framework built in the first step. In this section, we discuss the framework of the JASMINE simulator.

2.1. *Implementation of proto-model*

JASMINE Simulator will be mainly used for assessing the scientific goals of the mission, instrument design investigation, data treatment preparation, and algorithm study such as stellar image detection or evaluating astrometric parameters from raw data. The need for a clear, modular and easily communicable structure for the simulator led us to choose an object-oriented language for its implementation.

In the simulator, each component of JASMINE is treated as an "object" which has specifications as attributes. For example, an object which represents a CCD detector has attributes such as pixel size, number of pixels, quantum efficiency, flatness, charge transfer rate, etc. Such attributes affect final scientific outputs.

Within the context of system investigations, some attributes of some components may depend on other attributes of other components. We can draw diagrams which represents dependencies of each component (object) by analyzing the dependencies. This is called a "Data Flow Diagram". The framework is expected to provide the function of creating and editing such "data flow" diagrams.

The "data flow" during JASMINE observation begins with astronomical objects, is related to an optical system, detectors, and data-processing units, and ends as a communication system. One simple implementation is the specifying of a relation of abstract classes which represent such components beforehand. Then the framework can deal with detailed investigation easily by making concrete classes that are inherited from the abstract classes. The following calculation is performed in the flow of the JASMINE specification study,

$$f = \frac{N_{\mathrm{PSF}} w D}{\lambda},$$

where f, w, D, λ are total focal length of the optical system, pixel size of the CCD, diameter of the primary mirror, and wavelength, respectively, and N_{PSF} is a constant. An

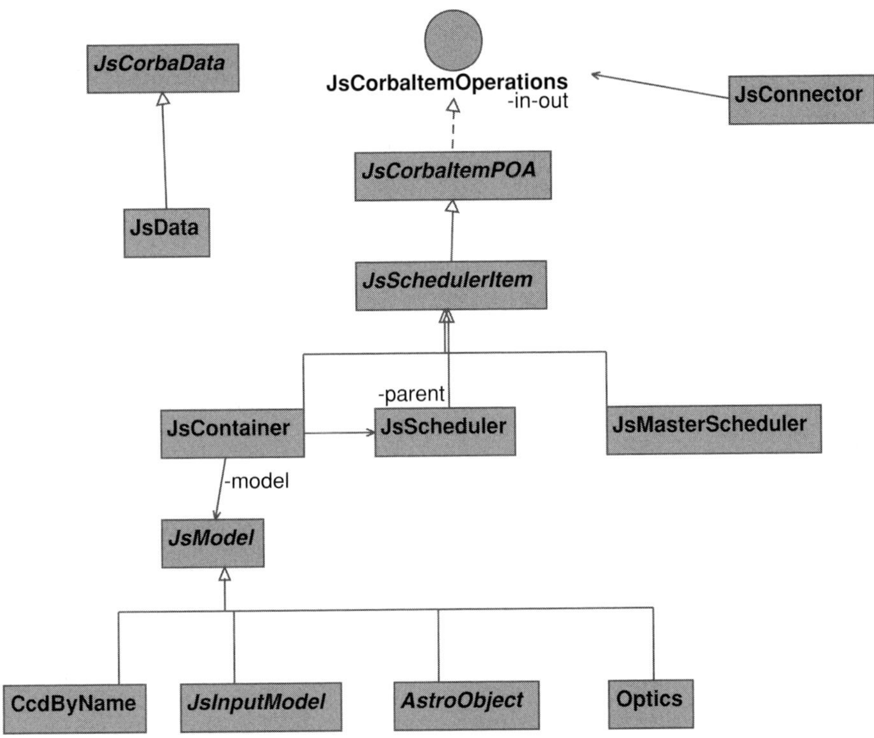

Figure 1. UML diagram of the framework of prototype simulator code. Names of classes JsModel, JsContainer, JsConnector, and JsData correspond to "model", "container", "connector", and "data" of the text.

attribute of the optical system, f, depends on an attribute of the detector, w. These dependencies ("data flow") have opposite direction of a real "data flow" in observation. One of requirements of our simulator is that we can specify direction of the "data flow" in our diagram as both the same and counter direction as the "data flow" of real observations.

From above considerations, the framework of the JASMINE simulator may have functions for managing DAG (directed acyclic graph). To summarize the requirements,

(a) computational modules are extendible,

(b) data types are extendible,

(c) open source (if using existing software),

(d) some simple graphical tools are provided,

(e) platform independent,

(f) availability of treating distributed computation via network.

Requirements (a)–(c) are critical and (d)–(f) are optional.

A schematic UML diagram of our prototype simulator is shown in Fig. 1. In our implementation of the prototype simulator, each of the components, such as optical systems or detectors are inherited from an abstract class "model". A class "container" which manages the dependency graph has the class "model". The structure ensures not also extendibility of modules but the arbitrariness of the direction of dependency. Each directed arc is represented as a class "connector", which encapsulates flowing "data" between two "models". The abstraction "data" ensures extendibility of data. The model can be easily extendible by making subclass of "model".

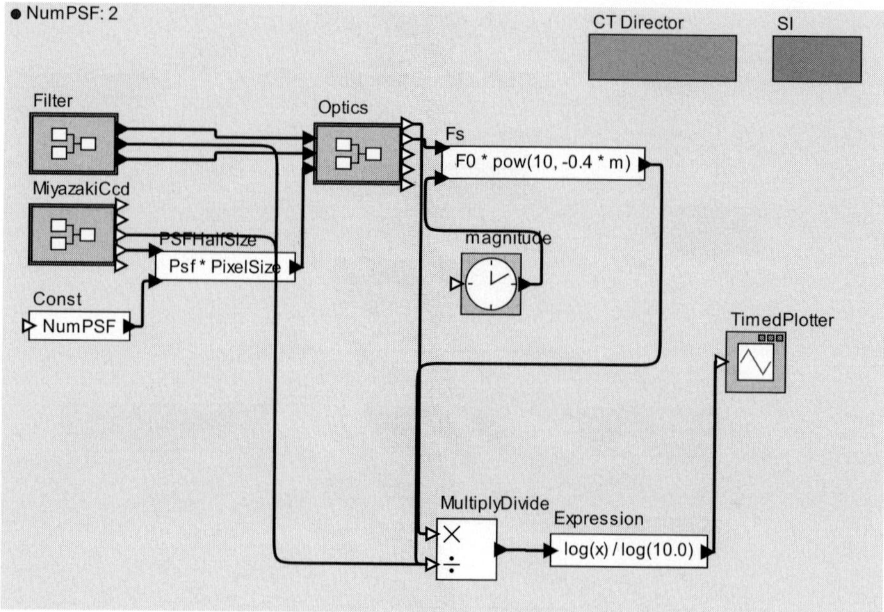

Figure 2. Framework of the prototype. Filter, MiyazakiCcd, and Optics modules are implemented as composite actors.

2.2. *Ptolemy II for framework*

Many implementations of such simple data flow simulators exist. Good examples are the visualization tools of numerical simulations. The methods of extending a calculation model are also provided in some tools. However, faults of such visualization tools are difficulties of data type extension. We concluded that building a JASMINE simulator as a module of AVS or OpenDX was not feasible.

The Ptolemy II project (http://ptolemy.eecs.berkeley.edu/ptolemyII/) carried out at UC Berkeley provides the framework which fills all of our above demands. Ptolemy II is designed as concurrent modelling and design tools under the fairly liberal UC Berkeley copyright (requirements a–d). It is implemented by Java language (requirement e). It also provides a "JXTA" class loader, which is JAVA PtoP implementation (requirement f). Thus, Ptolemy II is adopted as the framework of our simulator. A sample graph for the specification study implemented on Ptolemy II is shown in Fig. 2.

3. Contents of our simulator

In this section, we show implemented results and implementation plans of JASMINE specific components. We have already performed several items:

- observational method and accuracy,
- data reduction scheme,
- stellar image detection,
- data compression scheme,
- star number counts,
- optical system performance,
- satellite attitude and accuracy,
- orbits.

In this section, we show two topics.

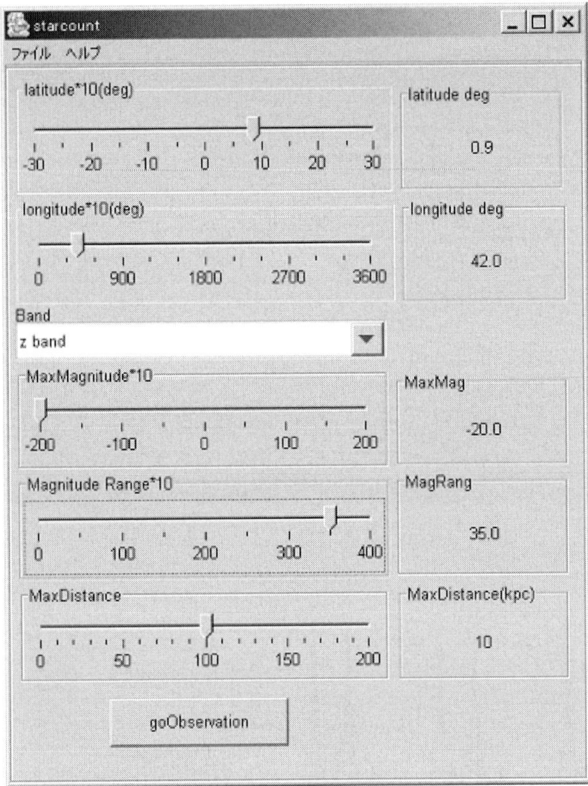

Figure 3. StarCount Simulator

3.1. *Galaxy model and stellar number counts*

The main targets of the JASMINE project are stars in the Galactic disk and bulge. There is no complete observation which shows number density of stars brighter than specified magnitude in the z-band and contained in the Galactic disk and bulge. Thus it is important to forecast the number of stars, data rate, etc. using a model Galaxy which is well-confirmed from observations. We use Wainscoat *et al.* (1992) and Cohen (1994) for the model of the stellar number distribution in the galaxy. This model reproduces well 2MASS results. We also take the motion of stars from the Evans model (1994), and the distribution of dust from COBE DIRBE data into account. An outline of the model is shown in appendix.

We implemented Star Count Simulator (SCS) first as Fig. 3. This is the tool for calculating THE number of stars with arbitrary direction, arbitrary distance, and arbitrary spectral types. This enables us evaluating the number of stars observed at every moment. We also implemented a three-dimensional viewer of the galaxy model (Fig. 4).

Fig. 5 shows the time variation of the number of stars observed by JASMINE. We cannot estimate the amount of processing by average stellar density because the variability of numbers is very large. Such information is very important for designing the data processing unit and data recorders.

3.2. *Modelling of CCD and focal plane simulation*

We have also started implementing the CCD model. The CCD can be modelled as the module which as many accumulators (pixels), data transfer mechanisms, and reading mechanisms. Efficiency of accumulation (quantum efficiency) is about 0.9 and depends

Figure 4. 3D Galaxy view.

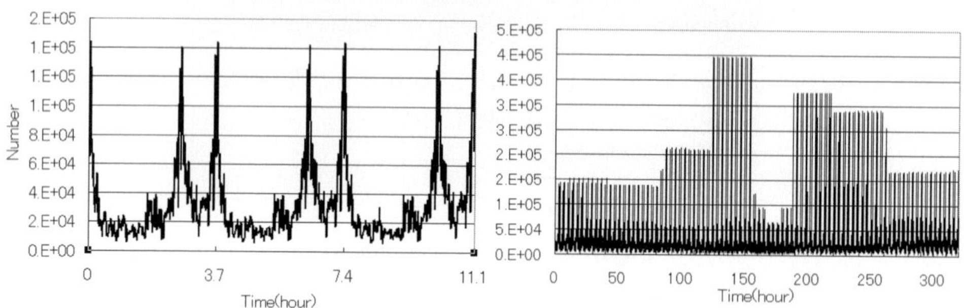

Figure 5. Time variation of Star Number.

on the wavelength. Data (charge) transfer efficiency is extremely close to unity, but is not unity. Readout of the data has also some errors. These effects are implemented as a probability process on the CCD model. Satellite environmental effects such as cosmic rays may affect such probabilities.

Using our galaxy model and satellite attitude model, a list of stars contained in two fields of view are generated. According to attributes of each star such as distance, spectral types, etc., photons are generated using a probability process. The astrometric optical system has a very long focal length. Point spread functions are broad compared with pixel size. To generate images on the focal plane, it is also required to use a probability process. From the above, several probability process (times, locations of detector plane) pairs belonging to each photon are generated. Each "photon" may be the trigger of the CCD model. By each trigger, the CCD model accumulates "charge" in the appropriate "pixel". Continuing pseudo-exposure during an appropriate interval, we can get a pseudo-image of a JASMINE observation.

4. Conclusions

We chose Ptolemy II as a framework of the JASMINE simulator. We have already preformed several investigations on the JASMINE system design using Fortran, C++

and JAVA codes. We will reconstruct such codes in a form suitable for the Ptolemy II framework.

Appendix A. Milky Way model

In this model, 87 spectral types of stars are considered. We can estimate the number of stars classified by types independently. The original Cohen (1994) model provides information on the number of stars within seven bands, $B, V, H, I, J, 12\,\mu m, 25\mu m$. We extend it to z (by inserting Cohen data), L, M (see Tokunaga 2000) bands.

We consider five components of the galaxy: disk, bulge, halo, spiral arm, molecular ring. The number density $\rho(S)$ with given spectral type S and distance r is denoted as

$$\rho(S) = \rho(S)_{\text{disk}} + \rho(S)_{\text{bulge}} + \rho(S)_{\text{halo}} + \rho(S)_{\text{arm}} + \rho(S)_{\text{ring}}. \tag{A1}$$

Number densities of each component are normalized by number density of S type stars near solar systems, $\rho_0(S)$. Density distributions of each component are as follows.

(a) exponential disk

$$\rho(S)_{\text{disk}} = \rho_0(S) f_D(S) \exp\left[-\frac{r - r_0}{h} - \frac{|z|}{h_z(S)} \right] \tag{A2}$$

where r_0, h, $h_z(S)$ are the distance to the Galactic center (8 kpc), the radial scale length (3.5 kpc), and scale height for each S type star, respectively,

(b) axisymmetric bulge component

$$\rho(S)_{\text{bulge}} = \rho_0(S) f_B(S) g_B x^{-1.8} e^{-x^3} \tag{A3}$$

where $x = \sqrt{r^2 + k_1^2 z^2}/r_1$ (k_1: the bulge axis ratio = 1.6; r_1: the bulge radius = 2 kpc). g_B: the ratio of densities of bulge at the point $x^{-1.8} e^{-x^3} = 1$ and disks near solar system − 3.6,

(c) Halo with $r^{1/4}$ law

$$\rho(S)_{\text{halo}} = \rho_0(S) f_H(S) g_H \exp\left[-7.669 \left(\frac{r}{r_e}\right)^{1/4} \right] \frac{r}{r_e}^{-7/8} \tag{A4}$$

where r_e: the effective radius of the halo = 2.83 kpc, $g_H = 0.002$,

(d) spiral arm pair which shown in table A,

$$\theta(r) = a \log\left(\frac{r}{r_{\min}}\right) + \theta_{\min} \tag{A5}$$

where

$$\rho(S)_{\text{arm}} = \rho_0(S) f_A(S) g_A \exp\left[-\frac{(r - r_0)}{h} - \frac{|z|}{h_z(S)} \right] \tag{A6}$$

where g_A: ratio of the total area of disk and the area occupied by arm (=5),

(e) and molecular ring at $r \sim 0.45 r_0$

$$\rho(S)_{\text{ring}} = \rho_0(S) f_r(S) g_r \exp\left[\frac{-(r - r_r)^2}{2\sigma_r^2} - \frac{|z|}{h_z(S)} \right] \tag{A7}$$

where r_r: the ring radius = $0.45 r_0$, $\sigma_r = 0.064 r_0$, g_r: ratio of the total area of disk and ring (=25).

We use dust absorption with the result from the DIRBE/IRAS dust Map 9

Table 1. parameter of spiral arm

Arm	a	$r_{\min}$ (kpc)	$\theta_{\min}$ (radian)	Extent (radian)	width (pc)
1	4.25	3.48	0.000	6.0	750
1'	4.25	3.48	3.141	6.0	750
2	4.89	4.90	2.525	6.0	750
2'	4.89	4.90	5.666	6.0	750
L	4.57	8.10	5.847	0.55	300
L'	4.57	7.59	5.847	0.55	300

(http://astron.berkeley.edu/dust/). To convert from the integration value provided by DIRBE to the real distribution, we assumed an exponential distribution of dust. The absorption δA_λ within the distance r and $r + \delta r$ is written as

$$\delta A_\lambda = A_{\lambda 0} \exp\left[-\frac{r - r_0}{h_{da}} - \frac{|z|}{h_{za}}\right]\delta r. \tag{A 8}$$

where $A_{\lambda 0}$: absorption near solar system, h_{da}: the radial scale length of the absorption $=$ 3.5 kpc, h_{za}: the scale height of the absorption $=$ 100 pc. The value of $A_{\lambda 0}$ at each (ℓ, b) is determined by

$$A_\lambda(\mathrm{DIRBE}) = A_{\lambda 0} \int_0^\infty \exp\left[-\frac{r - r_0}{h_a} - \frac{|z|}{h_{za}}\right] dr. \tag{A 9}$$

The simulation of astrometric observations such as JASMINE also requires a model of stellar motion. The axisymmetric Evans (1994) model is simple for discussing the motion. The potential in the Evans model is as follows,

$$\Psi = \frac{\Psi_a R_c^\beta}{(R_c^2 + R^2 + z^2 q^{-2})^{\beta/2}}, \qquad \beta \neq 0, \tag{A 10}$$

where q, Ψ_a, R_c are axis ratio of equipotential, central potential, core radius respectively. We can get mass density from Poisson's equation.

$$\rho = \frac{v_a^2 R_c^\beta}{4\pi G q^2} \frac{R_c^2(1 + 2q^2) + R^2(1 - \beta q^2) + z^2[2 - q^{-2}(1 + \beta)]}{(R_c^2 + R^2 + z^2 q^{-2})^{(\beta+4)/2}} \tag{A 11}$$

Using these models, we can compute the motion of stars in each component.

Acknowledgements

We would like to acknowledge Y. Kawakatsu, A. Noda, A. Tsuiki, M. Utashima, A. Ogawa, N. Sakou, H. Ueda for their collaboration in the investigations on the JASMINE spacecraft system. Furthermore, we would like to thank Miyashita Hisashi at IBM Tokyo Research for useful advice on general code construction technique. This work has been supported in part by the Grant-in-Aid for the Scientific Research Funds (15340066) and Toray Science Foundation.

References

Cohen, M. 1994, *ApJ* **107**, 582

Evans, N.W. 1994, *MNRAS* **267**, 333

Tokunaga, A.T. 2000 *in Allen's Astrophysical Quantities, 4th edition*, ed. A.N. Cox, Springer-Verlag (New York), p. 143

Wainscoat, R. J., Cohen, M., Volk, K., Walker, H. J., Schwartz, D.E. 1992, *ApJS* **83**, 111

Transits of Venus: New Views of the Solar System and Galaxy
Proceedings IAU Colloquium No. 196, 2004
D.W. Kurtz, ed.

© 2004 International Astronomical Union
doi:10.1017/S1743921305001651

Nano-JASMINE: a nano size astrometry satellite

Yukiyasu Kobayashi[1], Taihei Yano[1], Naoteru Gouda[1], Yoshiyuki Yamada[2], Naruhisa Takato[1], Satoshi Miyazaki[1], Masahiro Suganuma[1], Seiji Ueda[4], Shin'ichi Nakasuka[3], and the JASMINE working group

[1]National Astronomical Observatory, 2-21-1 Osawa Mitaka Tokyo 181-8588, Japan
email:yuki@merope.mtk.nao.ac.jp,t.yano@nao.ac.jp,naoteru.gouda@nao.ac.jp,
takato@subaru.naoj.org, satoshi@naoj.org, suganuma@merope.mtk.nao.ac.jp

[2]Department of Physics, Kyoto University, Kyoto 606-8502, Japan
email:yamada@amesh.org

[3]The Graduate University for Advanced Studies, 2-21-1 Osawa Mitaka Tokyo Japan
email: seiji.ueda@nao.ac.jp

[4]Faculty of Engineering, The University of Tokyo, 7-3-1 Hongo Bunkyo-ku Tokyo Japan
email:nakasuka@space.t.u-tokyo.ac.jp

Abstract. We present the outline and current status of the nano-JASMINE project. Nano-JASMINE is a nano size astrometry satellite (the total payload mass is between 1 kg and 10 kg), which is expected to be launched in 2006. The main purpose of the project is to prove and demonstrate the key technologies required for JASMINE (Japanese Astrometry Satellite Mission for Infrared Exploration) in a real space environment Gouda et. al. (2002). Nano-JASMINE will measure annual parallaxes of bright stars (7 mag) with an accuracy greater than 1 milliarcsecond after two years of operation. This is comparable to the accuracy of the Hipparcos catalog. Currently, the subject of major research and development of the project is to build optical and telescope structures using new material, and to establish a data acquisition and control system.

1. Mission Target

Nano-JASMINE is a very small satellite that will be used to demonstrate and verify new technologies developed for the JASMINE project. JASMINE will carry out astrometry observations for stars located in a narrow strip within 4° of the Galactic plane at the z-band (central wavelength of 0.95 μm). JASMINE will measure the position, annual parallax and proper motion of these stars with accuracies greater than ten microarcseconds. By measuring the motion of millions of stars, both deep in the Galactic plane and in the central bulge region, JASMINE will reveal the dynamic evolution of the Galaxy. Its launch is planned within the next ten years, and vigorous technical research and development is currently underway. These includes TDI control of an array detector, ultra-light mirror, electronic components for space radiative environment, techniques on thermal deformation and ultra precision position measurements. Nano-JASMINE will demonstrates and verify these key technologies in real space environments. At the same time, it will make clear the future problems that JASMINE might face. Thus, this would be a good exercise for the JASMINE team. Using nano-JASMINE, we can test new technologies in a real space environment during the development stage itself. The advantages of using a very small satellite, such as a nano-sized satellite is that its cost is not very high and the results can be realized in a short span of time. Furthermore, we can use the unoccupied space of a launcher; this makes the chance of a launch high and its cost low.

491

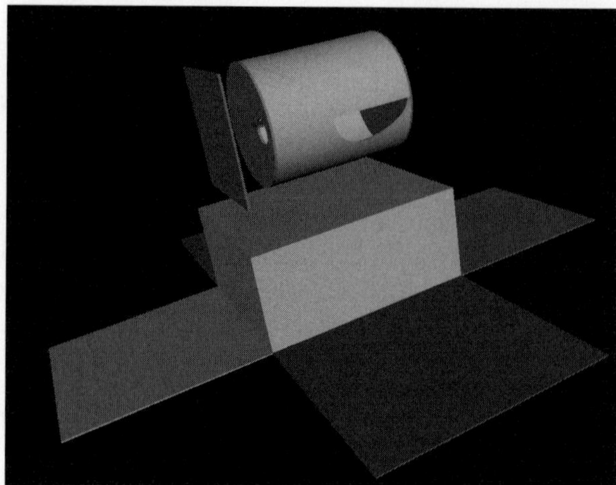

Figure 1. Overall view of nano-JASMINE.

Table 1. General Features of nano-JASMINE

Orbit	Sun-synchronous
Spin rate	36 minutes
Height	300 km – 800 km
Diameter	5 cm
Focus Length	167 cm
Field of View	$30' \times 30'$
Basic Angle	$99°\!.5$
Detector	Full depletion CCD
Pixel Format	1024×1024
Pixel Size	$15\,\mu$m

2. Mission Plan

The first nano-JASMINE satellite is expected to be launched in 2006 into a sun-synchronous orbit at an altitude of between 300 km and 800 km. The total payload mass of nano-JASMINE is a few kilograms. A satellite that weighs between 1 kg and 10 kg is usually called a nano-satellite. Thus, we have named this satellite nano-JASMINE. We will employ a nano-satellite, because of its high chance of launch and because it is capable of supporting an appropriate level of astrometry observations. In order to obtain very precise position information, we simultaneously observe two fields in a great circle with a large separation angle. We expect an accuracy of one milliarcsecond for stars in a certain great circle after one year of operation. The selection of the great circle depends on the launch conditions. The maximum data rate will be of the order of 5000 kbps. We use the z-band filter (central wavelength of $0.95\,\mu$m), the center of which is the longest part of the silicon CCD's sensitive wavelength region. The satellite will spin at a rate of one revolution every forty minutes in order to conduct a survey along the concerned great circle.

3. Telescope and Optics

The nano-JASMINE telescope has a diameter of 5 cm with a beam combiner placed in front of the primary mirror. We adopt the Korsh optical system. This system consists

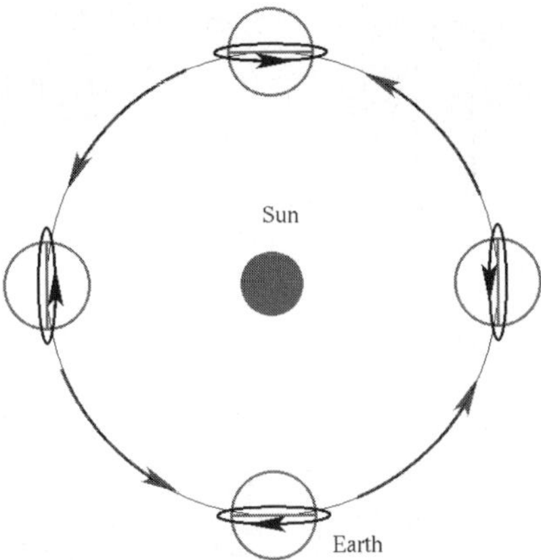

Figure 2. Sun-synchronous Orbit.

of three mirrors that provide a flat focal plane and a wide field-of-view. With a focal length of 167 cm the telescope provides a diffraction limited image which fills two pixels at the focal plane. Basically, the optical system used is a scaled down version of that of JASMINE; however, some modifications are required to solve the problems that arise when optical components become very small. Fig. 1 shows the complete view of the optical layout. Two flat mirrors, angled at 49°, and placed in front of the primary mirror allow the system to simultaneously observe two fields separated by an angle of 95°. Deformation of the optical components due to thermal variations in the environment would make the achievement of high precision astrometry observations almost impossible. The structure and optics are constructed using the same material to reduce thermal deformation. Currently, we concentrate on the research and development of two kinds of materials. The first material is aluminum. Employing modern super-precise machines, it is possible to construct a telescope using aluminum optics and structure. This type of telescope is suitable for nano-JASMINE. The other material under consideration is reaction sintered silicon carbide developed by Toshiba Tsuno et. al. (2004). Reaction sintered silicon carbide is manufactured by reaction sintering molten silicon to spray dried SiC Carbon powder. It has high thermal stability and favorable properties for space telescope material. The reason behind developing the nano-JASMINE telescope using reaction sintered SiC is that this is one of the best candidate materials used to construct the JASMINE telescope.

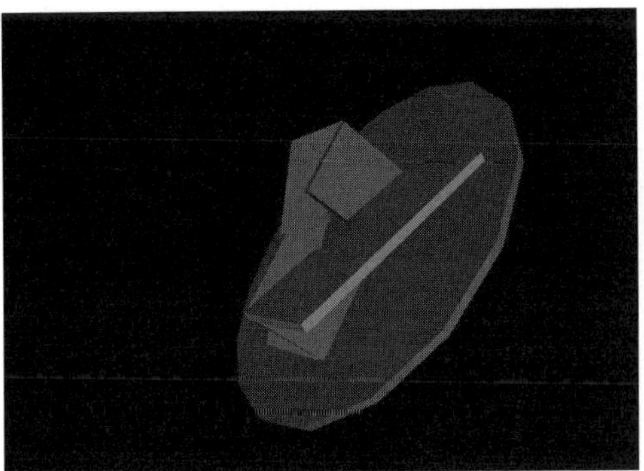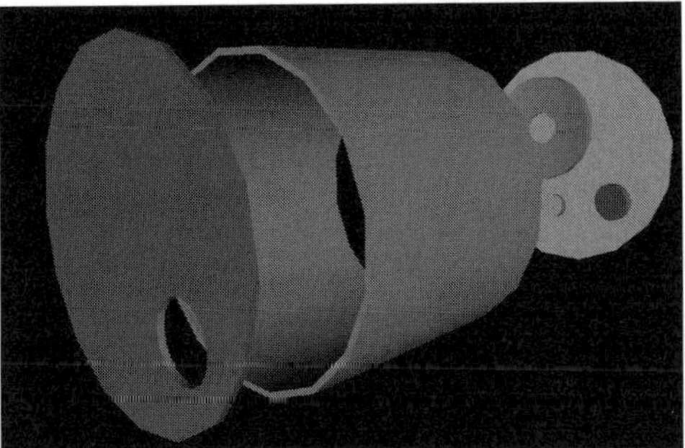

Figure 3. Telescope structure design.

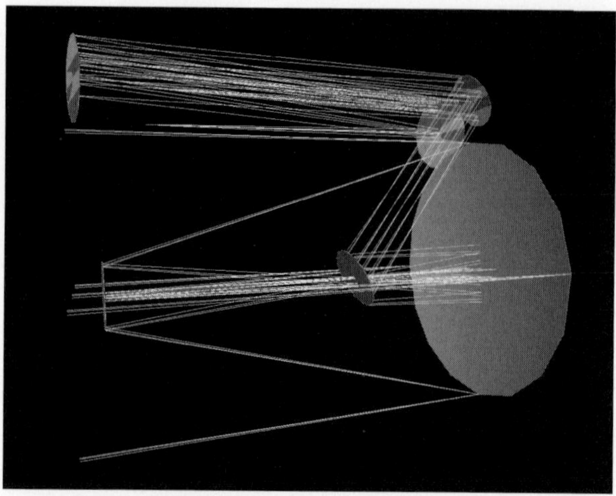

Figure 4. Optical layout.

4. CCD and Control System

For the purpose of survey observations, the CCD and control system of nano-JASMINE use a 1024×1024 format, full depletion CCD, operated in TDI mode. The pixel size is $15\,\mu$m, which is equivalent to $1''.8$ projected on the sky. We are developing our own readout and controlling system. The full depletion CCD has a high response at longer wavelengths. The National Astronomical Observatory and Hamamatsu Photonics are developing the same type of CCD for the new Subaru instrument. We will manufacture the same type of CCD with a smaller format for nano-JASMINE, with a total field of view of $30' \times 30'$. In order to make the readout noise low at a high read out frequency, the CCD has two output channels. A 3-s integration time is employed under the TDI operation. Fig. 2 shows the readout circuit board, which has two 16 bit AD channels with a 3 mus sampling rate. The TDI scan rate must be adjusted according to the satellite spin rate; the CCD controller is equipped to accept a fine adjust control command to change the scan rate. The digitized data is picked up and saved as single stellar information. This part of the operation is realized as hardware using FPGA. This is because it requires high speed processing; however, it has a relatively simple processing procedure. After receiving data from FPGA, the system controller carries out the calibration and compresses the data to reduce the data size. In order to synchronize the TDI scan rate with the satellite

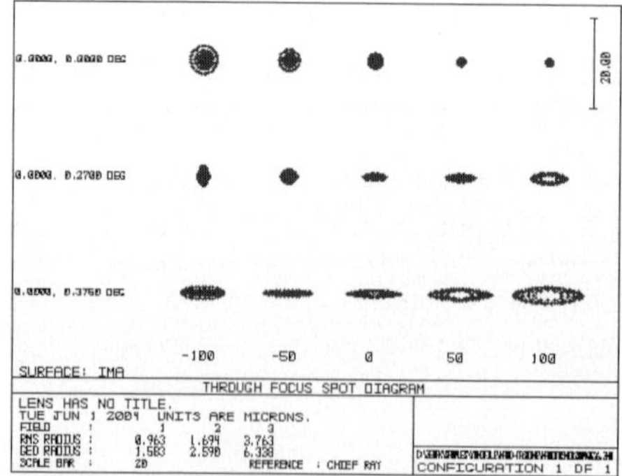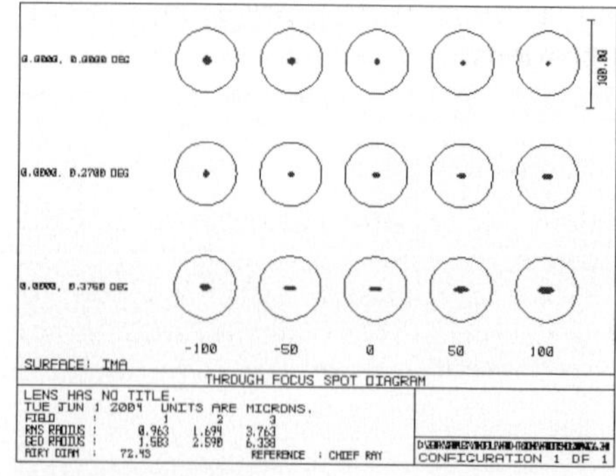

Figure 5. Optical system spot diagram.

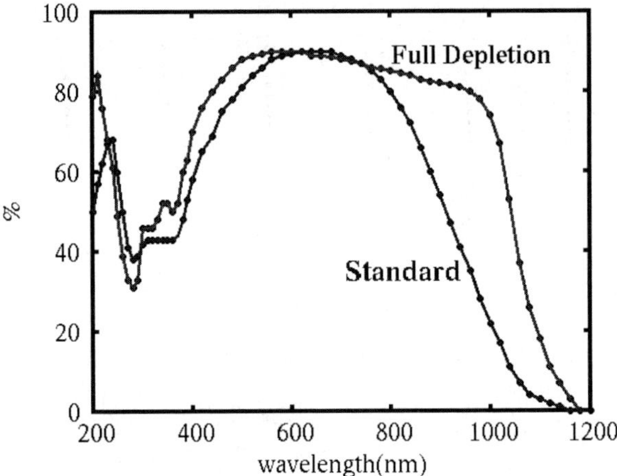

Figure 6. Spectral response of the full depletion CCD.

spin rate, stellar image data is analyzed by the system controller and feedback is sent to the posture control system and a scan shift command is sent to the CCD controller.

5. Bus System

The new bus system, based on the PRISM (Pico-satellite for Remote sensing and Innovative Space Missions) system, will be developed for nano-JASMINE. PRISM is a system that has been under development at the University of Tokyo as a second generation system based on the Cube-Sat XI-V. PRISM will realize a higher performance satellite mission based on a very small satellite system. The system offers a 3-axis controller, equipped with a magnetic torquer and a reaction wheel. It also includes a command and data processing system, power supply and a telecommunication system. Although it is a simple system, it is capable of performing several tasks.

6. Conclusion

As a facility to demonstrate and verify new technologies in real space environments, a nano-sized satellite is very useful. It provides the opportunity of conducting experiments in a short span of time at low costs. In fact, realizing a real space environment is difficult given the existing ground-based facilities. If the equipment commonly used for such small satellites is fully prepared, it would serve as a powerful platform for many other applications.

References

Gouda N., Tsujimoto T., Kobayashi Y., Nakajima T., Matsuhara H., Yano T., Yasuda N., Kan-Ya Y., Yamada Y., Ueno M. 2002 *The Proceedings of the IAU 8th Asian-Pacific Regional Meeting Volume II, Edited by S. Ikeuchi, J. Hearnshaw, and T. Hanawa* , 5–6

Tsuno Bearman P.W., Graham J.M.R. 2004 *Proceedings of 5th Internatinal Conference on Space Optics 2004, Toulouse.*

Transits of Venus: New Views of the Solar System and Galaxy
Proceedings IAU Colloquium No. 196, 2004
D.W. Kurtz, ed.

© 2004 International Astronomical Union
doi:10.1017/S1743921305001663

The VERA project
(VLBI Exploration of Radio Astrometry)

Hideyuki Kobayashi[1], Noriyuki Kawaguchi[1], Seiji Manabe[1],
Toshihiro Omodaka[2], Katsunori Shibata[1], Mareki Honma[1],
Yoshiaki Tamura[2], Osamu Kameno[2], Tomoya Hirota[2], Hiroshi Imai[1]

[1]National Astronomical Observatory of Japan, Mitaka, Tokyo, 181-8588, Japan

[2] Department of Physics, Kagoshima University, Kagoshima, Japan

Abstract. VLBI – Very Long Baseline Interferometry – is a radio interferometry technique which provides the highest spatial resolution observations to human kind. But at the present, the accuracy of the astrometry observations is limited by the atmospheric light path variations and instrumental phase errors, and only group delay measurements are used. To overcome these error factors, we have developed the VERA system, which has the first dual beam system. VERA is the first VLBI array to be free from the atmospheric phase fluctuations. It has four VLBI stations with 2300-km maximum baseline length in Japan. To compensate phase fluctuations of interferometer visibilities, which are mainly caused by the atmosphere, the VERA antenna observes two objects simultaneously. In order to do such observations, VERA has a two-receiver system, which tracks a focal plane according to a separation angle between observing objects. By comparing the visibility phase between two beams, simultaneous phase referencing VLBI will be achieved. The goal accuracy of astrometry observations is 10 micro-arcseconds, which makes annual parallax and proper motion measurements of galactic maser objects possible. 10 micro-arcsecond accuracy is equivalent to 10% distance accuracy for the galactic centre. This becomes 20% accuracy at the opposite side of the galaxy. The main scientific targets of VERA are to make a 3-dimensional maser object map of the galaxy and reveal the velocity field of the galaxy. This will show the mass distribution of the galaxy. Currently, construction of four stations is complete, and test observations are underway. We show the scientific goal, current results and instrumental accuracy of VERA.

ED BUDDING: Could you give some idea about the limiting flux density of point sources that you can receive with VERA? Is it something like 0.1 mJy?

HIDEYUKI KOBAYASHI: For the VLBI observations the structure is omitted, because we use only the interferometer. So it is only sensitive for very compact structures and we need some correction for the structure effect. We want to know the mapping, especially for the calibrator – from VLBI, or from ourselves.

ED BUDDING: The contours on the map that you showed: they are what kind of fluxes in mJy?

HIDEYUKI KOBAYASHI: Its a mJy per bin, and the bin is very small. This shows very high temperature regions; the usual extended structure vanished.

Part 7

MEETING SUMMARY

Hideyuki Kobayashi and Taihei Yano observe the transit

Coryn Bailer-Jones

Transits of Venus: New Views of the Solar System and Galaxy
Proceedings IAU Colloquium No. 196, 2004
D.W. Kurtz, ed.
© 2004 International Astronomical Union
doi:10.1017/S1743921305001675

Shades of the goddess: Venus in transit

Richard G. Strom

Radiosterrenwacht, ASTRON, Postbus 2, 7990 AA Dwingeloo
email: strom@astron.nl
and
Astronomical Institute 'Anton Pannekoek', University of Amsterdam, Kruislaan 403,
1098 SJ Amsterdam, the Netherlands

Abstract. I review the talks given during IAU Colloquium 196, sometimes in a revised order to suggest certain connexions. The AU now, its definition, value and uncertainty, and its modern determination are contrasted with the situation in 1640. While there are differences, not least in the value of the AU and its error, some things have not changed. As an enduring constant we require: a correct theoretical framework, precise observations, and accurate calculations. The history and context of Horrocks' transit observations are set against the backdrop of our own sightings during the 2004 event, and our journeys to Carr House and other sites in Much Hoole.

The apparent success of the subsequent 1769 world-wide effort belied the limitations imposed by the 'black drop' effect, now said to have two causes: finite resolution and limb darkening. Some mysteries surrounding Henderson's determination of the parallax of α Centauri were dispelled, which led to a discussion of modern astrometry, both from the ground and in space. A passionate plea for continuing ground-based astrometry was followed by results from satellite observatories, in particular discordant values for the parallax of the Pleiades. A graph of parallax determinations since 1769 illustrates the steadily increasing precision reminiscent of a 'Livingston curve,' with improvement by an order of magnitude every 50 years. This progression is expected to continue, as the next space missions (Gaia, JASMINE) should better Hipparcos by large factors. Time on the Earth and our very definition of the second are quite naturally related to motion of the planets, and the dynamical history of the solar system.

The 19th-century transit efforts were the last gasp in a 250-year endeavour linking Kepler with his Victorian heirs: From the viewpoint of determining solar parallax the Venus transit must have had its day. Discussion of its history, though, can be expected to continue. Finally, I trace the progress in determining the value of the AU over nearly 400 years, and suggest that more rapid advancement could have been facilitated by the introduction of other techniques. The danger of sticking to one strategy for too long is perhaps the best lesson which the Venus transits have to offer.

---

1. Introduction

"I came in with Halley's Comet in 1835. It is coming again next year, and I expect to go out with it. It will be the greatest disappointment of my life if I don't go out with Halley's Comet."

Mark Twain (1835-1910)

The reappearance of Halley's comet every three-quarters of a century is the astronomical event which best reflects the measure of a human lifespan. (It is perhaps fitting that the noted Dutch astronomer Jan Oort [1900-1992], whose contributions to our science included work on the origin of comets, could see two apparitions of P/Halley during his long life.) Total solar eclipses occur more frequently (though irregularly), but they are rare at any single location on the Earth. A transit of Venus (ToV), while less of a spectacle than the other two, does, by its rarity and its potential for probing the inner solar system, occupy a special place in astronomical legend, and practice. Indeed, the transits of the 17th, 18th and 19th Centuries can be seen as a bridge linking the 'new astronomy' of Kepler with modern astrophysics. This conference was an opportunity to celebrate the event (and experience the latest edition in an historical setting), cast an eye back to the three centuries preceding the one just completed, take stock of where we stand in the game of direct distance determination, and discuss where we might go from here.

It is in some sense poetic justice that a ToV involves the most mysterious (cloud-enshrouded) of the inner planets, the one associated with the goddess of love and beauty. On those rare occasions, Venus glides across our view of the Sun, providing both a visual spectacle and the potential for taking measure of the scale of our earthly orbit. With the poet we can affirm:

"Between the idea
 And the reality
 Between the motion
 And the act
 Falls the shadow"   from *The Hollow Men*, T.S. Eliot

2. The Astronomical Unit: now, and in 1639

2.1. *The modern situation*

Modern determination of the Astronomical Unit (AU) has progressed in the past 50 years from classical (optical) triangulation to direct radar measurement (Standish, C196 [=these proceedings]). This has resulted in an immense decrease in the uncertainty, from $\simeq 80\,000$ km in 1950, to a few meters anno 2004‡. Crucial to its value, however, is our definition of the AU. For the instantaneous Earth–Sun distance can be determined to high accuracy, but it constantly changes because of orbital ellipticity. For adopted values of the mass of the Sun, the gravitational constant (G) and the mean interval between vernal equinoxes (i.e., our definition of the tropical year), 1 AU is an average distance to the Sun for some assumed mean motion of the Earth. In the beautiful experiment we call the solar system, one sees Mother Nature integrating the equations of motion to perfection.

The resulting ephemerides have been used to test general relativity, the equivalence and Mach's principles, modern Newtonian dynamics, and to set limits to $\dot{G}$ and changes in the AU. Combining the very latest precision observations with numerical integrations of the equations of motion, including perturbations from the planets, the Moon and some 300 asteroids, highly accurate ephemerides are being calculated to carry out such tests (Pitjeva, C196). Possible changes in the AU, for example, can be investigated at the level of $\simeq 0.1\,\mathrm{m\,yr^{-1}}$.

2.2. *The AU in 1639*

When considering the work of Jeremiah Horrocks, it must be placed in the context of his time, with due consideration for its impact upon his contemporaries (Chapman, C196). Horrocks and his alter ego in things astronomical, William Crabtree, were 'grand amateurs,' quite unlike their continental examples Brahe and Kepler (who had noble and regal connections). Theirs was in a great English tradition of self-supporting savants, and they were, moreover, out to tackle the big questions in science: their interest was in true science, not mere phenomenology. Horrocks was widely read, and combined keen observing skills with the mathematical proficiency necessary to tackle Keplerian theory.

‡ The same accuracy is not, however, achieved for the outer planets, where traditional optical methods are still in use.

Figure 1. An impressive battery of telescopes prepares to capture first contact of the 2004 Transit of Venus.

When his calculations led him to conclude that the 1631 ToV was to be followed by a second one in 1639, he felt that the realization just a month before the event must have resulted from divine providence (a reflection of his Puritanism). When he first observed the disk of Venus, he was astonished at its diminutive size, in much the same way that Pierre Gassendi had been surprised to see how small Mercury was during its 1631 transit.

In the short time available after his prediction, Horrocks could only warn a few friends, including Crabtree, the only other person whom we know observed it. The latter (who described Horrocks as "my friend and second self") observed the transit with rapture and awe (Kollerstrom, C196). The pair, to whom we can trace the Keplerian tradition in England, complemented one another in many of their scientific endeavours. Crabtree's determination of the diameter of Venus ($1'\,03''$) was, one can say in retrospect, more nearly correct than Horrocks' (Chapman, C196).

2.3. *In three and a half centuries, we've come a long way*

The observations of 1639 revealed not only Venus' diminutive angular scale, but also showed that it was round and opaque (not self-luminescent), that is it had the qualities of a planet. Moreover, and perhaps most important at the time, according to Horrocks' argument the implied distance to the Sun (the AU) was much larger than previously imagined, some $4.5\times$ greater than Kepler's estimate. And yet Horrocks' value was barely two-thirds of the modern AU, or in absolute terms he was off by some 54 million km,

while today's determination (given the provisos discussed by Standish, C196) is believed
to be accurate at the level of a few meters. To achieve this, both now and in the 17th
century, required the correct theoretical framework, precision observations, and accurate
calculations.

3. 'T-day'

"The sight of a planet through a telescope is worth all the course on astronomy"

from Essays, Ralph Waldo Emerson

The day of 'our' ToV, 8 June 2004, was a memorable blend of observations, tourism
and historical context. Before 6 AM BST we journeyed from Preston to the University's
Alston Observatory well in time for first contact (05:19:46 UT), but under scattered cloud.
The only disappointment was that a cloud prevented direct viewing of the instant of first
contact, but observations (with an assortment of telescopes set up by the organizers)
were possible shortly thereafter (see Fig. 1). In fact, good observations could be (and
were) made during the entire transit (Figs 2 and 3).

3.1. *Carr House*

After a solid English breakfast, most of the conference attendees went on a tour of Horrocks sites in Much Hoole. This small community, quiet, obscure, and set in a "dark
corner of England" (in the eyes of 17th-century Puritans; Walton, C196) is where Horrocks, more by default than design, made his momentous observations. The likely location
was Carr House, which we were able to visit by the kindness of the present owners. The
stately brick edifice, some 400 years old, would have been familiar to Jeremiah Horrocks
and certainly delighted us (Fig. 4). We were able to peek through the very window from
which, plausibly (the Sun could be seen), the 1639 observations were made (Fig. 5).

3.2. *St. Michael's Church*

There followed a visit to St. Michael's of Much Hoole, with its memorials relating to the
1639 ToV. Then as now, it was the parish church, and would have been frequented by
Horrocks. His religion — Puritanism — should not be ignored when considering his scientific achievements. Its austerity (as illustrated by the sombre garb depicted in paintings
of Horrocks by Eyre Crowe, and of Crabtree by Ford Maddox Brown) may have been its
most tangible aspect from our perspective. Among their other qualities, Puritans were
serious yet flexible, they believed in predestination, and to be a Puritan required literacy
(Walton, C196) — and Jeremiah Horrocks *was* well read, consuming 15 or 16 books per
year (Chapman, C196). Not long after his early death in 1641, Lancashire and much
of England were engulfed in a bloody civil war pitting royalists against parliamentari-
ans, Catholics against Protestants (or roundheads). Or, as humorously summed up last
century, it was the

*"utterly memorable Struggle between the Cavaliers (Wrong but Wromantic) and the
Roundheads (Right and Repulsive)."*

from 1066 And All That, W.C. Sellar and R.J. Yeatman

3.3. *Hoghton Tower*

"A good time was had by all."

Stevie Smith

The day's events were fittingly (and exquisitely) rounded off with a superb dinner at
the 16th-century manor of Hoghton Tower, where James I, among others, was a famous
visitor. (The stately building and its entrance are shown in Fig. 6.)

Figure 2. "Mad dogs and Englishmen / Go out in the midday sun." — English women also!
from *Mad Dogs and Englishmen*, Noël Coward

4. The great 18th-century expeditions; Venus' planet credentials boosted

"I didn't go to the moon, I went much further — for time is the longest distance between two places."
from *The Glass Menagerie*, Tennessee Williams

The most direct way of measuring the distance to an inaccessible object is to use parallax. (It is the main reason we have two well-separated eyes: originally, no doubt, to avoid being eaten, though later it helped put meat on the menu.) For a Venus transit, this requires the determination of tiny angles. It was the great English astronomer/naturalist Edmund Halley who, reflecting upon his own attempts to determine the parallax of Mercury in transit, hit upon the idea of converting the measurement of an angle into the determination of a time interval: in particular, the elapsed time between 2nd contact (ingress), and 3rd (egress). He outlined a global attack on the problem involving observers at widely separated terrestrial latitudes timing the transit. Although he did not live to see its practical application, Halley's approach would set the agenda for the next two-and-a-half centuries.

Figure 3. And this is what they saw: Venus' shadow near the solar limb (the small dot on the right side of the disk). "Looks like someone punched a hole in the Sun," was my sister's comment.

Figure 4. Conference delegates approach Carr House. Tradition has it that Jeremiah Horrocks observed the ToV from the second storey central bay window (see Fig. 5).

Figure 5. The bay window through which Horrocks' 1639 observations might have been made.

Figure 6. Delegates from the conference approach the entrance to Hoghton Tower.

4.1. *French and English expeditions*

While many groups took up the challenge, France and England set the tone in the 18th century. Débarbat (C196) described the origins of the French efforts, from Gassendi's Mercury transit observations of 1631 (and an attempt to see if the 1631 ToV might be visible from Europe, despite Kepler's prediction to the contrary) through Delisle's systematic preparations (he had discussed the matter with Halley during a trip to England in 1724). Yet despite the investment of effort (and money) and the acquisition of 120 observing reports from 62 stations, the results were not generally felt to be satisfactory.

These measurements of the 1761 ToV were carried out against the backdrop of warfare (the Seven Years War, which involved England, France and most of Europe), with all the hardships it imposed upon expeditions trying to reach remote (and often unexplored) parts of the Indian and Pacific Oceans. The two 18th-century transits were to provide romantic tales of the exploits of many of the protagonists. Among the most gripping are the tragicomic adventures of Guillaume Le Gentil de la Galaisière, whose attempts to observe the 1761 event having been effectively thwarted by the war, decided to remain in the east for an additional 8 years to plan for (and observe) the 1769 transit. Finally settling upon Pondicherry, India, as a suitable observing site (after having been expelled from the Philippines as a suspected spy), he invested much effort in building an observatory and getting all the instruments set up, only to have the weather turn sour the very day of the transit. Even his return to Paris after over a decade would have pained all but the most zealous of masochists.

A much happier and successful lot would befall James Cook, whose voyage (and later ones of discovery to the Pacific) and exploration of Australia and New Zealand are the stuff of legend (particularly in the Anglo-Saxon world). By 1769 the war was over, and observing the Venus transit seems to have become something of a matter of national pride (war by other means...?). The British expedition organizers located a point in the Pacific as the ideal observing spot, and shortly thereafter as if by providence (and they weren't even Puritans! - § 2.2) Tahiti was discovered there (Orchiston, C196). Cook and his crew of nearly 100 sailed in the *Endeavour*, a former Whitby collier, making for Tahiti where they set up three observing sites as insurance against inclement weather. In the event, all the observations provided useful data, and the expedition can be adjudged to have been a success (even discounting the results of the subsequent explorations of the antipodes). When measurements from sites in Europe, Canada and the Pacific had been analysed, the resulting solar parallax ($\pi_\odot = 8''.78$) was in remarkably good agreement with the modern value ($\pi_\odot = 8''.794148$).

4.2. *Great observations! ... pity about that black drop*

Despite the success of their measurements, all of the observers in Cook's expedition discovered that their timing was limited by an effect first observed by Bergman in 1761: near the moments of 2nd and 3rd contact, the edge of Venus near the solar limb becomes distorted, shaped rather like a drop of water. The effect has been much discussed and speculated upon, being variously attributed to the atmosphere of Venus, that of the Earth, and instrumental resolution. During satellite (TRACE) observations of a recent transit of Mercury, Pasachoff, Schneider & Golub (C196) observed a similar black drop effect. Since Mercury has no atmosphere and the observations were done from space, any atmospheric cause is ruled out. By decomposing the TRACE images, the authors are able to show that solar limb darkening as well as the expected telescope point spread function both play a role in producing the black drop. The assumption is that this will also prove to be the case in the Venus transit, a conjecture which the team was testing even as the rest of us were celebrating Jeremiah Horrocks in Much Hoole.

4.3. *Planet Venus has an atmosphere*

The black drop effect may not provide evidence for a Venusian atmosphere, but there were 18th-century ToV observations which could be so interpreted. The Russian scientist Mikhail Lomonosov was involved in expeditions to time the 1761 transit (Russia, an ally of France, was also at war with England), but was himself more interested in observing associated physical phenomena (Marov, C196). He was thus among the first to regard the occasion as an opportunity to study something other than the geometry of the solar system. At ingress, just before second contact, Lomonosov saw a sudden, brief crescent of light encircling the side of Venus farthest from the Sun. He correctly interpreted this as sunlight refracted through a dense atmosphere surrounding the planet, and further made philosophical speculations about inhabitable worlds. Venus not only had the simplest properties of a planet (round, opaque) as had been shown by Horrocks and Crabtree (§ 2.3), but with an atmosphere was even beginning to look earth-like. And if the 'roaming stars' bound to the Sun were planets not unlike our Earth, couldn't other stars also have them; couldn't there be exoplanets?

4.4. *Why did they bother (to observe the transit, that is)?*

Intellectual quest? Technical challenge? Lust for adventure? All of these no doubt helped stimulate the 18th-century scientists and explorers who risked (and sometimes gave) their lives to observe Venus transiting the Sun. Perhaps it was simply the next step in the exploration of our world (in the broadest sense of the word). The voyages of discovery of the 15th – 18th centuries had negotiated uncharted seas. We now had the measure of our world, while the revolution spawned by Copernicus–Kepler–Galileo provided a model for the world beyond Earth — the solar system. Its structure was established, but its dimensions poorly determined. What better way to use our knowledge of the size of the Earth than as a sighting platform for determining the parallax of Venus?

5. Stepping stone to the stars

"We sit in the mud ... and reach for the stars."

from Enough, Ivan Sergeevich Turgenev

The parallax of Venus determined its distance from the Earth. Combined with Kepler's 3rd law, the distance to the Sun could be calculated, and hence the scale of the solar system established. The next rung on the distance-scale ladder would be to measure the distances to the nearest stars. Triangulation from the Earth being impossible (even with modern instrumentation, the parallax is too minute), the Earth's motion about the Sun could be used to give a baseline of 2 AU (some $23,000\times$ greater than the Earth's diameter). That is exactly what astronomers began to do in the first half of the 19th century.

5.1. *Henderson and the distance to α Centauri*

The observatory established at the Cape of Good Hope in the 1820s had a role similar to Greenwich Observatory: to obtain accurate stellar positions and provide time signals. Thomas Henderson, its second director (1831-1833), who hated the climate of the Cape, busied himself with astrometry, in particular to determine the declinations of 170 stars including α Cen (Warner, C196). Shortly before his departure for Scotland in 1833, Henderson learned that α Cen has a large proper motion. Concluding that it is likely to be nearby, he made additional observations during his last month at the Cape, and asked his successor, Thomas Maclear, to make still more. Although he clearly thought that α Cen must be nearby, why did he wait so long before analysing and publishing the data? Warner suggests that it was a combination of factors: the pressure of his new duties in Scotland, the fact that he didn't entirely trust the measurements he had made in 1832-1833 and so was waiting for Maclear's results, and the amount of work involved in the analysis itself. It was only Bessel's announcement of the parallax of 61 Cyg in 1838 that finally spurred him to action.

Though he was not the first to publish a fairly accurate stellar parallax (in fact his value was some 30% too large), Henderson is often credited with the first measurement. If we take the raw observations to be a 'measurement,' then this is, I suppose, strictly true. However, without calibration and data reduction, which would involve combining many observations taken at different times of the year, to separate the annual changes due to parallax from an even larger proper motion, the determination would be far from complete. As with the solar parallax results, from Horrocks to the present day (§ 2.3), data reduction is such an integral part of the 'measurement' that the observations alone are not sufficient to establish an outcome.

5.2. *For some passion: bring on the astrometrists!*

Anyone talking on a favourite topic can wax exuberant about it. However, passion is somehow not the first emotion most of us are likely to associate with astrometry (not in the way we might link it to cosmology, for example). Nonetheless, there was passion in Monet's plea for more ground-based astrometry, and in talks on the distance to the Pleiades Cluster.

5.2.1. *Parallax determinations from ground-based instruments*

The situation which astrometry from the ground finds itself in sounds rather precarious (Monet, C196). Although observations from space are immeasurably more expensive (and satellite observatories have much shorter lifetimes), there is plenty of money for space astrometry and practically none for terrestrial. If cost were the main factor, ground-based measurements would prevail, though space has the advantage when it comes to astrometric precision. A possible niche for future astrometry from the ground is, according to Monet, the combination of large aperture telescopes (2-m class) with wide fields of view (CCD detectors covering up to 0.5×0.5 m^2). In addition to the astrometry of faint ($R \simeq 24$) objects, such an instrument can detect fast-moving asteroids — indeed, searching for 'killer asteroids' may become its main selling point. Although terrestrial astrometry may be (all but) dead, there are many opportunities (but no funds).

5.2.2. *How far to the Pleiades?*

The Hipparcos Satellite has enabled us to extend the distance determination of stars to kiloparsec scales. But nearer to home, a well-known open cluster (known even to the ancients), the Pleiades, confronts us with a quandary. The expected distance from isochrone fitting to the main sequence is about 130 pc (or a parallax of, $\pi = 0\rlap{.}''0076$). Benedict explained how the Fine Guidance Sensors aboard the HST have been used to achieve astrometric precision to the $0\rlap{.}''0003$ level (Benedict & McArthur, C196). The technique has yielded distances for a dozen astrophysically interesting objects; for the Pleiades (based on six members) the distance is found to be 134.6 ± 3.1 pc, in good agreement with the 'expected' value.

The Hipparcos distance is significantly less than this: 120 pc ($\pi = 0\rlap{.}''0083 \pm 0\rlap{.}''0002$). In view of the discrepancy, van Leeuwen (C196) has reanalysed the Hipparcos data, evaluating and where necessary correcting for all possible sources of error. The result is not significantly different from the original value, with $\pi < 0\rlap{.}''008$ deemed highly unlikely. According to van Leeuwen, people should take the lower distance value seriously, and consider what its astrophysical consequences might be. In a subsequent talk, Southworth, Maxted & Smalley (C196) have reanalysed published data to derive a distance to the eclipsing binary and Pleiades member, HD 23642. They confirm the earlier value of slightly over 130 pc, but conclude that the smaller Hipparcos distance cannot be ruled out, noting that further observations could provide a determination accurate to ± 5 pc.

There is clearly a discrepancy in the Pleiades distance determination. Whether it is due to a problem in the Hipparcos calibration, or a misconception concerning the expected behaviour of Pleiades stars, was not settled at the conference (and is certainly unclear to this reviewer, although my impression is that the majority incline toward an instrumental problem). That the Hipparcos data contain many subtle effects was underlined by Pourbaix (C196), who discussed chromatic issues, in particular colour-induced position shifts, and how to correct for them. Problems have been reported for very red objects like Mira variables, although the influence on the distance scale is small. Correcting the achromatism results in a better period-luminosity relationship for the Miras.

Eclipsing binaries, also variable (but for reasons of geometry), have long been exploited for inferring stellar properties. Budding (C196) described the photometry required to extract accurate distances from apparent magnitudes, which can be an independent check on parallax determinations. The possible detection of planetary transits was noted.

Figure 7. Fr. Fintan O'Reilly, standing before the original observatory building, describes Stonyhurst and its telescope to an attentive group from the conference.

5.2.3. *The Ribble, Shires, Hobbits, a Dome Observatory and the Shireburn Arms*

After a day of discussing transits, orbits, astrometry and eclipsing binaries, what better way to wind down than to visit... an observatory? Stonyhurst College, a Catholic school with a long tradition, stands in the Forest of Boland where J.R.R. Tolkien (whose sons taught at the college) imagined much of his Hobbitarium. Stonyhurst also houses an historical telescope, still used by both students and a club. After an enthusiastic tour of the observatory (Fig. 7), we walked Dean Brook hoping to spot some of the small folk (but we're apparently better at looking up than down), ending up at the Shireburn Arms, overlooking the Ribble River Valley. (The observatory, it seems, was just a diversion to make the evening appear serious.)

5.3. *A digressionary comment on dark matter*

The effects of dark matter are well known on large scales (galaxies, clusters of galaxies), though they are also seen within individual galaxies (in flat rotation curves). Oort (1932 & 1960) suggested from the motions of stars perpendicular to the galactic plane that unseen material was required. The matter is, however, complex (Trimble 1987). Nonetheless, Bahcall, Flynn & Gould (1992) have found that there is only one chance in six that the local stellar motion can be explained without the need for dark matter. If this conclusion is correct, then it is interesting to note that while motion within the solar system (to scale sizes of perhaps 100 AU) can be explained to high precision without the need for dark matter, on a scale of 100 pc or so, a nonvisible component becomes indispensable. Perhaps with the next generation of astrometric satellites (Gaia and JASMINE), which will provide parallaxes for stars at tens of kpc, it will be possible to define a scale size for dark matter in the galactic disk (or discount the existence of such small scales).

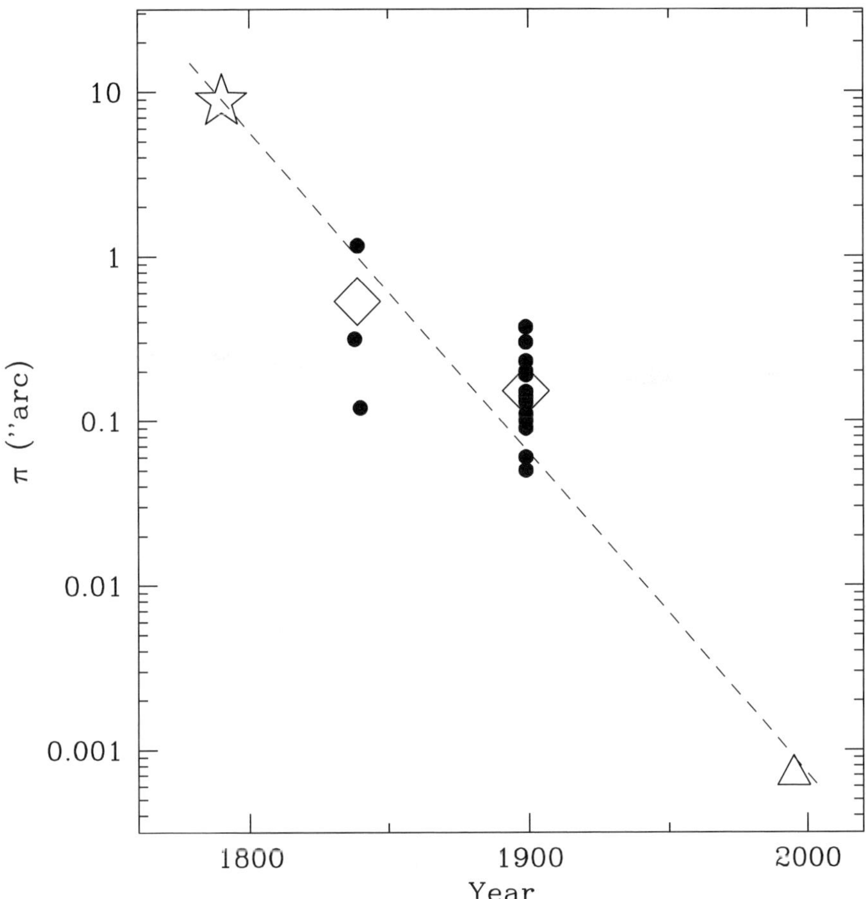

Figure 8. Parallax determinations over the past 200 years. The earliest point (star) is the solar parallax value after the 18th-century Venus transits (§ 4.1). The three points around 1840 are the earliest (successful) stellar parallaxes (§ 5.1), while those around 1900 come from Newcomb (1899). (Diamonds indicate averages of each group of points.) The most recent point (triangle) is based upon the smallest estimated Hipparcos error, σ_π (Perryman, Lindegren, Kovalevsky *et al.* 1997), and taken to be $2.5\sigma_\pi$. The dashed line is a fit through the large symbols.

5.4. *How far have we come with stellar astrometry, and where next?*

To get a sense of the progress made, Fig. 8 shows parallax determinations, beginning in the late 18th century with the ToV observations. There follow the first stellar parallaxes in the late 1830s, typical values around 1900, and finally the kind of measurements delivered by Hipparcos at the end of the last century. The result is a steady progression not unlike a 'Livingston curve' (originally used for comparing particle accelerators by M.S. Livingston in 1954; Panofsky 1997), with a slope suggesting improvement by a factor of 10 every 50 years. It is said that if a discipline's technical capability improves exponentially, then that is a sign of good health. On this basis, parallax determinations seem to have been vigorous for much of the past two centuries, and given new satellite missions on the planning board, should remain so for the foreseeable future.

During an evening lecture, Perryman (C196) illustrated in a lively manner the remarkable success of Hipparcos, describing the mission with the aid of three-dimensional sky images. In the coming decade, two successors to Hipparcos are planned: Gaia (an optical mission) and JASMINE (infrared). There is also a ground-based (radio) instrument, VERA. All three were described at the conference.

Gaia, the direct successor to Hipparcos, will improve upon its performance by several orders of magnitude in all dimensions of parameter space (Bailer-Jones, C196). It will perform an all-sky astrometric and photometric survey to $V = 20^{\mathrm{m}}$, and will detect galaxies, quasars and solar-system objects, in addition to stars, during its five-year mission. It will discover and investigate thousands of exoplanetary systems, and be able to detect planets of several $M_\oplus$ in nearby systems. Gaia also has a spectrograph which will provide radial velocities for $V < 17^{\mathrm{m}}$ objects (Cropper, C196). It is particularly sensitive to unusual objects (rapid rotators, accreting stars, etc.), and could locate the nearby 'killer star' whose close encounter with the solar system might have disrupted the Oort cloud, unleashing cometary activity which could have been responsible for terrestrial mass extinctions.

The Japanese satellite JASMINE will have astrometric capability similar to Gaia but will operate at $0.9\,\mu$m, or z-band (Gouda, C196). It will observe stars with $Z < 14$. At this infrared wavelength it will obtain useful parallaxes for $10\,000\times$ more stars than Gaia in the direction of the galactic centre. In the galactic bulge it will observe some $700\,000$ stars. Its main goal is to study the structure and evolution of the disk and bulge of the Milky Way, with launch planned in a decade. (In addition to Gouda's talk, there were also several poster presentations related to JASMINE.) VERA, a Japanese ground-based radio project, also has galactic structure as its primary objective (Kobayashi, Kawaguchi, Manabe *et al.*, C196). The four VERA antennas, which were recently completed, will simultaneously observe H_2O masers and reference sources, thereby eliminating atmospheric fluctuations. At its 22-GHz operating frequency, VERA will achieve an astrometric precision of $10\,\mu$arcsec, enabling the measurement of parallax and proper motion of water masers throughout the Galaxy. The ultimate goal is to determine the kinematics and mass distribution of the Milky Way.

6. Planets around stars

"Observe how system into system runs,
  What other planets circle other suns."   from *An Essay on Man*, Alexander Pope

Both Gaia and JASMINE will have the astrometric precision to detect more remote extra-solar planets by the recoil motion they induce in their 'sun,' adding to the impressive tally of recent years (see `exoplanets.org/almanacframe.html`). In fact, most of the exoplanet discoveries have resulted from the small Doppler shifts induced in the spectrum of the host star. Planetary transits *à la* Venus will occur in edge-on planetary systems, an effect which has been observed in HD 209458b (Henry *et al.* 2000; Charbonneau *et al.* 2000), and which the Kepler satellite hopes to exploit. In a mission already in orbit, the MOST satellite uses precision photometry to detect acoustic oscillations in Sun-like stars (Matthews, C196). It has, moreover, the capability of detecting reflected light from near-in exoplanets, and can thereby usefully constrain their radii and atmospheric properties. MOST achieves a sensitivity of a few millimag over periods of several days.

In addition to the (partial) occultation which a transiting exoplanet produces, its atmosphere (if any) can induce spectral changes, which have already been observed in the case of HD 209458b (Charbonneau *et al.* 2002). Such changes are unobservable from the ground, however, where the Earth's atmosphere limits photometric accuracy. However, Snellen (C196) outlined a method which exploits the Rossiter effect, whereby the star's rotation produces observable spectral changes.

The very first exoplanets were detected, in ground-based radio observations, around a millisecond-pulsar (or rapidly-rotating neutron star; the perfect clock for observing small Doppler shifts), PSR 1257+12 (Wolszczan & Frail 1992). Brisken (C196) described recent developments in pulsar astrometry, not so much for planet detections as for distances and proper motions. Although dispersion is a rough guide to pulsar distance, the scatter is large. VLBI observations can provide parallax as well as proper motion determinations to explore neutron star galactic distribution, kinematics, birthplace and age.

7. Earth and solar system

"Time is the measure of movement." (Medieval expression)
"What is actual is actual only for one time. And only for one place." T.S. Eliot

To carry out the accurate timing needed for the historical ToV measurements, as well as the determination of longitude (not to mention the precise frequencies required for the Doppler shifts just discussed), an accurate time standard is indispensable. Even Cook, Le Gentil, and their contemporaries took precision chronometers on their expeditions (§ 4.1). The Earth's rotation has, since life appeared on the planet, fulfilled the fundamental role of simple harmonic oscillator and provided our basic time interval. However, when it comes to precision timing, the Earth is not very accurate as is well demonstrated by historical solar eclipse records, for example.

McCarthy (C196) sketched the historical development of chronometry, which has seen accuracy improve by some eight orders of magnitude in ten centuries. A fundamental problem today is that the second, which has become our basic unit of time, is defined as a fraction (1/86,400) of the mean day in 1900, which is actually based upon observations of Earth rotation made in the 19th century, and from which atomic time is derived. Contemporary (VLBI) observations of the Earth's rotation, however, show that it is decelerating by $\simeq 1.7\,\mathrm{ms\,d^{-1}\,cy^{-1}}$ (due to geophysical processes: tidal friction, glaciation, etc.), requiring the regular insertion of leap seconds. As the rate at which this needs to be done may increase still further in the future, there are some who favour getting away from leap seconds altogether, or redefining the SI second. However, any change has implications: for physical equations; for navigation ($1\,\mathrm{s} \simeq 460\,\mathrm{m}$ at the equator); and there may be legal questions. In any event, the tendency is to move away from Earth rotation as the basis of our system of time.

Time also plays a fundamental role in observations of occultations of the moons of Jupiter and Saturn (Noyelles, Lainey & Vienne, C196). Why carry out such measurements? Because the timing can provide positional accuracies to better than 30 mas. Moreover, the events can be readily observed with small (50 cm) telescopes, and can provide improved ephemerides for studying tidal effects and predicting stellar occultations. Orbit determinations were also the subject of several other talks. De Saedeleer & Henrard (C196) have calculated the orbits of artificial satellites around the Moon. The approach is analytical, the important perturbations arising from lunar mass concentrations and the Earth's gravitational attraction. The results have obvious application to future space missions, as does an analytical model of Mercury's spin-orbit resonance (D'Hoedt & Lemaitre, C196). Using a Hamiltonian formalism, they calculate equilibrium states for four different configurations, concentrating upon the stable situation which prevails now. The work is being extended to include perturbations of Venus and the giant planets.

Near-resonance came into Message's (C196) exposition on the arithmetic of Venus transits. He showed, for the past millennium, the pattern of transits occurring in pairs at an eight-year interval, with each brace separated by 105 or 122 years. Approximate orbital resonance plays a role in this configuration. The more frequent Mercury transits have a similar pattern with some differences, the causes of which were discussed. The determination of orbital parameters from observations, particularly of asteroids, has been investigated by Gronchi and colleagues (C196). Classical methods, such as that due to Gauss, are often not applicable to modern data sets, which may only consist of a position and projected velocity vector. Under a set of reasonable assumptions, the distance and velocity can be constrained to an admissible region of parameter space, leading to a small number of possible orbits.

Finally, more complex orbital calculations have been undertaken by Tsiganis, Morbidelli & Levison (C196) to model the dynamic evolution of the solar system. In its formation and evolution, the Sun, planets and asteroids pass through several stages: shortly after the Sun forms, any remaining gas in the disk is driven out leaving proto-planets; the asteroid belt loses 99% of its mass; the outer (major) planets begin to migrate, and the Kuiper Belt (KB) forms. It is this last stage of migration and KB formation which is particularly addressed. The migration (Jupiter inwards, Saturn, Uranus and Neptune out) proceeds very slowly at first, but then accelerates by a dynamical mechanism. As this happens, KB members also migrate, and some are scattered away. This can account for three kinetic elements of the KB (stable, resonant [like Pluto], and scattered). From the timescales involved, it is suggested that the scattered KB objects may be responsible for the Late Heavy Bombardment of the inner solar system some 3.9 Gyr after it formed (and to which the lunar surface bears witness).

8. Wrapping up the historical Venus transit observations

"Most of us spend too much time on the last twenty-four hours and too little on the last six thousand years."   Will Durant

While England and France set the agenda for the 18th-century ToV expeditions (even while at war; § 4.1), they were by no means the only participants. Russian endeavours have already been noted in passing (§ 4.3), and the conference also heard about a Dutch and an Italian effort, as well as further British observations, in Ireland. The 19th century would see indigenous missions from the Americas too, and continued interest from the Old World as well, witness Lord Lindsay's 1874 expedition to Mauritius described in a poster (Brück, C196).

8.1. *Other 18th-century measurements*

In addition to the efforts of Delisle, Cook, Le Gentil *et al.* to execute Halley's grand plan (§ 4), lesser figures also contributed, in locations like Jakarta (as it now is), Rome and Ireland. At Batavia, the main Dutch colony on Java, the German-Dutch clergyman J.M. Mohr was initiated into the high art of 'transitry' by the 1761 event, during which he assisted the observers G. de Haan and P.J. Soele (van Gent, C196). With the zeal of the converted (and the wealth of his second wife), he built a stately mansion crowned by an observatory (total bill: 200 000 florins). No cost was spared in equipping it with the finest instruments shipped from Europe — for decades its quality would be unsurpassed in the region. Mohr succeeded in observing the 1769 Venus transit (although he was only able to see egress), and published the results. Their use in the greater — parallax determination — scheme was limited by uncertainty in the longitude of Batavia.

Longitude was also to play a role in a dispute which arose over observations of the 1761 transit by G.-B. Audiffredi. Pigatto (C196) sketched the history of the Santa Maria sopra Minerva Monastery in Rome, where Audiffredi, a Dominican, became the librarian in 1749. Being much enamoured of astronomy, he constructed an accurate meridian line in an upper loggia of the monastery to carry out observations. There, in 1761, he timed the transit of Venus and published the results. The French astronomer, A.-G. Pingré, described Audiffredi's data as "useless" since they seemed to conflict with Pingré's own determination of the longitude difference between the Observatoire de Paris and Saint Peter's in Rome. What Pingré had failed to grasp was that the transit observations had not been made from Saint Peter's, a fact which Audiffredi made clear in a refutation he later published.

After Mason (& Dixon)'s successful ToV observations in 1761 (& their famous M-D Line‡), Maskelyne implored him to once again set up a distant observing station for the 1769 transit (Butler, C196). Mason, by then tired of travelling to remote continents, refused to return to North America, but did finally agree to go to the Emerald Isle. Why observe from Donegal? Its northerly latitude would mean the Sun could be viewed at a higher elevation. The weather was (just in the nick of time) clear, and the measurements were a success. The problem was to determine their longitude, which took Mason months (and much encouragement from Maskelyne to persevere, and not abandon prematurely). In the end, ironically, the North American observations were of such high quality that Mason's would hardly figure in the parallax determination. But there was a spin-off from this, for it enhanced Ireland's reputation astronomically, and arguably (with not a modicum of assistance from Maskelyne) led to the establishment of at least one of Ireland's main observatories. Indeed, in the following century, Eire would (for a time) boast both the world's largest reflector *and* refractor.

8.2. *1874 & 1882: the last hurrah*

The 19th-century transits were to see 'new' nations of the western hemisphere actively enter the fray for the first time (they hadn't even existed as independent states in 1769). America, through the US Naval Observatory, mounted what were certainly the most professional campaigns of the period — and perhaps of all time (Dick, C196). Armed with a congressional appropriation of over a quarter-million dollars (for both transits), eight expeditions were sent out each time, fully trained and equipped with the latest instrumentation. Photographic plates were heavily relied upon, with hundreds exposed in 1874 (and over 1000 in 1882). Despite all the meticulous planning (to the point of erecting commemorative markers at the sites), the black drop effect (§ 4.2) still limited the ultimate parallax accuracy. Nonetheless, the final value derived by W. Harkness was within Halley's hoped-for 1/500th (giving an AU 0.17% smaller than the modern value, though the the estimated error exceeded the difference by over 2×). It was all written up but never published (galley proofs exist). The result did enter S. Newcomb's AU determination a few years later, but was given low weight (aberration was deemed 20× more important).

The American expeditions may have been among the most lavishly financed; Mexico participated in the 1874 transit on a shoestring (Allen, C196). It was a hard time for the country financially, but the Mexican president became convinced of the value of

‡ The Mason-Dixon line is the traditional boundary between the North and the South in the United States.

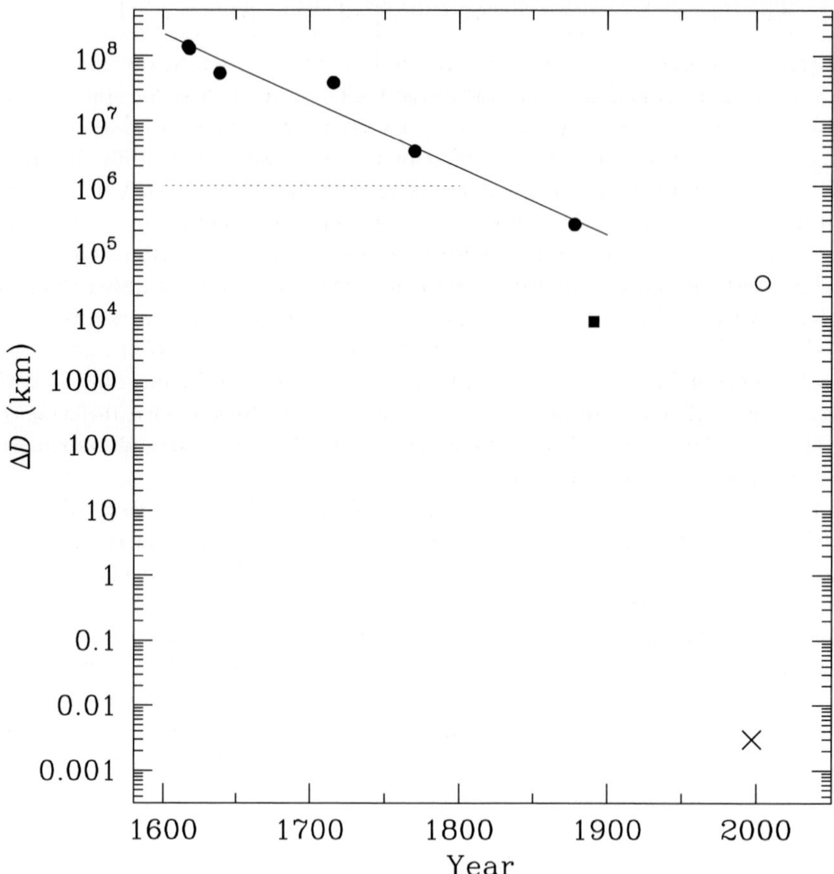

Figure 9. Absolute value of the deviation of AU measurements from the modern value. The first six points (solid circles) are based on classical methods, mainly involving the transit of Venus, and the solid line is a fit to them. The dotted horizontal line shows roughly where the black-drop effect begins to play a significant role. The most recent point (open circle) is based upon a provisional determination from the 2004 ToV (`www.vt-2004.org/central/cd-observers/obs-tim.html`). Newcomb's value from the 1890s (solid square) is an average based on several methods, including the ToV (though with low weight). The point with the smallest ΔD ($\times$) shows the estimated uncertainty in the modern (radar-determined) value.

the enterprise and gave the go ahead for an observing party to head for Japan. Armed with the necessary telescopes and clocks, two stations were set up near Yokohama, and both produced usable results. They were published within a year (the first of the 1874 measurements to appear). While the Mexican expedition may be little more than a footnote in transit history, it had significant impact within the country. As in Ireland a century before (§ 8.1), it facilitated the establishment of a national observatory. It also strengthened relations with Japan, leading to the first equal treaty with that country. Mexican observations of the 1882 Venus transit from Guadalajara were described in a poster by de Alba Martínez (C196).

9. And where has it all brought us?

"...and history came to a."‡   from *1066 And All That*, W.C. Sellar and R.J. Yeatman

At this point, a conference reviewer would probably conclude with words to the effect that the subject is vigorous and healthy, and (s)he looks forward to significant progress in the field before the next meeting in a few years time. Here however, I think I can safely state that we have come to the end of Venus transit observations for the purpose of parallax determination (in fact, the end came in 1882; §8.2). Fascination with the phenomenon will no doubt continue to inspire many, especially amateur, observations (and one website — www.vt-2004.org/central/cd-observers/obs-tim.html — reports, on the basis of thousands of [mainly amateur] timings of this ToV, a preliminary determination of the AU within 30 000 km of the current [radar] value). Further study of the history may also yield new insights. Allen (C196) noted the existence of a Mayan wall painting from the period 1200-1350, which might show a transit of Venus, although the interpretation seems far from unambiguous. Nevertheless, naked-eye transit observations are feasible, so a pre-Horrocks sighting is certainly possible. Oriental and Babylonian records would seem to be the likeliest sources.

How much progress did two-and-a-half centuries of observing Venus in transit achieve? Fig. 9 shows the deviation ($\Delta D = |AU_{\mathrm{now}} - AU_{\mathrm{old}}|$) in determinations of the AU from the modern value since the time of Kepler. We again see the steady progression typical of a Livingston curve: after each ToV campaign, the estimated Earth–Sun distance moves closer to the modern value. A fit to the values based upon measurements up to 1882 indicates that ΔD decreased by a factor of 10/century. The point in the 1890s is from Newton (§8.2) using several methods. Both it and the lowest value — based upon the estimated uncertainty in today's AU determination — lie well below the fit. This suggests that ToV measurements of the AU quickly became moribund (and remained so for over 200 years). The steepest possible exponential improvement shown by a Livingston curve occurs when new technologies are regularly introduced. The two lowest points, relying on completely different techniques (see §§2.1 & 8.2), suggest how much more rapid progress might have been. The plot also shows where the ΔD values should become limited by the black-drop effect, suggesting that not much more progress could have been expected from ToV determinations. (The preliminary 2004 value is also shown.)

Finally, what lessons can we learn from the transit of Venus experience? It seems ironical that although the black drop effect was well-documented by 1769, with a century to prepare for the 1874 transit (and a second chance eight years later), no effective strategy could be developed to mitigate its effect — even its cause remained obscure. But most likely the problem was (is?) insurmountable. If, as the Livingston curve (Fig. 9) suggests, new technologies were required for a breakthrough, then the weakness of pursuing only one scheme is exposed. However, it is difficult to see what alternatives there were. The ingenuity of the timing method (§4) lies in its conversion of a tiny angle into a small (but more readily measurable) time interval. Halley's hoped-for accuracy of 1/500th corresponds to an angular uncertainty in π of $\simeq 0\rlap{.}''05$, a value stellar astrometry would only reliably achieve around 1900. The one alternative I can imagine might have been precision timing of planets occulting stars. Clearly, a breakthrough like radar was essential. Perhaps the coming eight years will also teach us lessons we can discuss at the next (?) Venus transit conference in the antipodes.

‡ For those unfamiliar with British nomenclature, '.' = 'full stop' ["and history came to a *full stop*"]. Hence in their foreword [or "COMPULSORY PREFACE (*This Means You*)"] to *1066 And All That*, the authors note with some pride, "History is now at an end...; this History is therefore final."

518   Richard Strom

"The party's over, it's time to call it a day."

from *The Party's Over*, Betty Comden and Adolph Green

Acknowledgements

I am grateful to the SOC for their invitation to attend this superb colloquium, and to the IAU for financial support. I also wish to thank my sister, Janet, for the wonderful photographs she took, and for letting me use some of them to illustrate this summary (Figs 1-7).

References

Bahcall, J.N., Flynn, C. & Gould, A. 1992 *ApJ* **389**, 234-250.

Charbonneau, D., Brown, T.M., Latham, D.W., Mayor, M. 2000 *ApJ* **529**, L45-L48.

Charbonneau, D., Brown, T.M., Noyes, R.W., Gilliland, R.L. 2002 *ApJ* **568**, 377-384.

Henry, G.W., Marcy, G.W., Butler, R.P., Vogt, S.S. 2000 *ApJ* **529**, L41-L44.

Newcomb, S. 1899 *AJ* **20**, 1-6.

Oort, J.H. 1932 *BAN* **6**, 249-287.

Oort, J.H. 1960 *BAN* **15**, 45-53.

Panofsky, W.K.H. 1997 *Beam Line* **27**(1), 36-44.

Perryman, M. A. C., Lindegren, L., Kovalevsky, J., Høg, E., Bastian, U., Bernacca, P. L., Crézé, M., Donati, F., Grenon, M., van Leeuwen, F., van der Marel, H., Mignard, F., Murray, C. A., Le Poole, R. S., Schrijver, H., Turon, C., Arenou, F., Froeschlé, M., Petersen, C. S. 1997 *A&A* **323**, L49-L52.

Trimble, V. 1987 *ARA&A* **25**, 425-473.

Wolszczan, A. & Frail, D.A. 1992 *Nature* **355**, 145-147.

Robert van Gent and Richard Strom

Part 8
8 JUNE 2004
TRANSIT OF VENUS
AND
CONFERENCE BANQUET

Sunrise on transit day: St. Walburge's Church, Preston

Viewing the transit began at the University of Central Lancashire's Alston Observatory.

Alston Observatory's planetarium-lecture hall was set up for a web broadcast of the transit from sunnier climates in the event of cloudy skies, but, as is obvious in this picture, it was sunny yet again in Lancashire, as it was 24 November 1639 for Jeremiah Horrocks (there have been other sunny days between); meeting participants were all outside at the telescopes observing the real thing.

Gordon Bromage, head of the LOC, put in months of work guiding the large team that set of the Alston Observatory facilities for the transit. His pleasure in the clear sky on transit day is evident.

June Kurtz and Don Kurtz

Some time after second contact people settled down to individual activities

Janet Strom and Richard Strom

Phil Cox on security duty

Once the transit was well underway, participants went to breakfast at Alston Hall,
which neighbours the Observatory

Vladimir Elkin, Rick Collins, Steven Chapman, Arthur Missira
and Andreas Papageourgiou

Malcolm McVicar, Gordon Bromage and Barbara Bromage

Allan Chapman and Brian Warner

Following breakfast the conference moved to Much Hoole to continue observing the transit from Carr House, believed to be the site of Jeremiah Horrocks's 1639 transit of Venus observations

left: The BBC was broadcasting the transit live from Carr House, including an interview with Allan Chapman; right: Ed Budding and Clive Elphick in the room where Horrocks is believed to have made his 1639 observations

Groups of school children were invited to the transit at Carr House where they view the transit and talk to the astronomers

For the final stages of the transit and for third and fourth contact, the conference moved on the Much Hoole church.

The public turned out in large numbers at Much Hoole church to view the transit through telescopes of the University of Central Lancashire, along with explanations from university staff and PhD students.

The long entrance to Hoghton Tower

Hoghton Tower

Arrival for the banquet through the gatehouse

Announcement of the start of a tour of the house

left: Don and June Kurtz
right: Gordon Bromage, David Clarke and Suzanne Débarbat

Vladimir Elkin, Jon Riley, Jaymie Matthews, June Kurtz and Mikhail Marov

Bridget Bailey, Jackie Cunningham, Emma Woodward

Takuji Tsujimoto, Hideyuki Kobayashi, Naoteru Gouda and Yoshiyuki Yamada

John Southworth, Jon Riley, Steven Chapman, Mike Marsh and Mark Northeast

The high table in the background is the very one where King James I reputedly was served a loin of beef he so enjoyed that he drew his sword and knighted it "Sir Loin".

Allan Chapman and Paul Marston

Michael Perryman and Jaymie Matthews near the end of the evening

Robert Walsh, Clive and Jane Elphick

Clockwise from 12.00. Yukiyasu Kobayashi, Yoshiyuki Yamada, Seiji Ueda, Takuji Tsujimoto, Naoteru Gouda, Taihei Yano, Vladimir Elkin and Hideyuki Kobayashi

Author Index